Mechanics of Asphalt

Microstructure and Micromechanics

Linbing Wang

New York Chicago San Francisco
Lisbon London Madrid Mexico City
Milan New Delhi San Juan
Seoul Singapore Sydney Toronto

Cataloging-in-Publication Data is on file with the Library of Congress.

Mechanics of Asphalt

2 3 4 5 6 7 8 9 0   DOC/DOC   1 9 8 7 6 5 4 3 2 1 0

ISBN 978-0-07-149854-8
MHID 0-07-149854-0

The pages within this book were printed on acid-free paper.

Sponsoring Editor
Larry S. Hager

Acquisitions Coordinator
Michael Mulcahy

Editorial Supervisor
David E. Fogarty

Project Manager
Virginia Howe, Lone Wolf Enterprises, Ltd.

Copy Editor
Jacquie Wallace

Proofreader
Afton Woodson

Production Supervisor
Richard C. Ruzycka

Composition
Lone Wolf Enterprises, Ltd.

Art Director, Cover
Jeff Weeks

Contents

Acknowledgments

I want to take this moment to thank Dr. James Lai and Dr. David Frost, my former advisors, who have always inspired me to work innovatively. I would like to acknowledge my former and current graduate students and research associates for their contributions: Cristian Druta, Yanrong Fu, Yang Lu, Chris Harris, Xue Li, Tongyan Pan, Jin-Young Park, Wenjuan Sun, Dong Wang, Yongping Wang, Bing Zhang, Guoqing Zhou, and Yu Zhou. The support by the National Science Foundation, the Innovations Deserving Exploratory Analysis of the National Cooperative Highway Research Program, the Louisiana Transportation Research Center, and the Federal Highway Administration Turner-Fairbank Research Center is also acknowledged. I also would like to acknowledge many of my colleagues for the courtesy provision of some of the figures and equations. I would especially acknowledge Dr. Xingran Wang for his contribution to the interface model and the implicit implementation sections. Ms. Katie Thacker has provided professional editing, which is sincerely acknowledged. Last but not least, I am grateful to Larry Hager at McGraw-Hill, Virginia Howe and Roger D. Woodson at Lone Wolf, Ent. for their encouragement, support and patient work that finally made the book, and to all those who have helped me.

This book is dedicated to my parents, my wife Yan, and my sons George and Charlie, whose love has always inspired me and given me the strength to work hard.

—Linbing Wang

The author would also like to express his sincere thanks to the publishers of the following journals for their permissions to re-use contents in some papers published there. The co-authors of these papers are also sincerely acknowledged for their original efforts.

Computers and Geotechnics, Elsevier
Geotechnical Special Publication, American Society of Civil Engineers
International Journal of Geomechanics, American Society of Civil Engineers
International Journal of Pavements, Taylor & Francis
International Journal of Pavement Engineering, Taylor & Francis
International Journal of Pavement Research and Technology, Chinese Society of
    Pavement Engineering
Journal of Computing in Civil Engineering, American Society of Civil Engineers
Journal of Construction and Building Materials, Elsevier
Journal of Engineering Mechanics, American Society of Civil Engineers
Journal of Materials in Civil Engineering, American Society of Civil Engineers
Journal of Mechanics of Materials, Elsevier
Journal of Road Materials and Pavement Design, Lavoisier SAS
Transportation Research Record, Transportation Research Board

About the Author

Linbing Wang, Ph.D., P.E., is a leading researcher on asphalt mechanics, the coeditor of four proceedings, and the author of many papers on asphalt concrete. He is a member of several National Cooperative Highway Research Program (NCHRP) Project Panels and the Federal Highway Administration (FHWA) Expert Task Group on Fundamental Properties and Advanced Modeling of Asphalt.

CHAPTER 1

Introduction and Fundamentals for Mathematics and Continuum Mechanics

1.1 General Introduction

In this chapter, the fundamental properties that make asphalt concrete (AC) different from other materials are discussed. Logically, this discussion will serve as the guiding principles for the entire book.

Asphalt concrete is a unique material in four aspects: 1) its composition; 2) its manufacturing process; 3) its use environment; and 4) its performance or failure modes. These unique characteristics make AC different from other materials, and make this book valuable and uncommon. There are many textbooks on the mechanics of different materials and many books on the general topics of mechanics as well. What makes AC different from other materials is addressed in detail and serves as the focus of this book.

1.1.1 Unique Material Structure and Complexity

AC consists mainly of three constituents: asphalt binder, aggregates, and air voids. Other constituent/constituents may include additives such as fillers, modifiers, and fibers. These constituents are often used to enhance the properties of AC for specific aspects, such as moisture damage. The three major constituents have great differences in their properties and variety. While most of the binders are residuals of petroleum refinery, different petroleum resources and refining procedures will end up with binders of different molecular structure and compositions. Aggregates come with a variety of minerals that have different reactions with binders. The stiffness of the three constituents is so drastically different that tremendous localizations make properties of this material highly variable. While there are other distinct properties, three important properties of binder—thermal sensitivity, loading rate sensitivity, and moisture sensitivity—make this material much more challenging than most other materials in characterization, modeling and simulation, and especially predicting its behavior or performance under complex use environments including temperature, loading magnitude, and loading rates. The unique material structure and complexity requires a microscopic view.

In addition, due to the significant differences in chemical composition between binder and aggregates, the bonding between asphalt binder and aggregates is weak and has not been well understood. So far, there are not enough experimental and theoretical backgrounds pinpointing the behavior of interfaces between aggregates and binder, nor between fillers and binder.

1.1.2 Construction

Unlike other materials such as polymer and metals, the production of AC for real roads presents difficulties in controlling its quality in the areas of: 1) aggregate moisture; 2) the fines stuck at the aggregate surfaces; 3) temperature; 4) wind and moisture conditions at compaction; and 5) roadbed conditions. The variations in these conditions affect the bonding between aggregates and binder, binder content, void distribution, and even segregation. More importantly, a typical lift thickness is around three times the nominal maximum size of the aggregates, smaller than the required size for a representative volume, making compaction difficult and uniformity hard to achieve. These factors significantly limit understanding or predicting the behavior of asphalt pavement.

1.1.3 Environment Exposure

Roads must be in a natural environment without protection or isolation. Variations in temperature, moisture conditions, sunshine exposure, and the loading spectrum (loading magnitude, speed, and interval) are difficult to characterize, making predicting the performance of AC much more challenging than most other engineering materials. In addition, the "strength" of AC is usually degrading (due to aging, sometimes also increasing due to healing) with time, and since the degrading mechanisms are very complicated to explore, they are not well understood.

1.1.4 Failure Modes

Unlike many other engineering structures, roads are not designed against failures such as fracture or limited elastic deformation that can be accurately determined. They are designed against several distresses such as rutting (permanent deformation), fatigue cracking, thermal cracking, and roughness. These distresses accumulate with the repetitions of loads and thermal cycles. Due to the complexities described in previous sections, the performance characterized in the laboratory can hardly be used to predict road performance. This makes empirical methods, such as the mechanistic-empirical (M-E) method, necessary.

1.1.5 Moisture Damage

A special distress of AC is moisture damage. Moisture entrapped in AC produces large excess pore water pressure. When asphalt film is soaked in water, its light components may be dissolved and become weak. Moisture infused between aggregates and binder interface will also weaken this interface. Moisture damage causes the removal of fine materials, stripping the binder from the aggregate surface, and thus changing the material structure. This phenomenon is not well understood.

1.1.6 Complex Coupling

Experiments have also discovered that rutting induces fatigue cracking and vice versa. Moisture damage, weakening the mastics and binder-aggregate interface, leads to acceleration of fatigue and rutting.

The complexities described above can be summarized as the effect of non-linearity, non-uniformity, rate dependency, temperature dependency, large deformation, and complex coupling in terms of mechanics. The overall consequence is that the prediction of the pavement performance based on properties characterized in laboratories may be several magnitudes different from the performance observed in the field.

While the author began with an objective to document most of the studies relevant to the mechanics of AC, the study identified thousands of articles and reports. It is not the intent of this book to document all these studies, which may require an independent book of several hundred pages. Instead, this book will focus on the more rigorous theories of mechanics and their related applications. Laboratory discoveries, empirical correlation, relatively simple models such as isotropic elasticity models, and viscoelasticity models are not discussed in this book. Since systematic and fundamental research looking into the mechanisms of deformation and failure, stress-path dependency, and anisotropy is not available, it is difficult to present the book in such a way where concepts in different chapters are tied together to develop the various distress models.

1.2   Phenomenological Behavior of Asphalt

Phenomenologically, AC demonstrates behavior of non-linearity, rate dependency, temperature sensitivity, anisotropy, heterogeneity, irrecoverable deformation, fracturing, healing, hardening, softening, dilatancy, etc. Theories to cover all these phenomena and their couplings are too complicated to be realistic. Phenomenological models that describe AC behavior are typically elasticity, anisotropic elasticity, non-linear elasticity, elastoplasticity, linear or non-linear viscoelasticity, viscoplasticity, fracture mechanics, and continuum damage mechanics.

1.3   Need for This Book

The need for a book such as this one is two-fold: industrial and educational. Historically, the characterization of AC has been dominated by empirical approaches. In the last two decades an (M-E) approach for pavement design has been developed. Under this driving force, more rigorous theories have been developed or applied to solve problems encountered in asphalt industries. The variety of theories is so large that even researchers in this field would become lost seeking the links among these theories and understanding their capabilities and limits. An introduction to the fundamentals of these theories would be helpful to industrial researchers and practicing engineers to identify viable methods. The educational needs are relatively straightforward, as was indicated in the preface. So far, there is no such textbook. It is anticipated the book could serve as a textbook for characterization, modeling, and simulation courses on AC and pavements.

1.4   Logical Link of Chapters

There are many books in mechanics dealing with topics such as stress, strains, kinematics, tensor analysis, and fundamental laws of conservation and objectivity. Excellent books are available on these topics. There are also excellent textbooks on elasticity, viscoelasticity, plasticity, viscoplaticity, fracture mechanics, continuum damage mechanics molecular dynamics, and numerical implementations in Finite Element Method (FEM), Discrete Element Method (DEM) and Boundary Element Method (BEM). In the interest of being concise, only a brief description of these fundamentals is included in Chapter 1, Chapter 5, Chapter 6 , Chapter 8, Chapter 9 and Chapter 13 for completeness. Readers are assumed to have a background in elasticity and tensor analysis and the required mathematics background (such as Laplace Transform) to read these textbooks if needed. With this understanding, the book is organized into six sections with a total of 13 chapters with both straightforward fundamentals and applications so that practicing engineers and graduate students can learn this topic with convenience.

The first section has two chapters (Chapters 1 and 2) to discuss the fundamental properties of AC, fundamentals of tensor analysis and continuum mechanics, and the constituent properties and characterization methods. The second section has three chapters (Chapters 3, 4, and 5) focusing on microstructure characterization, strain characterization at the microscopic level, and mixture theory and micromechanics applications. The third section has two chapters (Chapters 6 and 7) to discuss macro-continuum types of theories and viscoplasticity models developed for AC. The fourth section has two chapters (Chapters 8 and 9) to mainly focus on three major numerical methods including FEM, BEM, and DEM. Section five has three chapters (Chapters 10, 11, and 12) to discuss topics of active research areas with applications of the fundamental theories and techniques. Section six has one chapter (Chapter 13) on multi-scale modeling and topics of current interest and less matured. Appendices document the Laplace Transform, Eshelby Tensors for elliptic inclusions, and fundamental solutions for BEM.

A list of the books recommended for further reading is presented at the end of this chapter. Again, the philosophy is not to perform a literature review or evaluation of these books. Instead, the goal is to assist the readers of this book to pick up the backgrounds quickly, so only a few books in each category are listed. One of the most important selection criteria for these books was simplicity, conciseness, and description of the fundamental physical mechanisms.

1.5   Fundamentals of Mathematics

1.5.1   Scalar

A scalar represents physical or geometrical quantities that require only a magnitude to represent. Quantities such as time, temperature, density, and distance can be represented by scalars. It is also called zeroth-order-tensor, requiring $3^0 = 1$ basis (independent direction). In this chapter, lower-case Greek letters such as α, β and γ are used to represent scalars. In fact, one may judge whether a quantity in an equation is a scalar, a vector, or a tensor from its physical meaning and dimensional analysis. The algebra operations such as + and − can be applied to scalars.

Geometrically, scalars are one-dimensional (1D) quantities that have zero as a reference point and positive and negative "directions" along only one axis.

1.5.2   Vector

Physical quantities such as force and displacement require both a magnitude and a direction to represent. These quantities can be represented as vectors, generally having three independent bases to represent in the three-dimensional (3D) space. If one compresses one or two of the three dimensions to zero, one can have two-dimensional (2D) vectors and a 1D vector (a scalar). A vector is also called a first-order tensor, having $3^1 = 3$ independent bases. Mathematically, a set of mutually independent N variables can be considered as an N dimensional vector. Lower-case characters such as v and u are used to represent vectors in this chapter. Vectors can be also represented in indicial format as v_i and u_i ($i = 1{\sim}3$). Vectors in rigorous sense (2D and 3D) follow the triangle rule for "add" and "subtract." Abstract extension to an N dimensional vector may not follow such rules.

A vector can be represented in its indicial format as follows:

$$v = v_1 e_1 + v_2 e_2 + v_3 e_3 = \sum_{i=1}^{3} v_i e_i \tag{1-1}$$

Where e_i are the three bases (with unit length). By following the Einstein convention (the dummy index or the repeaded index will sum running from 1 to 3), it can be simply represented as Equation 1-2:

$$v = v_i e_i \tag{1-2}$$

$\|v\| = \sqrt{v_1^2 + v_2^2 + v_3^2}$ is the magnitude of a vector or the "normal/mod/length" of a vector (see Figure 1.1a). The vector can be generally represented as scalar (magnitude) in an orientation where e is the orientation. It is this orientation that needs a reference system to measure it. The above representation (Equation 1-1) is for a Cartesian reference system.

Addition or subtraction of vectors will follow the following rules (Figure 1.1b):

$$w = u \pm v, \ w_i e_i = (u_i \pm v_i) e_i \tag{1-3}$$

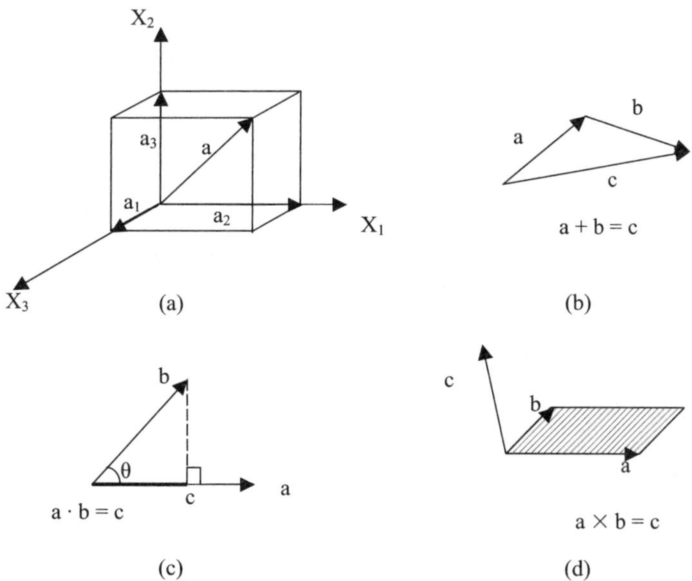

FIGURE 1.1   Illustration of typical vector operations.

Multiplication of a vector by a scalar is equivalent to extension or contraction of the vector along its original orientation (may reverse its direction, or make it a zero vector, which may not have a physical meaning).

$$\lambda v = \lambda v_i e_i \tag{1-4}$$

The dot (scalar) product of two vectors results in a scalar (the magnitude of the projection of one vector on the other) (Figure 1.1c).

$$u \bullet v = v \bullet u = \|u\|\|v\|\cos\theta = u_i v_i \tag{1-5}$$

Where θ is the angle between the two vectors.

The cross product of two vectors is a vector with its magnitude equal to $\|u\|\|v\|\sin\theta\|$ and its orientation normal to the plane formed by the two vectors and following the right-hand convention (Figure 1.1d). The magnitude is equal to the area of the parallel-isms as shown in Figure 1.1d.

$$u \times v = -v \times u = \|u\|\|v\|\sin\theta\|e \tag{1-6}$$

The indicial format of the cross-product will be presented in the next section as it requires the use of some tensors.

1.5.3 Tensor

Many quantities in mechanics require more than three independent components to represent. The stresses and strains are good examples. They can be represented as second-order tensor (rank of two), with a total of nine components ($3^2 = 9$). The nine components may not necessarily be independent. In this chapter, capital characters such as T, Q, and C are used to represent tensors. The indicial format of the tensors (second-order) may be represented as T_{ij} and Q_{ij}.

Two important tensors, the Kronecker Delta and the Permutation Tensor, are first defined as they are often used in other vector and tensor operations.

1.5.3.1 Kronecker Delta

Kronecker Delta is equivalent to the unit (often represented as unit isotropic tensor I) or one in scalar. It also serves as an operator (some tensors such as the transformation tensor also serve as operators).

$$\delta_{ij} = \begin{vmatrix} 1 & \text{for } i = j, i = 1\text{-}3; j = 1\text{-}3 \\ 0 & \text{for } i \neq j, i = 1\text{-}3; j = 1\text{-}3 \end{vmatrix} \tag{1-7}$$

It can be conveniently verified that:

$$e_i \bullet e_j = \delta_{ij} \ (i, j = 1,2,3)$$

$$\frac{\partial x_i}{\partial x_j} = \delta_{ij}$$

$$\delta_{ij} e_j = e_i$$

The last property represents the replacement operation.

Using the Kronecker Delta, the scalar product can be represented as:

$$u \bullet v = v \bullet u = u_i v_j e_i \bullet e_j = u_i v_j \delta_{ij} = u_i v_i = u_j v_j = u_1 v_1 + u_2 v_2 + u_3 v_3$$

1.5.3.2 Permutation Tensor

The permutation tensor, ξ_{ijk}, is a rank three tensor. It has $3^3 = 27$ components as defined in the following:

$$\xi_{ijk} = \begin{vmatrix} 1 & \text{if } ijk \text{ appears as in the sequence } 123123 \\ -1 & \text{if } ijk \text{ appears as in the sequence } 321321 \\ 0 & \text{if } ijk \text{ appears as in any other sequences} \end{vmatrix}$$ (1-8)

One can conveniently verify that: $\xi_{miq} \xi_{jkq} = \delta_{mj} \delta_{ik} - \delta_{mk} \delta_{ij}$ and $\xi_{ijk} = - \xi_{jik}$

It can also be conveniently verified that the cross product of the bases can be represented as:

$$e_i \times e_j = \xi_{ijk} e_k \quad (i, j, k = 1, 2, 3)$$

Therefore, the cross product of two vectors can be expressed as:

$$u \times v = (u_i e_i) \times (v_j e_j) = u_i v_j (e_i \times e_j) = u_i v_j \xi_{ijk} e_k$$

and $u \times v = -v \times u$, making use of $\xi_{ijk} = -\xi_{jik}$.

1.5.3.3 Dyadic Product of Two Vectors

The dyadic product of two vectors, $a \otimes b$, is defined as a linear operation that makes two vectors into a tensor. It has the following features (a, b, and c are vectors; α is a scalar).

$$(\alpha a) \otimes b = a \otimes (\alpha b) = \alpha (a \otimes b)$$ (1-9)

$$a \otimes (b + c) = a \otimes b + a \otimes c$$ (1-10)

$$(b + c) \otimes a = b \otimes a + c \otimes a$$ (1-11)

$$a \otimes b = (a_i e_i) \otimes (b_j e_j) = a_i b_j e_i \otimes e_j$$ (1-12)

In more rigorous mathematics, a tensor is defined as a linear combination of the basis dyads (Equation 1-13).

$$A = A_{ij} e_i \otimes e_j$$ (1-13)

1.5.3.4 Product of Dyad with Vector

The inner product of a dyad with a vector is defined as $(a \otimes b) \bullet c = a(b \bullet c)$. Its indicial format is as follows:

$$(a \otimes b) \bullet c = a_i e_i (b_j e_j \bullet c_k e_k) = a_i e_i (b_j \bullet c_k \delta_{jk}) = a_i b_k c_k e_i = a_1 (b_1 c_1 + b_2 c_2 + b_3 c_3) e_1 +$$ (1-14)

$$a_2 (b_1 c_1 + b_2 c_2 + b_3 c_3) e_2 + a_3 (b_1 c_1 + b_2 c_2 + b_3 c_3) e_3$$

Obviously, the inner product ends up with a vector. The physical meaning is the projection of a tensor on the surface to give the surface traction (see the stress-surface traction relationship).

$$a \bullet (b \otimes c) = (a_i e_i \bullet b_j e_j) c_k e_k = a_i b_j \delta_{ij} c_k e_k = a_i b_i c_k e_k = (a_1 b_1 + a_2 b_2 + a_3 b_3) c_1 e_1 +$$

$$(a_1 b_1 + a_2 b_2 + a_3 b_3) c_2 e_2 + (a_1 b_1 + a_2 b_2 + a_3 b_3) c_3 e_3 \tag{1-15}$$

Other vector dyad products may include:

$$a \times (b \otimes c) = (a_i e_i \times b_j e_j) c_k e_k = \xi_{ijq} a_i b_j c_k e_q \otimes e_k \tag{1-16}$$

$$(a \otimes b) \times c = a_i e_i (b_j e_j \times c_k e_k) = \xi_{jkq} a_i b_j c_k e_i \otimes e_q \tag{1-17}$$

1.5.3.5   Dyad-Dyad Product

The dyad-dyad product of two dyads is defined as:

$$(u \otimes v) \bullet (w \otimes s) = u_i e_i (v_j e_j \bullet w_k e_k) s_q e_q = u_i v_j w_j s_q e_i \otimes e_q \tag{1-18}$$

It is a dyad.

1.5.3.6   Vector-Tensor Products

$$v \bullet A = v_i e_i \bullet A_{jk} e_j \otimes e_k = v_i A_{jk} \delta_{ij} e_k = v_i A_{ik} e_k \tag{1-19}$$

$$A \bullet v = A_{ij} e_i \otimes e_j \bullet v_k e_k = A_{ij} e_i \delta_{jk} v_k = A_{ij} v_j e_i \tag{1-20}$$

If A is symmetric, it can be conveniently verified that the above two products are the same.

This product carries the meaning of the projection of a tensor (stress) to the surface as surface traction. In that case, A is the stress tensor and v is the direction normal of the surface.

1.5.3.7   Tensor-Tensor Product

$$A \bullet S = A_{ij} e_i \otimes e_j \bullet S_{pq} e_p \otimes e_q = A_{ij} S_{jq} e_i \otimes e_q \tag{1-21}$$

The product is a tensor. It follows the same rule as the dyad-dyad product in 1.5.3.5.

1.5.4   Frame Transformation

A scalar will not change its value when the frame is changed, or it is frame independent. However, vectors and tensors will follow certain rules when the frame is changed.

1.5.4.1   Vector Transformation

Considering two-frame systems (x_1, x_2, x_3) and (x'_1, x'_2, x'_3) (Figure 1.2), their basis vectors are corresponding (e_1, e_2, e_3) and (e'_1, e'_2, e'_3). From algebra geometry, one can represent the relationship of the two basis systems, or the coordinates of the basis vectors in the original frame in the following equation:

$$\begin{Bmatrix} e'_1 \\ e'_2 \\ e'_3 \end{Bmatrix} = \begin{Bmatrix} \cos(1'-1), \cos(1'-2), \cos(1'-3) \\ \cos(2'-1), \cos(2'-2), \cos(2'-3) \\ \cos(3'-1), \cos(3'-2), \cos(3'-3) \end{Bmatrix} \begin{Bmatrix} e_1 \\ e_2 \\ e_3 \end{Bmatrix}$$

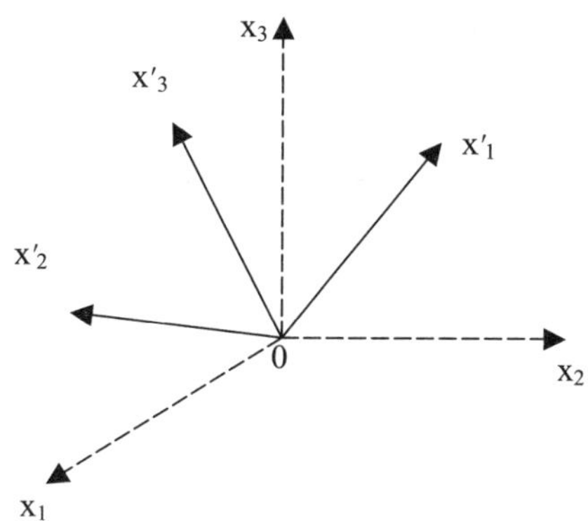

FIGURE 1.2 Illustration of frame rotation (0 x_1 x_2 x_3 rotated to 0 x'_1 x'_2 x'_3).

or

$$e' = Te \qquad e'_i = T_{ij}e_j \tag{1-22}$$

Considering the two basis vectors $e'_i = T_{ir}e_r$ and $e_j = T_{js}e_s$, the scalar product of the two basis vectors is:

$$e'_i \bullet e'_j = T_{ir}e_r \bullet T_{js}e_s = T_{ir}T_{js}e_r \bullet e_s = T_{ir}T_{js}\delta_{rs} = T_{ir}T_{jr} = \delta_{ij} = \delta'_{ij}$$

By the nature of the directional cosines, one can verify that $T_{ir}T_{jr} = \delta_{ij}$.

It can be deduced that any vector v can be transformed into v' following frame transformation.

$$\begin{Bmatrix} v'_1 \\ v'_2 \\ v'_3 \end{Bmatrix} = \begin{Bmatrix} \cos(1'-1), \cos(1'-2), \cos(1',3) \\ \cos(2'-1), \cos(2'-2), \cos(2',3) \\ \cos(3'-1), \cos(3'-2), \cos(3',3) \end{Bmatrix} \begin{Bmatrix} v_1 \\ v_2 \\ v_3 \end{Bmatrix}$$

or

$$v' = Tv \quad v'_i = T_{ij} \tag{1-23}$$

1.5.4.2 Tensor Transformation for Different Coordinate Systems

Considering a tensor $A = A_{ij}e_i \otimes e_j$ in (e_1, e_2, e_3) and its representation in the frame (e'_1, e'_2, e'_3).

$$A'_{ij}e'_i \otimes e'_j$$

Clearly, one has $e'_i = T_{ir}e_r$ and $e'_j = T_{js}e_s$ and therefore:

$$A'_{ij}T_{ir}e_i \otimes T_{js}e_s = A'_{ij}T_{ir}T_{js}e_r \otimes e_s$$

Therefore,

$$A_{rs} = A'_{ij}T_{ir}T_{js} \tag{1-24}$$

Isotropic Tensor

A second order isotropic tensor can be represented as:

$$I = \delta_{ij}e_i \otimes e_j$$

$$\delta_{ij}T_{ir}T_{js}e'_r \otimes e'_s = T_{jr}T_{js}e'_r \otimes e'_s = \delta_{rs}e'_r \otimes e'_s \tag{1-25}$$

$$\delta'_{ij} = \delta_{pq}T_{pi}T_{qj} = T_{qi}T_{qj} = \delta_{ij}$$

Therefore, it does not vary with the frame changes. In analogy, it is equivalent to one of the scalars.

Fourth Order Isotropic Tensor

Any of the fourth order isotropic tensors can be represented as the combinations of the following three fourth order istropic tensors:

$$\delta_{ij}\delta_{kl}e_i \otimes e_j \otimes e_k \otimes e_l$$

$$\delta_{ik}\delta_{jl}e_i \otimes e_j \otimes e_k \otimes e_l$$

$$\delta_{il}\delta_{jk}e_i \otimes e_j \otimes e_k \otimes e_l$$

or

$$(\lambda\delta_{ij}\delta_{kl} + \mu\delta_{ik}\delta_{jl} + v\delta_{il}\delta_{jk})e_i \otimes e_j \otimes e_k \otimes e_l \tag{1-26}$$

1.5.5 Matrix Operations

The second order tensors are often represented as 3×3 matrices and all the matrix operations are valid for the tensors. Among many of the operations, the operation to find the principal directions and principal values of matrices has significances in determining the principal directions and values of stresses. Considering a tensor A, its projection onto the u direction is a vector v.

$$A_{ij}u_j = v_i \text{ or } A \bullet u = v \tag{1-27}$$

A special case is that the projection on to the n direction ends up with a vector that is in line with n direction.

$$A_{ij}n_j = \lambda n_i \text{ or } A \bullet \hat{n} = \lambda\hat{n} \tag{1-28}$$

or

$$(A_{ij} - \lambda\delta_{ij})n_j = 0 \text{ or } (A - \lambda I) \bullet \hat{n} = 0 \tag{1-29}$$

Therefore, one has the following equations:

$$(A_{11} - \lambda)n_1 + A_{12}n_2 + A_{13}n_3 = 0$$

$$A_{21}n_1 + (A_{22} - \lambda)n_2 + A_{23}n_3 = 0$$

$$A_{31}n_1 + A_{32}n_2 + (A_{33} - \lambda)n_3 = 0 \tag{1-30}$$

The solution to λ is equivalent to the following matrix equation:

$$\left| A_{ij} - \lambda\delta_{ij} \right| = 0 \tag{1-31}$$

or

$$\lambda^3 - I_1\lambda^2 + I_2\lambda - I_3 = 0 \tag{1-32}$$

Where I_1 (first invariant, the trace), I_2 (second invariant), and I_3 (third invariant) are the invariants of the tensor (matrix).

$$I_1 = A_{ii} = trA \tag{1-33a}$$

$$I_2 = \frac{1}{2}\left(A_{ii}A_{jj} - A_{ij}A_{ji}\right) = \frac{1}{2}\left[(trA)^2 - tr(A)^2\right] \tag{1-33b}$$

$$I_3 = \xi_{ijk}A_{1i}A_{2j}A_{3k} = \det A \tag{1-33c}$$

It can be proved that I_1, I_2, and I_3 are invariant when the tensor is transformed into its representation into another rotated coordinate system.

Typically, it has three real roots.

$$\left[A_{ij} - \lambda_{(q)}\delta_{ij}\right]n_i^{(q)} = 0, (q = 1,2,3) \tag{1-34}$$

and

$$n_i^{(q)}n_i^{(q)} = 1, (q = 1,2,3)$$

Replacing A with the stress tensor σ or strain tensor ε in the above operations, one can find the principal stresses or principal strains. Due to the invariant nature of the three invariants, they are often used in constitutive laws to avoid the violation of objectivity principle.

1.5.6 Typical Operators

Operator ∇ (del) of vector calculus can be represented as:

$$\nabla = \frac{\partial}{\partial x_1}e_1 + \frac{\partial}{\partial x_2}e_2 + \frac{\partial}{\partial x_3}e_3 = \frac{\partial}{\partial x_i}e_i \tag{1-35}$$

If ϕ is a scalar function then $\nabla\phi = grad\,\phi$ is the gradient of the function. It has applications in the plasticity theory for representing the flow rule. The gradient of a scalar function is a vector.

The gradient of a vector field v is represented as:

$$\nabla v = \frac{\partial v_j}{\partial x_i}e_j \otimes e_i \tag{1-36}$$

It is a tensor.

The divergence of a vector field v is also scalar. It can be represented as:

$$\nabla \bullet v = \frac{\partial v_j}{x_i}e_i \bullet e_j = \frac{\partial v_j}{x_i}\delta_{ij} = \frac{\partial v_i}{\partial x_i} \tag{1-37}$$

The curl of a vector is a vector and is defined as:

$$\nabla \times v = \xi_{ijk}\frac{\partial v_j}{\partial x_i}e_k \tag{1-38}$$

1.5.7   Divergence Theorem

The Integral Theorems of Gauss and Stokes (or the divergence theorem) is widely used in continuum mechanics. It presents the relationship between the volume integral and the surface integral of any quantity. For a tensorial quantity $T_{ij...k}$ and its gradient

$$T_{ij...k,q} = \frac{\partial T_{ij...k}}{\partial x_q}$$

The theorem states:

$$\int_S T_{ij...k} n_q \, dS = \int_V T_{ij...k,q} \, dV \tag{1-39}$$

For a scalar field λ, it ends with:

$$\int_S \lambda n_q \, dS = \int_V \lambda_{,q} \, dV \tag{1-40}$$

For a vector field v:

$$\int_S v \bullet n \, dS = \int_V div v \, dV \quad \text{or} \quad \int_S v_q n_q \, dS = \int_V v_{q,q} \, dV \tag{1-41}$$

$$\int_S n \times v \, dS = \int_V curl v \, dV \quad \text{or} \quad \int_S \xi_{ijk} n_i v_j e_k \, dS = \int_V \xi_{ijk} \frac{\partial v_j}{\partial x_i} e_k \, dV \tag{1-42}$$

In 2D spaces, it reduces to the area integral and curve (line) integral.

1.6   Fundamentals of Continuum Mechanics

1.6.1   Concept of a Continuum and Representative Element

With unaided eyes, metals, glasses, other solids, and liquids look like a continuum; there is no space in them. One can imagine that the gases are filled with molecules everywhere and can also be considered a continuum. The physical concept of continuum is very complicated. There are always voids and minute cracking in solids, and voids in liquids. There are also spaces between atoms and molecules. Even within the atom there are tremendous spaces. However, there might be force fields in these spaces which bond the molecules and atoms. The concept of a continuum in continuum mechanics is both concrete and abstract. In the abstract sense, when the effects due to any discontinuity are negligible, the material may behave like a continuum. Mathematically, if the displacement field of a material can be represented as a continuous function, the material can be considered a continuum. While there is no restriction on the homogeneity assumption of the material, this assumption is usually implied as typically one addresses the material with the same set of constitutive equations. AC is, by nature, a heterogeneous material comprised of asphalt binder, aggregates, and voids. Each of the constituents can be considered as homogeneous. Using a large representative volume element (RVE), the thus obtained properties of RVE can be used to represent the properties of the material at any sizes, including the infinitesimal elements.

1.6.2   Description of Kinematics-Deformation Gradient and Finite Strain Tensor

1.6.2.1   Coordinate Representation

Figure 1.3 presents the reference configuration and the current configuration of a deformable body. A segment in the reference configuration is represented as:

$$dX = dX_A I_A \tag{1-43}$$

Therefore:

$$(dX)^2 = dX \bullet dX = dX_A dX_A \tag{1-44}$$

Where dX_A are the three components in the three orthogonal directions represented by I_A for the reference configuration.

In the current configuration the segment is represented as:

$$dx = dx_i e_i \tag{1-45}$$

Therefore:

$$(dx)^2 = dx \bullet dx = dx_i dx_i \tag{1-46}$$

Where dx_i are the three components in the three orthogonal directions represented by e_i for the current configuration.

For a continuum, one can assume a continuous and one-to-one mapping as:

$$x_i = \chi_i(X) \tag{1-47}$$

Therefore:

$$dx_i = \frac{\partial \chi_i}{\partial X_A} dX_A = \chi_{i,A} dX_A \tag{1-48}$$

$$\chi_{i,A} \equiv F_{iA} \tag{1-49}$$

is the deformation gradient and

$$dx = F \bullet dX \tag{1-50}$$

$$dX = F^{-1} \bullet dx \tag{1-51}$$

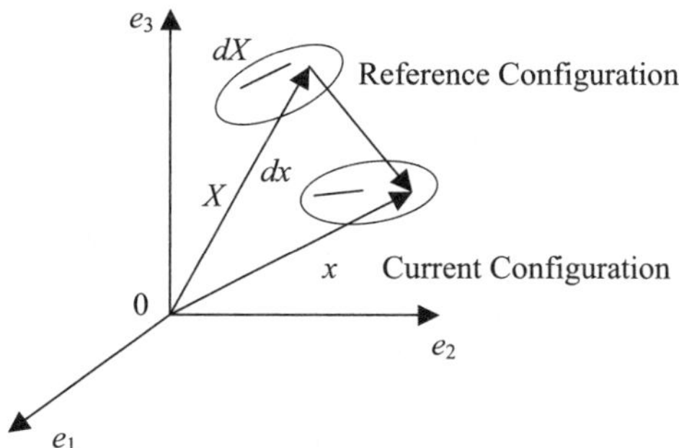

FIGURE 1.3   Illustration of deformation kinematics.

Considering:

$$(dx)^2 - (dX)^2 = dx_i dx_i - dX_A dX_A = (x_{i,A}dX_A)(x_{i,B}dX_B) - \delta_{AB}dX_A dX_B = (x_{i,A}x_{i,B} - \delta_{AB})dX_A dX_B$$

$$= (C_{AB} - \delta_{AB})dX_A dX_B \tag{1-52}$$

Where $C_{AB} = x_{i,A}x_{i,B}$ or $C = F^T \bullet F$ in matrix or tensor representation.

The Lagrangian Finite Strain Tensor is defined as:

$$2E_{AB} = C_{AB} - \delta_{AB} \tag{1-53}$$

or

$$2E = C - \delta \tag{1-54}$$

and

$$(dx)^2 - (dX)^2 = (C_{AB} - \delta_{AB})dX_A dX_B = 2E_{AB}dX_A dX_B \tag{1-55}$$

For the one-to-one mapping, one can represent X as a function of x. Therefore:

$$(dx)^2 - (dX)^2 = \delta_{ij}dx_i dx_j - (X_{A,i}dx_i)(X_{A,j}dx_j) = (\delta_{ij} - X_{A,i}X_{A,j})dx_i dx_j = (\delta_{ij} - c_{ij})dx_i dx_j$$

$$c_{ij} = X_{A,i}X_{A,j} \text{ or } c = (F^{-1})^T \bullet (F^{-1}) \tag{1-56}$$

The Cauchy Deformation Tensor or Eulerian Finite Strain Tensor is defined as:

$$2e_{ij} = \delta_{ij} - c_{ij} \text{ or } 2e = I - c \tag{1-57}$$

Considering two segments $dX^{(1)}$ and $dX^{(2)}$ in the reference configuration, $dx^{(1)}$ and $dx^{(2)}$ represent the two segments in the current configuration.

$$dx^{(1)} \bullet dx^{(2)} = F \bullet dX^{(1)} \bullet F \bullet dX^{(2)} = dX^{(1)} \bullet F^T \bullet F \bullet dX^{(2)}$$

$$= dX^{(1)} \bullet C \bullet dX^{(2)} = dX^{(1)} \bullet (I + 2E) \bullet dX^{(2)} = dX^{(1)} \bullet dX^{(2)} + dX^{(1)} \bullet 2E \bullet dX^{(2)} \tag{1-58}$$

If there is no strain, one can have:

$$dx^{(1)} \bullet dx^{(2)} = dx^{(1)}dx^{(2)}\cos\theta = dX^{(1)} \bullet dX^{(2)} = dX^{(1)}dX^{(2)}\cos\theta \tag{1-59}$$

It means the angles between any two segments remain unchanged.

1.6.2.2 Displacement Representation

Obviously, the displacement of point X can be represented as:

$$u_i(X_A) = x_i(X_A) - X_i$$

$$u_A(x_i) = x_A - X_A(x_i) \tag{1-60a, 1-60b}$$

Therefore the Langrangian Finite Strain Tensor is:

$$2E_{AB} = x_{i,A}x_{i,B} - \delta_{AB} = (u_{i,A} + \delta_{i,A})(u_{i,B} + \delta_{i,B}) - \delta_{AB}$$

or

$$2E_{AB} = u_{A,B} + u_{B,A} + u_{i,A}u_{i,B} \tag{1-61}$$

The Eulerian Finite Strain Tensor is:

$$2e_{ij} = \delta_{ij} - X_{A,i}X_{A,j} = \delta_{ij} - (\delta_{Ai} - u_{A,i})(\delta_{Aj} - u_{A,j})$$

$$2e_{ij} = u_{i,j} + u_{j,i} - u_{A,i}u_{A,j} \tag{1-62}$$

By neglecting the high order terms, one has:

$$2E_{AB} = u_{A,B} + u_{B,A}, \; 2e_{ij} = u_{i,j} + u_{j,i} \tag{1-63a, 1-63b}$$

Both reduce to the small strain representation.

1.6.2.3 Rate of Deformation

The previous sections have focused on the deformation gradients and strains that are based on displacements or locations. They are spatial gradients. This section will discuss time-dependent characteristics of deformation. Instead of using the displacement, the velocity v_i will be the field variables for the study.

The spatial velocity gradient is defined as:

$$L_{ij} = \frac{\partial v_i}{\partial x_j} \tag{1-64}$$

This tensor can be decomposed into a symmetric part and a skew-symmetric part:

$$L_{ij} = D_{ij} + W_{ij} \tag{1-65}$$

$$D_{ij} = \frac{1}{2}\left(\frac{\partial v_i}{\partial x_j} + \frac{\partial v_j}{\partial x_i}\right) \qquad W_{ij} = \frac{1}{2}\left(\frac{\partial v_i}{\partial x_j} - \frac{\partial v_j}{\partial x_i}\right) \tag{1-66a, 1-66b}$$

The symmetric part D_{ij} is the rate of deformation tensor and the skew-symmetric part W_{ij} is the spin tensor representing rigid rotation.

The spatial velocity change can be represented as:

$$dv_i = \frac{\partial v_i}{\partial x_j}dx_j \text{ or } dv = L \bullet dx \tag{1-67}$$

Since

$$\frac{\partial v_i}{\partial x_j} = \frac{\partial v_j}{\partial X_A}\frac{\partial X_A}{\partial x_j} = \frac{d}{dt}\left(\frac{\partial x_i}{\partial X_A}\right)\frac{\partial X_A}{\partial x_j} \tag{1-68}$$

One has:

$$L = \dot{F} \bullet F^{-1} \tag{1-69}$$

or

$$\dot{F} = L \bullet F \tag{1-70}$$

This relationship is often used for deriving other relationships in the following sections.

1.6.2.4 Stretch Ratio

The stretch ratio α is defined as the ratio between the length dx and that of dX measured in the reference direction N. It is a scalar.

$$\alpha = \frac{dx}{dX} \tag{1-71}$$

One may use a definition that can be more conveniently implemented.

$$\alpha^2 = (\frac{dx}{dX})^2$$

$$(dx)^2 = dx \bullet dx = F \bullet dX \bullet F \bullet dX = dX \bullet C \bullet dX$$

$$\alpha^2 = (\frac{dx}{dX})^2 = \frac{dX}{dX} \bullet C \bullet \frac{dX}{dX} = N \bullet C \bullet N \tag{1-72}$$

One can define the stretch ratio in the n direction.

$$\frac{1}{\beta} = \frac{dX}{dx} \tag{1-73}$$

$$(dX)^2 = dX \bullet dX = F^{-1} \bullet dx \bullet F^{-1} \bullet dx = dx \bullet c \bullet dx$$

$$\frac{1}{\beta^2} = (\frac{dX}{dx})^2 = \frac{dx}{dX} \bullet c \bullet \frac{dx}{dX} = n \bullet c \bullet n \tag{1-74}$$

Due to the deformation, n may not be a unit vector and therefore typically $\alpha \neq \beta$.

1.6.2.5 Material-Time Derivative

For any quantity (may be a scalar, vector, or tenor),

$$B_{ij\ldots} = B_{ij\ldots}(X,t) \text{ or } B_{ij\ldots} = B_{ij\ldots}(x,t) \tag{1-75a, 1-75b}$$

Its time variation rate can be represented either as:

$$\frac{d}{dt}\left[B_{ij\ldots}(X,t)\right] = \frac{\partial}{\partial t}\left[B_{ij\ldots}(X,t)\right] \tag{1-76}$$

or in the spatial reference as:

$$\frac{d}{dt}\left[B_{ij\ldots}(x,t)\right] = \frac{\partial}{\partial t}\left[B_{ij\ldots}(x,t)\right] + \frac{\partial}{\partial x_k}\left[B_{ij\ldots}(x,t)\right]\frac{dx_k}{dt} \tag{1-77}$$

$$\frac{d}{dt}\left[B_{ij\ldots}(x,t)\right] = \frac{\partial}{\partial t}\left[B_{ij\ldots}(x,t)\right] + \frac{\partial}{\partial x_k}\left[B_{ij\ldots}(x,t)\right]v_k \tag{1-78}$$

Therefore, one has the following operator:

$$\frac{d}{dt} = \frac{\partial}{\partial t} + v_k \frac{\partial}{\partial x_k} \tag{1-79}$$

or

$$\frac{d}{dt} = \frac{\partial}{\partial t} + v \bullet \nabla_x \tag{1-80}$$

1.6.2.6 Changes of Segment, Area, and Volume

1.6.2.6.1 Line Segment Change

$$\dot{dx} = \dot{F} \bullet dX \tag{1-81}$$

$$\dot{F} = L \bullet F \tag{1-82}$$

$$\dot{dx} = L \bullet F \bullet dX = L \bullet dx \tag{1-83}$$

The material derivative of the dot product of dx:

$$(dx \bullet \dot{dx}) = 2dx \bullet \dot{dx} = dx \bullet 2L \bullet dx = dx \bullet 2(D+W) \bullet dx = dx \bullet 2D \bullet dx \tag{1-84}$$

1.6.2.6.2 Area Change

The area change can be evaluated through an area element composed of two segments $dX^{(1)}$ and $dX^{(2)}$. In the reference configuration the area is:

$$dS_A^o = \xi_{ABC} dX_B^{(1)} dX_C^{(2)} \tag{1-85}$$

In the current configuration:

$$dS_i = \xi_{ijk} dx_j^{(1)} dx_k^{(2)} = \xi_{ijk} x_{j,B} dX_B^{(1)} x_{k,C} dX_C^{(2)}$$

$$x_{i,A} dS_i = \xi_{ijk} x_{i,A} x_{j,B} x_{k,C} dX_B^{(1)} dX_C^{(2)} = \xi_{ijk} F_{iA} F_{jB} F_{kC} dX_B^{(1)} dX_C^{(2)}$$

$$\xi_{ijk} F_{iA} F_{jB} F_{kC} = \xi_{ABC} \det F = \xi_{ABC} J$$

$$x_{i,A} dS_i = \xi_{ABC} J dX_B^{(1)} dX_C^{(2)}$$

Multiply both sides of the above equation by $X_{A,q}$, one obtains:

$$x_{i,A} X_{A,q} dS_i = \xi_{ABC} J dX_B^{(1)} dX_C^{(2)} X_{A,q}$$

$$x_{i,A} X_{A,q} = \delta_{iq}$$

$$dS_q = X_{A,q} J dS_A^o \tag{1-86}$$

Considering the identity $\dot{\det B} = tr(\dot{B} \bullet B^{-1}) \det B$ and replace B with F, one can obtain:

$$\dot{\det F} = tr(\dot{F} \bullet F^{-1}) \det F = J tr L \text{ or } \dot{J} = J divv \tag{1-87}$$

This relationship is important for defining stress in the reference configuration:

$$dS = J(F^{-1})^T \bullet dS^o$$

$$dS \bullet F = J dS^o \tag{1-88}$$

Therefore:

$$\dot{dS}\bullet F + dS \bullet \dot{F} = \dot{J}dS^{o} = JtrLdS^{o} \text{ Multiply both sides by } F^{-1}$$

$$\dot{dS} + dS \bullet \dot{F}\bullet F^{-1} = JtrLdS^{o} \bullet F^{-1} = trLdS$$

$$L = \dot{F}\bullet F^{-1} \text{ One has:}$$

$$\dot{dS} = trLdS - dS \bullet L \tag{1-89}$$

1.6.2.6.3   Volume Change

In the reference configuration:

$$dV^{0} = dX^{(1)} \bullet dX^{(2)} \times dX^{(3)} = \xi_{ABC}dX_{A}^{(1)}dX_{B}^{(2)}dX_{C}^{(3)} \tag{1-90}$$

In the current configuration:

$$dV = dx^{(1)} \bullet dx^{(2)} \times dx^{(3)} = \xi_{ijk}dx_{i}^{(1)}dx_{j}^{(2)}dx_{k}^{(3)}$$

$$dx^{(1)} = F \bullet dX^{(1)} \; dx^{(2)} = F \bullet dX^{(2)} \; dx^{(3)} = F \bullet dX^{(3)}$$

$$dx^{(1)} = x_{i,A}dX_{A}^{(1)} \quad dx^{(2)} = x_{i,B}dX_{B}^{(2)} \quad dx^{(3)} = x_{i,C}dX_{C}^{(3)}$$

$$dV = \xi_{ijk}x_{i,A}x_{i,B}x_{i,C}dX_{A}^{(1)}dX_{B}^{(2)}dX_{C}^{(3)} = JdV^{0}$$

$$dV = \dot{J}dV^{o} = J(trL)dV^{o} = (trL)dV \tag{1-91}$$

If the deformation is isochoric, then $dV = 0 \quad trL = 0$

1.6.2.6.4   Rate of Stretch Ratio

Considering the identity:

$$dxn_{i} = x_{i,A}dXN_{A} \tag{1-92}$$

$$n_{i}\alpha = x_{i,A}N_{A}$$

or

$$n\alpha = F \bullet N$$

So:

$$\dot{n}\alpha + n\dot{\alpha} = \dot{F}\bullet N = L \bullet F \bullet N = L \bullet n\alpha$$

$$\dot{n} + n\dot{\alpha}/\alpha = L \bullet n$$

$$n \bullet \dot{n} + n \bullet n\dot{\alpha}/\alpha = n \bullet L \bullet n$$

$$(n \bullet n = 1 \quad n \bullet \dot{n} = 0)$$

$$\dot{\alpha}/\alpha = n \bullet L \bullet n$$

or

$$\dot{\alpha}/\alpha = L_{ij}n_i n_j$$

$$L_{ij}n_i n_j = (D_{ij} + W_{ij})n_i n_j = D_{ij}n_i n_j$$

$$\dot{\alpha}/\alpha = n \bullet D \bullet n \qquad (1\text{-}93)$$

or

$$\dot{\alpha}/\alpha = D_{ij}n_i n_j$$

1.6.2.7   Rotation Tensor and Stretch Tensor

The deformation gradient tensor F can be decomposed into a rotational component R followed by a stretch component U or stretch component tensor V followed by a rotational R. It is named as polar decomposition. Figure 1.4 illustrates this concept. PQ is deformed into pq but can be considered as first stretched to pq' (in parallel to PQ) and then rotated to pq. R is the orthogonal rotational tensor and U and V are symmetric, positive definite right and left stretch tensors.

dX is stretched to dx' and is then rotated to dx. Then one can have:

$$dx' = U \bullet dX \qquad (1\text{-}94)$$

$$dx = R \bullet dx' \qquad (1\text{-}95)$$

Therefore $dx = R\bullet U\bullet dX$ and $F = R\bullet U$. Similarly, one can obtain $F = V\bullet R$.

Therefore:

$$F = R \bullet U = V \bullet R \qquad (1\text{-}96)$$

Since R is purely rotational, one has $dx \bullet dx = (R \bullet dx') \bullet (R \bullet dx') = dx'R^T R dx' = dx' \bullet dx'$ and $R^T R = I$.

Through this relationship and the dot product and strain tensor definitions:

One can prove that:

$$U\bullet U = C \text{ and } F\bullet F^T = V\bullet V \qquad (1\text{-}97a,\ 1\text{-}97b)$$

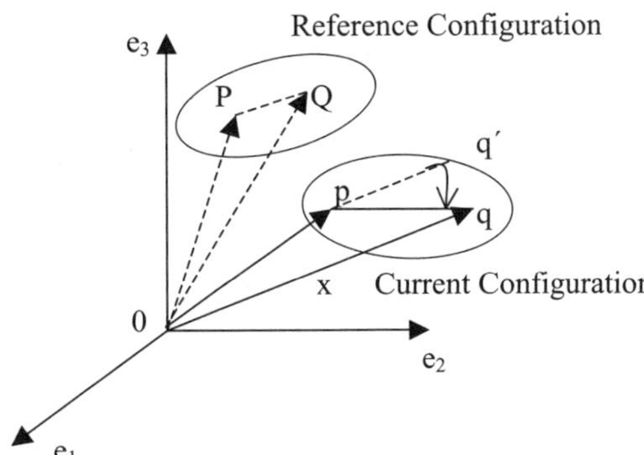

FIGURE 1.4   Illustration of polar decomposition.

1.6.3   Stresses

1.6.3.1   Cauchy Stress

Considering an area element ΔS of current configuration, having applied forces Δf_i and momentum ΔM_i on it, the Cauchy stress principle asserts that the following limits (as the area ΔS approaches zero) exist:

$$\lim_{\Delta S \to 0} \frac{\Delta f_i}{\Delta S} = \frac{df_i}{dS} = t_i^{(n)} \tag{1-98}$$

It is named the stress vector or traction vector.

$$\lim_{\Delta S \to 0} \frac{\Delta M_i}{\Delta S} = 0 \tag{1-99}$$

This means that the distributed momentum is equal to zero.

1.6.3.2   Stress Tensor

Consider an infinitesimal tetrahedron in Figure 1.5 at point P in the current configuration, the three stress vectors can be represented as:

$$t^{(e_1)} = t_1^{(e_1)}e_1 + t_2^{(e_1)}e_2 + t_3^{(e_1)}e_3$$
$$t^{(e_2)} = t_1^{(e_2)}e_1 + t_2^{(e_2)}e_2 + t_3^{(e_2)}e_3$$
$$t^{(e_3)} = t_1^{(e_3)}e_1 + t_2^{(e_3)}e_2 + t_3^{(e_3)}e_3 \tag{1-100}$$

Using summation convention, these stress vectors can be represented as:

$$t^{(e_i)} = t_j^{(e_i)}e_j, \ (i=1,2,3)$$

In Figure 1.5, the minus sign denotes the opposite direction of the stress vectors.

The Newton second law for the tetrahedron can be represented as:

$$t_i^{(n)}dS - t_i^{(e_1)}dS_1 - t_i^{(e_2)}dS_2 - t_i^{(e_3)}dS_3 + \rho b_i dV = \rho a_i\, dV \tag{1-101}$$

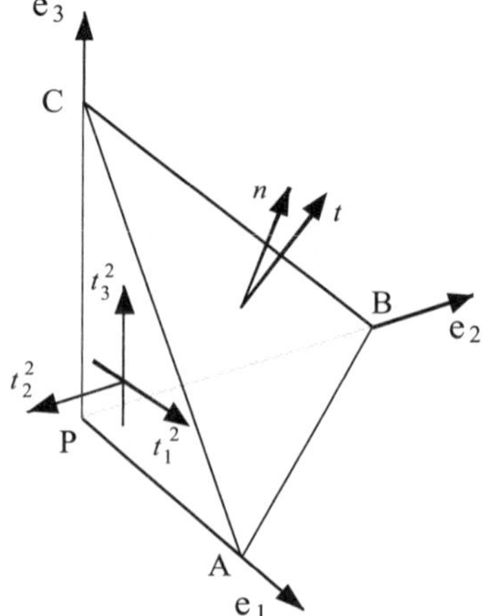

Figure 1.5   Free-body diagram for illustrating stress tensor.

Where dV is the volume of the infinitesimal tetrahedron.

Dividing both sides of the above equation by dS, and considering $n_i = \dfrac{dS_i}{dS}$ and $\dfrac{dV}{dS} = 0$ when dS approaches zero, one obtains:

$$t_i^{(n)} = t_i^{(e_1)}n_1 + t_i^{(e_2)}n_2 + t_i^{(e_3)}n_3 \tag{1-102}$$

Where $i = 1,2,3$, therefore the above equation actually represents three equations. The nine stress vector components $t_i^{(e_j)}$ as a set of quantities represent the Cauchy stress tensors. It is represented as $\sigma_{ij} = t_i^{e_j}$ $(i = 1,3; j = 1,3)$. Figure 1.6 is an illustration of the nine components of a stress tensor.

The stress vector on the plane with normal direction n can then be generalized as:

$$t_i^{(n)} = \sigma_{ji}n_j \tag{1-103}$$

or

$$t^{(n)} = n \bullet \sigma \tag{1-104}$$

1.6.3.3 Stress Tensor Transformation in Different Coordinate Systems

For the same stress vector, it can be represented using the components in two different coordinate systems, the e_i and e_i' systems. Logically (it is the same vector but different representations), the following relationship will hold.

$$t^{(n)} = t_i^{(n)}e_i = t_r^{(n')}e_r'$$

Considering $t_i^{(n)} = \sigma_{ji}n_j$ and $t_r^{(n)} = \sigma'_{rs}n'_s$

One has the following:

$$t^{(n)} = \sigma_{ji}n_je_i = \sigma'_{rs}n'_re'_s$$

$$n_r' = T_{rl}n_l \text{ and } e_s' = T_{sk}e_k$$

After some manipulations of the summed indices:

$$\sigma_{ij}n_je_i = T_{rl}T_{sk}\sigma'_{rs}n_le_k$$

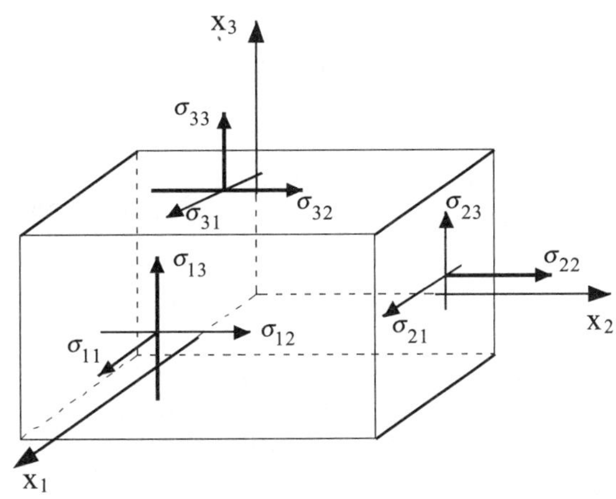

FIGURE 1.6 Schematic diagram for components of stress tensor.

Please note that $n_l e_k$ represents the nine independent bases and can be replaced as $n_j e_i$. By replacing l with j and k with i, one has:

$$\sigma_{ij} n_j e_i = T_{rj} T_{si} \sigma'_{rs} n_j e_i$$

$$\sigma'_{rs} = T_{sr} T_{rj} \sigma_{ij} \tag{1-105}$$

In matrix format, it can be represented as:

$$\sigma' = T \sigma T^T \tag{1-105a}$$

1.6.3.4   Stress Invariants
Following Section 2.5, the three invariants of the stress tensor are:

$$I_1 = \sigma_{ii} = tr\sigma \tag{1-106}$$

$$I_2 = \frac{1}{2}(\sigma_{ii}\sigma_{jj} - \sigma_{ij}\sigma_{ji}) = \frac{1}{2}\left[(tr\sigma)^2 - tr(\sigma^2)\right] \tag{1-107}$$

$$I_3 = \xi_{ijk}\sigma_{1i}\sigma_{2j}\sigma_{3k} = \det\sigma \tag{1-108}$$

1.6.3.5   Symmetry
Due to the non-existent distributed momentum (the second principle of Cauchy Stress), the stress tensor thus defined will be symmetric. Section 3.4.5 will present a proof.

1.6.3.6   Principal Stresses
For symmetric real valued tensors, there exist three directions where the stress vectors are coincident with the direction of the normal of the planes. These three stresses are the principal stresses. One can use Equations 1-31, 1-32, and 1-34 to find the principal directions and principal stresses.

1.6.3.7   Maximum and Minimum Stresses
As the stress vector is $t^{(n)} = n \bullet \sigma$, its component along the normal of the surface is:

$$\sigma^N = t^{(n)} \bullet n = n \bullet \sigma \bullet n = \sigma_{ij} n_i n_j \tag{1-109}$$

Note that σ_N is the magnitude of the stress vector projected on the normal direction. The magnitude of the stress vector projected on the tangent direction is:

$$\sigma_T^2 = t^{(n)} \bullet t^{(n)} - \sigma_N^2 \tag{1-110}$$

The maximum or minimum are applicable to σ_N or σ_T. Taking σ_N as an example, to find maximum or minimum σ_N is equivalent to find the maximum or minimum of the function:

$$f(n_i) = \sigma_{ij} n_i n_j \tag{1-111}$$

The Lagrangian Multiplier method can be used to find the solution. One can constitute an equation $g(n_i) = \sigma_{ij} n_i n_j - \zeta(n_i n_i - 1)$ with $n_i n_i - 1 = 0$. With the necessary condition of the extremal conditions, one will have:

$$\frac{\partial g}{\partial n_k} = \sigma_{ij}\delta_{ik}n_j + \sigma_{ij}\delta_{jk}n_i - 2\zeta\delta_{ik}n_i = 0 \tag{1-112}$$

Considering the symmetry of the stress tensor and the replacing function of the Delta operator, one can have:

$$\sigma_{ij}\delta_{ik}n_j = \sigma_{kj}n_j = \sigma_{ij}\delta_{jk}n_i = \sigma_{ik}n_i$$

Therefore:

$$(\sigma_{ik} - \zeta\delta_{ik})n_i = 0 \tag{1-113}$$

Comparing it with Equation 1-34, it is the same equation. Therefore, the principal stresses are actually the maximum and minimum stresses. Following the similar procedure, one can determine the maximum shear stresses.

1.6.3.8 Deviatoric and Hydraulic Stresses

The mean of the sum of the three normal stresses ($I_1 = \sigma_{ii} = tr\sigma$) is the hydraulic stress or spherical stress. It can be represented as an isotropic stress tensor:

$$\sigma_{ij}^M = \frac{1}{3}I\delta_{ij} \tag{1-114}$$

The difference between the stress tensor and its mean stress tensor is called the deviatoric stress tensor.

$$S_{ij} = \sigma_{ij} - \sigma_{ij}^M \tag{1-115}$$

1.6.3.9 Octohedral Stress

The plane whose normal makes equal angle to the directions of the principal stresses is called the octahedral plane. The directional cosines of this plane are equal to $1/\sqrt{3}$ and therefore $\sigma^N = \frac{1}{3}\sigma_{ii}$

$$\sigma_T = \sigma_{oct} = \frac{1}{3}\sqrt{\left[\sigma_1 - \sigma_2\right]^2 + \left[\sigma_2 - \sigma_3\right]^2 + \left[\sigma_3 - \sigma_1\right]^2} \tag{1-116}$$

Where $\sigma_1\, \sigma_2\, \sigma_3$ are the three principal stresses.

1.6.3.10 Piola-Kirchhoff Stresses

In calculating the engineering stresses, one often uses the force applied on a surface normalized with the surface area in the reference configuration (the original surface). The first Piola- Kirchhoff (P-K) stress is such a measure.

$$\lim_{\Delta S^o \to 0} \frac{\Delta f_i}{\Delta S^o} = \frac{df_i}{dS^o} = p_i^N \tag{1-117}$$

It should be noted that the direction N is also a measure in the reference frame. The first P-K stress is defined in such a way that:

$$p_i^N = P_{Ai}^{1st}N_A \tag{1-118}$$

$$df_i = \sigma_{ji}n_j dS$$

$$df_i = P_{Ai}^{1st}N_A dS^o$$

Considering the area transform identity:

$$n_i dS = X_{A,i} J N_A dS^o$$

$$df_i = \sigma_{ji} n_j dS = \sigma_{ji} X_{A,j} J N_A dS^o$$

$$P_{Ai}^{1st} = \sigma_{ji} X_{A,j} J$$

Since

$$X_{A,j} = F^{-1}$$

One has

$$P_{Ai}^{1st} = \sigma_{ji} F_{Aj}^{-1} J \tag{1-119}$$

The second Piola-Kirchhoff stress tensor refers a measure of the forces in the reference frame to the area in the reference frame.

$$df_B^o = P_{AB}^{2nd} N_A dS^o$$

Considering the unit normal n versus N as vector element in the domain, one would have:

$$df_i = F_{iB} \bullet df_B$$

$$df_i = \sigma_{ji} n_j dS = \sigma_{ji} X_{A,j} J N_A dS^o = F_{iB} \bullet df_B = F_{iB} \bullet P_{AB}^{2nd} N_A dS^o$$

Therefore:

$$P_{AB}^{2nd} = F_{iB}^{-1} \sigma_{ij} F_{Aj}^{-1} J \tag{1-120}$$

1.6.4 Fundamental Continuum Mechanics Equations

1.6.4.1 General Concepts

There is a set of equations that a continuum must observe during its deformation process. These equations include the conservation equations involving mass, momentum, angular momentum, energy; the second Law of Theromodyanmics; free energies; the objectivity assumptions; the strain compatibility conditions; and constitutive laws.

The divergence theorem is widely used in derving the differential format of the conservation laws. It is represented in the following three formats in terms of a scalar field, a vector field, and a tensor field.

$$\iiint_V \eta_i dV = \iint_{\partial V} \eta n_i dS \quad \iiint_V \frac{\partial a_i}{\partial x_i} dV = \iint_{\partial V} a_i n_i dS \quad \iiint_V \frac{\partial A_{ij}}{\partial x_i} dV = \iint_{\partial V} A_{ij} n_i dS \tag{1-121a, b, c}$$

1.6.4.2 Density Definition

Considering a volume element ΔV surrounding point P, the amount of mass contained in ΔV is Δm, the average density of this volume element is:

$$\rho_{ave} = \frac{\Delta m}{\Delta V} \tag{1-122}$$

When ΔV approaches infinitesimal, it becomes the density ρ at point P. In reality, the infinitesimal may not be meaningful when the volume element is smaller than a few molecules or atoms.

The density is a scalar function of position and time and thus may vary from point to point within a given body.

$$\rho = \rho(x_i, t)$$

or

$$\rho = \rho(x, t)$$

1.6.4.3 Mass Conservation (Material Derivative)

For any domain, the mass cannot be destroyed or produced. Therefore, the mass flowing in or out of a specific volume and its local mass change (within the volume) will sum up to zero.

$$\iint_S \rho v \bullet n dS + \frac{\partial}{\partial t} \iiint_V \rho dV = 0 \tag{1-123}$$

Considering that:

$$\iint_S \rho v \bullet n dS = \iiint_V \frac{\partial(\rho v_i)}{\partial x_i} dV$$

If the above equations hold for any domain, one has the following equation:

$$\frac{\partial(\rho v_i)}{\partial x_i} + \frac{\partial}{\partial t}\rho = 0$$

$$\frac{\partial(\rho v_i)}{\partial x_i} = \frac{\partial \rho}{\partial x_i} v_i + \rho \frac{\partial v_i}{\partial x_i}$$

Considering the material derivative in Section 1.6.2.5, one has:

$$\frac{d\rho}{dt} = \frac{\partial \rho}{\partial t} + \frac{\partial \rho}{\partial x_i} v_i$$

Therefore, the mass conservation or the continuity equation can be represented as:

$$\frac{d\rho}{dt} + \rho \frac{\partial v_i}{\partial x_i} = 0 \tag{1-124}$$

1.6.4.4 Linear Momentum Equilibrium Equation

For any finite domain (body), the force equilibrium equation can be represented as:

$$\iint_S t^{(n)} dS + \iiint_V \rho b dV = 0 \tag{1-125}$$

Its indicial format:

$$\iint_S n_i \sigma_{ij} dS + \iiint_V \rho b_j dV = 0$$

By applying the divergence theorem, one has:

$$\iiint_V \left(\frac{\partial \sigma_{ij}}{\partial x_i} + \rho b_j\right) dV = 0$$

If the above equation holds for any of the domain, the following will also hold:

$$\frac{\partial \sigma_{ij}}{\partial x_i} + \rho b_j = 0 \tag{1-126}$$

1.6.4.5 Angular Momentum

For any finite domain (body), the angular momentum equilibrium equation can be represented as:

$$\iint_S x \times t^{(n)} dS + \iiint_V \rho x \times b \, dV = 0 \tag{1-127}$$

The indicial format can be represented as:

$$x \times t = e_{pqr} x_p t_q e_r = e_{pqr} x_p n_j \sigma_{jq} e_r$$

$$x \times b = e_{pqr} x_p b_q e_r$$

Using the divergence theorem for surface integral one obtains:

$$\iiint_V e_{pqr} \left\{ \frac{\partial}{\partial x_j}(x_p \sigma_{jq}) + \rho x_p b_q \right\} dV = 0$$

$$e_{pqr} \left\{ \frac{\partial(x_p \sigma_{jq})}{\partial x_j} + \rho x_p b_q \right\} = 0$$

$$e_{pqr} \{ x_p \left(\frac{\partial \sigma_{jq}}{\partial x_j} + \rho b_q \right) + \sigma_{pq} \} = 0$$

Since:

$$\frac{\partial \sigma_{jq}}{\partial x_j} + \rho b_q = 0$$

$$e_{pqr} \sigma_{pq} = 0$$

or

$$\sigma_{pq} = \sigma_{qp} \tag{1-128}$$

This actually means that stress tensor is symmetric. The condition is that there is no distributed angular momentum.

1.6.4.6 Energy Conservation

The energy conservation principle can be stated as: for any domain of interest, the rate of change of the kinetic and internal energy is equal to the rate of work done by the surface tractions and body forces, plus other rates of energies entering or leaving the surface. Other energies may include all types of energies but are limited to thermal and mechanical energies in this book. While a small section may focus on some polar medium, the majority of the book deals with non-polar media.

The kinetic energy in any domain: $K(t) = \dfrac{1}{2} \int_V \rho v_i v_i \, dV$ $\hspace{2em}$ (1-129)

The rate of work by surface traction: $\int_{\partial V} t_i v_i \, dS$ $\hspace{2em}$ (1-130)

The rate of work by body force: $\int_V \rho b_i v_i dV$ $\qquad$ (1-131)

The power of mechanical force is: $P(t) = \int_V \rho b_i v_i dV + \int_{\partial V} t_i v_i dS$ $\qquad$ (1-132)

The rate of thermal energy includes heat supply r (heat supply per unit mass per unit time) and heat flux q (heat flux per unit surface area per unit time). It can be represented as:

$$Q(t) = \int_V \rho r dV - \int_{\partial V} q_i n_i dS \qquad (1\text{-}133)$$

The rate of internal energy U accounts for all the other energies stored in the medium. It is abstract at this stage. The specific internal energy u can be introduced so that:

$$U(t) = \int_V \rho u dV \qquad \dot{U}(t) = \int_V \rho \dot{u} dV \qquad (1\text{-}134a, b)$$

Using the above terms, the energy conservation law can be represented as:

$$\dot{U}(t) + \dot{K}(t) = P(t) + Q(t) \qquad (1\text{-}135)$$

$$\dot{K}(t) = \frac{d}{dt}\int_V \frac{1}{2}\rho v_i v_i dV = \int_V \rho \dot{v}_i v_i dV$$

By using the divergence theorem and the stress-surface traction relationship, one can convert the surface integral $\int_{\partial V} t_i v_i dS$ into a volume integral:

$$\int_{\partial V} t_i v_i dS = \int_{\partial V} \sigma_{ij} n_j v_i dS = \int_V \frac{\partial(\sigma_{ij} v_i)}{\partial x_j} dv = \int_V (\frac{\partial \sigma_{ij}}{\partial x_j} v_i + \sigma_{ij}\frac{\partial v_i}{\partial x_j}) dV$$

$$\int_{\partial V} q_i n_i dS = \int_V \frac{\partial q_i}{\partial x_i} dV = \int_V div(q) dV$$

$$\int_V \rho \dot{u} dV + \int_V \rho \dot{v}_i v_i dV = \int_V \rho b_i v_i dV + \int_V (\frac{\partial \sigma_{ij}}{\partial x_j} v_i + \sigma_{ij}\frac{\partial v_i}{\partial x_j}) dV + \int_V \rho r dV - \int_V div(q) dV$$

$$\int_V \rho \dot{u} dV = \int_V (-\rho \dot{v}_i + \rho b_i + \frac{\partial \sigma_{ij}}{\partial x_j}) v_i dV + \int_V \sigma_{ij}\frac{\partial v_i}{\partial x_j} dV + \int_V \rho r dV - \int_V div(q_i) dV$$

By using the momentum equation, $\left(-\rho \dot{v}_i + \rho b_i + \frac{\partial \sigma_{ij}}{\partial x_j}\right) = 0$, and

$$\int_V \sigma_{ij}\frac{\partial v_i}{\partial x_j} dV = \int_V \sigma_{ij}(D_{ij} + W_{ij}) dV = \int_V \sigma_{ij} D_{ij} dV$$

The energy equation will reduce to:

$$\rho \dot{u} - \sigma : D - \rho r + div(q) = 0 \qquad (1\text{-}136)$$

1.6.4.7 Objectivity

The objectivity of constitutive modeling includes the Principle of Equipresence, which states that an independent variable assumed to be present in one constitutive equation of a material should be assumed to be present in all constitutive equations of the same material, unless its presence contradicts an assumed symmetry of the material, or contradicts the principle of material frame-indifference or some other fundamental principle.

There are three other fundamental postulates including the Principle of Determinism for stress, the Principle of Local Action, and the Principle of Material Frame-Indifference (Malvern, 1969). The Principle of Material Frame-Indifference is described as follows.

The material frame-indifference principle actually states that an event $\{x,t\}$ (location and time) in frame x should be observed the same by the observers in another frame x^* as (vector transformation law):

$$\left.\begin{aligned} x^* &= c(t) + Q(t) \bullet x \\ t^* &= t - a \end{aligned}\right\} \tag{1-137}$$

Vectors v

$$v^* = Q(t) \bullet v \tag{1-138}$$

Second-order tensors T or S regarded as linear vector transformations:

$$u = T \bullet v \text{ or } u = v \bullet S$$

$$\left.\begin{aligned} T^* &= Q(t) \bullet T \bullet Q(t)^T \\ S^* &= Q(t) \bullet S \bullet Q(t)^T \end{aligned}\right\} \tag{1-139}$$

Deformation gradient $F(X,t)$

$$F^* = Q(t) \bullet F \tag{1-140}$$

(This two-point tensor transforms like a vector under change of frame at time t)

Motion of a medium:

$$x = \chi(X,t)$$

$$x^* = \chi^*(X,t^*) = c(t) + Q(t) \bullet \chi(X,t), \ t^* = t - a$$

$$x^* = c + Q \bullet x \tag{1-141}$$

Velocity:

$$v^* = \frac{dx^*}{dt} = \dot{c} + \dot{Q} \bullet x + Q \bullet \frac{dx}{dt} \tag{1-142}$$

1.6.5   Green Function

If L is a linear operator, a partial (or ordinary) differential equation may be written as:

$$L\,u(x,y,z) = f(x,y,z)$$

As L is a linear operator, it satisfies the following conditions for linear operator:

For two functions $f_1(x,y,z)$ and $f_2(x,y,z)$, and their corresponding solutions $u_1(x,y,z)$ and $u_2(x,y,z)$, and two scalars α_1 and α_2, the solution to $L\,u(x,y,z) = \alpha_1 f_1(x,y,z) + \alpha_2 f_2(x,y,z)$ will be $\alpha_1 u_1(x,y,z) + \alpha_2 u_2(x,y,z)$. This is an important feature. If it is known the solution u^* (Green Function) to a unit source at point $(x_0,\ y_0,\ z_0)$, that is $f(x,y,z) = \delta(x - x_0)\delta(y - y_0)\delta(z - z_0)$, then one can integrate the solution u^* to obtain the solution to any source $z(x_0,\ y_0,\ z_0)$.

Green functions are widely used in the Boundary Element Method. Interested readers may find more detailed description in Qin (2007).

Recommended Books for Further Reading

The organization of the fundamental theory part of this book was very much inspired by the following books. If readers need more detailed descriptions of the fundamental theories, these books will be very helpful.

Anderson, T.L. (1995). Fracture Mechanics, Fundamentals and Applications. 2nd Edition. *CRC Press LLC*, Boca Raton.

Betten, J. (2004). Creep Mechanics. 2nd Edition. *Springer*, New York.

Brebbia, C.A., Telles, J.C.F. and Wrobel, L.C. (1984). Boundary Element Techniques, Theory and Application in Engineering. *Springer-Verlag*, New York.

Bower, F.A. (2010). Applied Mechanics of Solids. *CRC Press, Taylor & Francis Group*, Boca Raton.

Christensen, R.M. (1982). Theory of Viscoelasticity. *Academic Press, Inc*, New York.

Cristescu, N. (1982). Viscoplasticity. *Martinus Nijhoff Publishers*, The Netherland.

Findley, W.N., Lai, J. and Onaran, K. (1989). Creep and Relaxation of Nonlinear Viscoelastic Materials, with an Introduction to Linear Viscoelasticity. *Dover*, New York.

Hertzberg, R.W. (1996). Deformation and Fracture Mechanics of Engineering Materials. Fourth Edition, *John Wiley & Sons*, New York.

Lemaitre, J. (1996). A Course on Damage Mechanics. 2nd Edition. *Springer*, Berlin.

Liu, W.K., Karpov, E.G. and Park, H.S. (2006). Nano Mechanics and Materials: Theory, Multiscale Methods and Applications. *John Wiley & Sons, Ltd*, New York.

Lubliner, J. (1990). Plasticity Theory. *Macmillan Publishing Company*, New York.

Malvern, L.E. (1969). Introduction to the Mechanics of a Continuous Medium. *Prentice-Hall Series in Engineering of Physical Sciences*, New Jersey.

Mase, G.E. and Mase, G.T. (1991). Continuum Mechanics for Engineers. *CRC Press*, Boca Raton.

Maugin, G.A. (1993). Material Inhomogeneities in Elasticity. *Chapman & Hall*, London.

Munjiza, A. (2005). The Combined Finite-Discrete Element Method. *John Wiley & Sons, Ltd.*

Nemat-Nasser, S. and Hori, M. (1999). Micromechanics: Overall Properties of Heterogeneous Materials. 2nd Edition. *Elsevier*, Amsterdam.

Oda, M. and Iwashita, K. (Eds.) (1999). An Introduction to Mechanics of Granular Materials. *Elsevier*, New York.

Paris, F. and Canas, J. (1997). Boundary Element Method. *Oxford University Press*. Oxford.

Phan-Thien, N. and Kim, S. (1994). Microstructures in Elastic Media, Principles and Computational Methods. *Oxford University Press*, Oxford.

Qin, Q.H. (2007). Green's Function and Boundary Elements of Multifield Materials. *Elsevier Ltd.* Oxford.

Qu, J.M. and Cherkaoui, M. (2006). Fundamentals of Micromechanics of Solids. *John Wiley and Sons.*, Hoboken, NJ.

Rapaport, D.C. (2004). The Art of Molecular Dynamics Simulation. 2nd Edition. *Cambridge University Press*, Cambridge.

Scarpas, T. (2004). A Mechanics based Computational Platform for Pavement Engineering. *Delft University of Technology*, the Netherland.

Segel, L.A. (1977). Mathematics Applied to Continuum Mechanics. *Dover*, New York.

Suresh, S. (2003). Fatigue of Materials. 2nd Edition. *Cambridge University Press*, Cambridge.

Timoshenko, S.P. (1951). Theory of Elasticity. 3rd Edition. *McGraw-Hill*, New York.

Ting, T.C.T. (1996). Anisotropic Elasticity. *Oxford University Press*, New York.

Wriggers, P. (2002). Computational Contact Mechanics. *John Wiley & Sons, Inc.*, New York.

Weiner, J.H. (2002). Statistical Mechanics of Elasticity. *Dover,* New York.

Zienkiewicz. O.C. (1977). The Finite Element Method. 3rd Edition. *McGraw-Hill UK Limited,* New York.

Zohdi, T.I. and Wriggers, P. (2005). Introduction to Computational Micromechanics. *Lecture Notes in Applied and Computational Mechanics,* Vol.20, *Springer-Verlag,* Berlin.

Suggested Readings

This chapter is mainly based on the teaching notes of the author. These notes are prepared following the presentations from two excellent textbooks by Malvern (1969) and Mase and Mase (1991). Malverm (1969) gave excellent descriptions about the history of the development of continuum mechanics and the physics of the concepts. The book by Mase and Mase presented the major concepts in continuum mechanics in a concise and mathematic manner. The problems for excises are also excellent. If readers need more backgrounds, please read these two books. For convenience and connections, the symbols adopted in this chapter are consistent with those used in these books.

References

Malvern, L.E. (1969). Introduction to the Mechanics of a Continuous Medium. *Prentice-Hall Series in Engineering of Physical Sciences,* New Jersey.

Mase, G.E. and Mase, G.T. (1991). Continuum Mechanics for Engineers. *CRC Press,* Boca Raton.

Qin, Q.H. (2007). Green's Function and Boundary Elements of Multifield Materials. *Elsevier Ltd.* Oxford.

CHAPTER 2

Mechanical Properties of Constituents

From the microscopic view, the properties of asphalt mixtures are determined by the properties of the constituents and the spatial arrangement of the constituents. This chapter will focus on the properties of the constituents and the implication of how these properties affect the mixture behavior. For consistency, the binder-filler mixture is defined as Mastics. The mastic is modeled as a homogeneous and isotropic viscoelastic medium. This chapter consists of five sections. Section 1 focuses on the properties of asphalt binder; Section 2 focuses on the properties of aggregates; Section 3 focuses on the properties of mastics; Section 4 focuses on the characterization of mixture properties; and Section 5 discusses nano-characterizations. There are various kinds of properties such as chemical, physical, and mechanical properties. For the interest of this book, discussion will focus on the mechanical properties only, although chemical and thermal properties affect mechanical properties in a complex coupling mode. In addition, as far as pavement performance is concerned, only properties at pavement temperatures regularly encountered will be addressed. In order to study the compaction properties, the properties at the compaction temperature and below will be the focus of this book. In this temperature range, asphalt binder is viscous, viscoelastic, or viscoplastic.

Asphalt concrete (AC) consists of approximately 5% air void by volume (not porous asphalt pavement) and 5% asphalt binder by mass, and the volumetric composition of AC is roughly 5% air void, 12% binder/mastics, and 83% aggregates plus fillers.

2.1 Binder Properties

Binder properties are very much related to temperatures. They can behave like a viscous fluid, a viscoelastic solid, or a viscoplastic solid. Depending on the temperature range, the parameters to represent these properties are significantly different. In addition, due to the variability in properties of asphalt binder as a result of composition and sources, these properties are quite different as well. Therefore, for binder properties, only models are presented. The model constants (material properties) should be characterized for any specific type of binder. These properties mainly fall into the category of deformation. Other properties such as toughness (fracturing), anisotropy, and failure/healing have not been well studied.

2.1.1 Modeling of Asphalt Binder

Asphalt binder at relatively high temperatures is usually considered a viscous fluid. Viscosity models in either linear, shear thinning, or shear hardening can be used to model their behavior. At low temperatures, it behaves like an elastic medium. In the temperature range encountered in pavement, it is a viscoelastic/viscoplastic material. Several models that are often used in AC include the Kelvin Model and the General Maxwell model for this range of temperatures. For viscoplastic behavior, one may adapt the viscoplastic models presented in Chapter 7.

Asphalt binder is a polymeric material. It consists of several constituents including asphaltenes, resins, aromatics, and saturates. Resins, aromatics, and saturates are collectively named as maltenes. These constituents have different properties (Whiteoak, 1990). Depending on the volume fractions of these constituents, asphalt binder could have quite different physical, chemical, and mechanical properties. The determination of the chemical composition of polymers can be achieved by polarity analysis (Petersen, 1984), differential thermal analysis, and differential scanning calorimetry. A detailed literature review on binder composition analysis and the binder composition mechanical properties relationship can be found in (Cheung and Cebon, 1997a; Anderson et al., 1994). For simplification, only a brief description is presented here so that readers can have some working knowledge about composition and properties, which link the entire book, from micro-structure to macro-behavior.

2.1.2 Glass Transition of Asphalt Binder

Glass transition is an important concept in understanding the properties of asphalt binder. It is a temperature at which a material transitions from a liquid to solid or vice versa. The glass transition temperature of asphalt binder ranges from −40°C to 0°C. It increases with the volume fraction of asphaltenes for the binder from the same source (Wada and Hirose, 1960; Hirose et al., 1963; Schmidt and Barral, 1965). The empirical concept of high or low temperatures is actually related to this temperature.

2.1.3 Linear Viscous Model

At temperatures well above the glass transition temperature, asphalt binder behaves like a viscous fluid. Its stress is related to the strain rate in the following equation:

$$\sigma = \eta \dot{\varepsilon} \qquad (2\text{-}1)$$

Where η is the viscosity.

Generally, this relationship is not followed. The viscosity usually varies with strain rates. If the viscosity decreases with the strain rate, it is shear thinning. The opposite is named as shear thickening. The power law model better describes the behavior of binder.

2.1.4 The Power Law Model (PLM)

At a particular temperature, the power law equation can be written as:

$$\frac{\dot{\varepsilon}}{\dot{\varepsilon}_r} = \left(\frac{\sigma}{\sigma_r} \right)^n \qquad (2\text{-}2)$$

Where n is the creep exponent, and $\dot{\varepsilon}_r$ is the strain rate at a reference stress σ_r.

If $n = 1$, the PLM model degrades into the linear viscosity model. When $n > 1$, it describes the shear thinning behavior while $n < 1$ describes the shear thickening behavior.

By including the temperature dependence (Arrhenius Equation) in the Equation 2-2, the following equation can be obtained:

$$\frac{\dot{\varepsilon}}{\dot{\varepsilon}_{op}} = \left(\frac{\sigma}{\sigma_{op}}\right)^{n_p} \exp\left(-\frac{Q_p}{RT}\right) \tag{2-3}$$

Where Q_p is the thermal activation energy, R is the universal gas constant, and T is the temperature.

The power law model is applied at relatively higher stress levels. It is confirmed by numerous experimental studies at stress levels of 100 kPa to 1 MPa, where bitumen was found to behave as a power-law material.

2.1.5 Williams-Landel-Ferry Shift Function

The viscosity of binder at different temperatures varies. Experimental results indicate that the viscosity at two temperatures T_1 and T_2 can be related by:

$$\eta(T_2) = \eta(A(T_1,T_2)T_1) \tag{2-4}$$

Where A is given by:

$$\log[A(T_1,T_2)] = -\frac{C_1(T_2 - T_1)}{C_2 + (T_2 - T_1)} \tag{2-5}$$

This is called the Williams-Landel-Ferry (WLF) shift function. C_1 and C_2 are material constants.

2.1.6 The Modified Cross Model (MCM)

The MCM is to model the transition from linear behavior to power law behavior. It is expressed as:

$$\bar{\eta} = \bar{\eta}_\infty + \frac{\bar{\eta}_0 - \bar{\eta}_\infty}{1 + \alpha \dot{\gamma}^{n_c}} \tag{2-6}$$

Where $\bar{\eta}$ is the shear viscosity,
    $\dot{\gamma}$ is the shear strain rate,
    $\bar{\eta}_\infty$ is the shear viscosity when $\dot{\gamma} \to \infty$,
    $\bar{\eta}_0$ is the shear viscosity when $\dot{\gamma} \to 0$, and α is constant.

For bitumen $\bar{\eta}_\infty$ is usually small compared with $\bar{\eta}_0$ and is rarely observable (except for non-residual bitumens). Ignoring $\bar{\eta}_\infty$, Equation 2-6 can be re-written for tensile deformation as:

$$\eta = \frac{\eta_{oT}}{1 + \beta \dot{\varepsilon}^{n_c}} \tag{2-7}$$

Where η is the tensile viscosity,
    $\dot{\varepsilon}$ is the tensile strain rate,
    η_{oT} is the limiting viscosity when $\dot{\gamma} \to 0$,
    β is a constant.

The considerable evidence strongly suggests the suitability of the modified cross model in describing the pseudoplastic flow behavior of pure bitumen at a wide range of stress levels.

2.1.7   Linear Viscoelasticity

At other temperatures, the binder may be described as linear elastic or non-linear elastic material.

Based on an experimental study by Cebon and his colleagues, they generalized the deformation mechanism in Figure 2.1. While this information is very useful, it also indicates the complexity of the deformation mechanism of asphalt binder. This complexity presents tremendous challenges for modeling and simulating the behavior of asphalt binder and AC. In other words, a universal model that covers all the temperature ranges and strain rate ranges may not be realistic. Transitional properties and viscoplastic properties were also modeled by Cebon and his colleagues (Cheung and Cebon, 1995, 1997a, 1997b).

2.1.8   Film Thickness Effect

When asphalt binder is sandwiched between two aggregate particles, its mechanical properties will be affected by the particle-binder interface friction. This can be modeled as plane strain problem as illustrated in Figure 2.2 (Cheung and Cebon, 1997b). For the

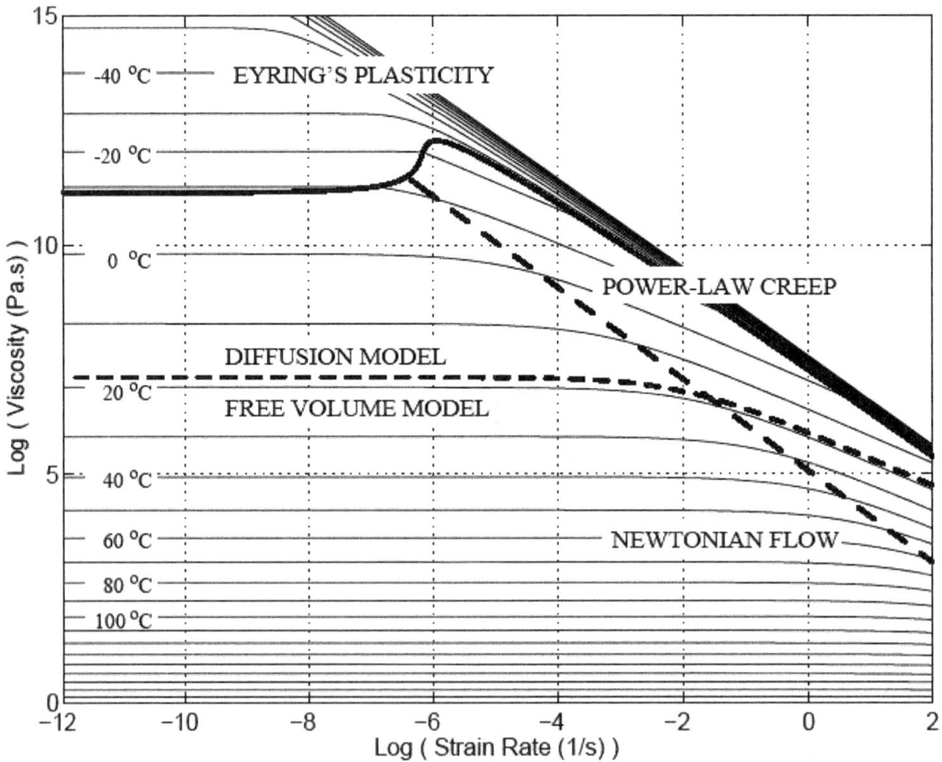

Figure 2.1   Deformation mechanism of asphalt binder (Cheung and Cebon, 1997a).

elastic contact behavior, an approximate solution to the compression of an incompressible elastic material between two rigid plates under full friction conditions for plane strain and axisymmetric deformations was obtained by Nadai (1963). Nadai suggested the use of a biharmonic stream function to obtain an approximate solution. If the width, thickness, and length are $2a$, $2h$, and $2l$, the elastic solution for the equivalent elastic modulus of compression by Nadai is presented as Equation 2-8.

$$E_n^{ai} = \frac{E_m}{8} A^2 \qquad (2\text{-}8)$$

Where $A = \frac{a}{h}$ and E_m is the Young's modulus of the material sandwiched. Obviously the equivalent modulus will increase when the film thickness is reduced. This is consistent with the observation by Wang (2007).

For viscous binder material at ambient temperatures, the steady-state deformation behavior of pure bitumen can be expressed using a generalized power law relationship such as:

$$\frac{\sigma'_{ij}}{\sigma_0} = \frac{2}{3} \left(\frac{\dot{\varepsilon}_e}{\dot{\varepsilon}_0} \right)^{\frac{1}{n}} \frac{\dot{\varepsilon}_{ij}}{\dot{\varepsilon}_e} \qquad (2\text{-}9)$$

Where σ'_{ij} is the deviatoric stress tensor, $\dot{\varepsilon}_{ij}$ is the deviatoric strain rate tensor, $\dot{\varepsilon}_e$ is the effective deviatoric strain rate, and σ_0 and $\dot{\varepsilon}_0$ are the reference deviatoric stress and deviatoric strain rate, n is material constant. If $n = 1$, the material is linear viscous (Newtonian). For many bitumen at modest stress levels, $n > 2$, indicating power law creeping behavior. Equation 2-8 can be used to represent the viscous behavior of the film material under multi-axial stress states (Cheung and Cebon, 1997a, 7b).

Consider the deformation of a plane strain contact in compression, for which the two rigid plates approach each other at a constant velocity of $-2\dot{h}$. The governing equation can be expressed as:

$$\left(\frac{\sigma_n}{\sigma} \right) = \left[\left(\frac{n}{2n+1} \right)(n+2)^{\frac{1}{n}} \left(\frac{A}{\sqrt{3}} \right)^{\frac{n+1}{n}} \right] \left(\frac{\dot{\varepsilon}_n}{\dot{\varepsilon}_0} \right)^{\frac{1}{n}} \qquad (2\text{-}10)$$

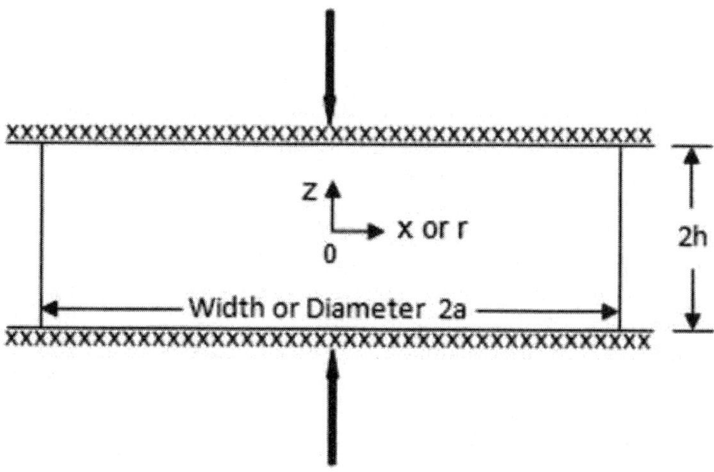

  Thin binder film model (Cheung and Cebon, 1997b).

F is compression load; A is aspect ratio of the film; 2h is thickness of the film; a is width or diameter of the plate; and l is length of the plate.

For small changes in the thickness of the film, both h and a can be assumed constant. Therefore, the deformation behavior of a plane strain contact in compression can be expressed as:

$$\left(\frac{\sigma_n}{\sigma}\right) = \left[\left(\frac{n}{2n+1}\right)(n+2)^{\frac{1}{n}}\left(\frac{A}{\sqrt{3}}\right)^{\frac{n+1}{n}}\right]\left(\frac{\dot{\varepsilon}_n}{\dot{\varepsilon}_0}\right)^{\frac{1}{n}} \tag{2-11}$$

Where $\sigma_n = F/4al$ is the nominal compressive stress and $\dot{\varepsilon}_n = -\dot{h}/h$ is the nominal compressive strain rate.

For larger changes in the thickness of the film, h and a can no longer be assumed constant.

If the initial dimensions of the contact are denoted by h_0 and a_0, the change of these dimensions can be assumed to be a constant volume process. The governing equation can be expressed as:

$$\frac{dh}{dt} = -h\dot{\varepsilon}_0\left(\frac{1}{n+2}\right)\left(\frac{2n+1}{n}\right)^{n}\left(\frac{\sqrt{3}}{A}\right)^{n+1}\left(\frac{F/4al}{\sigma_0}\right)^{n} \tag{2-12}$$

For a viscoelastic contact in which both the elastic and the viscous constitutive properties of the film material are to be included, the exact constitutive equation can be obtained by using a spring in series with a non-linear dashpot (Cheung and Cebon, 1997b).

2.1.9   Characterization of Mechanical Properties of Asphalt Binders and Mastics

This section presents the methods for characterizing the mechanical properties of asphalt binder and mastics. Only brief descriptions are given for readers as convenient referencing.

2.1.9.1   Absolute (Dynamic) Viscosity (Pa.s or Poise)

Viscosity is the resistance to flow of a fluid. The viscosity grading of asphalt binders is based on measurements at 60°C. This temperature was selected for testing due to the fact that it is the assumed maximum asphalt pavement temperature during summer in the United States. Two capillary tube viscometers are used to perform viscosity tests. They are the Cannon-Manning and Asphalt Institute vacuum viscometers (Roberts et al., 1996; Asphalt Institute, 2007). These viscometers measure the asphalt consistency at 60°C using a procedure described in ASTM D2171. They are placed in a thermostatically controlled water/oil bath at constant temperature (60°C) and their tubes are filled with binder that is forced to flow by applying a partial vacuum. The time in seconds required for it to flow between two timing marks is measured and then multiplied by the calibration factor of the viscometer tube to obtain the viscosity in poises. These test methods were introduced in the early 1960s by the FHWA, AASHTO, State DOTs, and Industry.

2.1.9.2   Kinematic Viscosity (cm²/s or Stoke)

Kinematic viscosity is equal to the absolute viscosity divided by density (for asphalt it is 1 g/cm³). It is a consistency measurement performed at 135°C using a Zeitfuchs Cross-

Arm viscometer placed in a thermostatically controlled oil bath at constant temperature. A slight vacuum is applied to the small opening of the viscometer to induce the flow of asphalt binder over the siphon section. As for the dynamic viscosity, the kinematic viscosity is determined (in centistokes) by multiplying the time (in seconds) the asphalt flows between two timing marks by the calibration factor provided with the viscometer. This method was proposed by the FHWA and AASHTO and is covered under ASTM D2170.

2.1.9.3 Zero Shear Viscosity (ZSV)

Zero shear viscosity of asphalt binder was first studied by Puzinauskas (1967, 1979) as he extrapolated the calculated apparent coefficients of viscosity to a zero strain rate. It was also assumed that the steady-state strain rate was reached at a series of shear stress levels. Zero shear viscosity has been also studied by various researchers (Anderson, et al., 2000, 2002; Shenoy 2001b, 2004; Bahia et al., 2001) since the inception of the Strategic Highway Research Program (SHRP) as an alternative to the high-temperature specification parameter for evaluating rutting resistance $G^*/\sin \delta$. In his work, Shenoy (2001a, 2004) has developed a performance-based specification parameter, $|G^*|/(1 - (1/\tan\delta \sin\delta))$ that describes the unrecovered strain in the binders more accurately as it is more sensitive to the variations in the phase angle (δ) than the original Superpave specification parameter ($|G^*|/\sin\delta$).

Dongre and D'Angelo (2003) proposed the following steady-state viscosity equation, based on the Carreau model, which can be used to calculate the zero shear viscosity from data obtained using the dynamic shear rheometer:

$$\eta = \frac{\eta_0}{\left\{1+[\lambda(d\gamma/dt)]^2\right\}^{[(1-n)/2]}} \qquad (2\text{-}13)$$

where η = steady state viscosity
η_0 = zero-shear viscosity
$d\gamma/dt$ = shear rate at steady state
λ, n = constants

The cross model is based on four parameters to describe the flow curves of pseudoplastic fluids:

$$\frac{\eta^* - \eta_\infty^*}{\eta_0^* - \eta_\infty^*} = \frac{1}{1+(K\omega)^m} \qquad (2\text{-}14)$$

where η^* = complex viscosity
η_0^* = complex zero shear viscosity
η_∞^* = limiting viscosity in the second Newtonian region
ω = angular frequency (rad/s)
K, m = constants

For dynamic shear rheometer (DSR) frequency sweep performed in the 0.1 rad/s and 100 rad/s interval it can be assumed that $\eta^* \gg \eta_\infty^*$, thus the above equation becomes:

$$\eta^* = \frac{\eta_0^*}{1+(K\omega)^m} \qquad (2\text{-}15)$$

2.1.9.4   Material Volumetric-Flow Rate (MVR)

Shenoy (2001b, 2002) showed that the material's volumetric-flow rate (MVR) in cc/10 min can be used to correctly determine the high temperature performance grade specification of asphalt binders. The MVR can be determined using a relatively inexpensive and portable flow measurement device (FMD). Through determination of the volumetric-flow rate, the fundamental viscoelastic data curves obtained from the DSR testing could be unified. The MVR is the material volume (in cubic centimeters) that is extruded in 10 min through the capillary die of a certain diameter and length of a FMD. The extrusion takes place by applying "dead weight" pressure under certain temperature conditions. The volume-flow rate is easy to measure and does not need the material density for the calculations. The flow measurement device consists of a steel cylinder provided with a calibrated thermocouple surrounded by insulated heaters that can control the temperature. The heaters can maintain a constant temperature at 10 mm above a steel die during the test. A 9.5 mm steel piston is used to extrude the material through the die, which has an internal diameter of 2.1 mm.

A modified frequency can be calculated with the following formula for $|G^*|/\sin\delta = 1$ kPa:

$$\omega\left(\frac{L^{1/n}}{MVR}\right) = 0.245 \tag{2-16}$$

where ω = modified frequency
$\quad L$ = load (kg)
$\quad n$ = power-law index corresponding to the slope of the load L

2.1.9.5   Handling and Pumping–*Asphalt Flow Characteristics*

The Brookfield rotational viscometer was adopted in Superpave to determine the viscosity of asphalt binders at construction temperatures above 100°C (Roberts et al., 1996). The test measures the original or modified binder viscosity for mixing and pumping operations. Rotational viscosity is determined by measuring the torque that maintains a constant rotational speed (20 rpm) of a cylindrical spindle submerged in asphalt at 275°F (135°C). Superpave binder specifications limit the viscosity to 3 Pa.s at 135°C. The method was proposed by the FHWA and AASHTO and is covered under ASTM D4402 or AASHTO TP48.

2.1.9.6   Permanent Deformation and Fatigue Cracking—*Rutting and Structural Cracking*

These performance parameters can be evaluated using the DSR. Samples from original binder, rolling thin film oven (RTFO) tests, and pressure aging vessel (PAV) tests are placed between a fixed base plate and an oscillating plate (Asphalt Institute, 2007). All Superpave DSR asphalt tests are performed at 10 rad/s (1.59 Hz or cycles/s). There are two types of DSRs: controlled stress and controlled strain. Controlled stress rheometers work by applying a fixed torque to oscillate the plate while the controlled strain rheometers oscillate the plate at a specified frequency and measure the required torque. Depending on the binder's stiffness, the torque necessary to move the plate at the recommended frequency will vary. DSR testing method characterizes both viscous and elastic behavior of asphalt at high and intermediate temperatures by measuring its complex shear modulus (G^*) and phase angle (δ). Results are automatically reported by the rheometer software. The complex shear modulus, G^*, is the ratio of total shear stress to total shear strain and is given by the following formula:

$$G^* = \frac{\tau_{max} - \tau_{min}}{\gamma_{max} - \gamma_{min}} \tag{2-17}$$

Where τ is the shear stress and γ is the shear strain. Permanent deformation is controlled by limiting $G^*/\sin\delta$ to values greater than 1 kPa for original binder and 2.2 kPa for RTFO aged binder. For fatigue cracking, the $G^*\sin\delta$ parameter is desired to be less than 5000 kPa at test temperatures for PAV aged material. In general, multiple single frequencies or frequency sweep (ranging from 0.0001 to 1000 Hz) are used within the DSR to construct the master curves in order to determine the rheological properties of binders and mastics. Testing procedures are covered by ASTM D 7405 or AASTHO T 315 or TP70.

2.1.9.7 Low Temperature Cracking Resistance—*Thermal Cracking*

To determine binder properties at low pavement temperatures, SHRP researchers proposed two devices: the Bending Beam Rheometer (BBR) and the Direct Tension Tester (DTT).

Bending Beam Rheometer (BBR)

The BBR apparatus is used to measure how much deflection or creeping a binder can sustain under a constant load at a constant temperature (Anderson et al., 2001; Asphalt Institute, 2007). For this purpose, asphalt samples from RTFO and PAV aging tests are formed into rectangular beams and tested at low pavement service temperatures for thermal cracking. Beam theory is used to calculate the stiffness of an asphalt beam under a creep load. During testing a constant creep load of 100 g is applied in bending mode at constant temperature for 240 s and creep stiffness $S(t)$ and creep rate (m-value) are provided by the rheometer software. The creep load simulates thermal stresses that gradually build up in a pavement when temperatures get very low. Creep stiffness (S) is the resistance of the asphalt binder to creep loading and m-value is the change in asphalt stiffness with time during loading. Creep stiffness is given by the following equation:

$$S(t) = \frac{PL^3}{4bh^3\delta(t)} \tag{2-18}$$

where P = applied constant load (N)
L = distance between beam supports (102 mm)
b = beam width (12.5 mm)
h = beam thickness (6.25 mm)
$\delta(t)$ = deflection (mm)

The Superpave specification requires that the creep stiffness not exceed 300 MPa and m-value be greater than or equal to 0.3 at 60 s. Testing procedure for BBR is covered by ASTM D 6648 or AASTHO T 313.

Direct Tension Tester (DTT)

The amount of binder strain before failure at very low temperatures is measured by the Direct Tension Tester (DTT) (Asphalt Institute, 2007). The test is performed at the temperature range where asphalt generally exhibits brittle behavior (0°C to –36°C). Furthermore, the test is performed on binders that have been aged in both a RTFO and the

PAV. Dog-bone shaped asphalt samples are tested in tension (pulled slowly) at low temperatures and constant rate (1 mm/min) to address the stiff and ductile characteristics of binders with a creep stiffness between 300 and 600 MPa. The elongation at failure is used to calculate the failure strain, which is an indication of whether the binder will behave in a brittle or ductile manner at the low test temperature. The failure strain (ε_f) is the change in length (ΔL) divided by the effective gauge length (L_e). For this test, failure is defined as the load when the stress reaches its maximum value, not necessarily the load when the specimen breaks. Failure stress (σ_f) is the failure load divided by the original cross section area of the specimen (36 mm^2). The specification requires that the failure strain must be a minimum of 1% at failure. Test specimens have a mass of approximately 3g and are 100 mm long, including the end inserts. Testing procedure is covered by ASTM D 6723 or AASTHO T 314.

2.1.9.8 Fracture (Tensile) Strength in Cleavage

Double Cantilever Beam (DCB)
The DCB consists of two loading tabs on which two very thin asphalt beam specimens with thicknesses from 0.5 to 3 mm can be attached (Harvey and Cebon, 2003, 2005). Specimens are compressed for 30 min at 30°C and then tested at temperatures ranging from −30°C to 30°C. This test (ASTM D3433) is used to obtain the mode *I* fracture energy of the asphalt bonds, which is a measure of the fracture toughness of the asphalt in the presence of flaws. Similar to a wedge test, a crack is initiated first by inserting a wedge. The specimen is then loaded by pulling apart the two beams at a certain rate; the increasing load results in increased deflection of the two beams. The applied forces are converted to stresses. At a certain critical load, the crack begins to propagate resulting in a slight drop in the load (due to the increased compliance). At this point, the beams are stopped from moving apart, thus keeping the deflection constant. Due to increasing crack length, the load and the crack length decrease. After the equilibration of the crack, the specimen is repeatedly unloaded and then loaded until total cleavage of the specimen. The data collected at different times consist of load, deflection, crack length, and compliance. Mode *I* strain energy release rate, G_I, is calculated using the following formula:

$$G_I = \frac{12F^2a^2}{EB^2H^3}\left(1 + \frac{(1+v)H^2}{4a^2}\right) \tag{2-19}$$

where F = load measured in a constant speed test at crack initiation

a = initial crack length

B, H = beam's width and depth

E and v = Young's modulus and Poisson's ratio of the adherend

Butt Joint
The butt joint test follows the same procedure as for the DCB but the asphalt adhesive area is 500 mm^2. In this test, two aluminum alloys adherends (loading tabs) with cross-sections of 22 × 25 mm are placed very close to each other and kept apart by a rubber membrane. Constant velocities were applied to separate the specimens (Harvey and Cebon, 2005). These corresponded to nominal strain rates in the range 0.03 – 100 s^{-1}. The

resulting forces are measured and converted to stresses. G_I is calculated from the area under the load displacement curve to failure using the following formula:

$$G_I = \frac{\partial W}{B \partial a} \qquad (2\text{-}20)$$

where ∂W = area under the load-displacement curve
$B \partial a$ = specimen area in the plane

The test procedure is covered by ASTM D 2094.

2.1.9.9 Monotonic Test

Monotonic, recovery, and cyclic behaviors of pure asphalt bitumens were studied by Ossa et al. (2005a) to develop a simple phenomenological model. These tests are described below. Mean stress, axial strain, and accumulated strain are determined in relation to permanent deformation.

The monotonic stress versus strain behavior of AC was determined from the constant strain-rate and constant stress creep tests. In the constant strain-rate test, a certain uniaxial tensile strain rate, $\dot{\varepsilon}$ is applied to the specimen and the resulting nominal tensile stress σ and strain ε are recorded. In the constant stress creep test, a constant nominal tensile stress is applied instantly to the specimen and the nominal tensile strain ε is recorded as a function of time, t.

For this study, a hydraulic testing machine was employed for the tensile tests on dumbbell-shaped specimens. The nominal stress in the specimen was defined by the load measured with a 2-kilonewton load cell, while the load line displacement defined the nominal strain. The specimen grips were diametrically split cups whose inner surfaces were shaped to match the heads of the dumbbell specimens. Tests have been performed at –5, 0, 10, and 20°C, which are representative operating temperatures in the United Kingdom. Samples were conditioned for 2 hours in an environmental chamber prior to testing.

2.1.9.10 Creep Recovery Test

In the creep recovery tests, a series of single load/unload tests are applied to the pure bitumen or mastic to investigate their creep recovery behavior. A stress, σ, is applied very quickly to the specimen and then held constant, and the material is allowed to creep to a specified total nominal tensile strain, ε^T. At this strain, the loading stress is released with the tensile strain monitored until the strain rate is zero, $\dot{\varepsilon} \approx 0$. The strain at this point, $\varepsilon^{pl} = \varepsilon^T - \varepsilon^r$, is the irrecoverable strain and ε^r is the recoverable strain. Such tests are repeated for a series of strains, ε^T, and creep stresses, σ, using the same specimens as for monotonic tests.

2.1.9.11 Continuous Cyclic Test

In the continuous cyclic tests, the nominal tensile stress σ is varied between σ_{min} and σ_{max}. The load levels are calculated with the formulae provided below:

$$R = \frac{\sigma_{min}}{\sigma_{max}}; \quad \sigma_m = \frac{\sigma_{min} + \sigma_{max}}{2} \qquad (2\text{-}21)$$

In this test, the loading rate is defined by the frequency, f, of the triangular waveform. The nominal tensile strains are measured as a function of time and the tests are repeated for a series of values of R, σ_m and f. The same specimens as for creep recovery tests were used and the same conditioning was applied.

2.1.9.12 Pulse Train Tests

To simulate pavement loading conditions, intermittent identical tensile stress pulses tests with a trapezoidal shape function of time are performed on dumbbell-shaped specimens. The purpose of this test is to investigate the relationship between the single load/unload behavior analyzed through the creep and creep recovery tests. During the test, a constant maximum stress σ_p in each trapezoidal stress pulse is applied for a time period Δ_p with a loading and unloading rate $\dot{\sigma} = 4\sigma_p/\Delta_p$. Also, tests at –5, 0, 10, and 20°C are performed with varoius time periods Δ_p between consecutive trapezoidal pulses at a predetermined σ_p.

2.1.9.13 Bulge Test

The bulge test was first introduced for the study of thin film mechanical properties (Vinci and Vlassak, 1996). In the first version of the test, a thin film was secured over a circular orifice and a uniform pressure applied to one side of the film to measure the deflection of the film as a function of the applied pressure. The pressure-deflection data were then converted into a stress-strain curve for the film. In this test, the results are very sensitive to small variations in the dimensions of the sample as the occurrence of small compressive stresses causes the film to buckle or wrinkle due to its small bending stiffness.

2.1.9.14 Spherical Indentation Test

The spherical indentation response of bitumen was investigated by Ossa et al. (2005b) using small cylindrical specimens (60 × 50 mm) tested in a hydraulic machine. Spherical indenters with diameters of 15 or 40 mm were used and indentation was performed to a depth less than 2 mm in a temperature-controlled chamber at 0°C. Cyclic and creep recovery indentations were conducted to model the asphalt response based on effective stress (σ^{eff}) and effective strain (ε^{eff}) concepts. Measured indentation load (F) versus indentation depth (h) and the indentation depth (h) versus time response were used for comparison with model predictions.

2.2 Aggregates Properties

2.2.1 Mechanical Properties

As indicated at the beginning of the chapter, aggregates are about 80% by volume of AC. The properties of aggregates certainly have an important influence on AC. In addition, the mineral composition and surface characteristics of aggregates also affect the bonding between aggregates and binder. In this section, typical mechanical properties of aggregates including Young's modulus, Poisson's ratio, and strength are presented. Realistic mechanical models of aggregates are also discussed. Methods to quantify surface characteristics such as aggregate shape, angularity, and texture are briefly discussed. Abrasion properties using different methods are also discussed. The abrasion properties affect the surface characteristics and the friction behavior of pavements.

Considering that typical aggregates (of high quality) have a strength of about 3.0 MPa and above, and the stress that pavement (highway pavement) can experience is about 0.5 MPa, it can be concluded that most of the aggregates should be in the linear elastic range. It might be possible that zones at aggregate-aggregate contacts may experience some plastic deformation. For computing efficiency, aggregates may be considered as linear elastic and brittle with a stress-strain relationship as linear elastic-perfect plastic. While generally aggregates are anisotropic, the random orientation effects and the material anisotropy may cancel out. For the sake of simpler material characterization and computational efficiency, the anisotropy behavior may be neglected in current literature. Strengths of rocks are typically considered to follow Mohr-Coulomb criteria.

More complicated models such as the continuum damage model, the viscoplasticity model, and the elastoplasticity model with back stresses, while more accurate, may not be necessary for modeling the behavior of aggregate. During the compacting, aggregate breakage may be modeled as brittle, linear elastic until fracture. For more advanced modeling, rate independent elastic-plasticity may be considered.

2.2.1.1 Concepts of Mechanical Properties

Aggregate strength, usually crushing strength, represents maximum compressive or tensile stress that an aggregate particle can carry before breaking. Unconfined compressive strength (UCS), aggregate crushing value (ACV), and aggregate impact value (AIV) provide strength data.

Aggregate stiffness is an aggregate particle's resistance to deformation and is based on its modulus of elasticity. The modulus of elasticity is calculated from the applied compressive or tensile stress divided by the corresponding recoverable strain.

Toughness is the resistance of aggregates to crushing and impact fracture during transport, placing, compaction, and under traffic. British tests such as aggregate crushing value (ACV) and aggregate impact value (AIV) are usually employed to determine the aggregates' toughness.

Durability or resistance to weathering is the ability of an aggregate particle to resist disintegration due to wetting and drying, heating and cooling, and freezing and thawing. Common tests such as the soundness test or the unconfined freeze-thaw test are used to assess the aggregates' susceptibility to weathering.

Hardness or wear resistance is the aggregate surface's ability to resist polishing or wearing due to rubbing and/or friction produced by externally applied forces, such as vehicles or foot traffic. Thus, the aggregate surface should wear non-uniformly to maintain a high level of surface friction. Hence, a high percentage of hard, well-bonded mineral grains is desired in the aggregate to resist the abrasive smoothing action of tires. Tests used to evaluate aggregate's hardness are Los Angeles abrasion, micro-deval abrasion, and polished stone value tests.

Degradation resistance is an aggregate's ability to resist fractionation when subjected to applied forces such as those from mixer blades, compaction, heavy wheel loads, and grinding action.

2.2.1.2 Typical Mechanical Properties from Literature

Table 2.1 presents the typical mechanical properties of mineral rocks. It should be noted that the rock properties represent those of intact rocks. When intact rocks (with distributed discontinuity) are crushed into aggregates of different sizes, the porosity of aggregates becomes smaller and therefore the modulus of aggregates are typically larger for aggregates of smaller size. Some relatively simple micromechanics approaches, such as Mori-Tanaka, self-consistent method, and Hsih-Strikman method, may be used to estimate these properties.

Rock Type	Elastic Modulus (GPa)			Standard Deviation
	Maximum	Minimum	Mean	
Granite	100	6.41	52.7	24.5
Diorite	112	17.1	51.4	42.7
Gabbro	84.1	67.6	75.8	6.69
Diabase	104	69.0	88.3	12.3
Basalt	84.1	29.0	56.1	17.9
Quartzite	88.3	36.5	66.1	16.0
Gneiss	82.1	28.5	61.1	15.9
Sandstone	39.2	0.62	14.7	8.21
Limestone	89.6	4.48	39.3	25.7
Dolostone	78.6	5.72	29.1	23.7

TABLE 2.1 Typical values of elastic modulus of intact rocks (from AASHTO HB-17, 2002).

While rocks may be modeled as isotropic/anisotropic linear elasticity, linear visco-elasticity, plasticity, and viscoplasticity, how these different models affect the mixture properties (especially the anisotropy) needs assessment.

Table 2.2 presents the typical Poisson's ratios for these rocks. For the same type of rocks, porosity may vary. In addition, when particle sizes become smaller, the porosity typically decreases as well. Table 2.3 presents the ratios between the maximum and minimum moduli of different rocks. Table 2.4 presents the typical values of the uncon-

Rock Type	Poisson's Ratio			Standard Deviation
	Maximum	Minimum	Mean	
Granite	0.39	0.09	0.20	0.08
Gabbro	0.20	0.16	0.18	0.02
Diabase	0.38	0.20	0.29	0.06
Basalt	0.32	0.16	0.23	0.05
Quartzite	0.22	0.08	0.14	0.05
Gneiss	0.40	0.09	0.22	0.09
Sandstone	0.46	0.08	0.20	0.11
Limestone	0.33	0.12	0.23	0.06
Dolostone	0.35	0.14	0.29	0.08

TABLE 2.2 Typical values of Poisson's ratio of intact rocks (from AASHTO HB-17, 2002).

Rock	Anisotropy ratio R_E
Rothbach sandstone	5.31
Diatomite	3.81
Bentheim sandstone	1.45
Hornfel	1.31
Adamswiller sandstone	1.26
Sandstone	1.23

TABLE 2.3 Deformability anisotropy ratio R_E for different rocks (from Zhang, 2005).

General Description	Rock	Unconfined Compressive Strength σ_c [1](MPa)
Carbonate rocks with well-developed crystal cleavage	Dolostone	33–310
	Limestone	24–290
	Carbonatite	38–69
Arenaceous rocks with strong crystals and poor cleavage	Conglomerate	33–221
	Sandstone	67–172
	Quartzite	62–379
Fine grained igneous crystalline rock	Andesite	97–179
	Diabase	21–572
Coarse grained igneous and metamorphic crystalline rock	Amphibolite	117–276
	Gabbro	124–310
	Gneiss	24–310
	Granite	14–338
	Quartz diorite	10–97
	Quartz monozonite	131–159
	Syenite	179–427

[1]Range of unconfined compressive strength reported by various investigators.
[2]Not including oil shale.

TABLE 2.4 Typical range of unconfined compressive strength of intact rocks (from AASHTO HB-17 2002).

Rock	ϕ_b Dry (deg)	ϕ_b Wet (deg)
Sandstone	26–35	25–34
Limestone	31–37	27–35
Basalt	35–38	31–36
Fine-grained granite	31–35	29–31
Coarse-grained granite	31–35	31–33
Gneiss	26–29	23–26
Dolerite	36	32
Amphibolite	32	

TABLE 2.5   Basic friction angles ϕ_b for different rocks (from Barton and Choubey, 1977).

fined compressive strengths of various rocks. Table 2.5 lists the basic friction angles ϕ_b for different rocks (Barton and Choubey, 1977). Table 2.6 presents the typical peak cohesion, c_i, and internal friction angle, ϕ_i, for different rocks (Goodman, 1989). Table 2.7 presents the strength anisotropy ratio R_c for different rocks without confining compression (expanded from Ramamurthy, 1993).

Rock	Porosity (%)	c_i (MPa)	ϕ_i (°)	Range of Confining Pressure (MPa)
Berea sandstone	18.2	27.2	27.8	0–200
Bartlesville sandstone		8.0	37.2	0–203
Pottsville sandstone	14.0	14.9	45.2	0–68.9
Sioux quartzite		70.6	48.0	0–203
Wolf Camp limestone		23.6	34.8	0–203
Indiana limestone	19.4	6.7	42.0	0–9.6
Hasmark dolomite	3.5	22.8	35.5	0.8–5.9
Inada biotite granite	0.4	55.2	47.7	0.1–98
Stone Mountain granite	0.2	55.1	51.0	0–68.9
Nevada Test site basalt	4.6	66.2	31.0	3.4–34.5
Schistose gneiss				
30° to schistocity	0.5	46.9	28.0	0–69
90° to schistocity	1.9	14.8	27.6	0–69

TABLE 2.6   Typical peak cohesion c_i and internal friction angle ϕ_i for different rocks (from Goodman, 1980).

It should be noted that the above mechanical properties are referred to as the properties of bulk materials. For individual granular particles these properties may be slightly different from those of bulk materials. Nevertheless, it is very difficult to characterize these properties for individual particles.

2.2.2 Morphological Properties

Shape, angularity, and texture are three related concepts in morphology analysis. They represent spatial variations (irregularities) in different dimension scales. Shape represents variation in the large dimension; angularity represents variation in the medium dimension super-imposed on shape; and texture represents variation in the small dimension super-imposed on angularity. In the frequency domain, large spatial variations are related to low frequency and small spatial variations are related to large frequency. Due to the different dimension scales, shape, angularity, and texture affect material properties in different ways. For example, the local curvature of particles at the contact is important in load transfer among the skeleton, while the texture at the interface between aggregates and binder is important for durability. There is a need to separate the shape, angularity, and texture. One of the most efficient methods is the Fourier morphological analysis.

Rock	β for σ_{cmax}	Anisotropy Ratio R_c
Fractured sandstone	90°	6.37
Diatomite	90°	3.74
Green River shale	0°, 90°	1.37
Kota sandstone	0°	1.12
Arkansas sandstone	0°	1.10
Sandstone-A (fine grained)	90°	1.75
Sandstone-B (fine grained)	90°	1.62
Sandstone-C (fine grained)	90°	1.15
Sandstone-D (medium grained)	90°	1.34
Sandstone-E (medium grained)	90°	1.23
Siltstone-A	90°	1.94
Siltstone-B	90°	2.30
Quartzitic	90°	2.19
Carbonaceous	90°	2.19

Table 2.7 Strength anisotropy ratio R_c for different rocks at unconfined compression (from Ramamurthy, 1993 and Zhang, 2005).

There are various image-based direct quantification methods for shape, angularity, and texture. These quantities are often used in modeling the contacts between aggregates, and the interfaces between aggregates and binder. As there are no consistent models that directly incorporate these quantities into modeling, no details are presented on this topic.

2.3   Mastics

Mastics are the mixtures of binder and fillers, and their properties are often characterized using similar methods as for asphalt binder. Many theories applicable to binder and mixtures are often used to characterize the mechanical properties of mastics. There are also theories that use micromechanics to evaluate mastic properties from binder properties and the filler properties. The testing methods for binder are often applicable to those of mastics as well.

2.4   Characterization of Mixture Properties

Due to the variety of asphalt binder, fillers, and aggregates, the mixture properties vary widely. It is difficult to determine the typical values of these properties. One of the important branches of pavement mechanics is to predict the mixture properties from the properties of their constituents. As a result, only the characterization methods are presented in this section.

2.4.1   Marshall Stability Test

The Marshall Stability Test (ASTM D 1559) was developed by the Corps of Engineers in the 1940s and is used to measure the strength of an asphalt mixture sample compacted to a standard laboratory compactive effort. The stability is the maximum load which the sample can withstand or the load that must be applied to cause the sample to fail. In this test procedure, a compressive load is applied to a 101.6 mm diameter and 63.5 mm thick cylindrical specimen through semicircular testing heads. The displacement is applied at a rate of 5 mm/min after the specimen is brought to a temperature of 60°C. This temperature represents the assumed maximum pavement temperature in the summer in the United States, thus being the weakest condition for the HMA mixture.

2.4.2   Hveem Stabilometer Test

The Hveem or stabilometer test is an empirical measure of the internal friction within an asphalt mixture commonly used in the United States. In this test, a vertical axial load is applied to an AC specimen that is 101.6 mm in diameter and 63.5 mm high and the displacement is measured. The specimen temperature has to be 60°C at the time of the test to simulate the most critical field condition of the mixture. Values of stability can vary between 0 (for a liquid) and 100 (incompressible solid). For asphalt mixes, values could range from 30 to 40.

2.4.3   Resilient Modulus (Repeated Load Indirect Tensile) M_R Test

Resilient modulus is the most common parameter of measuring stiffness for AC samples that are field cored or laboratory prepared. Mr test is a repeated-load indirect tension test method covered in ASTM D4123. The specimen used for the resilient modulus test is 62.5 mm by 101.6 mm diameter. The applied stress is calculated the same way as for tensile strength, but the specimen is not loaded to failure. For this test, the specimen is loaded to a stress level between 5 and 20% of indirect tensile strength. The load is typically applied for 0.1 s along with a rest period of 0.9 s. This loading sequence is repeated 100 times and the resilient axial strain is measured by linear variable displacement transducers (LVDT). The following equation is used to calculate the resilient modulus:

$$M_R = \frac{P}{Ht}(\mu + 0.27) \tag{2-22}$$

where P = applied load (lb)

H = horizontal deformation

t = sample thickness (in)

μ = Poisson's ratio

2.4.4   Dynamic Complex Modulus Test

The dynamic (complex) modulus is determined by applying sinusoidal vertical loads to cylindrical samples while measuring their deformation (Roberts et al., 1996). The test procedure is provided in ASTM D3497. The height-to-diameter ratio of the samples has to be at least 2 to 1 to minimize the effect of friction at the top and bottom of the samples. Also, the diameter of the sample must be at least twice the maximum aggregate size. The most commonly used sample sizes are 101.6 × 203.2 mm and 152.4 × 304.8 mm. The applied load is around 35 psi (241.5 kPa) and is usually defined as a percentage of the mixture strength. Specimens are tested at five different temperatures—10, 5, 21, 38, and 55°C and at five loading frequencies—25, 10, 5, 1, 0.5, and 0.1 Hz. The dynamic modulus is calculated with the following formula:

$$|E^*| = \sigma_0/\varepsilon_0 \tag{2-23}$$

where σ_0 = amplitude of applied sinusoidal loading

ε_0 = amplitude of resulting sinusoidal strain

Also, a phase angle, which is a measure of the viscous properties of the material, is calculated using the following formula:

$$\phi = \frac{T_l}{T_p} \cdot 360 \tag{2-24}$$

where T_l = time lag (s)

T_p = period of sinusoidal loading (s)

2.4.5 Static Creep (Flow Time F_t) Test

In the static creep test, a quick axial load (0.05 s) is applied to a test specimen until an axial stress, usually between 69 and 207 kPa, is achieved (Asphalt Institute, 2007). This axial stress is then maintained until the total axial strain reaches approximately 2% (20,000 microstrain) or tertiary flow (strain rate increase zone) occurs in the test specimen. In the primary zone the strain rate decreases, while in the secondary zone the strain rate is constant. The flow time is the time when shear deformation occurs under constant volume. It is also the minimum point where the rate of creep compliance changes as a function of time is zero. Typically, higher values of flow time (F_t) are associated with a better rutting resistance of the HMA mixture in the pavement.

2.4.6 Dynamic Creep (Flow Number F_n) Test

Dynamic creep testing is conducted using repeated axial stress loading cycles (usually 10,000 cycles at 63–207 kPa) and the resulting axial strain (cumulative permanent strain) is measured (Asphalt Institute, 2007). The test lasts for approximately 3 h or until excessive tertiary deformation is observed in the specimen. The axial stress is applied to cylindrical specimens, 101.6 mm in diameter and 150 mm in height, for 0.1 s (loading) followed by a rest period of 0.9 s. The starting point, or cycle number, at which tertiary flow occurs is referred to as the flow number (F_n).

2.4.7 Flexural Fatigue Test

The flexural fatigue test is conducted to evaluate the fatigue life of a small HMA beam specimen (380 mm long x 50 mm thick x 63 mm wide) by subjecting it to repeated flexural bending until failure. The beam specimen is prepared either in the laboratory or is sawed from field compacted HMA. The test is conducted in accordance to the procedures in AASHTO T 321 as repeated haversine loads (0.1-second load and 0.4-second unload) are applied at the third points of the beam specimen. Deflection is measured at the center of the beam. The flexural fatigue test can be conducted in controlled stress or controlled strain mode.

Flexural stiffness (S) can be calculated with the following formula:

$$S = \sigma_t / \varepsilon_t \tag{2-25}$$

where σ_t = maximum tensile stress (Pa)
ε_t = maximum tensile strain

2.4.8 Shear Modulus Test

The shear modulus test, provided by the AASHTO T 320 standard, is used to describe the viscoelastic behavior of an asphalt mix. In this test, a sinusoidal shear strain of 0.01% (peak-to-peak) is applied to a Superpave-gyratory-compacted specimen at 10 frequencies (0.031 to 10 Hz). Specimens are usually 150 mm in diameter and 50 mm in height. A diameter-to-height ratio of 3:1 or greater should be used for all specimens. The obtained shear stress and strain are then used to calculate the complex shear modulus (G^*) of the asphalt mixture. This test can be used to determine the effect of loading frequency (traffic speed) and temperature on an asphalt mixture. The results of this test are used as an input into the mechanistic-empirical design guide for pavement thickness design.

2.4.9   Indirect Tensile Test (IDT)

The IDT procedure is provided by the AASHTO T 322 standard and measures the creep compliance and strength of asphalt mixtures using indirect tensile loading techniques at intermediate to low temperatures (<20°C). It analyzes mixtures for low-temperature cracking. In this test, a compressive load is applied across the diametrical axis of a cylindrical specimen. Due to the mechanics of the test, a nearly uniform state of tensile stress is placed across the diametrical plane. The testing machine consists of a closed-loop servo-hydraulic, or mechanical screw system capable of applying static loads as low as 5 N. A complete analysis of the test results requires that creep compliance and tensile strength be measured at temperatures varying from –10°C to 20°C.

2.4.10   Four-Point Bend Beam Test for Anisotropic Properties

This test is similar to the standard three-point bend test except the locations of two concentrated loads are not at one-third of the span, but 95 mm away from the support (Li and Wang, 2008). The location of the concentrated load allows enough space to attach the extensometers between loading rollers. During a test, load (measured with a 2-kilonewton cell), compressive deformation (strain) on the top of the beam, and tensile deformation (strain) at the bottom of the beam are recorded every 0.1 s. These parameters are measured with strain-gauge extensometers with a gauge length of 25 mm attached at the mid-span of the beam. The load is applied to the beam at a rate of 5 N/s so that the target peak load can be reached in a reasonable amount of time without introducing significant creep deformation.

2.4.11   True Triaxial (Cubical) Test

This flexible boundary type of device was originally developed by for multiaxial testing of rock materials. The stress-controlled cubical testing device consists of six main components: (1) a steel frame, (2) six wall assemblies, (3) a deformation measuring system, (4) a stress application and control system, (5) six rigid membranes, and (6) a data acquisition and process control system (DA/PCS) (Wang et al., 2005). The frame functions are: (1) to apply external multiaxial load to the cubical asphalt concrete specimen, and (2) to serve as the reaction structure for the application of the total normal stresses to the top, bottom, and lateral faces of the cubical asphalt concrete specimen. Pressure limits are 132.5 MPa for uniaxial pressures and 64.52 MPa for a hydrostatic loading condition. The deformation of the cubical HMA specimens (100 mm long) is measured at three points on the top, bottom, and each of the four lateral faces using three LVDTs per face. Typically, the cubical specimens are loaded in isotropic compression (IC) in each direction followed by triaxial compression (TC), triaxial extension (TE), simple shear (SS), conventional triaxial compression (CTC), conventional trixaial extension (CTE), and cyclic CTE. Elastic moduli and Poisson's ratios are calculated from experimental data.

2.4.12   Gyratory Testing Machine (GTM)

The GTM was developed by the Corps of Engineers as an effective tool in the evaluation of asphalt mixture quality (Roberts et al., 1996). It can also be used for mix design and shear strength measurement during compaction. This machine can compact asphalt mixtures using a kneading process that simulates the action of rollers during pavement compaction. Parameters such as the vertical pressure, gyration angle, and number of gyrations can be adjusted to simulate field compaction equipment and traffic conditions. The applied vertical pressure is 828 kPa, which is similar to truck tire inflation pressures. For strength measurement, the pressure required for producing the desired gyration angle is determined and then converted to shear strength. Typically, one degree is used for the gyration angle and 300 revolutions for compaction. Also, the gyratory shear index (GSI), which is determined by dividing the intermediate gyration angle by the initial angle, is a measure of mixture stability as it is related to permanent deformation.

GSI values around 1 indicate stable mixtures, while values above 1.1 indicate unstable mixtures.

2.4.13   Continuum Damage Mechanics Approach

Continuum damage mechanics is used for modeling the changes in the stress-strain behavior of AC. In this approach, the state of a damaged body, represented as a homogeneous continuum, is quantified based on internal state variables (ISVs). The scale for this representation is much larger than the defect sizes in the damaged body. Typically, a suitable damage evolution law is controlling the growth of damage. The material's stiffness is determined by fitting the theoretical model to the existing experimental data, being a function of the internal state variables and depending on the extent of the damage.

To describe the deformation behavior and performance of AC, a viscoelastoplastic continuum damage (VEPCD) model was developed that is based on the theory of viscoelasticity, continuum damage mechanics, and the theory of viscoplasticity. Through this model the AC behavior is studied in tension and compression.

2.4.14   VEPCD Model Calibration in Tension

The VEPCD model is based on the strain decomposition principle developed by Schapery in 1999, which states that the total strain can be comprised of viscoelastic and viscoplastic strain (Kim, 2009). Also, linear viscoelasticity (LVE), viscoelastic continuum damage (VECD) model and viscoplastic (VP) models are employed for the calibration of the VEPCD model. The LVE characterization is based on complex modulus testing at various temperatures and frequencies. The time-temperature shift factor from this testing is used in converting the physical time to reduced time, while the relaxation moduli, obtained from the dynamic modulus and phase angle master curves, are used in the pseudostrain calculation.

To calibrate the VECD model, constant crosshead rate monotonic test results are used as stress-strain data with minor viscoplasticity required. The data are obtained for four loading rates at 5 and 40°C. The calibration procedure involves calculation of pseudostrains and relaxation modulus in the Prony series as well as damage parameter (S) and normalized pseudo-secant modulus (C). The C-S relationship represents the damage characteristic relationship on which the VECD model is based.

For VP model calibration, the permanent strains after the rest periods in monotonic cycling loading can be considered viscoplastic strain. Monotonic testing was performed at temperatures ranging from $-10°C$ to $40°C$ and strain rates of $10^{-5}/s$ to $0.1/s$.

The VEPCD model (developed for the Maryland Superpave mixture) is validated by the thermal strain restrained specimen tensile (TSRST) tests performed on AC beam specimens ($250 \times 50 \times 50$ mm) at the Turner-Fairbanks Highway Research Center.

2.4.15   VEPCD Model Calibration in Compression

In this case, compression data are used under same calibration procedure for tension modeling regarding the LVE and VECD models (Kim, 2009). The tension-compression tests were performed at a strain amplitude below 70 micro-strains. Also, for viscoelastic damage characterization, constant crosshead rate monotonic test results at 5°C were used. Pseudo-secant modulus (C) and damage parameter (S) were calculated based on stresses and pseudostrains. At peak stresses acquired from different tension rate tests, C values were between 0.3 and 0.35.

Repetitive creep and recovery tests with fixed time and stress are performed for the VP model calibration. As in the tension calibration, the permanent strains at the end of rest periods are used to determine the p, q, and Y (VP) model coefficients. Creep compliance obtained from the complex modulus, damage characteristic relationship (C-S), and VP model coefficients (p, q, and Y) are used to verify the VEPCD model in compression.

2.4.16   Permanent Deformation Continuum Damage

The constant rate of strain tests to failure is used to calibrate the uniaxial continuum damage model parameters, $C(S)$ (damage function), α (material constant), and $a(T)$ (shift factor). Compression tests are conducted at multiple strain rates and at low temperatures to minimize the viscoplastic effects (Kim, 2009). A nonlinear optimization procedure is used to calibrate the damage function $C(S)$ and the material constant, α. Also, the uniaxial form of the damage model can be generalized to confined triaxial conditions using a dual pseudostrain energy density function of axial pseudostrain and confining pressure (Ha and Schapery, 1998).

Confined strain rate to failure tests, conducted at different conditions from those used for the model calibration, were used to validate the multi-dimensional damage model (viscoelastic, viscoplastic, and damage).

2.5   Nanoscale Characterization

In recent years, nanoscale modeling and simulation offer significant insights into the behavior of asphalt binder, mastics, and asphalt binder-aggregate interfaces. Researchers in the AC field have also started applying these advanced tools. The following presents just a few of these methods.

2.5.1   Atomic Force Microscope (AFM)

The atomic force microscope (AFM) or scanning force microscope (SFM) is a very high-resolution type of scanning probe microscopy, with resolution of fractions of a nanometer (Bonnell, 2001). It provides 3D, real-space images of various materials based on detecting the local interaction between a small probe tip and a surface. Depending on the separation between probe tip and sample, a variety of tip-surface interactions can be measured with the force microscope. In the AFM, the tip is attached to a spring in the form of a cantilever and as the tip moves over the surface, the cantilever bends back and forth in the z direction. A laser beam is directed onto the cantilever and as the cantilever bends, the movement of the reflected beam is detected by a photo diode. The movement of the tip over the surface is controlled by a piezoelectric ceramic, which can move in the x, y, and z directions in response to applied voltages. Sample scanning is carried out by the movement of the piezo ceramic in the x and y directions. A feedback circuit integrates the photo diode signal and applies a feedback voltage to the z piezo to exactly balance the cantilever bending. Since the probe force is proportional to the cantilever bending, this is constant. There are two operating modes for the AFM: contact-mode and intermittent-contact mode. Contact-mode cantilevers work well with reasonably hard materials such as metals, ceramics, and most polymers and the surface topography should not have abrupt edges or tall, steep features. Intermittent-contact cantilevers generally work well on all surfaces. They are more expensive and a bit more difficult to use, but they can accurately image surfaces that are very soft (e.g., organics, polymer coatings) and surfaces with steep features. Both types of cantilevers have the ability to detect the elastic and adhesive properties of surface materials.

When using the vibrational characteristics of the cantilever (classical 1-D, fixed-free beam) its mechanical resonant frequency can be determined using the dimensions of the structure and material properties from which it is made. Thus, this frequency is related to the cantilever spring constant (k) and material density (ρ) according to the following equation:

$$\omega_0 = \frac{t}{l^2}\sqrt{\frac{E}{\rho}} = \sqrt{\frac{k}{m_{eff}}} \qquad (2\text{-}26)$$

where m, t, and l are cantilever parameters, and E is the elastic modulus

2.5.2 Scanning Tunneling Microscope (STM)

A scanning tunneling microscope (STM) is a powerful instrument for imaging surfaces at the atomic level based on electron tunneling (Bonnell, 2001). Electron tunneling is a phenomenon that relies on quantum mechanics which allows a finite number of electrons to cross a very thin barrier between two closely spaced metals or semiconductors. Images are obtained by "rastering" a tip over the surface of a sample. For an STM, good resolution is considered to be 0.1 nanometer lateral resolution and 0.01 nanometer depth resolution. With this resolution, individual atoms within materials are routinely imaged and manipulated. The STM can be used not only in ultra high vacuum, but also in air, water, and various other liquid or gas ambients, and at temperatures ranging from near zero Kelvin to a few hundred degrees Celsius. During scanning, when a conducting tip is brought very near to the surface to be examined, a bias (voltage difference) applied between the two allows electrons to tunnel through the vacuum between them. The resulting "tunneling current" is a function of tip position, applied voltage, and the local density of states (LDOS) of the sample. Images are acquired by monitoring the current as the tip's position scans across the surface.

STM can be operated in two modes: constant current mode where a constant bias is applied between the sample and the tip, and constant height operation mode where a constant height and bias are simultaneously maintained. The tip motion for the three directions (x, y, and z) is controlled by piezoelectric elements. STM requires extremely clean and stable surfaces, sharp tips, very good vibration control, and sophisticated electronics.

For a metal tip brought very close to a flat sample surface, the resulting current (I) is given by the following equation:

$$I = C\rho_t\rho_s e^{z*k0.5} \qquad (2\text{-}27)$$

where ρ_t and ρ_s are the sample and tip electronic structures
$\quad z =$ the tip-sample separation
$\quad k =$ the wave number, and
$\quad C =$ a constant

2.5.3   Nanoindenter

The nanoindenter is a popular nanomechanical test instrument for measuring mechanical properties such as hardness, elastic modulus, fracture toughness, wear resistance, coefficient of friction, and viscoelastic properties of thin films, coatings, and particles with nanometer spatial resolution using controlled indentation of surfaces (Oliver and Pharr, 1992; Vinci and Vlassak, 1996; Fischer, 2004; Kelsall, et al., 2005). Nanoindenter dimensions are very small, less than 50 μm in some cases. They are made with precise angular geometry in order to achieve the highly accurate readings required for nanoindentation. Nanoindentation, also called *depth sensing indentation* or *instrumented indentation*, is using instruments that can continuously record small loads and displacements with high accuracy and precision. Using a sample stage controlled by piezoelectric drivers, a nanoindenter can be positioned with submicron accuracy within the microstructure of a sample. Useful data such as Young's modulus and hardness can be obtained from the loading-unloading versus displacement curve after testing.

Many types of nanoindenters are currently in use based on their tip geometry. Some available geometries are three- and four-sided pyramids, wedges, flat, cones, cylinders, filaments, and spheres. Nanoindenters such as Berkovich, cube corner, Vickers, and Knoop have been used extensively for hardness testing. To meet the high precision required, nanoindenters must be inspected and measured with equipment and standards provided by the National Institute of Standards and Technology (NIST). Nanoindenters are made mostly of diamond and sapphire, although other hard materials can be used such as quartz, silicone, tungsten, steel, tungsten carbide, and almost any other hard metal or ceramic material. Diamond is the most commonly used material for nanoindentation due to its properties of hardness, thermal conductivity, and chemical inertness.

References

AASHTO (2002). Standard Specifications for Highway Bridges. 17th Edition, *American Association of State Highway and Transportation Officials*, Washington DC.

Anderson, D.A., Christensen, D.W., Bahia, H.U., Dongre, R., Sharma, M.G., Antle, C.E. and Button, J. (1994). Binder characterization and evaluation. *Volume 3: Physical characterization. Strategic Highway Research Program (SHRP) A369 Report*, National Research Council, Washington, DC.

Anderson, D.A., Marasteanu, M.O., Mahoney, J.M. and Stephens, J.E. (2000). Factors affecting variability in strategic highway research program binder tests. *Transportation Research Record*, No. 1728, pp.28–35.

Anderson, D.A., Lapalu, L., Marasteanu, M.O., Le Hir, Y.M., Planche, J.P. and Martin, D. (2001). Low-temperature thermal cracking of asphalt binders as ranked by strength and fracture properties. *Transportation Research Record*, No. 1766, pp.61–68.

Anderson, D. A., Le Hir, Y.M., Planche, J.P. and Martin, D. (2002). Zero-shear viscosity of asphalt binders. *Transportation Research Record*, No. 1810, pp.54–62.

Asphalt Institute (2007). The Asphalt Handbook, *MS-4*, 7th Edition.

Bahia, H. U., Hanson, D.L., Zeng, M., Zhai, H., Khatri, M.A. and Anderson, R.M. (2001). Characterization of modified asphalt binders in Superpave mix design, *NCHRP Report*, No. 459.

Barton, N. and Choubey, V. (1977). The shear strength of rock joints in theory and practice. *Rock Mech.* Vol.10, pp.1–54.

Bonnell, D.A. (2001). Scanning Probe Microscopy and Spectroscopy: Theory, Techniques, and Applications. 2nd Edition. *John Wiley and Sons Inc.*, Hoboken, NJ.

Cheung, C. Y. and Cebon, D. (1995). Experimental study of pure bitumens in tension, compression, and shear. *J. of Rheol.* Vol.41, No.1, pp.45–73.

Cheung, C. Y., and Cebon, D. (1997a). Deformation mechanisms of pure bitumen. *J. Mater. Civ. Eng.*, Vol.9, No.3, pp.117–129.

Cheung, C. Y. and Cebon, D. (1997b). Thin Film Deformation Behaviour of Power Law Creeping Materials. *Journal of Engineering Mechanics*, Vol. 123, No.11, pp.1138–1152.

Dongre, R. and D'Angelo, J. (2003). Refinement of Superpave high-temperature binder specification based on pavement performance in the accelerated loading facility. *Transportation Research Record,* No.1829, pp.39–46.

Fischer-Cripps, A. C. (2004). Nanoindentation. 2nd Edition. *Springer-Verlag,* New York.

Goodman, R. E. (1980). Introduction to Rock Mechanics. *John Wiley and Sons Inc.,* New York.

Ha, K. and Schapery, R.A. (1998). A three-dimensional viscoelastic constitutive model for particulate composites with growing damage and its experimental verification. *Intl. J. of Solids and Struct.,* Vol.35, No.26–27, pp.3497–3517.

Harvey, J.A.F. and Cebon, D. (2003). Failure mechanisms in viscoelastic films. *J. of Mater. Sc.* Vol.38, pp.1021–1032.

Harvey, J.A.F., and Cebon, D. (2005). Fracture tests on bitumen films. *J. of Materials in Civil Engr.* Vol.17, No.1, pp.99–106.

Hirose, H., Tsuyuki, U. and Wada, Y. (1963). Physical properties of asphalt at different stages of blowing process. *Oyobutsur,* Vol.32, pp.29–36.

Kelsall, R.W., Hamley, I.W. and Geoghegan, M. (2005). Nanoscale Science and Technology. *John Wiley and Sons Inc.,* Chichester, England.

Kim, Y.R. (2009). Modeling of Asphalt Concrete. *ASCE Press, McGraw Hill,* USA.

Li, X. and Wang, L. (2008). Simultaneous determination of bimoduli of asphalt material with single viscoelastic beam. *Intl. J. of Pavm. Res. and Techn.,* Chinese Society of Pavement Engineering, Vol.1, No.2, pp.57–63.

Nadai, A. (1963). Theory of Flow and Fracture of Solids, Vol.2. *McGraw Hill,* New York.

Oliver, W.C. and Pharr, G.M. (1992). An improved technique for determining hardness and elastic modulus using load and displacement sensing indentation experiments. *J. Mater. Res.,* Vol.7, No.6, pp.1564–1583.

Ossa, E.A., Deshpande, V.S. and Cebon, D. (2005a). Phenomenological model for monotonic and cyclic behavior of pure bitumen. *J. of Materials in Civil Engr.* Vol.17, No.2, pp.188–197.

Ossa, E.A., Deshpande, V.S. and Cebon, D. (2005b). Spherical indentation behavior of bitumen. *Acta Materialia,* Vol.53, pp.3103–3113.

Petersen, J.C. (1984). Chemical composition of asphalt as related to asphalt durability: state of the art. *Transportation Research Record,* No.999, pp.13–30.

Puzinauskas, V.P. (1967). Evaluation of properties of asphalt cements with emphasis on consistencies at low temperature. *Proc. Assoc. Asph. Pav. Technol.,* Vol.36, pp.489–540.

Puzinauskas, V.P. (1979). Properties of asphalt cements. *Proc. Assoc. Asph. Pav. Technol.,* Vol.48, pp.646–710.

Ramamurthy, T. (1993). Chapter 13: Strength and modulus response of anisotropic rocks. *Comprehensive Rock Engineering,* Pergamon, U.K., Vol. 1, pp.313–329.

Roberts, F.L., Kandhal, P.S., Brown, E.R., Lee, D-Y. and Kennedy, T.W. (1996). Hot Mix Asphalt, Mixture Design, and Construction. 2nd Edition. *National Asphalt Pavement Association Education Foundation Inc.,* Lanham, MD.

Schmidt, R.J. and Barrall, E.M. (1965). Asphalt transitions. *J. of the Institute of Petroleum,* Vol.51, pp.162–168.

Shenoy, A. (2001a). Refinement of the Superpave specification parameter for performance grading of asphalt. *J. of Transp. Engr.,* Vol.127, No.5, pp.357–362.

Shenoy, A. (2001b). High temperature performance grade specification of asphalt binder from the material's volumetric-flow rate. *Mat. and Struct.,* Vol.34, pp.629–635.

Shenoy, A. (2002). Estimating the unrecovered strain during a creep recovery test from the material's volumetric-flow rate (MVR). *The Intl. J. of Pavm. Engr.,* Vol.3, pp.1–13.

Shenoy, A. (2004). High temperature performance grading of asphalts through a specification criterion that could capture field performance. *J. of Transp. Engr.,* Vol.130, No.1, pp.132–137.

Vinci, R.P., and Vlassak, J.J. (1996). Mechanical behavior of thin films. *Annu. Rev. Mater. Sci.,* Vol.26, pp.431–62.

Wada, Y., and Hirose, H. (1960). Glass transition phenomena and rheological properties of petroleum asphalt, *Journal of the Physical Society of Japan,* Vol.15, No.10, pp.1885–1894.

Wang, L., Hoyos, L.R., Wang, J., Voyiadjis, G. and Abadie, C. (2005). Anisotropic properties of asphalt concrete: characterization and implications for pavement design and analysis. *J. of Mat. in Civil Engr.,* Vol.17, No.5, pp.535–543.

Wang, D. (2007). Binder Film Thickness Effect on Aggregate Contact Behavior, *M.S. Thesis,* Department of Civil and Environment Engineering, Virginia Tech.

Whiteoak, D. (1990). The Shell Bitumen Handbook, *Shell Bitumen,* UK, London.

Zhang, L.Y. (2005). Engineering Properties of Rocks. *Elsevier Geo-Engineering Book Series, Vol. 4, Elsevier Ltd.*

CHAPTER 3

Microstructure Characterization

When asphalt concrete (AC) is treated as a continuum composed of three micro-continuums of aggregates, asphalt binder, and voids, microstructure of AC refers to the distributions and configurations of the three micro-continuums. The implications of microstructure may be interpreted in three different perspectives—the aggregate skeleton view, the inclusion view, and the distributed solid view. In the aggregate skeleton view, aggregate particle shape, angularity and texture, and the film thickness between two particles in contact are the quantities of interest. Correspondingly, the modeling techniques in the skeleton category include the Discrete Element Method (DEM) and the methods used in granular mechanics. The inclusion view takes aggregates as individual inclusions and air voids as cavities and uses Ashelby mechanics. In this view, the aggregate/air void shape and size, the spacial locations, and orientations are important. The distances between two air voids/cavities are also important. The third view is the distributed solid view. This view is consistent with the mixture theories where the local solid (aggregates and binder) and air void volume fractions and their gradients are the parameters.

There are several destructive approaches available including the thin sectioning and serial sectioning approaches. These approaches are hardly useful to AC where heat produced during the removal of the materials softens the asphalt binder and smears the surfaces. Nevertheless, these sectioning approaches have been used to expose a few sections for observation of the microstructure and their changes during testing, using either a microscope, a camera, or a scanner. This technique has been replaced by more advanced nondestructive approaches, such as the X-ray computed tomography (XCT) approach. This chapter will discuss the fundamentals of XCT, image analysis, and stereology-based interpretation techniques. There are four helpful reference books on XCT and stereology (Hsieh, 2003; DeHoff and Rhines, 1968; Underwood, 1970; Russ and Dehoff, 2000).

3.1 X-ray Tomography Imaging

3.1.1 Fundamentals of X-ray Computerized Tomography (XCT)

X-ray was discovered by the German physicist Wilhelm Conrad Roentgen in 1895 during his study of electrical discharge. In performing the discharge tests, Roentgen noticed some invisible and unknown rays that could penetrate into metals, cause fluorescence, and affect photographic films. He named it X-ray, an unknown ray. The first X-ray picture (radiogram), according to story, was the hand of Mrs. Roentgen. Due to this important discovery, Roentgen received his Nobel Prize in 1901. The first picture (a human hand) also inspired various applications of X-ray radiography in medical diagnosis. X-ray radiography is basically a two-dimensional (2D) picture (overall projection of the attenuation properties on one plane). Due to the projection, the three-dimensional (3D) details of the materials were smeared. XCT is a revolutionary improvement over X-ray radiography. It was invented more than a half century after the discovery of X-ray. This invention initiated a revolution in medical diagnosis; both inventors, Allan M. Cormack and Godfrey N. Hounsfield, were awarded the Nobel Prize in 1979. The following section illustrates the fundamentals of XCT.

X-rays have very high penetrating capability and can penetrate into solids such as metals, stones, and asphalt mixtures. When X-rays penetrate into a material, its intensity (energy per unit area per unit time) becomes attenuated due to the absorption and scattering of X-rays by the atoms of the material. Different materials attenuate X-rays at different rates. Therefore, by penetrating X-rays into different materials and measuring their attenuations (the difference between the intensity before penetration and the intensity after penetration), the types of materials can be determined. Figure 3.1a and Equation 3-1 present the law governing the X-ray attenuation.

$$I = I_0 e^{-ut} \tag{3-1}$$

Where I_0 and I are respectively the X-ray intensity before and after the penetration; u is the attenuation coefficient of the material and t is the thickness of the material. By knowing I_0, I, and t, the attenuation coefficient u can be determined.

Generally, materials with higher densities have larger attenuation coefficients. For a material mixed with different constituents (i.e., aggregates, asphalt binder, and voids) the determination of the spatial distribution of the constituents is much more complex and requires the use of computerized tomography. For example, for a layered material such as the one illustrated in Figure 3.1b, the attenuation relation with different materials and thicknesses can be presented in Equation 3-2.

$$I = I_0 e^{-\sum_{i=1}^{N} u_i t_i} \tag{3-2}$$

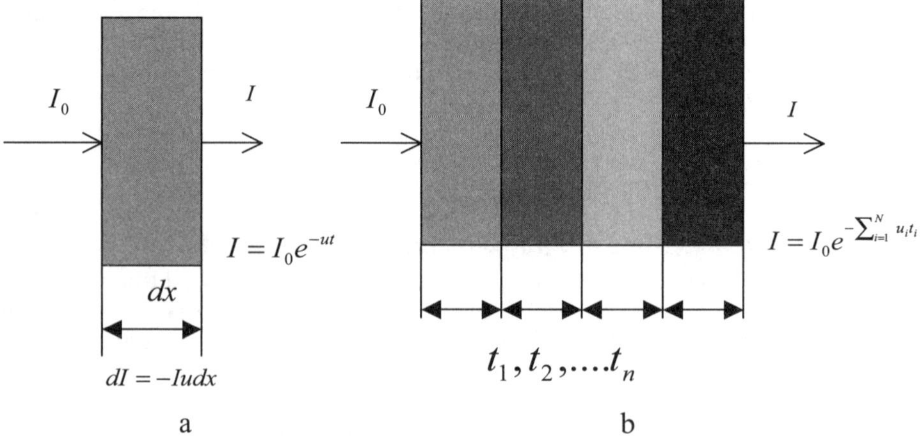

Figure 3.1 Illustration of the x-ray attenuation mechanisms.

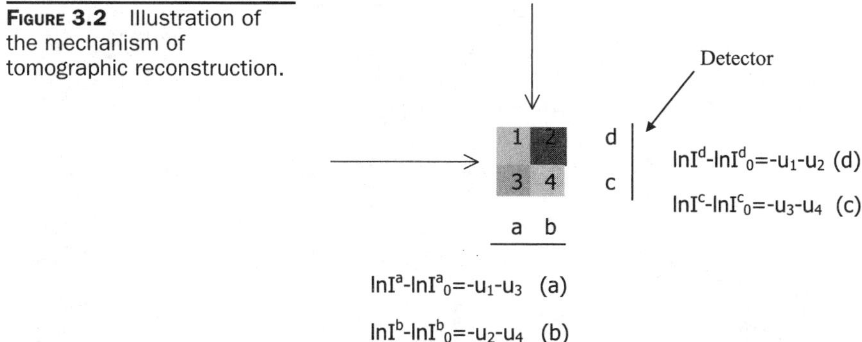

FIGURE 3.2 Illustration of the mechanism of tomographic reconstruction.

Where u_i is the attenuation coefficient of the ith material and t_i is the thickness of the ith material. By knowing I_0 and I in only one orientation, u_i cannot be uniquely determined. In this case and more complicated configurations, for example, with four different constituents (this could be the same constituent; it should be interpreted as a space divided into many small elements), each of the same unit length (1 pixel) arranged in Figure 3.2, it is required that X-rays be penetrated into two different orientations and their original and attenuated intensities be measured. In the 4-pixel case, X-rays may be penetrated in two orthogonal directions, measuring the intensities using a, b, c, and d detectors (a and b detectors can be used, but either the specimen or the detectors should be rotated 90 degrees). Four equations, as illustrated in Figure 3.2, can be obtained. By solving the four linear equations, the four attenuation coefficients can be obtained, and therefore it can be determined where different materials are located (it is actually the determinations of the attenuation coefficients corresponding to different locations). The solution to a general case is much more complicated and will not be presented.

Figure 3.3 presents the general working mechanism of X-ray tomography imaging and reconstruction. An X-ray tomography imaging system usually consists of four components including the X-ray source, the collimator (the window), the specimen manipulator, and the X-ray detector. The collimator can control the X-ray beam thickness and shape. By incrementally rotating the specimen 180 degrees, the attenuation of the X-rays in many orientations can be measured by the detector arrays or an image intensifier. By using more complicated reconstruction mechanisms, the attenuation coefficients

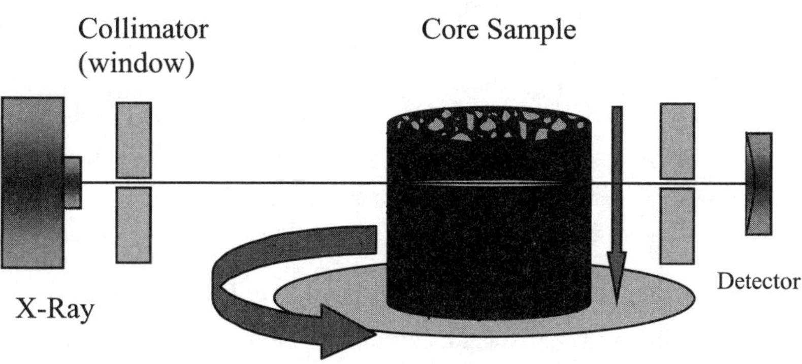

FIGURE 3.3 General mechanism of x-ray tomography scanning.

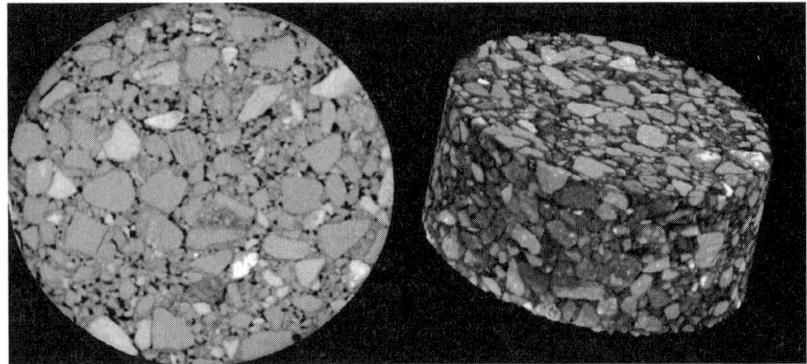

a. A Sectional Image   b. 3D Visualization

FIGURE 3.4   Sectional image and 3D image acquired through X-ray tomography imaging.

of different pixels (spatial location) can be obtained and visualized. Figure 3.4a presents one of the sectional images (mapping of the attenuation coefficients on the spatial locations). By moving the specimen down a little (Figure 3.3), another sectional image can be obtained, etc. Figure 3.4b presents the 3D visualization of 81 sectional images of an AC specimen. Unlike a conventional picture, this visualization contains details of the structure of the material.

Most current computed tomography (CT) systems include cone beam scanning, and sometimes spiral scanning. While many of the asphalt researchers use industrial CTs, some use medical CTs as well. Some more recent CT systems have reached resolutions up to 50 nm with specimen size limitations of 50 micrometers.

In recent years, XCT has also been used in the characterization of soil, cement concrete, wood, and composite materials. The next few sections describe some of the applications in AC.

3.2   Fundamental Stereology Principles and Interpretation

Stereology is a science that abstracts 3D information from 2D observations. Major resources come from Buffon Transform. A brief introduction to some of the fundamental principles is presented. The selection of these principles is based on the understanding that the geometric factors of interest to asphalt concrete researchers include the following:

1. Volume fraction of the constituents (aggregate, binder/mastics, and air voids).

2. Local volume fraction of the constituents and their gradients.

3. Specific surface area of aggregates and film thickness.

4. Spacing among the different entities (voids and aggregates).

Some fundamental concepts are presented as follows:

Random—In exact statistical meaning, "randomness" means equal chances for any individual member of a population. It is used in random sampling, or sampling without conscious choice of location and orientation. The results of any of the two random samplings are therefore independent.

Homogeneity—If the specimen is completely homogenous, then any random sectioning (or sampling) would reveal the same statistical information. However, since most materials are not completely homogenous, the section orientation should be perpendicular to the direction in which the feature (such as phase distribution) has the largest gradients.

Anisotropy—Anisotropy in this scenario may mean two aspects: the anisotropy of the morphology of particles or voids and the anisotropy of the orientation distribution of these entities.

Accessible parameters versus inaccessible parameters—Stereology should provide an unbiased estimate of the properties of higher dimensions from the observations made on lower dimensions. Some parameters can be estimated without bias (with no assumptions) and some cannot (an assumption must be made on the distributions). Volume fractions are one of the accessible parameters, while particle size, volume per particle, and number of particles per unit volume cannot be estimated without some assumptions.

3.2.1 Volume Fraction

In Figure 3.5, if dots are randomly scattered on the image and then counted within the various phases, it can be shown that if enough dots are used and the phase distribution is random, the dot density is equal to line segment fraction, the area fraction, and the volume fraction (Equation 3-3). This relationship is exact when the average is over the entire volume. Nevertheless, sampling over the entire specimen would take tremendous effort. Data based on sampling a portion of the specimen, however, are not accurate and depend on the uniformity of the microstructure.

$$P_P = \frac{N_\alpha}{N_T} = \frac{L_\alpha}{L_T} = \frac{A_\alpha}{A_T} = \frac{V_\alpha}{V_T} \qquad (3\text{-}3)$$

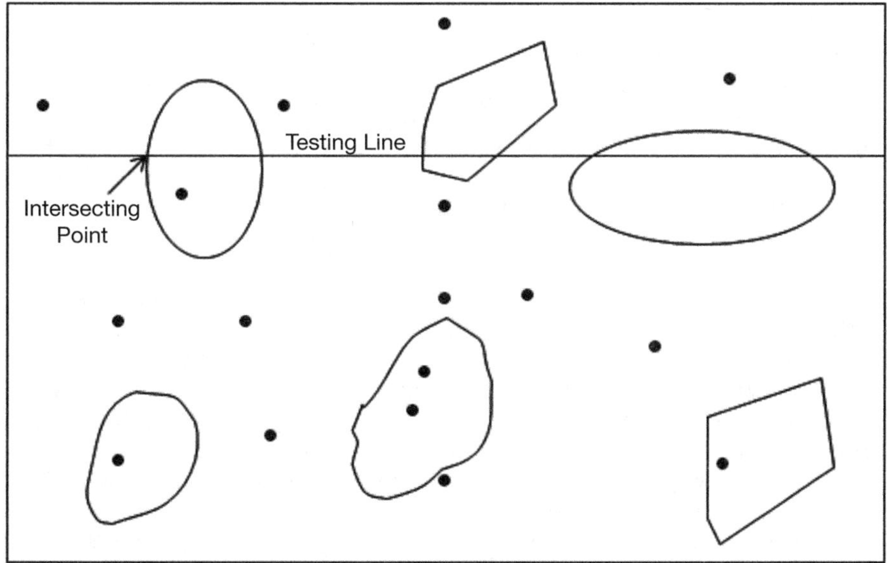

If the dots (points) represent pixels, $P_P = A_A$ (point fraction is equal to area fraction)

FIGURE 3.5 Testing line, testing dots, and intersecting points.

Equation 3-3 represents that the point fraction, line fraction, area fraction, and volume fraction are equal to one another. This concept is now understood by laymen when digital images enable people to directly talk about the length, area, and volume of images in pixels and voxels. A pixel represents a point, a length, and an area in 2D space. A voxel represents a point with volume in 3D space.

3.2.2 Specific Surface Area

A well-known relationship often used for estimating the specific surface area (Equation 3-4) of a phase is presented in Equation 3-5.

$$S_V = A/V \tag{3-4}$$

$$S_V = 2\overline{N}_L \tag{3-5}$$

Where S_V is the specific surface (area per unit area) of a phase, while $\overline{N}_L$ is the average number of intersecting points per unit length of test lines.

3.2.3 Average Spacing (Average Dimension of a Phase)

Equation 3-6 presents a relationship among $\overline{N}_A$ (number of testing planes intersecting with particles per unit area), N_V (number of particles per unit volume) and $\overline{D}_V$ (average size of all particles and in all directions).

$$\overline{N}_A = N_V \overline{D}_V \tag{3-6}$$

While Equation 3-6 is accurate and independent of the shape and size of particles, separate determinations of N_V and $\overline{D}_V$ require assumptions on the structural geometry of particles.

Even with CT methods, the above relationships may still be very useful in saving time in analyzing 3D datasets.

3.3 3D Image Reconstruction

The stereology approaches described above are very helpful in obtaining the quantities of certain parameters. They typically involve sectioning the specimen randomly or in a uniform sequence. XCT images are usually acquired in parallel sections at a fixed spacing. When the spacing is too large, directly stacking the images may not be so accurate. Frequently, the information in other planes is also needed. Both cases require the image interpolation technique. X-ray tomography sectional images can be used to reconstruct the 3D microstructure of a sample for computer simulation of material performance under various loading and environmental conditions. Meanwhile, the sample is still intact and can be used for other non-destructive or destructive macro-property tests such as rutting or fatigue tests. Therefore, X-ray tomography is one of the most effective tools to study the microstructure—macro-property relationship. However, tomography imaging has one disadvantage in that the spacing between adjacent image sections is sometimes much larger than the image resolution. For example, the spacing between the two images A and B in Figure 3.6 is 0.8 mm, while the resolution of the images is 0.3 mm/pixel. If the 3D microstructure is reconstructed by simply stacking the images, the resolutions in the transverse directions (i.e., horizontal and vertical) will be different. Figure 3.7a presents a virtual transverse cut through a stack of 76 real-world-acquired images. In this stack, the microstructure between the two sections

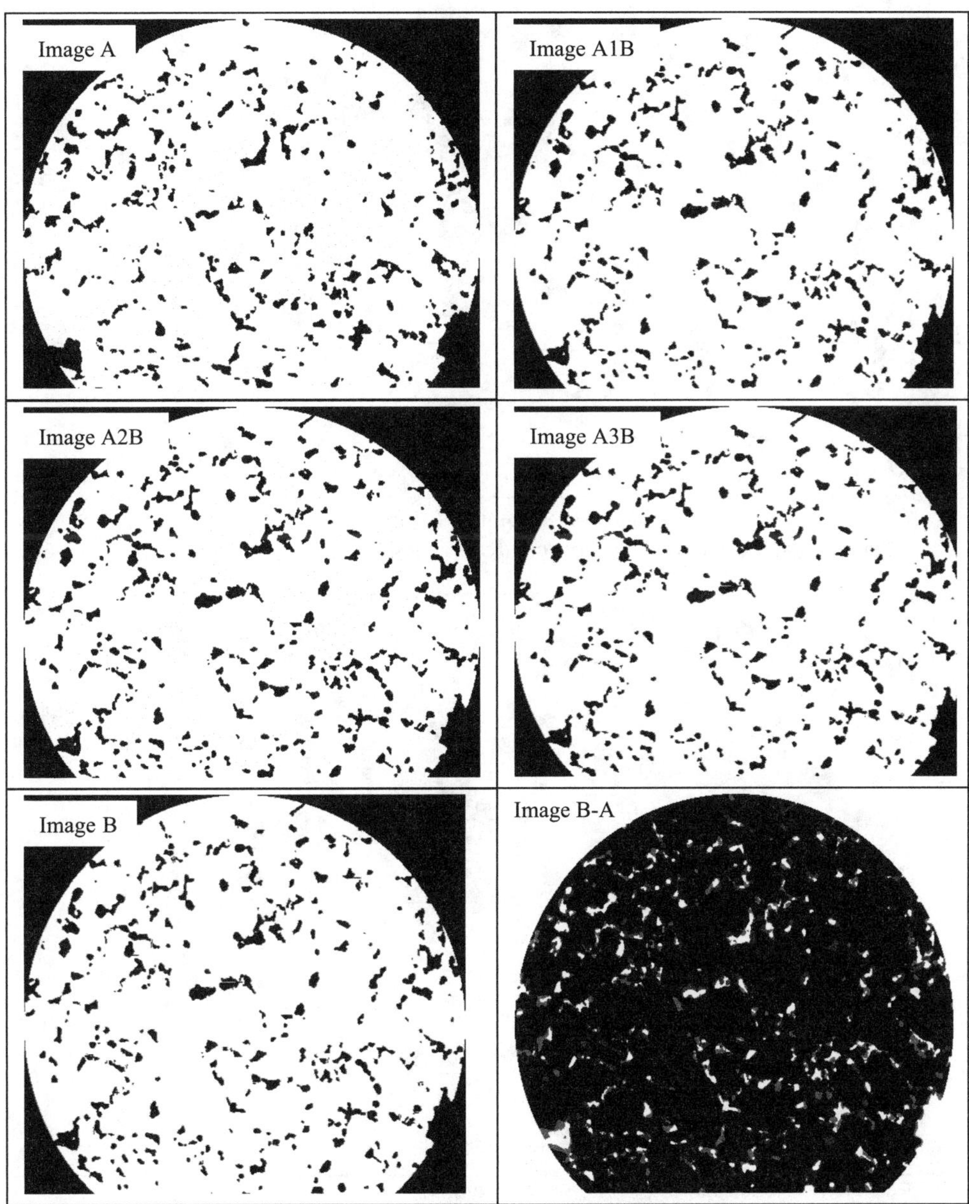

FIGURE 3.6   Illustration of the difference between image and image interpolation.

I and J is assumed to be the same as that represented by I. Obviously, direct stacking results in unrealistic microstructural features on the transverse sections and will lead to complicated formulations in feature abstraction computations such as the computation of stereology-based fabric tensors. Therefore, it is imperative to interpolate between the sectional images to provide smooth transitions and inherent connections between the images.

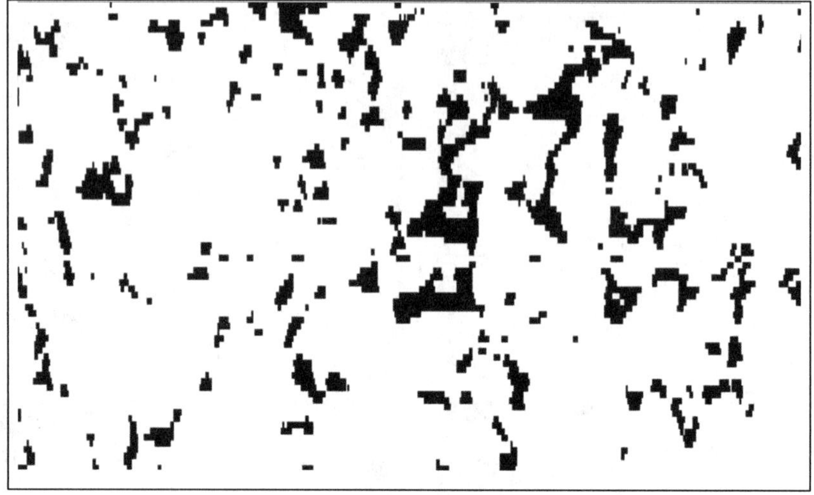

Horizontal Resolution 0.3 mm/Pixel

a. Direct Stack of the Real-World Acquired Images

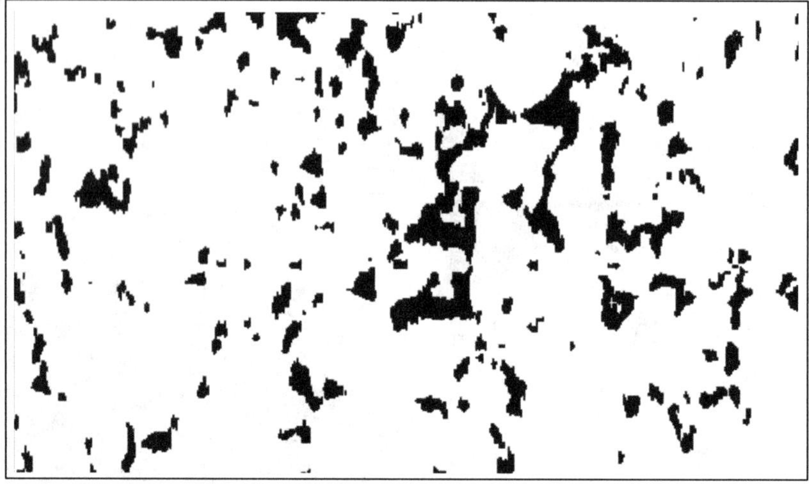

Horizontal, Resolution, 0.3mm/Pixel

b. After the Linear Proportional Erosion Interpolation

FIGURE 3.7   Illustration of the interpolation effect.

Various methods are used for interpolating tomography images in medical sciences. Methods such as Dynamic Elastic Interpolation (Lin et al., 1988), Shape-Based Interpolation (Raya and Udupa, 1990), and Morphology-Based Interpolation (Guo et al., 1995) are well accepted. Most of these methods involve complicated manipulations of images or intensive computations. However, the most influential factor is still the spacing between adjacent sections where images are acquired. With this in mind, Wang et al.

(2001b) developed a simpler method—Linear Proportional Erosion (LPE) method. The algorithm of the method is briefly summarized in the following steps:

Step 1: Obtain the difference image ($\Delta AB = B - A$) between two adjacent images A and B (image ΔAB in Figure 3.6.). For binary images, the pixel values of the difference images will have the following combinations: 1-1=0, 0-0=0, 1-0=1, 0-1=-1. The interpolation will be performed only on those pixels whose pixel value differences are equal to 1 or –1.

Step 2: The length, or the continuing counts of the pixels with the same pixel value on the different image, will then be linearly proportionally scaled according to the distance of the image to its base Image A (see Figure 3.8). This is achieved by assigning different pixel values (0, 1) to those pixels in a proportional length (Figure 3.8). The pixel values (0, 1) are determined according to their adjacent pixel values at both ends of the line sections. Figure 3.8 also presents the nine possible combinations of the adjacent pixel values, which can be further abstracted as five cases.

Step 3: Add the modified (to one phase it is erosion, to the other phase it is dilation) image resulting from Step 2 back to the base image to obtain the interpolated image.

Images $A1B$, $A2B$, and $A3B$ in Figure 3.6 are the three images interpolated between images A and B. Figure 3.7b is a virtual cut through the stack of the 76 images with three interpolated images between any two adjacent images of the 76 images. Compared with Figure 3.7a, the interpolated stack results in smoother curves. It should be noted that the images in Figure 3.7 have different resolutions in horizontal (0.3 mm/pixel) and vertical (0.2 mm/pixel) directions.

There are several commercially available programs (for example, VoxBlast) that can perform the interpolation conveniently.

3.4 Microstructural Quantities and Implications

3.4.1 Local Volume Fractions

As a heterogeneous material, the microstructure of AC is very complicated. It is related to the gradation of aggregates, the orientation and number of contacts of aggregate particles, the properties of aggregate-binder interface, and the void structure. Voids in asphalt mixtures exist in different shapes, sizes, and interconnectivity. The influence of voids on the properties of AC is related to not only the total void content, but also the spatial distribution of the voids. The Strategic Highway Research Program (SHRP) Phase I mix design actually determines the volumetrics of the constituents, including air void content, voids in the mineral aggregates (related to the binder volume), and aggregate volume and gradation. The volumetrics for all AC therefore is very similar. However, the performance of these mixes is sometimes quite different. This implies that other parameters, in addition to the overall volumetrics, may also play important roles. Local volume fractions of these constituents and their gradients may be of importance. Among the three constituents, air voids have the properties that are more drastically different from the other two. This can be illustrated through the comparisons of the stress fields of two inclusions: an inclusion of a solid and an inclusion of a cavity. The stress concentration also indicated the significant influence of voids (Wang et al., 2007).

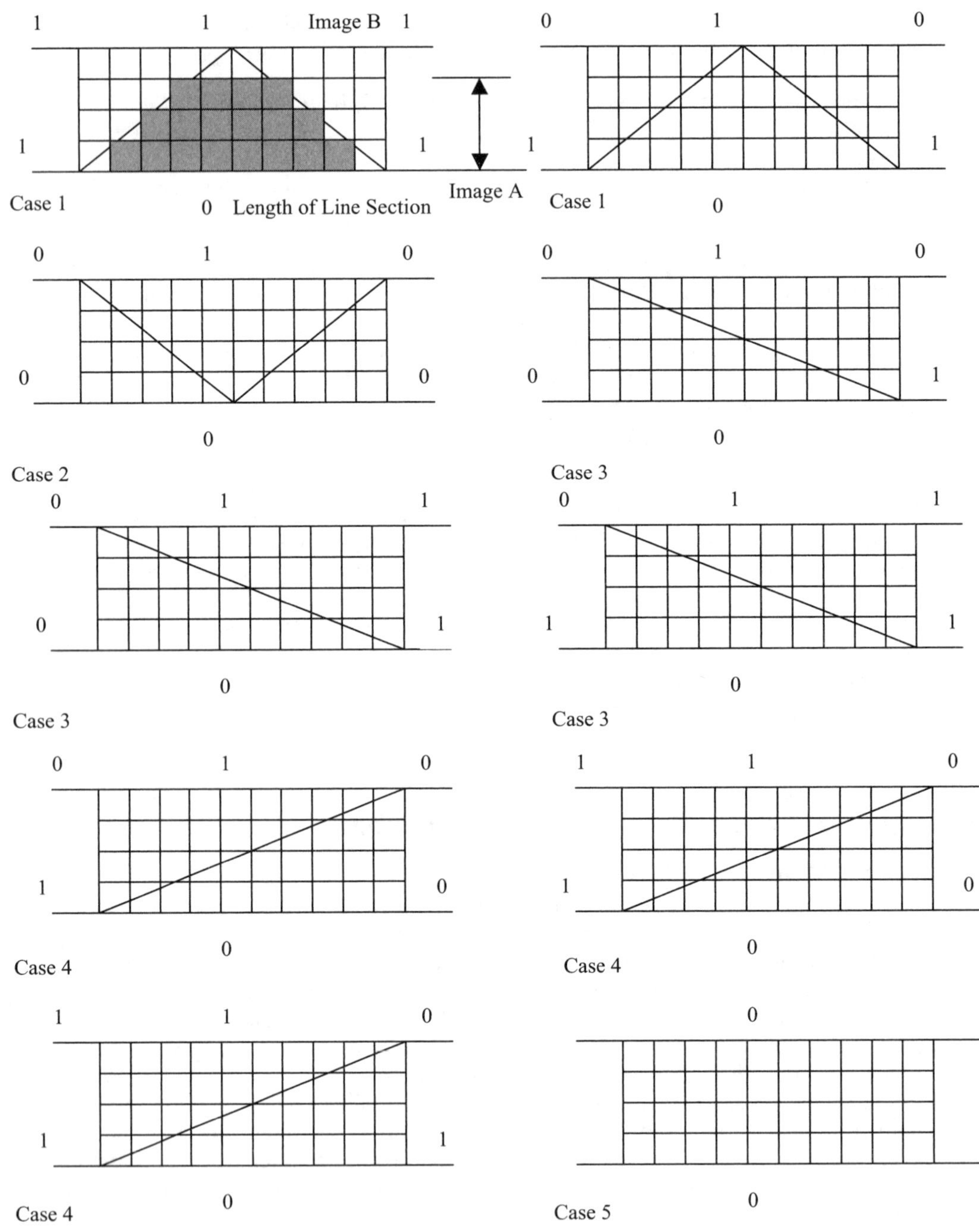

FIGURE 3.8   Illustration of the interpolation cases.

The spatial distribution of the voids may be characterized by the local volume fraction of voids and their spatial gradients. Equations 3-7 and 3-8 present the definitions of these two parameters:

$$\gamma = \frac{V_{sd}^a}{V_{sd}}$$

(3-7)

$$grad\phi = \frac{\Delta\gamma}{\Delta d}$$

(3-8)

γ = the local volume fraction of voids
V_{sd} = the volume of any sub-domain in the mixture
V_{sd}^a = the void volume in the sub-domain
$grad\phi$ = the gradient of the local volume fraction of voids
$\Delta\gamma$ = the difference of the local volume fraction of voids of two points
Δd = the distance between the two points

There are several theories including mixture theory, material force mechanics, and micro mechanics that relate the local volume fractions of the constituents and the mechanical properties of the mixture. These theories show that local volume fraction and the gradient of local volume fraction are related to the stress concentration or strain localization within the mixture. The stress concentration and strain localization are related to fatigue cracking and permanent deformation properties of the mixture, and therefore are related to the performance of the mixture. For example, in the continuum theory for granular materials (Goodman and Cowin, 1972) the stress in the mixture (granular particles and voids) under the applied strain D_{ij} can be expressed in the following equation:

$$T_{ij} = (\beta_0 - \beta v^2 + \alpha v_{,k} v_{,k} + 2\alpha v v_{,kk})\delta_{ij} - 2\alpha v_{,i} v_{,j} + \lambda D_{kk}\delta_{ij} + 2\mu D_{ij}$$

(3-9)

In Equation 3-9, α, β_0 and β are material constants, v represents the local volume fraction of solids, and $v_{,k} = \dfrac{\partial v}{\partial x_k}$ is the gradient of the local volume fraction of solids. Compared with classic linear elasticity, $T_{ij} = \lambda D_{kk}\delta_{ij} + 2\mu D_{ij}$, additional terms contribute to the stress tensor.

The second order terms of the gradients, $v_{,k} v_{,k}$, $v v_{,kk}$, and $v_{,i} v_{,j}$, are usually very small. If these terms are neglected, and D_{kk} and D_{ij} (macro variables) are considered as constants, then the stress difference between two points in the space can be expressed in the following equation:

$$T_{ij}^2 - T_{ij}^1 = \left[\beta_0 - \beta v_2^2 - (\beta_0 - \beta v_1^2)\right]\delta_{ij}$$

(3-10)

These two equations indicate that under the same macro-strain tensor D_{ij}, the stresses at two different locations would be different if their local volume fractions (local solid volume fractions) are different. This conclusion has an important implication to heterogeneous materials such as granular soil and AC: two specimens that have the same total void ratio (void content) may not have the same properties as the non-uniform void distribution will introduce additional stresses (inhomogeneity induced material force) in the material.

3.4.1.1 Local Volume Fraction Determinations

Local void volume fraction of granular soil was quantified by Frost and Kuo (1996) based on serial section and stereology principles. Wang et al. (2002) presented a method for the quantification of the gradient of local volume fraction based on XCT. It consists of five steps and is briefly described as follows.

Step 1: Obtaining XCT images and recording the thickness of slices and spacing.

Step 2: Converting gray images into binary images through determining the thresholding value, which is the value at which the overall void content of the specimen is the same as that determined through physical experimental measurements.

Step 3: Dividing the cross-section into squares. An inscribed square is constructed on the binary image to study the distribution of air voids in the sample. The square was divided into 25 small squares. Figure 3.9 shows the inscribed square and the divided small squares. The unit in Figure 3.9 is pixel; the size of the pixel is 0.3 mm.

Step 4: The areas of voids in every small square were measured. Then, the local void volume fraction and its gradient were calculated. By stereology principles (Equation 3-3), the volume fraction of the constituents is equal to their area fractions. Then, the volume fraction of voids in every small square is calculated using the following equation:

$$\gamma = \frac{a}{A} \tag{3-11}$$

Where a denotes the total area of voids in one square, A denotes the area of the square, and in the image, the area of every small square (A) is 449.4 mm^2.

Step 5: Calculating the gradient of the local void volume fraction. In every square, gradients in X, Y, and Z directions are calculated. However, the formulation is different for the small squares at the boundary and the small squares inside. For the small square at the boundary, the following equation is used to calculate the gradient:

$$grad\phi = \frac{\gamma_d}{d} \tag{3-12a}$$

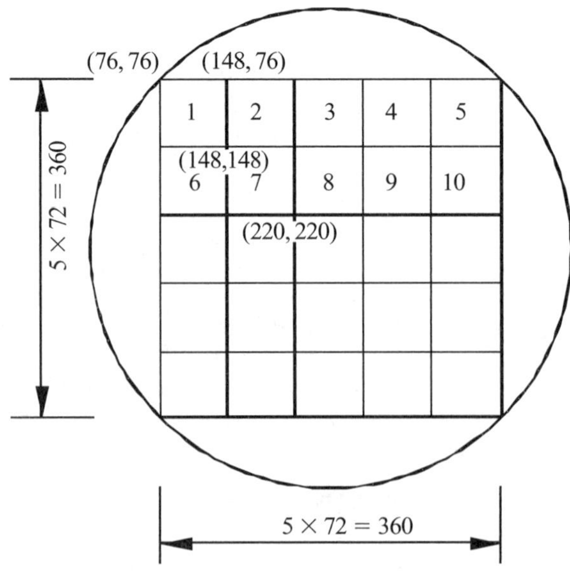

FIGURE 3.9   Inscribed square and square division.

Where γ_d denotes the absolute value of the difference of volume fractions of voids in the two adjacent small squares. For example, the following equation was used to calculate the gradient in X-direction for Square 1 in Figure 3.9:

$$grad\phi = \frac{|\gamma_2 - \gamma_1|}{d} \tag{3-12b}$$

Where $d = 21.2$ mm, γ_1 represents the void volume fraction of Square 1, and γ_2 represents the void volume fraction of Square 2. The gradient of Square 1 in Y-direction can be expressed as in the following equation:

$$grad\phi = \frac{|\gamma_6 - \gamma_1|}{d} \tag{3-12c}$$

Where γ_6 represents the void volume fraction of Square 6. However, the gradient of squares in the centered area is calculated using a different method. For example, the gradient of square 7 in X-direction is represented by the following equation:

$$grad\phi = \frac{|\gamma_6 - \gamma_8|}{d} \tag{3-12d}$$

Where $d = 42.2$ mm. The gradient in Z-direction is calculated as the ratio between the absolute of the difference of volume fractions of slices between the upper and lower position and the spacing between these two slices.

There is one gradient in each of the three directions for one point. The gradients will be related to the anisotropic properties of the materials. For simplification of data presentation, an effective gradient is defined by the following equation:

$$grad\phi = \sqrt{\phi_x^2 + \phi_y^2 + \phi_z^2} \tag{3-13}$$

Where ϕ_x, ϕ_y, and ϕ_z represent the gradients of local void volume fraction in X, Y, and Z directions.

Wang et al. (2002) presented a study on the WesTrack specimens. Six specimens, two for each of the three mixes, were studied. Specifically, TM06 and TM07 were the fine mix cores; TM08 and TM09 were the fine plus mix cores; and TM10 and TM11 were coarse mix cores. Eighty sectional images were captured for each specimen except for TM08, which had only 62 images. Table 3.1 and Table 3.2, respectively, show some statistical information of the local volume fractions and gradients.

TM06–TM11	Statistical Information			
	Standev	Mean	Maximum	Minimum
TM06	0.02146	0.015836	0.20327	0
M07	0.018812	0.018052	0.178246	0
TM08	0.05623	0.053774	0.353412	0
TM09	0.05288	0.051661	0.35322	0
TM10	0.049723	0.072603	0.338013	0.006545
TM11	0.060915	0.0770255	0.436568	0.004937

TABLE 3.1 Statistical information of local volume fraction of TM06–TM11.

TM06–TM11	Standev	Mean	Maximum	Minimum
		Statistical Information		
TM06	0.01	0.003881	0.179551	3.3E-05
TM07	0.004659	0.003301	0.095954	6.8E-05
TM08	0.008796	0.006858	0.13898	0.000112
TM09	0.007164	0.006482	0.120672	0
TM10	0.009126	0.011444	0.097332	0.000612
TM11	0.010233	0.011708	0.116603	0.000433

TABLE 3.2 Statistical information of the effective gradient of TM06–TM11.

From Table 3.1 it can be observed that the local volume fraction in the coarse mix (TM10 and TM11) is the largest of the three mixes. It means that void volume (larger than 0.3 mm in size) of the coarse mix is the largest among the three mixes although their total void volume contents are the same (about 8%). The fine plus mix (TM08 and TM09) is in the middle and the fine mix (TM06 and TM07) has the smallest void volume fraction. There are no apparent differences between the two specimens of the same mixes.

From Table 3.2 it can be noted that the mean value of the gradient and the standard deviation are apparently different for the three mixes. TM10 and TM11 have the largest effective gradient; TM8 and TM9 have the medium effective gradient; and TM6 and TM7 have the smallest effective gradient.

3.4.1.2 Data Interpretation

To allow for more intuitive understanding about how local void volume fraction and its gradient affect the stress in a heterogeneous material, Equation 3-10 is written in terms of local void volume fraction. If γ_1 and γ_2 represent the local volume fractions of voids at two points l distance away, then the local solid volume fractions are respectively $v_1 = 1 - \gamma_1$ and $v_2 = 1 - \gamma_2$. The stress gradient is then:

$$\frac{T^2{}_{ij} - T^1{}_{ij}}{l} = \beta \left[2 \frac{(\gamma_2 - \gamma_1)}{l} (\frac{\gamma_1 + \gamma_2}{2} - 1) \right] \delta_{ij} \tag{3-14}$$

Equation 3-14 indicates that the differences in stresses between two points in the material space are related to the gradient of local void fractions and the average local void fractions of the two points. When the two points are close, the stress gradient is proportional to the product of the gradient of the local void volume fraction and the mean local void volume fraction. This has an important implication: the stresses in the specimen are not uniformly distributed. The larger the average local void volume fraction and its gradient, the greater the internal stress difference or the more significant the stress concentration. The more severe stress concentration may directly lead to faster crack initiation and propagation and strain localization related to rutting or permanent deformation. Therefore, it may be deduced that the coarse mix may be the most prone to rutting and fatigue cracking and the fine mix may be the least prone to rutting and fatigue cracking. This theory based deduction is consistent with the ranking of the field performance of the three mixes. This deduction is also based on an implicit assumption that small voids may self-heal quickly and affect the fatigue and rutting properties to a lesser degree.

While the above method quantifies the void volume fractions in regular cubic or rectangular cells and is very convenient, definition of the cells may not be necessarily regular. Bagi (1996) introduced the concepts of material cell and spatial cell. For granular materials, the different cells may show different advantages. Other well-established cells, including Delaunay cells, can also be used.

3.4.2 Fabric Tensors and Their Implications

Particle orientation is an important concept for granular materials and weakly bonded granular materials such as AC at high temperatures. Masad et al. (2002) used the modified stress tensor (Satake, 1983; Nemate-Nasser and Mehrabadi, 1984; Tobita, 1989) for granular mechanics. While this relationship may not be valid for AC where restraint from binder or mastics may be significant, it offers a theoretical guide for understanding particle shape, particle orientation, and particle configuration on material anisotropy and stress concentration. General theories based on Eshelby mechanics may be more relevant where the matrix constraint can be considered. Certainly the computational mechanics approach (e.g., the digital specimen and digital test methods in a later chapter) can also deal with the orientation, shape, and distribution of particles. The following presents one of the approaches for quantifying the particle fabric tensors.

3.4.2.1 A Review on Fabric Quantities

Fabric quantities have been widely used in granular mechanics to describe the particle spatial geometric relationship. However, advances in characterizing these parameters in a 3D scenario have been slow until most recently when high-resolution X-ray tomography imaging has provided a viable tool to obtain 3D datasets. Different views, including the void view and the skeleton view, are used to look at the microstructure of a granular system or a weakly bonded granular system. The skeleton view emphasizes the load transfer through particle contacts while the void view emphasizes the weakening effects of voids to a continuum and void connectivity. The skeleton view better describes the mechanical behavior of unbonded granular materials. However, recent discoveries by Shashidhar (Figure 3.10) using photoelasticity indicate that load transfer in

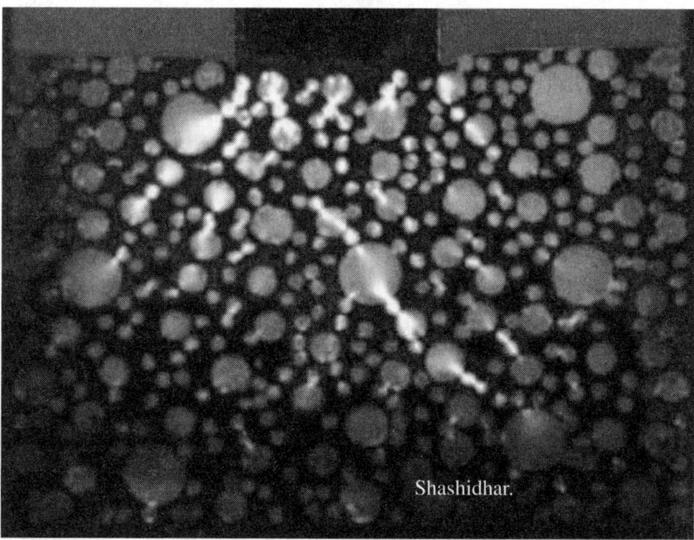

FIGURE 3.10 Load transfer of a weakly bonded granular material (Shashidhar).

AC is similar to that in a granular material. By the skeleton view, important parameters of the microstructure of a granular system include:

- The average number of contacts among the particles forming the particle skeleton
- The contact normal distribution (number of contacts in different orientations)
- The branch vector distribution (average length in different orientations)
- The particle orientation distribution (number of particles with the longest dimensions in different orientations)

These parameters have been shown to be related to the strength and deformation of granular materials. They have been used in micromechanics (Christoffersen et al., 1981; Chang, 1991; Rothenburg, 1992), doublet mechanics (Granik and Ferrari, 1993; Ferrari et al., 1997), and Distinct Element Method (Cundall, 1979). Applications of granular mechanics theory with assumed idealized particles and packing have successfully made qualitative predictions about granular behaviors such as dilatancy (Rowe, 1962) and shear band thickness (Mauhlhaus and Vardolarkis, 1987). These idealized theories cannot account for the effects of particle shape and surface texture, the contacting characteristics, and the actual particle configurations including clustering, and usually fail to predict the behavior of granular materials quantitatively. A general practice in granular mechanics is to introduce various fabric terms such as the contact normal tensor (Oda et al., 1982), the branch vector tensor (Mehrabadi et al., 1983), and the combined contact normal and branch vector tensor (Nemat-Nasser,et al., 1984) to describe the spatial arrangement of the particles and their inter-particle contacting characteristics.

Although significant efforts have been focused on the quantification of these fabric quantities since Oda's (1972a, 1972b, 1972c) study in the early 1970s, these studies were limited to 2D sections and thus inherently not correct for some of the quantities. For example, the number of contacts and the contact orientations are inaccessible parameters, which means they cannot be quantified without assumptions from a small number of sections. For two particles in contact, the chance for a random cross-section to cut through the contacting point (a small area) is very small. Figure 3.11 illustrates this concept, where the two particles pointed by arrows are not in contact, while they may actually contact each other at some other sections. Therefore, contact counts based on 2D sectional information are usually misleading. In addition, particle orientation, defined as the orientation of the longest axis of a particle, is also an inaccessible parameter. The measurement of orientation by Oda (1972a, 1972b, and 1972c) used the angle between the direction of the longest Feret diameter of a particle cross-section and the horizontal direction. The orientation based on 2D sectional observation varies with sections and is usually not unique and may not be coincident with the actual longest dimension in 3D space. Figures 3.11a and 3.11b illustrate how the irregular shape of the particles affects the orientation of the longest dimension on particle cross-sections. Generally, for irregular shapes, the orientation of the longest Feret diameter on a cross-section is not coincident with that of the longest dimension of the particle in 3D space.

The limitations of the current 2D methods have long been recognized; however, no general methods have been developed to quantify the branch vector, the contact normal distribution, or the particle orientation distribution in a true 3D scheme. A most recent development (Wang et al., 2001a), by use of the relation between the whole space and the sub-space probability distribution functions of random vectors, quantified the branch vector distribution from the sub-space distributions of projected random branch

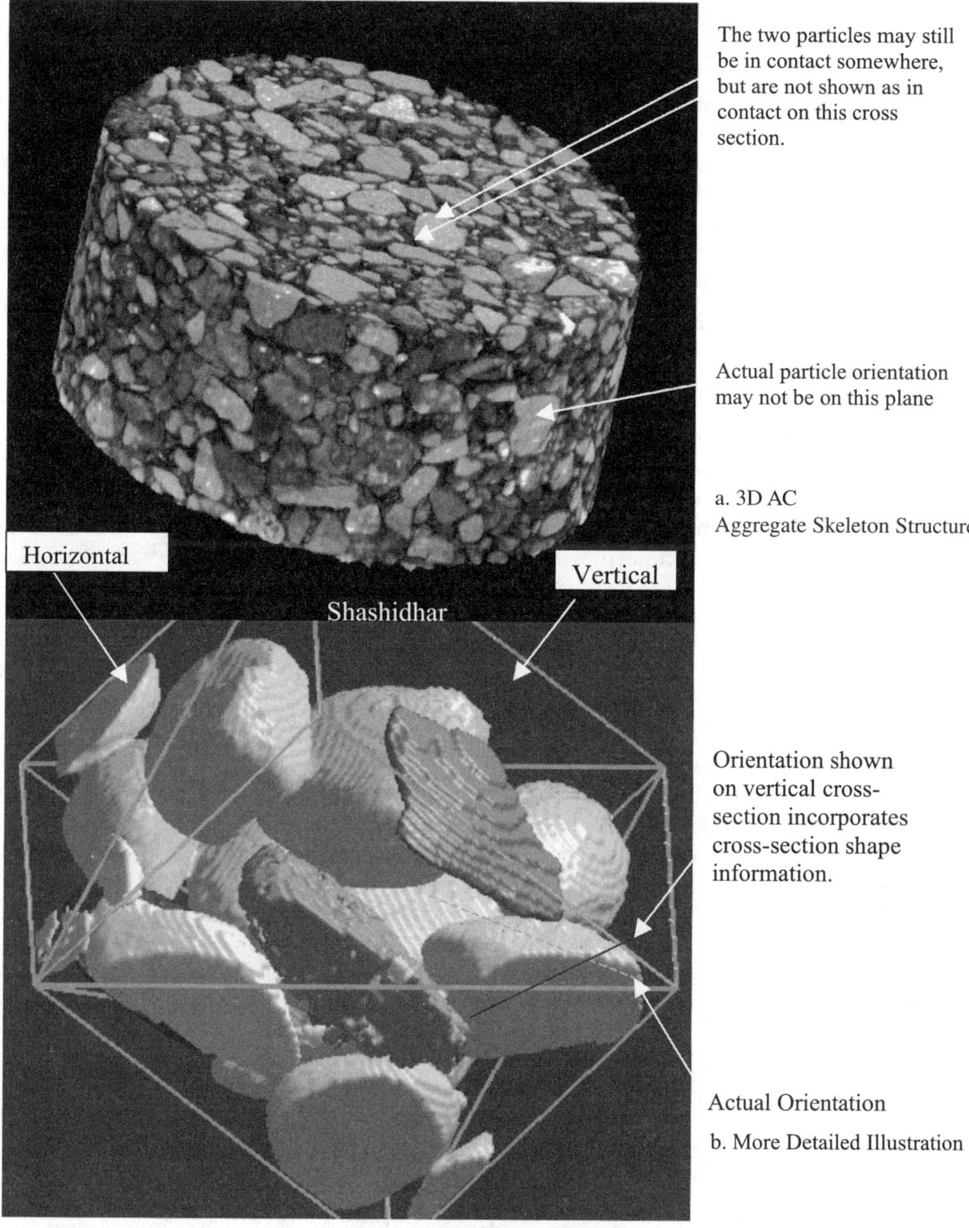

FIGURE 3.11 Illustration of misleading information introduced on 2D sections (the second image is provided by Naga Shashidhar).

vectors in three orthogonal sub-spaces. This method is valid only for the axial symmetric distribution. In general, the quantification of the contact normal and branch vector distribution requires the use of 3D datasets. XCT provides a unique capability to characterize these fabric quantities of a real granular system. This section presents the methods recently developed (Wang et al., 2010) to quantify the distribution of contact normal, branch vector, and particle orientation using XCT.

3.4.2.2   3D Quantification Methods

To accurately quantify the distribution of the contact normal and branch vector distributions, it is necessary to determine the boundary coordinates of the particle. Wang, et al. (2004) developed a method to reconstruct the 3D microstructure of granular materials using tomography imaging and image analysis. This method can be used to obtain the surface coordinates of each individual particle in a particulate system. A brief description of the method is presented below. The study by Wang et al. (2003) mainly focused on the methodology development, which directly used a real aggregate. The description of the method is directly tied into an application as an example. The aggregate is Louisiana sandstone passing a 12.5 mm sieve size but retained on 9.5 mm sieve size. The aggregate was placed in a 35 mm film cartridge for X-ray scanning. The size of the cross-section images was 512 × 512 pixels with a resolution of 0.07 mm/pixel. The interval of the cross-sections was 0.42 mm or 6.0 pixels. Using the method by Wang, et al. (2003), the particle cross-sections belonging to the same particle were identified and reconstructed into the 3D particle representation. The x, y, and z coordinates for 72 points on each of the particle cross-sections are determined. Figure 3.12 presents the 3D visualization of an assembly of 52 particles in the cartridge. With this representation, the mass center coordinates and the coordinates of surface points of each particle are available. This set of data places the foundation for the quantification of the contact normal, branch vector, and particle orientation distribution.

3.4.2.3   Determination of Particle Contacts

Particles in a granular specimen are in contact with their adjacent particles. In fabric analysis, the average number of contacts per particle is called coordination number. The contacts between particles can be observed using X-ray tomography imaging, which basically obtains the images of horizontal cross-sections of certain thickness and spacing non-destructively. If the particles are spheres, the contact occurs at a point and there is little chance that a cross-section image includes the contact point. However, the shapes of the real particles are irregular and contacting areas rather than contacting points are involved in the contact between two particles as shown in Figure 3.13. It is also possible for two particles to contact at more than one portion (see also Figure 3.13).

FIGURE 3.12   Rendering from surface Cartesian coordinates of particles.

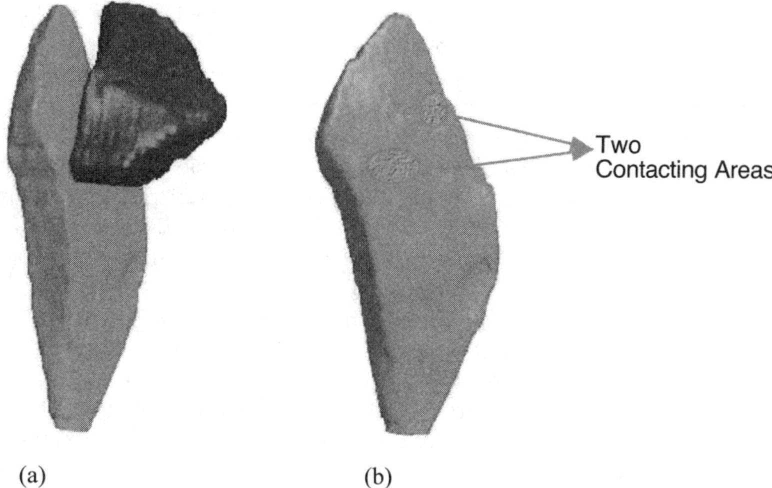

(a) Two Contacting Particles (b)Particles are Contacting at Many Points in Two Places

FIGURE 3.13   Illustration of the complexity of contact.

The determination of the existence of a contact can be made by the distances among the surface points of two particles. Theoretically, this distance should be zero for two particles in contact. However, due to errors involved in digitalization, the spacing and thickness of sections, and the image processing required to separate particle cross-sections, two particles with a certain small distance are considered contacting particles. This small distance is called threshold distance. The threshold distance must be determined based on an overall evaluation of the above errors and equilibrium considerations. Figure 3.14 presents the relation between the threshold distances and the corresponding

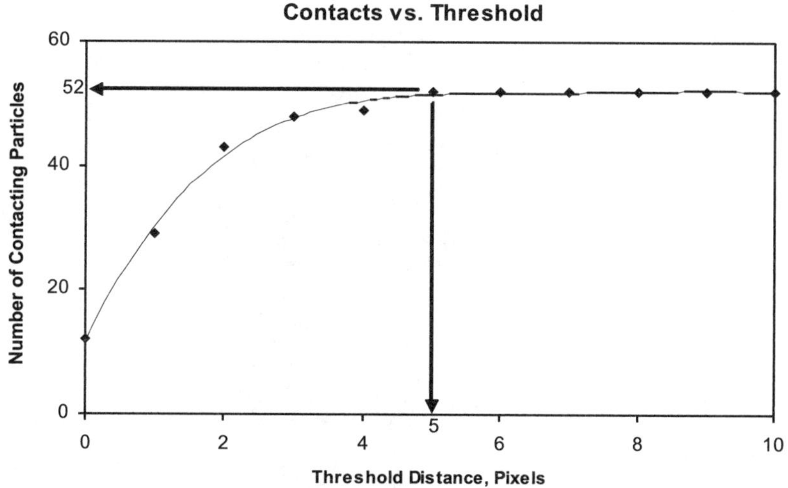

FIGURE 3.14   Number of particles in contact versus threshold distances.

number of particles in contact acquired from the specimen scanned. The specimen contains a total of 52 particles. At a 0-pixel threshold distance, just 12 particles (six pairs) are involved in contact. With so few particles in contact, the particulate system would not be able to achieve equilibrium. It may be reasonable to assume that all 52 particles are in contact with at least one other particle. The relation presented in Figure 3.14 indicates that when the threshold distance is 5 pixels, each particle will contact with at least one other particle. Therefore, the threshold distance is determined as 5 pixels. In other words, when two surface points on two different particles have a distance of 5 pixels or less, the two points are considered in contact. If two particles have more than one pair of points having a distance less than 5 pixels, the point pair that has the shortest distance will be selected as the contact point. When more than one pair of points between the two particles has the same shortest distance, the point closer to the central part of the particle surface (i.e., point O in Figure 3.15) is selected as the contact point. In some cases, when more than one portion of two particles has a distance smaller than the threshold distance, especially for two flat particles, only the portion that has a larger number of point pairs that have distance below the threshold distance is considered as the contact point. In other words, only one contact between two particles in contact is counted.

Based on the above criteria and the calculation and ordering of the distances between any point pairs on two particles, a total of 102 contact pairs were detected using a 5-pixel threshold distance for the specimen. Each contact is also visualized for verification of the contact. Correspondingly, the average coordination number for this specimen is 3.92 (102 × 2/52). It is anticipated that the coordination number would increase if particles at the boundaries were removed from the calculations.

3.4.2.4 Contact Normal Vectors and Their Distribution

Among the very many fabric terms for a granular system, the contact normal vector is one of the most important fabric quantities. Contact normal vector is defined as a unit vector of the normal to the tangent plane at the contact point as shown in Figure 3.16. The tangent plane should be tangential to the surfaces of both contacting particles at the

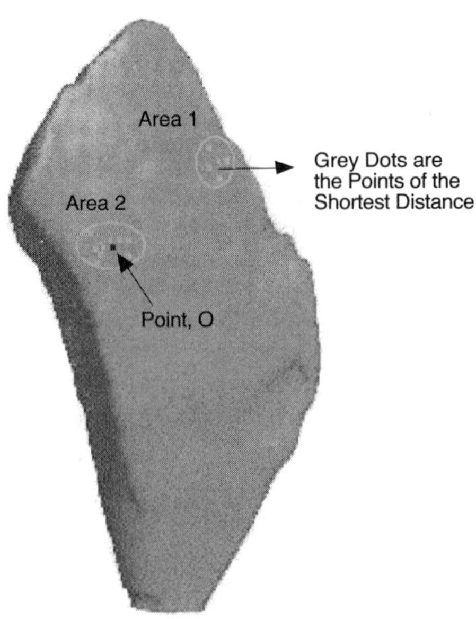

FIGURE 3.15 Illustration of the selection of the contact point.

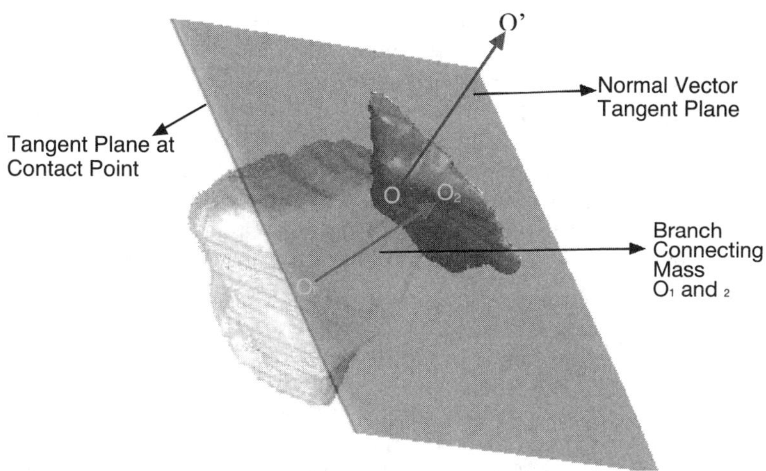

FIGURE 3.16 Tangent plane, contact normal vector, and branch vector.

contact point. Ideally, the tangent plane could be uniquely determined by using the curvature plane that passes through the contact point and best fits the surrounding points. In real implementation, the tangent plane is approximated by the plane that has its normal equal to the mean normal of the four planes comprised of the contact point and its four surrounding points (Figure 3.17). As demonstrated in Figure 3.17, the contact point (o_1) and its nearest left side point (a), its right side point (c), the closest point

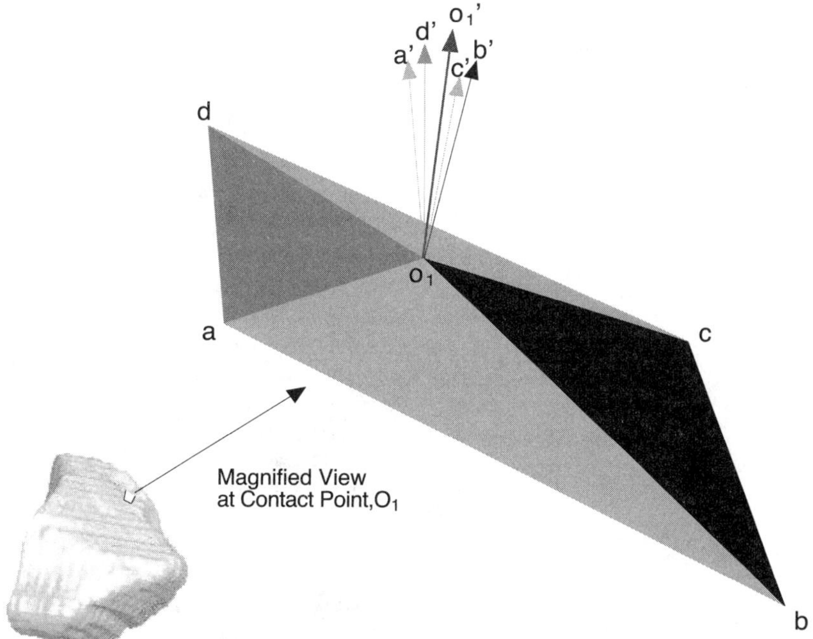

FIGURE 3.17 Definition of the contact normal vector O_1O_1' (obtained from the contact point O1 and its neighbor points, a, b, c, and d).

from its lower adjacent section (b), and the closest point from its upper adjacent section (d) are selected as the surrounding points. These four points and the contact point can form four triangular planes. Each plane, abo_1, bco_1, cdo_1, and dao_1 has a unique normal vector o_1a', o_1b', o_1c', and o_1d', respectively. The mean vector (o_1o_1') of the four normal vectors can approximately represent the normal vector of the tangent plane at the contact point. On the other hand, another normal vector from the other side of the contact point (on the surface of the other particle in contact) can be determined in the same manner. The mean of these two normal vectors is finally defined as the contact normal vector (oo' in Figure 3.18) for the contacting particles. In the analysis of subsequent sections, vectors v and $-v$ will be considered as the same vector in terms of the minus half-z space.

By using the above procedure, all the contact normal vectors for the 102 contact points detected for the specimen can be obtained. Figure 3.19 presents the 3D visualization of the contact normal vectors with their origins at the origin of the frame ($x=0$, $y=0$, *and $z=0$*). It should be noted that even though all the vectors are unit vectors, the lengths of the vectors in Figure 3.19 appear different due to the projection effect.

The orientational distribution of contact normal vectors is important to the study of the behavior of both bonded and unbonded granular materials. In order to evaluate the orientational distribution of these vectors, the half-space ($z > 0$) is evenly divided into 12 regions as shown in Figure 3.20a. Regions 1, 4, 7, and 10 are close to z axis and regions 2, 3, 5, 6, 8, 9, 11, 12 are close to x and y axes. To better illustrate the locations of the regions, a 3D view of Regions 1, 2, 3, 10, and 12 on a unit sphere is schematically shown in Figure 3.20b. The orientational distribution of the contact normal vectors can be obtained by counting the number of vectors that fall within the regions. Table 3.3 presents the distribution of the contact normal vectors in the 12 regions for the specimen studied.

FIGURE 3.18 Two normal vectors at O_1 and O_2 and their mean vector OO'.

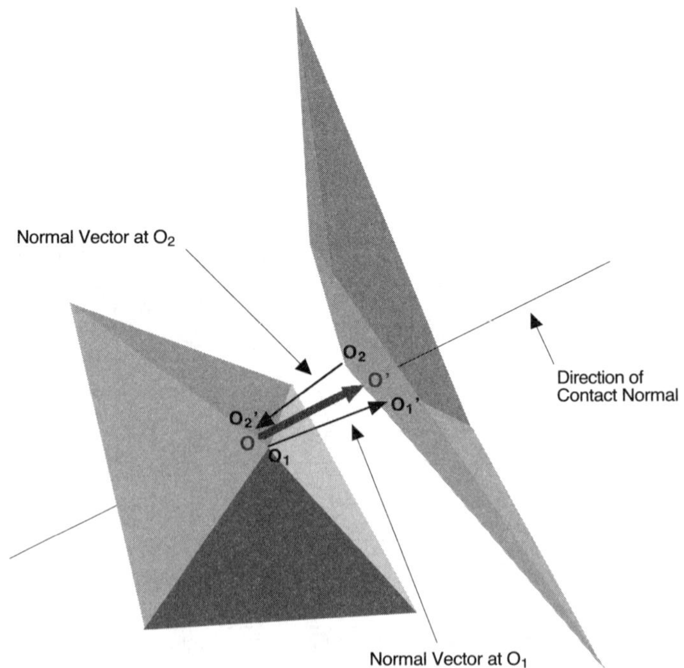

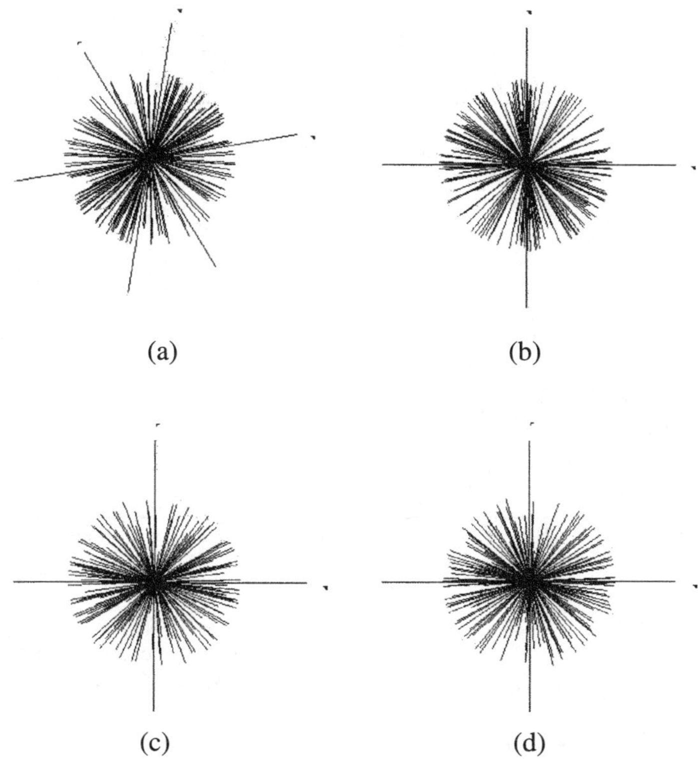

(a) Contact Normal Vectors in XYZ Coordinates, (b) Contact Normal Vectors Projected on XY Plane, (c) YZ Plane, and (d) ZX Plane

FIGURE 3.19   Orientational distribution of contact normal.

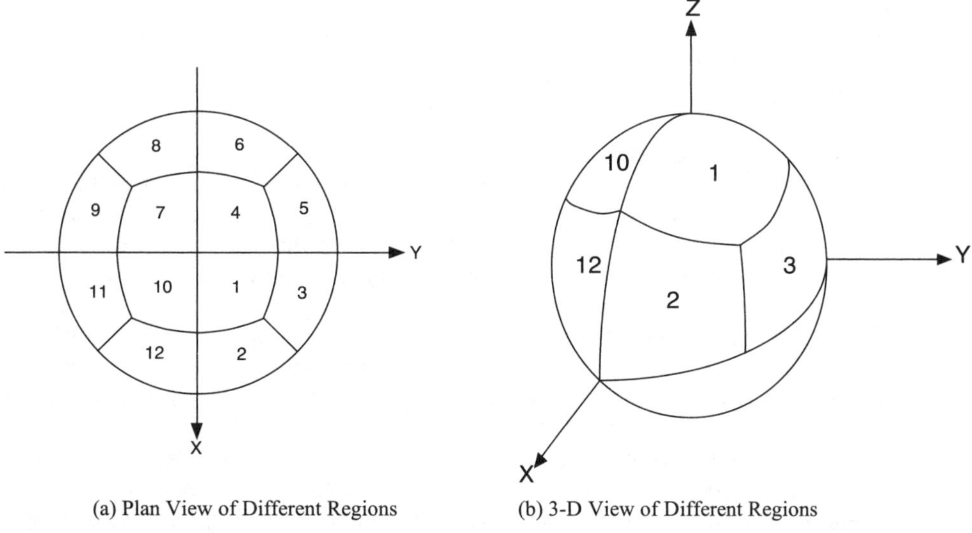

(a) Plan View of Different Regions   (b) 3-D View of Different Regions

FIGURE 3.20   Divided regions in 3D space.

Regions	n_1, or x	n_2, or y	n_3, or z	Contact Normal Vectors	Branch Vectors	Long Axis Orientation
1	0.2706	0.2706	0.9239	4	7	4
2	0.8536	0.3536	0.3827	8	8	3
3	0.3536	0.8536	0.3827	4	14	7
4	−0.2706	0.2706	0.9239	8	6	2
5	−0.3536	0.8536	0.3827	10	13	4
6	−0.8536	0.3536	0.3827	12	10	4
7	−0.2706	−0.2706	0.9239	6	9	3
8	−0.8536	−0.3536	0.3827	9	8	9
9	−0.3536	−0.8536	0.3827	12	7	5
10	0.2706	−0.2706	0.9239	4	5	4
11	0.3536	−0.8536	0.3827	13	8	6
12	0.8536	−0.3536	0.3827	12	7	1
Total	NA	NA	NA	102	102	52

TABLE 3.3 Distribution of contact normal, branch vectors, and long axis orientation.

Considering the location of the regions (e.g., Regions 1, 4, 7, and 10 and are close to the z axis), the numbers of contacts (Table 3.3) in different orientations are different, indicating the anisotropic distribution (it may require a larger specimen).

3.4.2.5 Branch Vectors

When two particles are in contact, the unit vector connecting the two mass centers of the particles is called the branch vector. Figure 3.16 also shows the branch vector of the two contacting particles. Since the 3D Cartesian coordinates of the mass center of each particle have already been identified, the branch vectors can be obtained conveniently. The distribution of the branch vectors can be obtained in the same manner as the distribution of contact normal vectors. Table 3.3 also shows the distribution of the branch vectors for the specimen. By comparing the data presented in Table 3.3, the differences of the two distributions should be noted. For spherical particles the two distributions should be the same.

3.4.2.6 Particle Orientation

The particle orientation has been defined as the orientation of the longest axis orientation of a particle cross-section for many 2D studies. The orientation of the longest orientation of a cross-section may be significantly different from the particle orientation. This problem can be solved by using the 3D surface datasets. Since the Cartesian coordinates of the surface of each particle are already determined, the real long-axis orientation of each particle in 3D can be determined. The method presented is simple but inefficient. It calculates the distance between any two points on the surface and selects the two points that have the longest distance. The line connecting the two points represents the orientation of the longest axis or the particle orientation. The orientational distribution of the particle orientation can be obtained in the same manner as the contact normal vector distribution was obtained. Table 3.3 also presents the distribution of the particle orientations for the specimen. Considering the location of the regions, it can be seen that more particles are oriented in the horizontal direction, consistent with common sense as gravity is in the vertical direction.

3.4.2.7 Tensor Analysis

Fabric tensor is often used as a means to describe the spatial arrangement of particulate systems. The orientational distribution of various microstructural quantities such as contact normals, branch vectors, and long-axis orientations of particles can be described by fabric tensors. A method to determine fabric tensors was developed by Hilliard (1962, 1967) based on stereological principles. Kanatani (1984a, 1984b, 1985) developed Cartesian tensor formulations to represent the distribution function $f(n)$ of any fabric quantities. An approximate formulation using only the second-order fabric tensor is presented as follows:

$$f(n) = \frac{C}{4\pi}\left[1 + \phi_{ij}n_i n_j\right] \tag{3-15}$$

where n_i is the unit vector; $f(n)$ is the distribution function representing the orientational distribution of any fabric quantities; ϕ_{ij} is a second-order fabric tensor; and denotes the spherical domain of integration.

There are six unknown terms with a second order symmetric fabric tensor. The observed distributions for the 12 directions, $f(n)_{ob}$, listed in Table 3.3 for contact normal vectors, branch vectors, and particle orientations can be used to obtain the six unknowns for each of the distributions by minimizing the square differences between the observed and predicted distributions (Equation 3-16). The unit vectors, n_i, representing the 12 regions are also listed in Table 3.3. The C value for contact normal and branch vectors is 102 and for particle orientation is 52. These values can be substituted into Equation 3-16 to generate six linear equations. Then, the six unknown terms of the fabric tensor can be solved using linear algebra.

$$\frac{\partial}{\partial \phi_{ij}}\left(\sum\left(f(n)_{ob} - \frac{C}{4\pi}\left[1 + \phi_{ij}n_i n_j\right]\right)^2\right) = 0 \tag{3-16}$$

The contact normal vector tensor, the branch vector tensor, and the particle orientation tensor for the specimen studied are calculated as follows:

$$\phi_{ij}\Big|_{\text{Contact Normal Vector}} = \begin{bmatrix} 0.398 & -0.359 & -0.206 \\ -0.359 & 0.296 & -0.265 \\ -0.206 & -0.265 & -0.437 \end{bmatrix}$$

$$\phi_{ij}\Big|_{\text{Branch Vector}} = \begin{bmatrix} -0.008 & 0.005 & -0.120 \\ 0.005 & 0.452 & 0.335 \\ -0.120 & 0.335 & -0.235 \end{bmatrix} \qquad \phi_{ij}\Big|_{\text{Long Axis Orientation}} = \begin{bmatrix} 0.009 & 0.449 & -0.266 \\ 0.449 & 0.510 & -0.106 \\ -0.266 & -0.106 & -0.296 \end{bmatrix}$$

It should be noted that the predicted distributions and the observed distributions have large differences (Table 3.4), indicating that higher order fabric tensors might be needed to represent the actual distributions. Although it is not rational to conclude that fabric quantities cannot be accurately represented by second-order tensors based on this dataset alone, similar doubts were cast by Tozeren and Skalak (1989) and Wang et al. (2001a). A much larger dataset should be used, ensuring a statistically valid conclusion.

3.4.2.8 Limitation of 2D Techniques

None of the three fabric quantities, the contact normal, the particle orientation, or the branch vector for irregular particles, can be quantified using rational mechanisms with a

Region	Contact Normal			Branch Vector			Orientation		
	$f(n)_{ob}$	$f(n)_{pre}$	Error, %	$f(n)_{ob}$	$f(n)_{pre}$	Error, %	$f(n)_{ob}$	$f(n)_{pre}$	Error, %
1	4	3.163	20.9	7	7.628	9.0	4	2.754	31.2
2	8	6.819	14.8	8	8.375	4.7	3	4.536	51.2
3	4	6.134	53.4	14	12.038	14.0	7	6.039	13.7
4	8	5.685	28.9	6	8.588	43.1	2	3.309	65.5
5	10	10.554	5.5	13	12.512	3.8	4	4.389	9.7
6	12	12.516	4.3	10	9.595	4.0	4	3.727	6.8
7	6	6.983	16.4	9	5.884	34.6	3	4.290	43.0
8	9	10.164	12.9	8	8.176	2.2	9	6.209	31.0
9	12	9.849	17.9	7	9.014	28.8	5	7.205	44.1
10	4	6.169	54.2	5	4.899	2.0	4	2.647	33.8
11	13	12.463	4.1	8	8.436	5.5	6	4.366	27.2
12	12	11.501	4.2	7	7.853	12.2	1	2.528	152.8
Total average	102	102	19.8	102	101	13.7			

TABLE 3.4   Predicted versus observed distributions of the fabric quantities.

2D dataset. It is therefore not meaningful to make any comparison between the 3D techniques and the 2D techniques. First, a particle in contact may not be shown as in contact in 2D. Similarly, for particle orientation, a cutting plane may not be consistent with the orientation of the longest Feret diameter of the particle. The only fabric quantity of the three that might be estimated with reasonable accuracy from 2D datasets (three orthogonal planes) is the branch vector distribution (Wang, et al., 2001a). Nevertheless, with particles of irregular shapes, the mass center coordinates may be significantly different from the centroids of a revealed cross-section (Wang, et al., 2001a).

It should be noted that many researchers have developed 2D techniques to quantify these fabric quantities not because the 2D techniques are reliable, but because of the lack of 3D techniques.

3.4.2.9   Fabric Quantities and Their Use in Constitutive Models

The characterized fabric quantities can be directly used in constitutive models. This section presents a review on how fabric tensors are quantitatively related to stress, strain, and yielding function.

Stress-Fabric Tensor Relation

A direct application of the fabric quantities is to relate average stress (effective stress) with contact force and fabric tensors including contact normal tensor (J_{ij}) and branch vector tensor (F_{ij}). Equations 3-17 and 3-18 (Mehrabadi and Nemat-Nasser, 1983) present the relationship of how the average stress of σ_{ij} in a granular assemble is related to the contact normal tensor and the branch vector tensor, respectively.

$$\overline{\sigma}_{ij} = \alpha_0 J_{ij} + \alpha_1 J_{ik} J_{kj} \tag{3-17}$$

$$\overline{\sigma}_{ij} = \beta_0 F_{ij} + \beta_1 J_{ik} J_{kj} \tag{3-18}$$

α_0, α_1 and β_0, β_1 are functions of the invariants of the stress tensor.

Other relations include those defined by Tobita (1989), Tobita and Yanagisawa (1992), and Bagi (1996).

Strain-Fabric Tensor Relation
In continuum mechanics, the deformation rate tenor (D_{ij}) and the spin tensor (W_{ij}) are used to represent the strain characteristics. Mehrabadi and Nemat-Nasser (1983) also derived the relations relating D_{ij} and W_{ij} to contact normal tensor J_{ij} and local strain rates E_{ij}.

$$D_{ij} = \frac{1}{2}(E_{ik}J_{kj} + E_{jk}J_{ki}) \tag{3-19}$$

$$W_{ij} = \frac{1}{2}(E_{ik}J_{kj} - E_{jk}J_{ki}) \tag{3-20}$$

Kanatani (1984a, 1985) offers relationships between strains and fabric evolution as well.

Other Relationships
Tobita (1988), through introducing the modified stress tensor, derived a new yielding function. Tobita and Yanagisawa (1992) systematically introduced the modified stress tensor that is explicitly related to the fabric tensors. Cowin (1985, 1986) introduced the relationships between fabric tensors, elasticity tensor, and strength.

Oda and Nakayama (1989), through modifying the first and second stress invariants by the following equations, related the stress invariants with the particle orientation tensor R_{ij} defined with 2D datasets.

$$\bar{I}_1 = (a_1\delta_{ij} + a_2 R_{ij})\sigma_{ij} \tag{3-21}$$

$$\bar{J}_2 = (2b_6\delta_{ik}\delta_{jl} + 4b_7 R_{ik}\delta_{lj})S_{ij}S_{kl} \tag{3-22}$$

Where a_1, a_2 and b_6, b_7 are functions of the orientation tensor constant of R_{ij}. S_{ij} is the deviatoric stress tensor.

The 3D non-destructive quantification of these fabric tensors provides a viable tool to verify these relations.

Significances
The quantification methods presented offer potential applications in characterizing the parameters needed in constitutive models that incorporate material microstructural quantities. They can also provide applications in correlating microstrutural quantities with anisotropic behavior of granular materials (or weakly bonded materials in discrete senses). More importantly, these methods place a foundation to better understand the deformation and strength mechanisms of granular materials on the basis of micromechanics, microstructure, and statistics. A rational elasto-plasticity constitutive model based on the fabric description can be conceived to include 1) the anisotropic elasticity base on the anisotropic fabric distribution; 2) anisotropic yielding strength that is based on material properties and fabric distributions; and 3) non-associated flow.

3.5   Microstructural Quantities in View of Damaged Continuum

Another set of parameters is related to the general concept of damage or weakness in a broader sense. For example, from a pure mechanics point of view, voids will weaken the material due to the stress concentration. The interface between aggregates and binder may have a larger concentration of micro-pores and therefore become weak. In addition, the spacing between two cavities, cracks, or solid inclusions is a measure of their interaction. The mean spacing and the spacing variation may serve as good measurements of how the weakening effects of two cavities can be magnified. This section presents methods to quantify a set of damage parameters including the specific damaged surface area tensor, the damage tensor, the average size of the cracks, the mean solid path among the damaged surfaces, and the mean solid path tensor. This section also presents the relationship between the quantified damage parameters and their applications in mechanical modeling. This work is applied, in particular, to quantify damage parameters of AC specimens of the WesTrack project.

3.5.1 Fabric Tensor and Stereology Method

Unlike the cases for contact normal, branch vector, and orientation tensor, which are actually inaccessible parameters, parameters in this section are accessible and can be evaluated using statistics within the framework of stereology. Kanatani (1984a, 1984b, 1985) presented a systematic approach using the Buffon transform and fabric tensors to represent the orientational distribution of geometric quantities. In these papers, Kanatani used a generic term "fabric" to represent the orientational distribution or arrangement of any geometric and/or microstructural quantities. This approach, in conjunction with automated image analysis, can be used to determine the damage parameters efficiently. Equation 3-23 presents the Buffon transform; Equation 3-24 presents the distribution function represented in tensorial format; Equation 3-25 is the expression of the Buffon transform in tensor format; and Equation 3-26 represents the inverse Buffon transform, through which the distribution function can be evaluated from the fabric tensors representing the orientational distribution of intersecting points. These equations were originally presented by Kanatani (1984b).

$$N(m) = \int |m \bullet n| f(n)dn \tag{3-23}$$

$$f(n) = \frac{C}{4\pi}[1 + D_{ij}n_i n_j + D_{ijkl}n_i n_j n_k n_l + ...] \tag{3-24}$$

$$N(m) = \frac{2\pi C}{4\pi}[1 + \frac{1}{4}D_{ij}m_i m_j - \frac{1}{24}D_{ijkl}m_i m_j m_k m_l + ...] \tag{3-25}$$

The inverse Buffon transform can be represented as:
If

$$N(m) = \frac{C}{4\pi}[1 + F_{ij}m_i m_j + F_{ijkl}m_i m_j m_k m_l + ...] \tag{3-26}$$

then

$$f(n) = \frac{C/2\pi}{4\pi}[1 + 4F_{ij}n_i n_j - 24F_{ijkl}n_i n_j n_k n_l + ...] \tag{3-27}$$

Where, $N(m)$ is the number of intersecting points per unit testing line in the m direction.

$f(n)$ is the distribution function so defined that $f(n)dn$ is the total quantity in orientation dn. The quantities could be curve length, surface area, or other geometric quantities.

D_{ij}, D_{ijkl}, F_{ij}, F_{ijkl}...are fabric tensors; C is the total quantity, and $C = \int f(n)dn$.

The general methodology for the quantification of the distribution function is to estimate F_{ij} through observed $N(m)$ in different orientations by using Equation 3-26 and then assessing the distribution function through Equation 3-27. If the distribution can be approximated by a second-order fabric tensor, the following equation results.

$$N(m) = \frac{C}{4\pi}[1 + F_{ij}m_i m_j] \tag{3-28}$$

If a cross-section with a unit normal I is revealed, the following integral over a circle $C(I)$ on the cross-section may be evaluated.

$$M(I) = \int_{C(I)} N(m)dm \tag{3-29}$$

$$M_{ij}(I) = \int_{C(I)} m_i m_j N(m)dm \tag{3-30}$$

If experimental observations are made on three orthogonal planes with basis vectors $e_1 = (1,0,0)$, $e_2 = (0,1,0)$, and $e_3 = (0,0,1)$, it can be shown that:

$$M(e_i) = \frac{C}{2}\left(1 - \frac{1}{2}F_{ii}\right) \text{ (not summed on } i) \tag{3-31}$$

$$M_{ij}(e_k) = \frac{C}{8}F_{ij} \tag{3-32}$$

$$C = \frac{2}{3}[M(e_1) + M(e_2) + M(e_3)] \tag{3-33}$$

With these relations, F_{ij}, and therefore $f(n)$, can be obtained if $M(e_i)$ and $M_{ij}(e_k)$ are known. $M(e_i)$ and $M_{ij}(e_k)$ can be evaluated by the following method:

1. Obtain three orthogonal sections (physically or virtually) from a specimen.

2. On each of the three cross-sections, a grid of parallel test lines in orientation θ_m is laid and the quantities $Q_L(\theta_m)$ over the circle are evaluated. In the case of a specific damaged surface area, this quantity is the number of intercepts per unit length of test lines. In the case of the mean solid path, this quantity is the mean solid path measured in θ_m orientation.

$$M(e_k) = 2\pi \sum_{m=0}^{N-1} \frac{Q_L(\theta_m)_{e_k}}{N} \tag{3-34}$$

$$M_{ij}(e_k) = \pi \sum_{m=0}^{N-1} \frac{Q_L(\theta_m)_{e_k} Sin\left(\dfrac{2\pi m}{N}\right)}{N} \tag{3-35}$$

As pointed out in the previous section, many quantities can be used to describe the damage status of a material, which may include the specific damaged surface area, the average spacing among the damaged surfaces, the area fraction of the damaged surfaces, and the size of the defects. However, most of these quantities are varying with orientations, therefore tensorial representations are required.

3.5.2 Specific Damaged Surface Area

The specific damaged surface area (damaged surface area per unit volume) is an overall evaluation of the severity of the damaged continuum. From a purely geometric point of view, the larger the specific damaged surface area, the more severe the damage. However, from the mechanical point of view, the influence of a naturally embedded void on the mechanical behavior of the material may significantly differ from that of an active crack. Generally speaking, the distribution of the damaged surface areas around the orientations is not uniform. If the distribution can be approximated by a second-order fabric tensor, the distribution may be represented as:

$$S(n) = \frac{S_v}{4\pi}(1 + S_{ij}n_i n_j) \tag{3-36}$$

Where $S(n)$ is the distribution density function of the specific damaged surface area. n is the unit normal to a unit sphere, and n_i and n_j are the directional cosines. S_v is the specific damaged surface area. S_{ij} is the second-order fabric tensor representing the orientational variation of the distribution. In brief, the distribution function has two parameters: the specific damaged surface area S_v and its anisotropic characteristics represented as the fabric tensor S_{ij}.

3.5.3 Mean Solid Path

Figure 3.21 illustrates a damaged continuum, where the damaged surface areas in Figures 3.21a and 3.21b are the same, while the average spacing between two adjacent damaged surfaces is different. The spacing among the damaged surfaces is an indication of the strength of interaction between two damaged surfaces and is an important parameter (Lacy et al., 1997, 1999). As the distribution of the average spacing among the damaged surfaces is usually not isotropic, tensors are needed to represent these distributions. Equation 3-37 represents the case where the anisotropy of the distribution can be represented by a second-order fabric tensor.

$$\lambda(n) = \lambda(1 + \lambda_{ij}n_i n_j) \tag{3-37}$$

Where $\lambda(n)$ is a function representing the orientational variations of the mean solid path among the damaged surfaces; n is the unit normal to a unit sphere; and n_i and n_j are the directional cosines; λ is the mean solid path in all orientations among the damaged surfaces. It can be obtained by placing test lines in different orientations and measuring the spacing among the cracked surfaces in different orientations and averaging them (Figure 3.21c). In addition, λ_{ij} is the second-order mean solid path tensor representing the orientational variations of the mean solid path.

Equations 3-36 and 3-37 represent only a second order approximation of the distribution density function, and the variation over orientations. Higher order tensors are usually needed for more accurately representing these functions. Nevertheless, Kanatani (1984b) indicated that for many fabric quantities the second-order fabric tensor representation could present approximations of reasonable accuracy.

3.5.4 Average Size and Shape Tensor of the Defects

Parallel to the mean solid path, the average diameter of the defects and its orientational variations can also be defined. Equation 3-38 presents this definition.

$$\delta(n) = \delta(1 + \delta_{ij}n_i n_j) \tag{3-38}$$

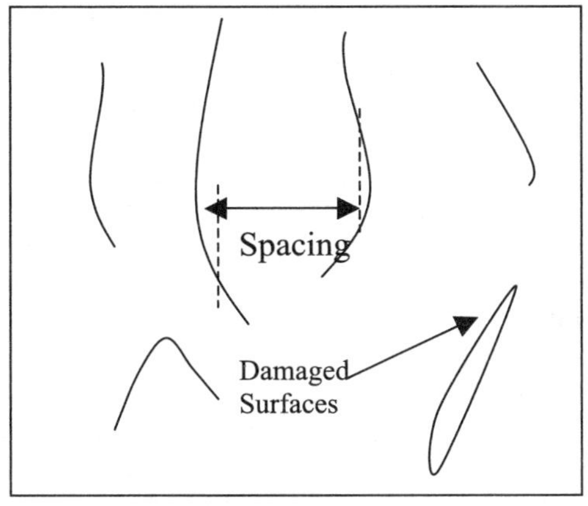

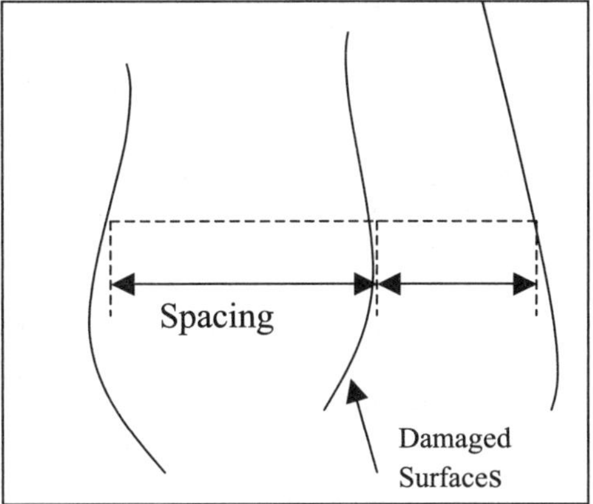

a) Small Spacing   b) Large Spacing

The Two Cases Have Similar Total Damaged Surfaces

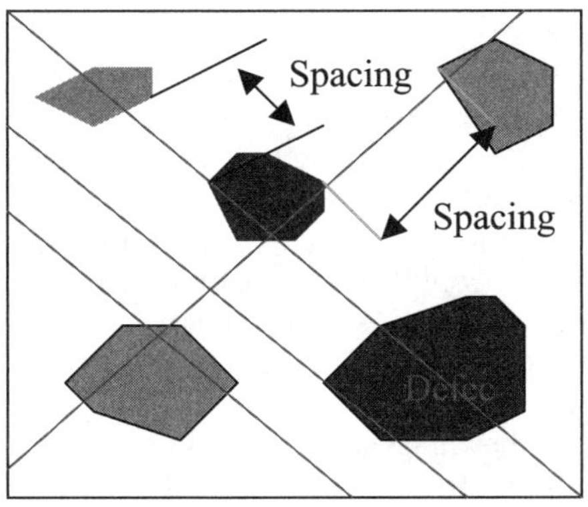

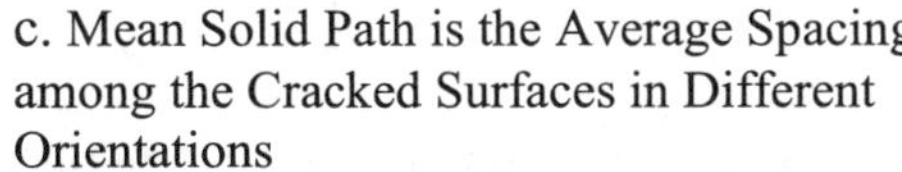

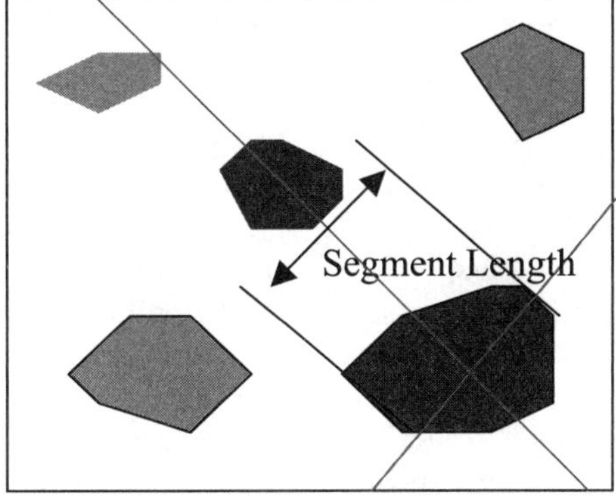

c. Mean Solid Path is the Average Spacing among the Cracked Surfaces in Different Orientations

d. Average Size of Defect is the Average Segment Length in Different Orientations of All the Defects

FIGURE 3.21 Damaged surfaces and the spacing among the surfaces and definitions of mean solid path and average size of defect.

Where $\delta(n)$ represents the average length of the intersecting line over all the defects in orientation n; δ_{ij} is the shape tensor and δ is the average size of all the defects over all the orientations (Figure 3.21d). The shape of the defects, such as cracks, tends to be so irregular that high order tensors should be used to represent it. In the example presented, only the average size will be quantified. The quantification method for the average size and the second-order tensor is the same as that for the mean solid path and mean solid path tensor as long as the intersecting line length in the solid phase is replaced with that in the void or crack phase.

The quantities defined in Sections 3.5.2, 3.5.3, and 3.5.4 and their corresponding tensors can be quantified using the general stereology methods (Kanantani, 1984a, 1984b, 1985) and will be discussed in the following section. The damage tensor, however, will be quantified in a different method.

3.5.5   Stereology-based Quantification Method Using Tomography Images

If the orientation data can be approximated by a second-order fabric tensor, the general approach by Kanatani (1984b) that uses observations made on three orthogonal planes can be used to obtain the required information. For higher order tensors (i.e., fourth-order), observations on more planes in different orientations are needed to determine the components of the tensors. Similar computational schemes can be developed.

Kuo, et al., (1998) presented an implementation scheme on the Q570 image analysis system based on Kanatani's method. However, the sampling scheme of this implementation used a square area, which may introduce bias at the corners. It is also platform-dependent (based on the Q570 system). In addition, Kuo's method needs to actually cut the specimens to reveal the surfaces in different orientations. If more than three orthogonal orientations are needed, this method becomes more difficult to implement as more than one statistically equivalent specimen may be needed. The method presented in this section uses X-ray tomography imaging and virtual cutting technique to conveniently obtain the cross-sections in different orientations. The general approach was developed by Wang et al. (2001b).

3.5.6   Image Interpolation and Analysis

As illustrated in Section 3.3, image interpolation may be required if the spacing between two adjacent sections is much larger than the in-plane resolution. Following the procedure illustrated in Section 3.3, images can be interpolated. The interpolation ends up with 0.3 mm/pixel resolution horizontally and 0.2 mm/pixel vertically.

The stereological quantification of these damage parameters and tensors requires the use of three representative images acquired from three orthogonal sections (Figure 3.22a) of a specimen. The different components or phases in the image should be separated by appropriate thresholds and converted into binary images for performing measurements. The major steps for determining the fabric quantities include the placement of parallel lines in different orientations on the images, the counting of intersecting points, and the computation of the mean solid paths or mean void paths (Figure 3.22b). Detailed algorithms can be found in Kuo et al. (1998). In this example application, Kuo's procedures were modified as follows:

1. Use one actually acquired image, and two orthogonal images obtained from virtual cuttings through the interpolated stack of the images (Figure 3.22a).

2. Account for the different resolutions in the vertical and horizontal directions by adopting a resolution of the common minimum cofactor of the two resolutions, in this case 0.1 mm/pixel for the two resolutions of 0.2 mm/pixel and 0.3 mm/pixel.

3. Use a circular area as the sampling area. A circular area will minimize the effect resulted from different sampling lengths in different orientations in a rectangular sampling area (Figure 3.22b).

3.6   Damage Tensor and Quantification Method

Damage tensor is usually defined as the tensorial representation of the area fraction of cracks or voids to the total area of representative cross-sections. For a representative sample, by stereology, the average area fraction over the slices in one orientation is the

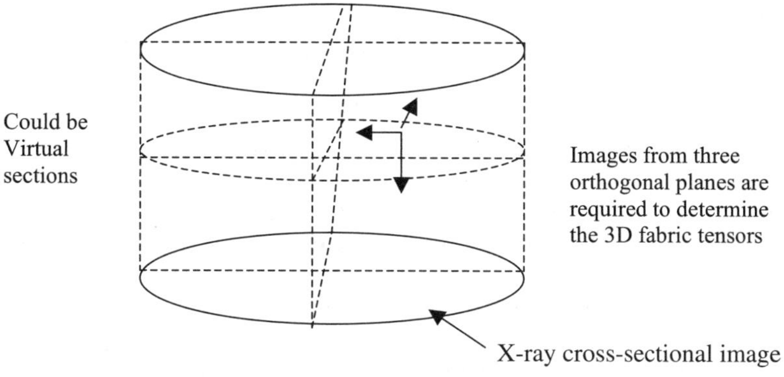

a. Orthogonal planes

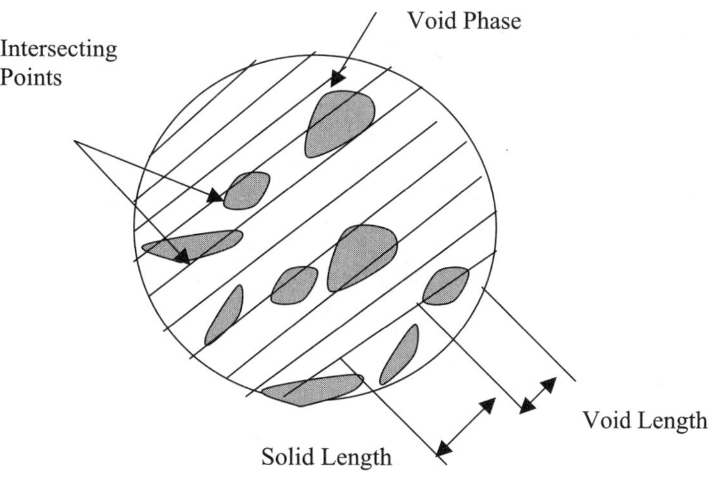

b. Placement of parallel lines, intersecting points and solid length

FIGURE 3.22   Illustration of the determination of stereological fabric tensors.

same as that in another orientation and equals to the volume fraction of cracks or voids in the specimen. Therefore, the damage tensor should be a field variable rather than a state variable. In damage mechanics, it is convenient to define the damage tensor. However, it is difficult to implement its quantification. For example, it is easy to define the area fraction as $\phi = \dfrac{dA_d}{dA}$ (where σ represents the area fraction of defects; dA_d represents the area of defects on a cross-section; and dA represents the total cross-section area of a representative cross-section), but how large should dA be in order to have a reasonable evaluation of the determination of the area fraction? In this section, the concept of a representative specimen is defined as a sphere centered at the point to be evaluated. If a cylindrical specimen is used, the point under evaluation should be the mass center of the cylinder and the diameter of the sphere should be equal to the diameter of the cylinder. To quantify the damage tensor, which is symmetric and therefore has six independent variables, the area fractions of voids and/or cracks in six different orientations are evaluated and fitted into the following equation to obtain the tensorial parameters.

$$\varphi_{ij} n_j = f_i \tag{3-39}$$

Where φ_{ij} is the damage tensor, n_j is the orientational cosine, and f_i is the observed area fractions in six orientations. The six orientations used in this development are (1,0,0), (0,1,0), (0,0,1), (1,1,1), (1,-1,1), and (-1,-1,1). By this method, the 3D dataset is re-sampled and the cross-section images are interpolated. Figure 3.23 presents a section in a certain orientation and the circular part that was used for the area fraction evaluation. This section was virtually cut from the 3D volume rendering.

3.6.1   Modeling Applications of the Damage Parameters

Damage tensors and damage parameters are widely used in damage mechanics. The following sections present some of these applications.

3.6.1.1   Specific Damaged Surface Area

Theoretically, the overall specific damaged surface area should be a pertinent quantity to represent the severity of the damage. However, the same total damaged surface can be contributed by a smaller number of large voids/cracks or a larger number of small voids/cracks. In addition, the volume average of the damaged surface area smears the information about the crack distribution and therefore the specific damaged surface area may not accurately represent the damage status. For brittle materials the specific damaged surface area is related to the energy required to generate the fractured surface.

Using the Griffith energy criterion, the following relationship shall hold.

$$\frac{d}{dS}(U - F + W) = 0 \tag{3-40}$$

or

$$\frac{d}{dS}(F - U) = \frac{dW}{dS} \tag{3-41}$$

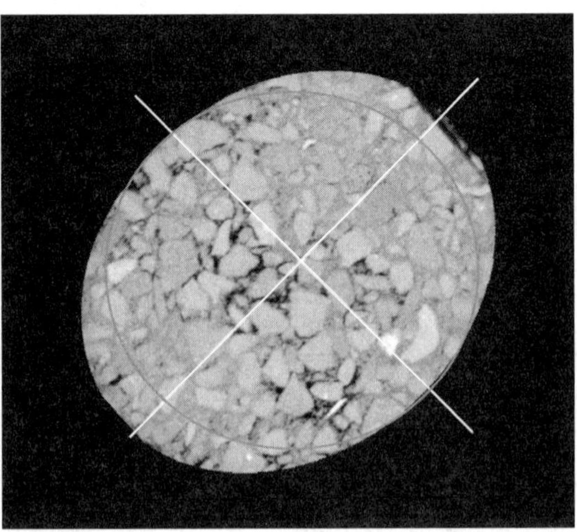

FIGURE 3.23   A section in the specimen and the circular area for area fraction evaluation.

Where U is the elastic energy contained in the volume; F is the work performed by external force; W is the energy for crack formation, and S is the cracked surface. $\frac{d}{dS}(F-U)$ is the energy release rate and $\frac{dW}{dS}$ is the crack resistance. If the material is truly brittle like glass, the energy for crack growth is the surface energy to form the new fractured surface, and $\frac{dW}{dS} = \gamma$ – the surface energy for the material. If the volume change is further assumed to be negligible during the crack growth process, then:

$$dS = dS_v V_0 \qquad (3\text{-}42)$$

Where S_v is the specific damaged surface area and V_0 is the constant volume. Therefore, the following evolution equation for S_v holds.

$$\frac{d}{dS_v}(F-U) = V_0 \gamma \qquad (3\text{-}43)$$

3.6.1.2 Damage Tensor

The damage tensor σ_{ij} thus defined is widely used in CDM. The typical application can be found in the following equation (Murakami, 1988):

$$\bar{\sigma}_{ij} = M_{ijkl}\sigma_{kl} \qquad (3\text{-}44)$$

Where σ_{ij}, σ_{kl} are the effective stress tensor and the Cauchy stress tensor respectively, while M_{ijkl} is a symmetric fourth rank tensor—the damage effect tensor. By using the vector formats of the tensors, Murakami (1988) showed that:

$$M = (I - \varphi)^{-1} \qquad (3\text{-}45)$$

Where I is the second-rank identity tensor, and φ is the damage tensor.

Voyiadjis and Kattan (1999) derived an explicit relation between M and φ. The diagonalized form involves the terms of $1/(1 - \varphi_i)$, and $1/(1 - \varphi_i)(1 - \varphi_j)$ (i,j=1,2,3, $i \neq j$). The averages of these terms and $1/(1 - \varphi_1)(1 - \varphi_2)(1 - \varphi_3)$ may serve as empirical quantity to represent the state of damage. Voyiadjis and Kattan (1999) also introduced a term $\varphi = \sqrt{\varphi_1^2 + \varphi_2^2 + \varphi_3^2}$ to represent the overall damage. This term was used to represent the damage states of the specimens obtained from the WesTrack project (Wang et al., 2001b).

3.6.1.3 Spacing-Size Ratio

The spacing between defects and the size of the defects are important variables in non-local damage theories. For modeling applications, the size of cracks or voids and the average spacing among the cracks offer two relevant parameters that are related to performance. Chudnovsky et al. (1987a, 1987b) developed a simple model addressing the interaction of two cracks, in which the two important model parameters are the spacing between the two crack tips and the sizes of the cracks. Figure 3.24 presents the configuration of two interacting cracks. The stress intensity (mode I) of the large crack is affected by the adjacent cracks. The ratio $\dfrac{K_1^{eff}}{K_1^0}$ (where K_1^{eff} is the effective stress intensity of a crack when subjected to the influence of the adjacent crack while K_1^0 is the stress intensity without the influence of adjacent cracks) decreases with the increase of λ/δ, which means, the larger the crack size and/or the smaller the spacing, the larger the

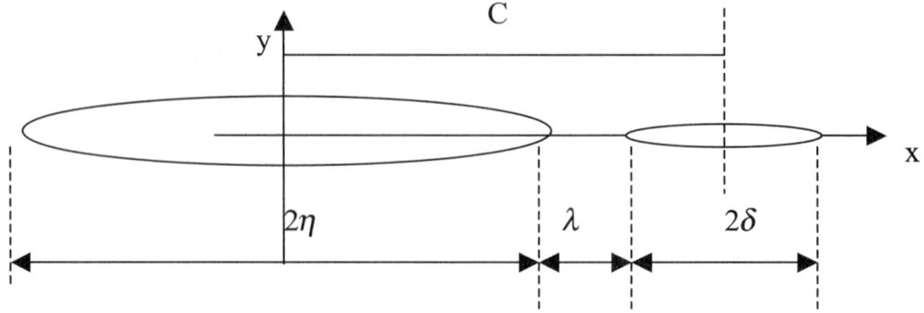

FIGURE 3.24 Illustration of the model for two interacting cracks.

effective stress intensity. Therefore, under the same external loading, the material with smaller λ/δ should have a larger stress intensity factor. Equations 3-46 and 3-47 present the relationship between the effective stress intensity and the crack configuration parameters illustrated in Figure 3.24.

$$\frac{K_1^{eff}}{K_1^0} = \frac{1}{1-q} \tag{3-46}$$

$$q = \frac{1}{\sqrt{2(\delta/\eta + \lambda/\eta)}} \int_{-1}^{1} \sqrt{\frac{1+t}{1-t}} \left(\frac{1}{\sqrt{1-(\frac{\delta}{\eta})^2/(t-C^2/\eta^2)}} - 1 \right) dt \tag{3-47}$$

As the sizes of interacting cracks are usually different and the configuration of the two cracks is complicated, the above relation is usually not followed quantitatively. However, it offers a guide for selecting rational damage quantities that may empirically relate to the effective properties of a damaged system. As it is impossible to develop any analytical models for a damaged system with randomly distributed cracks, empirical quantities are in many cases useful in materials evaluations. This is illustrated in the example application.

3.6.1.4 Application to the WesTrack Project

Wang et al. (2001b) applied these approaches to the WesTrack project and validated the applications of the above methods. The mean solid path tensor, the specific damaged surface area and damage tensor, and spacing-size ratio were quantified in their study to systematically evaluate the microstructure of the three mixes, and indicated that the average spacing (mean solid path) among the damaged surface, the comprehensive damage tensor quantity, and the spacing size ratio might represent the damage (weakening) state well. The specific damaged surface area may not represent the behavior of damaged materials accurately.

3.7 Other Studies

There are many other studies focusing on the use of microstructural information for characterizing, modeling, and simulating the macroscopic behavior of AC. The following presents a brief review of these studies. Kutay et al. (2007, 2009) studied the distribution of pore pressure and shear stress in AC with the considerations of asphalt pore structure. Arambula et al. (2007) and Masad and Button (2004) used XCT to study the influence of void distribution on compaction and moisture damage effects. Partl et al. (2007) carried out a detailed study to compare the microstructure of AC with different compaction methods. Song et al. (2005) made efforts to link XCT measurements with damage mechanics and micromechanics applications. While it is not necessary to cite all these applications, these efforts do indicate the consensus that researchers have reached in characterizing the microstructure quantities into the modeling and simulation of AC.

References

Arambula, E., Masad, E. and Martin, A.E (2007). Influence of air void distribution on the moisture susceptibility of asphalt mixes. *Journal of Materials in Civil Engineering,* Vol.19, No.8, pp.655–664.

Bagi, K.(1996). Stress and strain in granular assemblies, *Mechanics of Materials.* Amsterdam, North-Holland, Vol. 22, pp.165–177.

Chang, C.S. and Ma, L. (1991). A micromechanical-based micropolar theory for deformation of granular solids. *Int. J. Solids Structures.* Vol.28, No.1, pp.67–86.

Christoffersen, J., Mehrabadi, M.M. and Nemat-Nasser, S. (1981). A micromechanical description of granular material behavior. *Journal of Applied Mechanics,* Vol.48, pp.339–344.

Chudnovsky, A., Dolgopolsky, A. and Kachanov, M. (1987a). Elastic interaction of a crack with a microcrack array-I. Formulation of the problem and general form of the solution. *International Journal of Solids and Structure,* Vol.23, No.1, pp.1–10.

Chudnovsky, A., Dolgopolsky, A. and Kachanov, M. (1987b). Elastic interaction of a crack with a microcrack array-II. Elastic solution for two crack configurations (piecewise constant and linear approximations). *International Journal of Solids and Structure,* Vol.23, No.1, pp.11–21.

Cowin, S.C. (1985). The relationship between the elasticity tensor and the fabric tensor. *Mechanics of materials.* Vol. 4, pp.137–147.

Cowin, S.C. (1986). Fabric dependence of an anistropic strength criterion. *Mechanics of Materials.* Vol.5, pp.251–260.

Cundall, P.A. and Strack, O.D.L. (1979). A distinct element model for granular assemblies. *Geotechnique,* Vol.29, pp.47–65.

DeHoff, R.T. and Rhines, F.R. (1968). Quantitative Microscopy. *McGraw-Hill,* New-York.

Ferrari, M., Vladimir, T., Granik, A.L. and Nadeau, J.C. (Eds.) (1997). Advances in Doublet Mechanics. *Springer,* Berlin.

Frost, J.D. and Kuo, C.Y. (1996). Automated determination of the distribution of cohesionless soil subjected to cyclic and monotonic loading. *ASTM Geotechnical Testing Journal,* Vol.19, No.2, pp.107–117.

Goodman, M.A. and Cowin, S.C. (1972). A continuum theory for granular materials. *Arch. Rational Mechanics Analysis.* Vol.44, pp.248–266.

Granik, V.T. and Ferrari, M. (1993). Microstructural mechanics of granular media. *Mechanics of Materials.* Vol.15, pp.301–322.

Guo, J.F., Cai, Y.L. and Wang, Y.P. (1995). Morphology-based interpolation for 3D medical image reconstruction. *Computerized Medical Imaging and Graphics,* Vol.19, No.3, pp.267–279.

Hilliard, J.E. (1962). Specification and measurement of microstructural anisotropy. *Transactions, AIME.* Vol.224, pp.1201–1211.

Hilliard, J.E. (1967). Determination of structure anisotropy. *Proceedings, 2nd International Congress for Stereology.* New York, pp.219–228.

Hsieh, J. (2003). Computed Tomography: Principles, Design, Artifacts, and Recent Advances. *SPIE Optical Engineering Press,* Bellingham.

Kanatani, K.I. (1984a). Stereological determination of structural anisotropy. *International Journal of Engineering Science,* Vol.22, No.5, pp.531–546.

Kanatani, K.I. (1984b). Distribution of directional data and fabric tensors. *International Journal of Engineering Science,* Vol.22, No.2, pp.149–164.

Kanatani, K.I. (1985). Procedures for stereological estimation of structural anisotropy. *International Journal of Engineering Science,* Vol.23, No.5, pp.587–598.

Kuo, C.Y., Frost, J.D. and Chameau, J.L.A. (1998). Image analysis determination of stereology based fabric tensors. *Geotechnique,* Vol.48, No.4, pp.515–525.

Kutay, M.E., Aydilek, A.H. and Masad, E. (2007). Computational and experimental evaluation of hydraulic conductivity anisotropy in hot-mix asphalt. *International Journal of Pavement Engineering,* Vol.8, No.1, pp.29–43.

Kutay, M.E. and Emin, M. (2009) Pore pressure and viscous shear stress distribution due to water flow within asphalt pore structure, *Computer-Aided Civil and Infrastructure Engineering.* Vol. 24, pp.212–224.

Lacy, T.E., McDowell, D.L, Willice, P.A. and Talreja, R. (1997). On representation of damage evolution in continuum damage mechanics. *International Journal of Damage Mechanics.* Vol.6, pp.62–95.

Lacy, T.E., McDowell, D.L, Willice, P.A. and Talreja, R. (1999). Gradient concept for evolution of damage. *Mechanics of Materials.* Vol.31, pp.831–860.

Lin, W.C., Liang, C.C. and Chen, C.T. (1988). Dynamic elastic interpolation for 3D medical image reconstruction from serial cross section. *IEEE Trans. Med. Imaging.* Vol.7, No.3, pp.225–232.

Masad, E. and Button, J. (2004). Implications of experimental measurements and analyses of the internal structure of hot-mix asphalt. *Transportation Research Record, No. 1891,* pp. 212–220.

Masad, E; Tashman, L. Somedavan, N. and Little, D. (2002). Micromechanics-based analysis of stiffness anisotropy in asphalt mixtures. *Journal of Materials in Civil Engineering.* Vol.14, No.5, pp.374–383.

Mauhlhaus, H-B. and Vardoulakis, I. (1987). The thickness of shear bands in granular materials. *Geotechnique,* Vol.37, No.3, pp.271–283.

Mehrabadi, M.M. and Nemat-Nasser, S. (1983). Stress, dilatancy and fabric in granular materials. *Mechanics of Materials.* Vol.2, pp.155–161.

Murakami, S. (1988). Mechanical modeling of material damage. *Journal of Applied Mechanics,* Vol.55, pp.280–286.

Nemat-Nasser, S and Mehrabadi, M. (1984). Micromechanically based rate constitutive descriptions for granular materials. *Mechanics of Engineering Materials.* John Wiley & Sons Ltd., pp.451–463.

Oda, M. (1972a). Initial fabrics and their relations to mechanical properties of granular material. *Soils and Foundations.* Vol.12, No.1, pp.17–36.

Oda, M. (1972b). The mechanism of fabrics changes during compressional deformation of sand. *Soils and Foundations.* Vol.12, No.2, pp.1–18.

Oda, M. (1972c). Deformation mechanism of sand in triaxial compression tests. *Soils and Foundations.* Vol.12, No.4, pp.45–63.

Oda, M., Konishi, J. and Nemat-Nasser, S. (1982). Experimental micromechanical evaluation of strength of granular materials: effects of particle rolling. *Mechanics of Materials.* Vol.1, pp.269–283.

Oda, M. and Nakayama, H. (1989). Yield function for soil with anisotropic fabric. *Journal of Engineering Mechanics.* Vol.15, No.1, pp.89–104.

Partl, M.N., Alexander, F. and Jonsson, M. (2007). Comparison of laboratory compaction methods using x-ray computer tomography. *Road Materials and Pavement Design*, Vol. 8, No. 2, pp. 139–164.

Raya, S.P., Udupa, J.K. (1990). Shape-based interpolation of multidimensional objects. *IEEE Transactions of Medical Imaging.* Vol.9, No.1, pp.32–42.

Rothenburg, L. and Bathurst, R.J. (1992). Micromechanical features of granular assemblies with planar elliptical particles. *Geotechnique,* Vol.42, No.1, pp.79–95.

Rowe, P.W. (1962). The stress-dilatancy relation for static equilibrium of an assembly of particles in contact. *Proceedings of the Royal Society of London,* Vol.A269, No.1339, pp.500–527.

Russ, J.C. and Dehoff, R.T. (2000). Practical Stereology. 2nd Edition. *Kluwer Academic/Plenum,* New York.

Satake, M. (1983). Fundamental quantities in the graph approach to granular materials. *Mechanics of Granular Materials: New Models and Constitutive Relations. Elsevier,* New York, pp.9–19.

Song, I., Little, D., Masad, E. and Lytton, R. (2005). Comprehensive evaluation of damage in asphalt mastics using x-ray ct, continuum mechanics, and micromechanics. *J. of the Assoc. Asphalt Paving Technologists,* Vol. 74, pp. 885–920.

Tobita, Y. (1988). Yield condition of anisotropic granular materials. *Soils and Foundations.* Vol.28, No.2, pp.113–126.

Tobita, Y. (1989). Fabric tensors in constitutive equations for granular materials. *Soils and Foundations.* Vol.29, No.4, pp.91–104.

Tobita, Y. and Yanagisawa, E. (1992). Modified stress tensors for anisotropic behavior of granular materials. *Soils and Foundations.* Vol.32, No.1, pp.85–99.

Tozeren, A. and Skalak, R. (1989). Does fabric tensor exist for a fabric? *Journal of Materials Science.* Vol.24, No.5, pp.1700-1706.

Underwoood, E.E.(1970). Quantitative Stereology. *Addison-Wesley,* Reading.

Voyiadjis, G.Z. and Kattan, P. (1999). Advances in Damage Mechanics: Metals and Metal Matrix Composites, *Elsevier,* Oxford.

Wang, L.B., Frost, J.D. and Lai. J.S. (2001a). Quantification of the doublet vector distribution of granular materials. *Journal of Engineering Mechanics, Special Issue: Multiple Modeling of Damage and Material Characterization with Microstructure.* Vol.127, No.7, pp.720–729.

Wang, L.B., Frost, J.D. and Shashidhar, N. (2001b). Microstructure study of Westrack mixes from X-ray tomography images. *Transportation Research Record,* No.1767, pp.85–94.

Wang, L.B., Frost, J.D. and Lai. J.S. (2004). 3D digital representation of the microstructure of granular materials from X-ray tomography imaging. *Journal of Computing in Civil Engineering.* Vol.18, No.1, pp.28–35.

Wang, L.B., Wang, Y.P., Li, Q.B. and Flintsch, G.W. (2007). Stress concentration factor as a performance indicator for asphalt mixes. *Geotechnical Special Publication,* No.176, pp.1–13.

Wang, L.B., Wang, Y.P., Mohammad, L.N. and Harman, T.P. (2002). Voids distribution and performance of asphalt concrete. *International Journal of Pavements,* Vol.1, No.3, pp.22–33.

Wang, L.B., Park, J.Y., Zhou, C.B. and Fu, Y.R. (2010). Quantification of the contact normal, branch vector and particle orientation distribution using X-ray tomography imaging. Submitted to *Journal of Engineering Mechanics.*

<h2>CHAPTER 4</h2>

Experimental Methods to Characterize the Heterogeneous Strain Field

4.1 Introduction

Due to the heterogeneity of asphalt concrete (AC), its strain field is not uniform, even if it is subjected to a uniform stress field. Permanent deformations in AC pavements are not only a function of the magnitude and number of repetitions of external loading, but also are influenced by the characteristics and spatial arrangement of the particles in the mass. How individual particle motions (micro) relate to the global response (macro) is important in micro-mechanical studies that relate "micro" measures to "macro" response. Macro-response refers to the measurement that assumes that the AC is a homogeneous material whose response can be quantified by global measurements. However, since AC is composed of aggregates and asphalt binder, and the aggregates virtually do not deform under typical tire loading, interpretation using conventional strain measurements is inappropriate. Since the actual deformation takes place almost solely in the mastic (the mixture of asphalt binder and small particles), permanent deformation is localized in the mastic. In addition, fatigue cracking often initiates from the interfaces between the mastic and the larger aggregate particles. Accordingly, measurement of the permanent strains in the mastic may be a better measure for exploring the mechanisms of permanent deformation and fatigue cracking. The mastic strain measure is likely to be more relevant in that the mastic is more homogeneous than AC. In this chapter, methods which use imaging techniques and pattern recognition to measure the permanent strains in the mixture and the mastic are presented. Application of the procedure to the measurements of the strains developed in a specimen used in an asphalt pavement analyzer (Georgia Loaded Wheel Tester, GLWT) is described. The importance of these measurements is that they provide approaches to verify mechanics principles and computational simulation methods, even at the microscopic scales.

Strains are typically computed from measured displacements. For the measurement of the displacement of a point, a reference point and the displacement of the point relative to the reference point are required. The methods for measuring displacements generally fall into one of two fundamentally different techniques: invasive and non-invasive. For most invasive techniques, measurements tend to be limited to only a few points to minimize interference and cost. To obtain accurate displacement measurements, testers using invasive techniques must make sure that the sensors deform compatibly with the material being measured so that the interference of the sensors with the material being tested is minimized. For non-invasive measurements, there are no such limitations and thus techniques such as imaging offer the potential for non-invasive strain measurement of bonded granular materials such as AC and cement concrete. Non-invasive measurements have been widely applied in medical science (IAEA, 1995), in fluid flow measurement (Hanssan et al., 1991), and in granular particle tracking (Warr, 1994). These methods are presented first for the two-dimensional (2D) case using optical imaging and are then extended to the three-dimensional (3D) case using XCT imaging. A digital image correlation method, mainly applicable to the 2D case, will be discussed in the last section of this chapter.

4.2   2D Methodology

This section presents a method that permits non-invasive evaluation of the strains in the mastic using optical imaging of particle positions. It will be directly illustrated for specimens tested in the GLWT, while the methodology is equally applicable for use with other testing systems and specimens.

As described by Collins et al. (1996), the GLWT can be used to test specimens of AC and other pavement materials under accelerated simulated conditions in the laboratory. Specifically, a parallelepiped specimen measuring approximately $30 \times 13 \times 8$ cm is subjected to a simulated wheel load along the center of one of its long faces (Figure 4.1). The resulting rut depth is recorded as a function of the number of passes and is obviously a measurement of macro response as described earlier. To permit non-invasive micro measurements, the GLWT test procedure is modified as follows. The parallelepiped specimen of AC is prepared in the conventional manner; however, before being placed in the GLWT, the specimen is cut into two or more sub-specimens with a saw perpendicular to the long axis of the original specimen, and thus perpendicular to the direction of wheel travel (Figure 4.1). A reference grid is marked on the cut faces and images from these faces are captured using a charge coupled device (CCD) camera and stored for later analysis. The sub-specimens are then put together with the appropriate cut faces adjacent to each other in the GLWT. Although the GLWT specimen physically consists of sub-specimens, the cut faces are in intimate contact with each other and are held in that

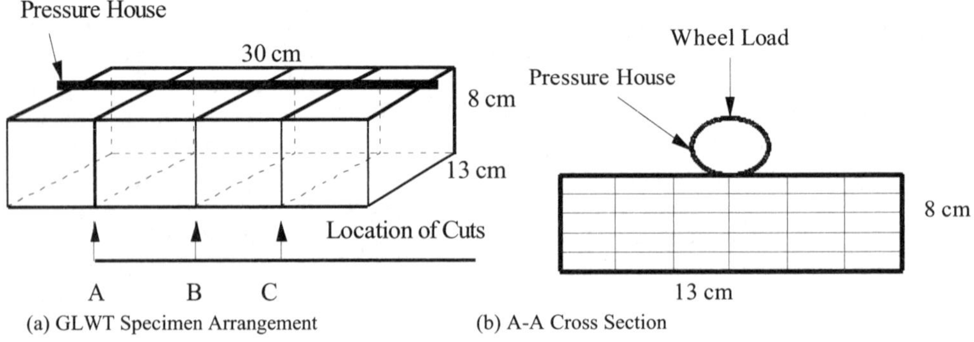

FIGURE 4.1   GLWT specimen configuration.

position by the clamping system of the GLWT. In this manner, the cut faces, although available for optical imaging at various stages during the rutting process, are located within the center of the specimen during actual application of the load by the rutting wheel, and thus are isolated from any boundary effects associated with the test system. The remainder of the rutting test procedure is implemented as with normal specimens.

When the desired number of passes of the simulated rutting wheel has been applied, the macro-response of the specimen is determined using the conventional procedures described by Collins et al. (1996) and Kandhal and Cooley (2003). The specimen is then separated into the sub-specimens to enable additional images to be captured from the cut faces for subsequent processing and analysis. Photographs of a complete section, prior to and after rutting, are shown in Figure 4.2. The indentation from the rutting wheel is evident in the post-rutting photograph. Images captured at this scale do not have the necessary resolution to permit quantitative measurement of the strains in the mastic. Clearly, selection of an appropriate magnification is dependent on the characteristic size of the aggregate in the material. Accordingly, the images captured for quantitative analysis need to be captured with a CCD camera mounted on a microscope. For the particular material being examined in the example study, a magnification of 7X was found to provide an appropriate level of resolution. Accordingly, the scheme shown in Figure 4.3 was selected for capturing images on a given cut section. Pre- and post-rutting images, along with their proposed digital counterparts, are shown in Figure 4.4. A subsequent montage consisting of eight images (5, 6, 9, 19, 13, 14, 17, 18) captured immediately below and adjacent to the location of the rutting wheel is shown in Figure 4.5. It is noted that some overlap exists between adjacent images to allow for appropriate referencing as required.

FIGURE 4.2   Cross-section of GLWT specimen (a) prior to and (b) after rutting.

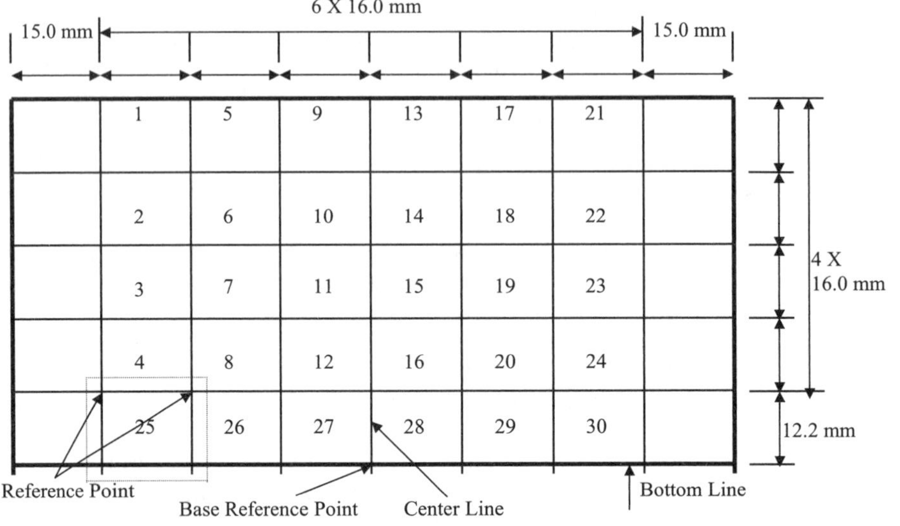

FIGURE 4.3   Location of images captured from cut-section.

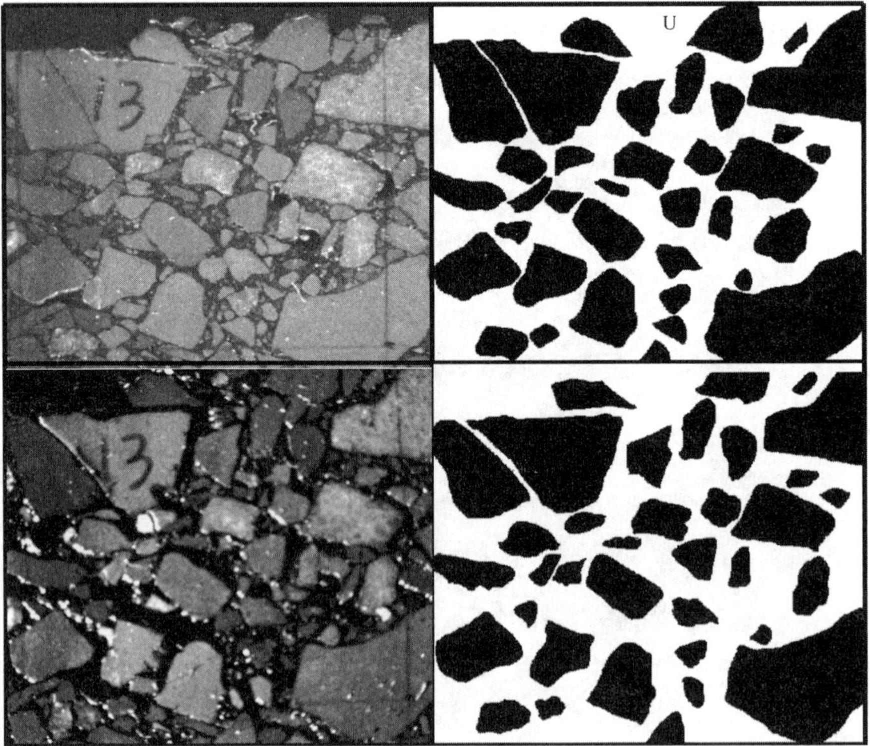

FIGURE 4.4   Images acquired at 7X magnification.

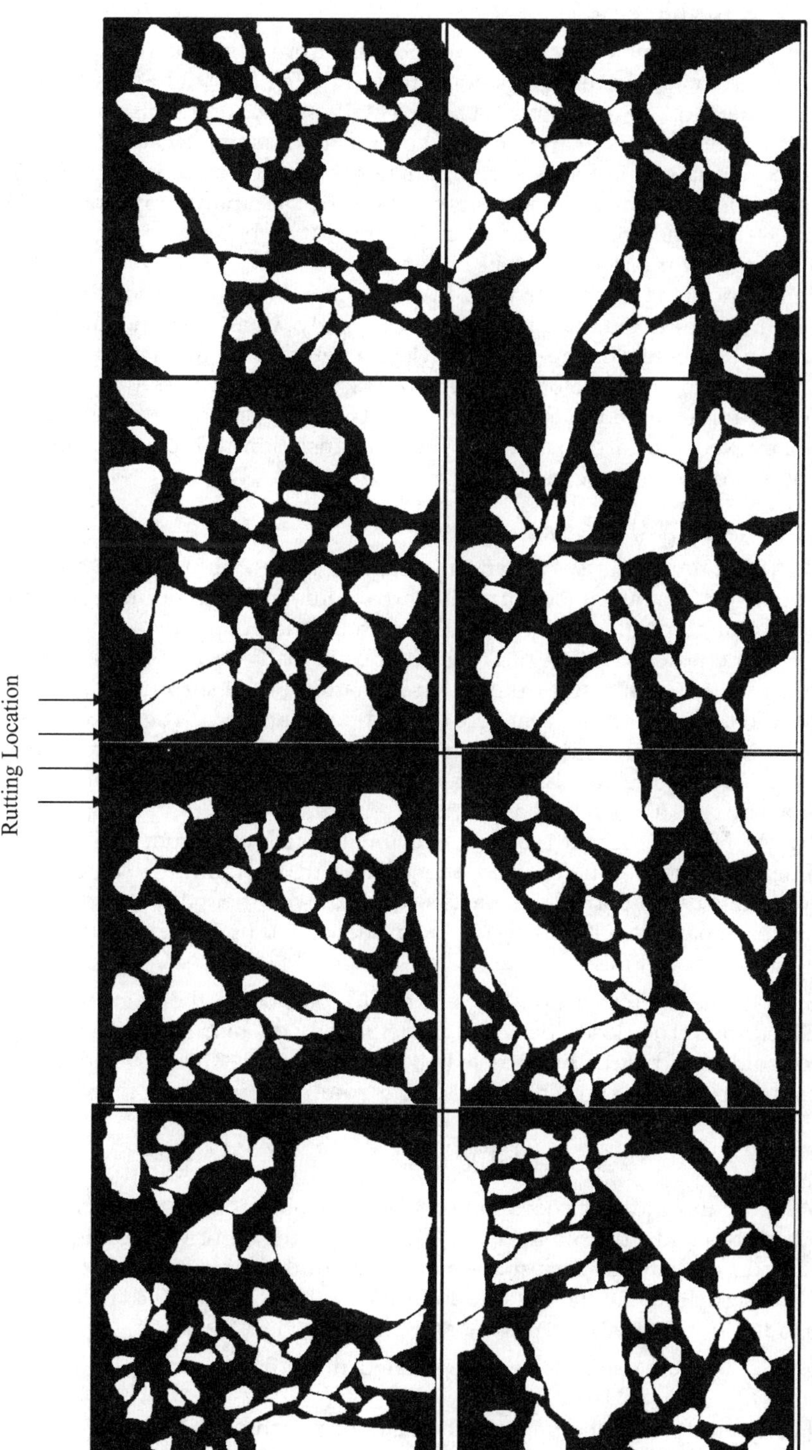

Figure 4.5   Montage showing particle configuration in eight images immediately adjacent to rutting hose.

4.2.1 Image Analysis

Most imaging systems use what is termed a feature count point (FCP) to identify all objects in an image. In many systems, the FCP is the last (right most) pixel of the last (bottom) line of pixels contained in an object. If all measurements are being conducted on a single image, this is an efficient way for the imaging system to identify objects. However, if analysis involves several images (e.g., successive images) where particles may be displaced relative to one another, as is the case in this application, then a given object's FCP in one image may not be the same in a subsequent image. For example, of the 35 objects in the image (Image 10) shown in Figure 4.6a, which was taken from a specimen prior to loading with the GLWT, 14 have different FCPs in the companion image (Figure 4.6b) taken after loading. Accordingly, an alternative approach to "track" particles in successive images must be used. A new procedure that uses pattern recognition based on the morphological characteristics of the particles was identified (Wang, 1999) and implemented in a program, "MATCH," as described below. After the particles are tracked, the particle motions, such as translation and rotation, can be computed, and therefore the strains can be estimated.

4.2.2 The MATCH Program

The MATCH program uses a pattern recognition algorithm to match the particles so that the displacement can be correctly and automatically computed. As previously indicated, under typical wheel loading, the deformation of aggregate particles is negligible. Permanent deformation of AC results principally from particle translations and particle rotations that result from the deformation of the surrounding mastic. In other words, the particles are only subject to rigid-body motions. Therefore, individual particle characteristics on a given cut plane such as their cross-section area, perimeter, and aspect ratio remain relatively constant. For relatively small displacements, their relative positions do not drastically change either.

The MATCH program requires two images acquired from the same region of a specimen cross-section before and after testing that contain the same particles. Small particles considered part of the mastic were removed manually during image processing. For each particle (*i*th particle) in the un-deformed image, the match procedure was implemented as follows:

Step 1: The five closest particles to the *i*th particle in the deformed configuration were identified. The closest particles are defined by the distances of their centroids from the centroid of particle *i*. The particle with the shortest distance to the *i*th particle was the closest particle, and so on. The distance was computed using the following formula:

$$d_j = \sqrt{(x_j^d - x_i^u)^2 + (y_j^d - y_i^u)^2} \tag{4-1}$$

Where x and y are the coordinates of the centroids of the particle cross-sections, the superscripts u and d denote the un-deformed and the deformed configurations respectively, and j is the index corresponding to the number of particles in the deformed image. The particles were then placed in order by this distance and the particle numbers were indexed in order to keep track of their properties.

Step 2: Particle similarity was evaluated through the **similarity index** that is defined as $|Q_j^d - Q_i^u|, j = 1-5$.

The quantity Q is the particle characteristic such as area, aspect ratio, etc. being compared.

Step 3: The particles were ranked by similarity index in ascending order and the top two were assigned a closeness value of 1.

Step 4: The previous two steps were repeated for different particle characteristics and the closeness values for each of the five particles was summed. The particle with the largest closeness value was identified as the matching particle.

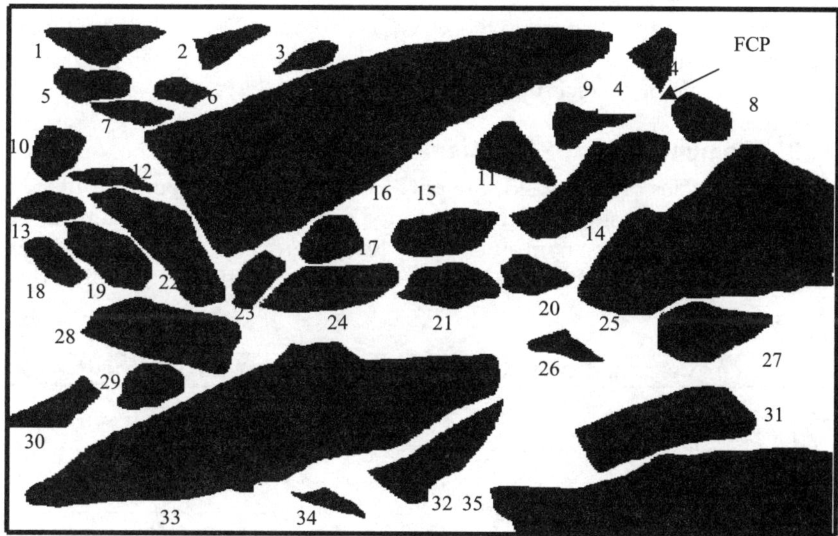

(a) Undeformed Configuration

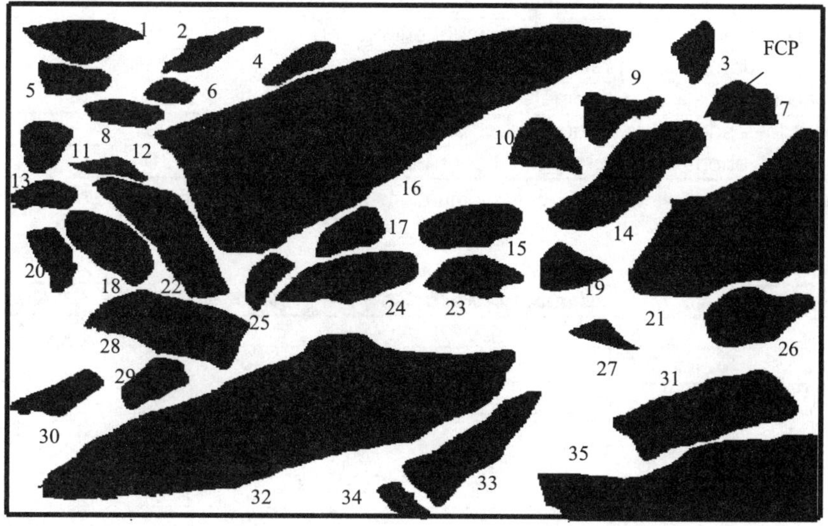

(b) Deformed Configuration

FIGURE 4.6 Numbering in undeformed and deformed configuration (Image 10).

The eight images (5, 6, 9, 10, 13, 14, 17, 18) were analyzed using the MATCH program. The rate of correct matching of particles using different combinations of the characteristics for all eight images is presented in Table 4.1. The matching results for each of the images using the optimum combinations of characteristics are presented in Table 4.2. The correct match was achieved through manual confirmation. Detailed test results can be found in Wang (1998). Due to the small deformation, the relative position of adjacent particles does not drastically change. Therefore, both locations and particle characteristics can be used to match the particles. At the present stage, only directly measured characteristics are used in the MATCH program. Derived characteristics and even Fourier techniques such as particle signature (Wang et al., 1997) can also be used to enhance the rate of correct matching.

4.2.3 Computation of Permanent Macro-Strain

Once the particles are matched, the particle translation can be computed from the differences of the centroid coordinates of the particles before and after deformation. The particle rotations can be computed from the differences of the orientations of the longest Feret diameters before and after deformation. With the computed displacements,

Test No.	Combination of Characteristics	No. Matched*	% Matched
1	Area	212	68.8
2	1 + perimeter	247	80.2
3	2 + aspect ratio	265	86.0
4	3 + breadth + length	275	89.3
5	4 + orientation + orthogonal orientation	286	92.9
6	Xfc + Yfc + Xmax + Ymin + Xmin + Xcen + Ycen = location	262	85.1
7	Location + area + perimeter	277	89.9
8	Location + breadth + length	281	91.2
9	Location + orientation + orthogonal orientation	275	89.3
10	Location + aspect ratio + roundness	268	87.0
11	Location + area + perimeter + breadth + length	289	93.8
12	11 + aspect ratio + roundness	297	96.4
13	12 + orientation + orthogonal orientation	302	98.1

TABLE 4.1 Correct matching of different combinations of (*total number of particles = 308).

Image No.	Number of Particles	Correct No.	Rate (%)
05	43	43	100
06	47	46	97.9
09	39	37	94.9
10	35	35	100
13	36	34	94.4
14	35	34	97.1
17	40	40	100
18	33	33	100

TABLE 4.2 Optimum matching results for the individual images.

macro-strains like those defined in the finite element method can be calculated. The calculation is illustrated as follows:

For each particle i, assume its two nearest neighbor particles are particles j and k. The displacements of the three particles (calculated from the differences between the coordinates of the centroids in un-deformed and deformed configurations) are u_i, u_j, u_k, v_i, v_j, v_k (i, j, and k are arranged counter-clockwise).

For a triangle, the valid displacement format can be assumed as:

$$u = a_0 + a_1 x + a_2 y \qquad (4\text{-}2)$$

$$v = b_0 + b_1 x + b_2 y \qquad (4\text{-}3)$$

Where x and y are coordinates in the image frame of the un-deformed configuration.

u = relative displacement in x direction.
v = relative displacement in y direction.

a_0, a_1, a_2 and b_0, b_1, b_2 are constants for a triangle but usually vary from triangle to triangle.

They can be represented conveniently by Equations 4-4 and 4-5.

$$\begin{vmatrix} 1, x_i, y_i \\ 1, x_j, y_j \\ 1, x_k, y_k \end{vmatrix} \begin{Bmatrix} a_0 \\ a_1 \\ a_2 \end{Bmatrix} = \begin{Bmatrix} u_i \\ u_j \\ u_k \end{Bmatrix} \qquad (4\text{-}4)$$

$$\begin{vmatrix} 1, x_i, y_i \\ 1, x_j, y_j \\ 1, x_k, y_k \end{vmatrix} \begin{Bmatrix} b_0 \\ b_1 \\ b_2 \end{Bmatrix} = \begin{Bmatrix} v_i \\ v_j \\ v_k \end{Bmatrix} \qquad (4\text{-}5)$$

For small deformations, the strains can be computed as follows:

$$\varepsilon_x = \frac{\partial u}{\partial x} = a_1 \qquad (4\text{-}6)$$

$$\varepsilon_y = \frac{\partial v}{\partial y} = b_2 \qquad (4\text{-}7)$$

$$\varepsilon_{xy} = \left(\frac{\partial v}{\partial x} + \frac{\partial u}{\partial y}\right) / 2 = (b_1 + a_2) / 2 \qquad (4\text{-}8)$$

4.2.4   Experimental Results and Analysis

The permanent macro strains for the eight images immediately beneath and adjacent to the rutting wheel were determined. As noted, the macro-strains reflect an averaged or homogenized strain because no distinction is made between the deformable mastic and the non-deformable aggregate particles. The x and y direct strains, the shear strains, and the volumetric strains for Image 9 are illustrated in Figure 4.7. The values of the strain

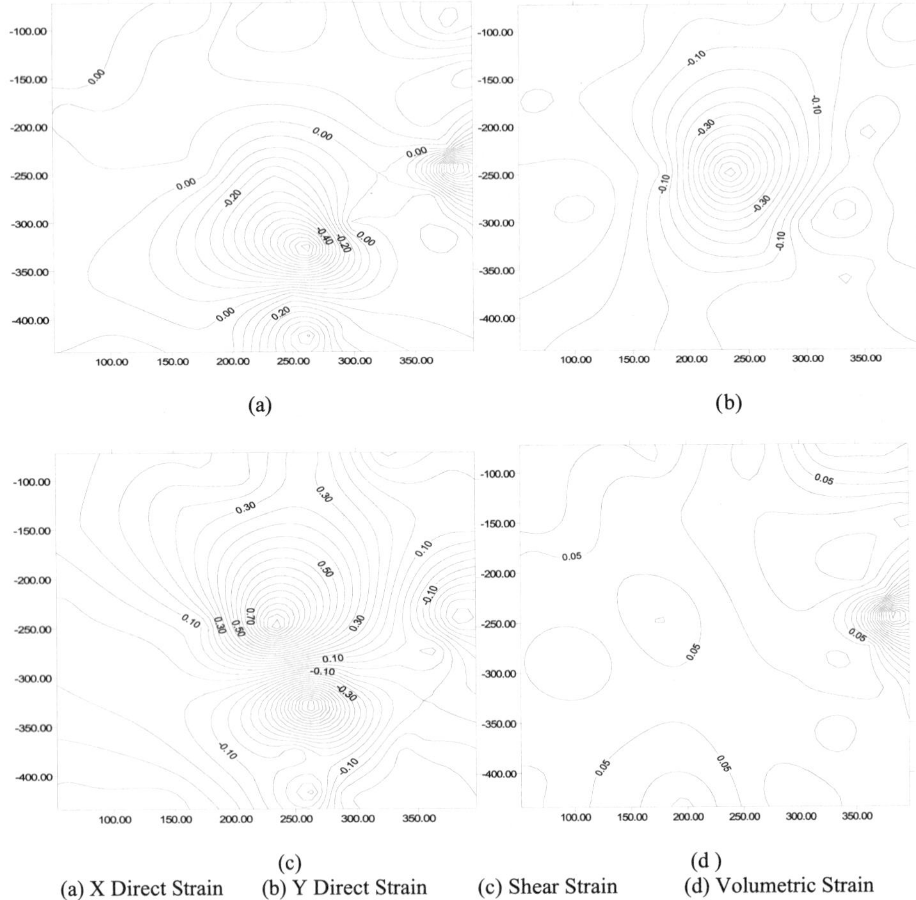

(a)

(b)

(c)   (d)

(a) X Direct Strain   (b) Y Direct Strain   (c) Shear Strain   (d) Volumetric Strain

(Note: The scales for different strains may not be the same due to the consideration of
optimizing the representation of different types of the strains)

Figure 4.7   Macro-strain contours for Image 9.

for one triangle (Figure 4.7a and Particles 9, 11, and 14 in Figure 4.6a) are summarized
in Table 4.3. In addition, the ratio of the mastic area to solid area or mastic solid area
ratio (MSAR) for each image, both before and after deformation, was also calculated
and is summarized in Figure 4.8. It is noted that in these calculations, the sign of the
strains follows the convention that tensile strain is positive. Thus, positive volumetric
strain indicates dilation and vice versa. From Figure 4.8, it can be seen that MSAR in the
zones close to the loading hose increased (Images 9, 10, 13, and 14), whereas MSAR in
the zones outside of these images decreased (Images 4, 6, 17, and 18). MSAR in the
plane strain case is equal to volumetric strains. Therefore, to a certain degree, the increase of this ratio is equivalent to dilation while the decrease of this ratio is equivalent
to contraction.

P No.	Xcen (p)	Ycen(p)	Orientation	u (p)	v (p)	Rotation	Strain
9 (U)	354.7	92.3	138.0				$\varepsilon_x = 0.0116$
11	310.3	123.2	53.0				$\varepsilon_y = 0.0231$
14	361.1	149.7	127.0				$\varepsilon_{xy} = 0.0525$
9 (D)	378.5	95.6	135.0	23.8	3.3	−3.0	$\varepsilon_v = 0.0347$
10	335.5	126.6	51.0	25.2	3.4	2.0	
14	389.4	151.8	127.0	28.3	2.1	0.0	

TABLE 4.3 Input for the mastic strain measurement and the global strains.

It should be also noticed from the volumetric strain contours (Figure 4.7) that dilation and contraction coexist within a few adjacent particles in the same image. This is quite reasonable as dilation and contraction are related to particle configurations that can be quite different among adjacent particles. In other words, tensile and compressive strains can take place within a few particles (a discrete element method [DEM] simulation also indicates this phenomena, see Chapter 9). From the plots of the volumetric strains, the direct strains, and the shear strains, it is evident that the macro-strain contours are quite complicated and indicative of the influence of the particle configurations.

4.2.5 Computation of Permanent Strain in the Mastic

As previously noted, the macro-strain measurements presented in the previous section were average strains, and while they can be useful in evaluating the overall deformation properties of AC, they mask the true micro-strain characteristics. For a better understanding of the microstructural behavior of AC, it is more appropriate to study

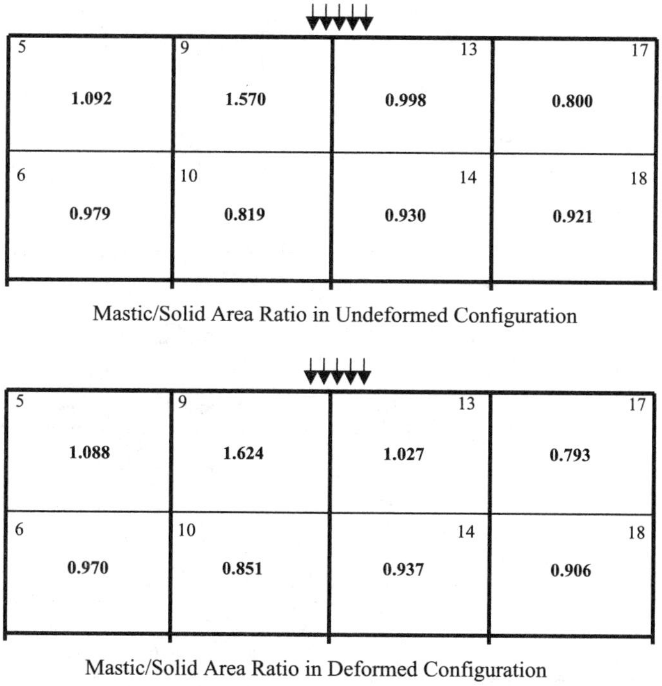

FIGURE 4.8 Mastic/solid area ratio evolution.

micro-strains. The strains, which occur in the mastic, are important parameters in the evaluation of rutting and fatigue resistance of AC. The procedure developed in the previous sections was thus extended to include the capability of measuring the average strains in the mastic only. Since mastic properties are both strain-rate and strain-level dependent (Brown et al., 1980), the measurement of the permanent strain in the mastic is only of interest in the analysis and experimental evaluation of the mechanical properties of AC.

The approach is best illustrated by considering the relative displacement between the gravity center (cross-section area center) of any two particles i and j, as shown in Figure 4.9. The displacement of particle i relative to particle j as a result of particle translation can be decomposed into two components:

The component along BA (see Figure 4.9):

$$\Delta u_n = (u_i - u_j)\cos\theta + (v_i - v_j)\sin\theta \tag{4-9}$$

and the component normal to BA:

$$\Delta u_t = -(u_i - u_j)\sin\theta + (v_i - v_j)\cos\theta \tag{4-10}$$

Where u_i, u_j, v_i, v_j are the x and y components of the displacements of the gravity center of the particles, while θ is the angle of BA (j to i) relative to the x axis.

Since the particles are essentially rigid, actual deformation takes place in the mastic only and therefore the relative displacement and the distance between A and B (Figure 4.9) should be used for computing the strain measure of the mastic between the two particles.

The average engineering strain along BA (normal strain) in the mastic is:

$$\varepsilon_n = \frac{\Delta u_n}{H} \tag{4-11}$$

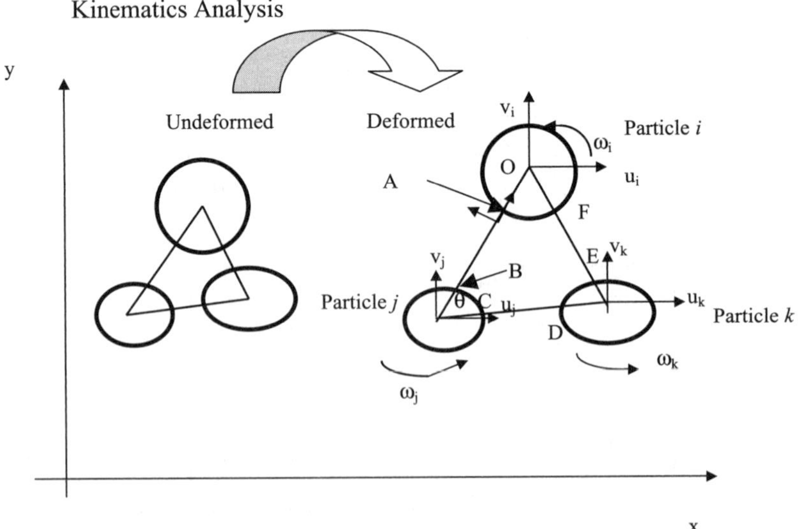

Figure 4.9 Illustration of particle kinematics.

The average engineering shear strain (tangent strain) in the mastic is:

$$\gamma = \frac{\Delta u_t - \omega_i r_i - \omega_j r_j}{H} \tag{4-12}$$

Where r_i is the distance from the center of the particle to a crossing point such as A; ω_i is the rotation of the particle (counter-clockwise rotation as positive); H is the distance between A and B.

As the vector connecting the centroids of two particles is not necessarily normal to the particle surface at crossing points such as A and B, the displacement normal to the vector, and therefore the shear strain, is only an approximation. Both direct and shear strain are average values between A and B. It should be noted that in the above formulations there is an implied assumption that there is no discontinuity between the particle and the mastic.

An interesting implication of the above formulation is that Equations 4-11 and 4-12 are related to particle translation and rotation (particle kinematics), respectively.

4.2.6   Experimental Results and Analysis

The above procedure was implemented to estimate the strains in the mastic for the eight images previously analyzed. The normal strains and the tangent strains for one of the images (Image 10) are presented in Figure 4.10. The micro-strain values for the same triangle whose macro-strain values are presented in Table 4.3 are presented in Table 4.4. From the results, several observations are made:

a. The mastic strain is very localized. The magnitude varies by several orders and is typically much larger than global strains.

b. In the zone dominated by compression, tensile strain does take place. This is consistent with the material properties of AC and with doublet mechanics predictions.

c. At a very large magnitude of strain, the interface between particles and the mastic might debond. The actual strain in the mastic may be smaller, however, before debonding the mastic likely was subjected to much larger strains than estimated using the conventional approach of measuring global strains. This has important implications in material property evaluations in that consistent strain levels should be used in experiments to evaluate the mechanical properties of AC.

The evolution of the mastic/solid area ratio indicated that the zone under the wheel demonstrated dilatancy, whereas the zone away from the wheel demonstrated contraction. The measured permanent strain field demonstrated strong localism. Thus, direct strains, shear strains, and volumetric strains all demonstrated particle configuration dependency.

The permanent strain in the mastic was also evaluated. It was found that this strain is generally much larger than the macro-strain. This measurement has important implications in the evaluation of rutting and fatigue resistance of AC because deformations in the mastic are the source of rutting and fatigue development and are related to particle rotations.

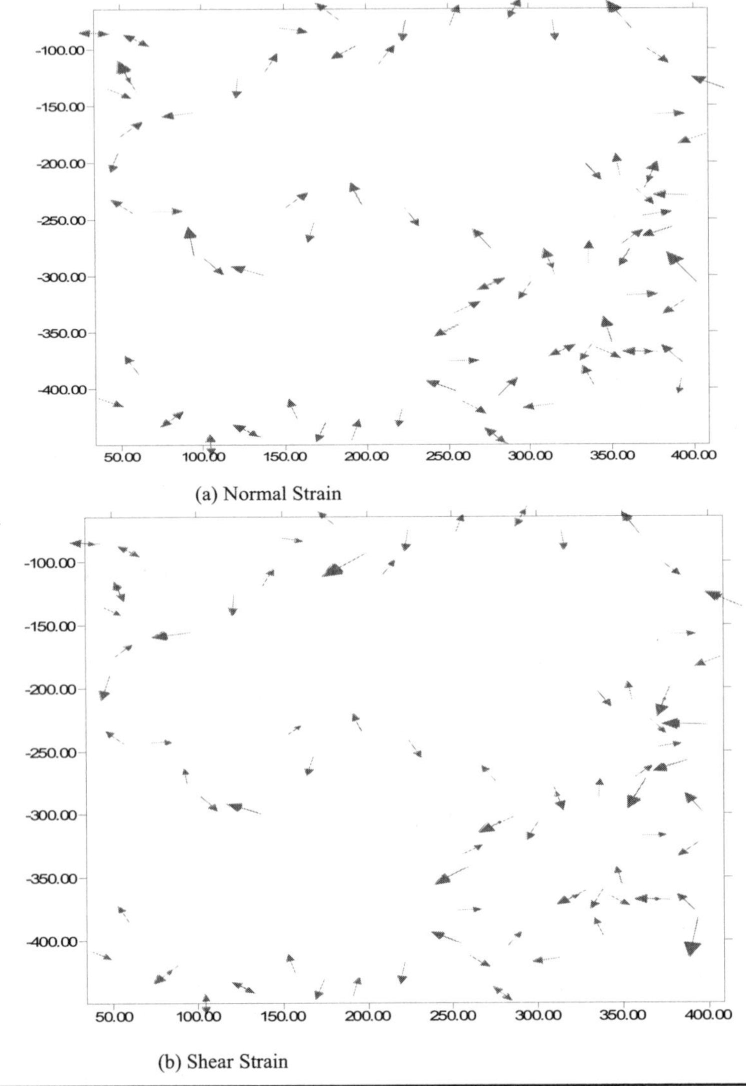

(a) Normal Strain

(b) Shear Strain

FIGURE 4.10   Normal and tangent strains in mastic in image 10.

Node	X (p)	Y(p)	u (p)	v (p)	Strain (ab,bc...)
a	56	24	24.4	3.2	$\varepsilon_n = 0.1378$
b	57	37	28.3	6.1	$\gamma = 0.1102$
c	32	54	28.3	6.1	$\varepsilon_n = -0.2325$
d	24	50	25.1	3.4	$\gamma = -0.0040$
e	26	31	25.1	3.4	$\varepsilon_n = 0.2193$
f	35	25	24.6	4.3	$\gamma = -0.2950$

TABLE 4.4   Strains in the mastic.

4.3   3D Methodology

For the 3D case, XCT imaging is used to obtain the internal structure of the materials non-destructively (as described in Chapter 3). The method similar to the 2D case will be presented in the following sections. Figure 4.11a shows the sectional images of a specimen. Figure 4.11b shows the reconstructed 3D specimens. In this study, before and after APA testing, the images acquired from the XCT scan were used to measure the strain in the mixture. Figure 4.12a shows the reconstructed 3D specimen before the test, while Figure 4.12b represents the specimen after the test. It should be noted that the images shown in Figure 4.12 are not simple digital pictures. They contain 200 sectional images.

4.3.1   Specimen Preparation

The size of the specimen is approximately $300 \times 125 \times 75$ mm. It was prepared using the conventional rolling compaction. The aggregates were limestone and the binder used was PG70-22.

4.3.2   Image Acquisition

Due to the energy limitation of the XCT system, the specimen could not be directly scanned in its original size. The specimen was cut into three pieces and the cutting direction was perpendicular to the long axis of the original specimen as shown in Figure 4.13. Before and after testing, only Piece B was scanned. The size of Piece B was $125 \times 75 \times 75$ mm. The image plane was consistent with the 75×75 mm cross-section. The two sets of images scanned before and after APA tests were used to measure the particle displacements for macro- and micro-strain quantification. The image size was

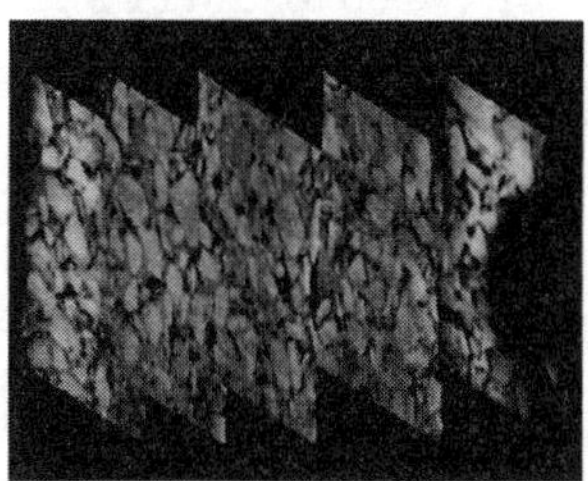

(a) Sectional Images

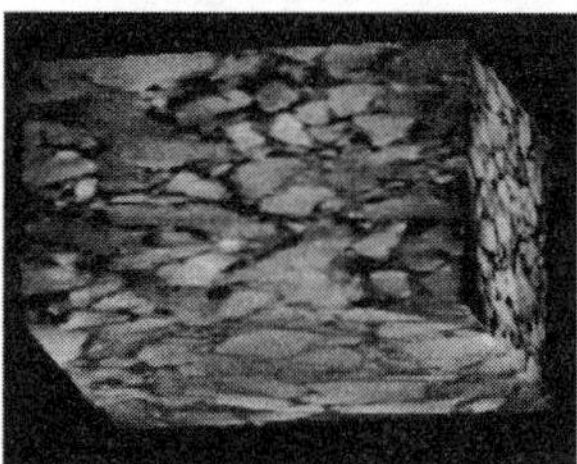

(b) Reconstructed Specimen

FIGURE 4.11   Reconstruction of 3D specimen from sectional images.

(a) Before Test

(b) After Test

FIGURE 4.12   Reconstructed 3D specimen before and after testing.

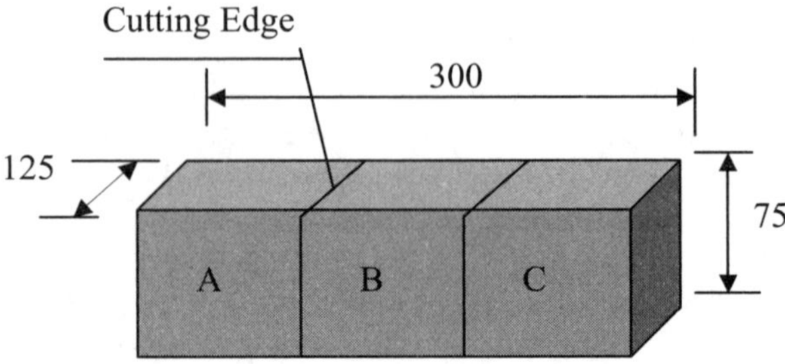

FIGURE 4.13 Specimen cutting position (mm).

512×512 pixels. One pixel represented 0.215 mm of physical size. There were 200 sectional images for one specimen. The spacing between two images was 0.54 mm. Due to the density differences of the constituents in the mixture, the aggregates have the lightest color compared with mastics and air voids. The density differences are the basis for the separation of the mixture phases.

4.3.3 APA Test

During testing, the three pieces were appropriately put together with the exact cutting face adjacent to one another in the APA. They were held in position by the clamping system of the machine. During the test, the specimen was subjected to a wheel load (100 lbs) by laying a linear hose along the centerline of its long edge. The cutting face was perpendicular to the direction of wheel travel. The resulting rut depth was a function of load cycles and was also the macro-response of the specimen. Parallel to the load direction, the thickness of the specimen was larger than 3 in, which is a representative size of an AC specimen. Therefore, it is anticipated that the test results from the centerpiece can still represent the properties of the materials. The permanent deformations along the wheel path were measured as 6.38 mm after 8000 loading cycles.

4.3.4 Image Processing

The objective of image processing is to separate the boundaries of aggregates from other phases. Image processing and analysis software, Image-Pro Plus, was used for processing the images. Each image was subjected to a threshold operation and different threshold values were assigned to each image. The values were determined by the brightness of the image in order to obtain the best particle boundary. By adjusting the threshold value, aggregates were separated from the matrix. Aggregates were assigned to black in binary images; Figure 4.14a shows a grayscale image, while Figure 4.14b shows the binary image where aggregates were separated from other phases.

4.3.5 Image Analysis

After the application of load, the aggregates had some rotation and translation within the mixture. In this study, a pattern recognition algorithm, based on the morphological characteristics of the particles, was applied to identify the particles in the images between adjacent slices before and after testing. Once the particles are tracked, the motions of the particles such as rotation and translation can be quantified and the strains can be further estimated. Since the particles are considered as rigid body, their deformation is negligible under the application of wheel load. Therefore, permanent deformation of AC mainly resulted from the rotation and translation of particles, which was actually induced by the deformation of mastics around the particles. In the deformation

FIGURE 4.14 Gray-scale image and separated aggregates.

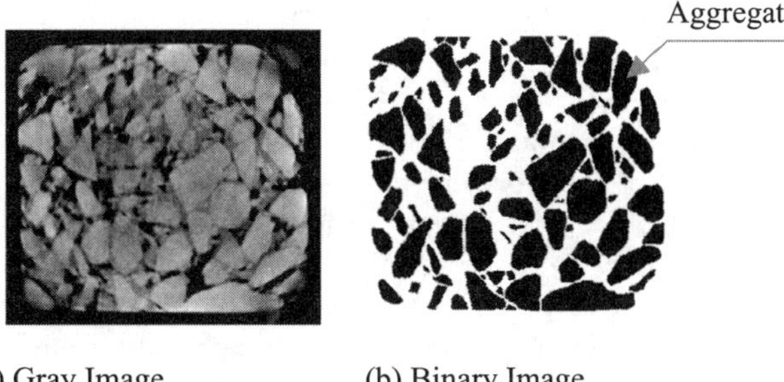

(a) Gray Image   (b) Binary Image

process, particle parameters such as volume, surface area, and aspect ratio remain constant. The recognition algorithm was based on the above assumption.

To identify the matching particles in the specimen, a method presented by Wang et al. (2004) was applied in this study. It involved two identification procedures. One is to identify particle cross-sections between adjacent slices; the other is to identify the particles before and after testing.

The first procedure is implemented through the following steps. The first step is to measure the sectional characteristics such as mass center coordinates Xi, Yi, and Zi, area, and perimeter of the particle i on the jth slice. Image-Pro Plus software was used to obtain these geometric parameters of the particles. The second step is to find the match particle cross-sections between adjacent slices based on comparing the similarity index (SI), which was defined in Equation 4-13.

$$SI = \left| x_{i,j} - x_{m,j+1} \right| + \left| y_{i,j} - y_{m,j+1} \right| + \left| A_{i,j} - A_{m,j+1} \right| + \left| P_{i,j} - P_{m,j+1} \right| \qquad (4\text{-}13)$$

Where $x_{i,j}$ and $y_{i,j}$ represent the mass center for particle i on j^{th} slice; $A_{i,j}$ represents the particle cross-sectional area for particle i on the jth slice; $P_{i,j}$ represents the particle cross-section perimeter for particle i on the jth slice; m is the dummy index representing the top five nearest neighbors for the particle cross-sections on $(j+1)^{th}$ slice. The nearest neighbors are defined by the distance between two cross-sections as shown by Equation 4-14.

$$D = \sqrt{(x_{i,j} - x_{m,j+1})^2 + (y_{i,j} - y_{m,j+1})^2} \qquad (4\text{-}14)$$

To find the nearest neighbors (denoted by dummy index m) in $(j+1)^{th}$ slice, distances from particle i in j^{th} slice to all the particles in $(j+1)^{th}$ slice are compared. Only the top five nearest neighbor particle cross-sections are used for SI matching. However, the differences of particle perimeters or areas between two adjacent slices could be relatively large while the mass center coordinate difference is small. Consideration of the area and perimeter may actually increase noises for particle cross-section identification. Research indicated that mass center was the most important factor for particle detection (Wang et al., 2004). Therefore, Equation 4-13 was modified as:

$$SI = \left| x_{i,j} - x_{m,j+1} \right| + \left| y_{i,j} - y_{m,j+1} \right| \qquad (4\text{-}15)$$

The section that has the minimum SI is considered the matching particle cross-section between two slices. Practically, SI cannot be the only factor that controls the particle cross-section identifications. There are some special cases that might confuse the detection. As was shown in Figure 4.15, particle *a* in the jth slice disappeared in the $(j+1)^{th}$ slice; and particle *b* in the $(j+1)^{th}$ slice appears in almost the same position as particle *c* in the jth slice. The above two conditions might also give a minimum *SI*, but the particles detected did not belong to the same particle in the adjacent slices. Therefore, computer intelligence and human judgment were interactively used to implement the particle recognition decisions. A FORTRAN program was developed to automatically execute the above steps with consideration of the above special cases.

The second procedure involves finding matching particles before and after testing. Since the particles were treated as rigid bodies, the volumes of the particles would remain the same before and after testing. The procedure to identify the particles before and after testing is briefly described as follows.

The first step is to compare the mass center coordinates of the particles. For bonded material, the displacement inside the mixture is relatively small, which means the mass center of the particle before and after testing will not change much. Therefore, the *SI* based on the mass center coordinates was used as the criterion to identify the particles. The second step is to compare the volumes of the particles before and after testing. The particles, which have the closest location and volume, belong to the same particle. The mass center (X_{mc}, Y_{mc}, Z_{mc}) and volume (V_i) of particle *i* are calculated using the following equations:

$$V_i \approx \sum \frac{A_{j-1} + A_j}{2} H_{j-1-j} \tag{4-16}$$

$$X_{mc} \approx \frac{\sum (A_{j-1} + A_j) H_{j-1-j} (x_{j-1} + x_j)/4}{V_i} \tag{4-17a}$$

$$Y_{mc} \approx \frac{\sum (A_{j-1} + A_j) H_{j-1-j} (y_{j-1} + y_j)/4}{V_i} \tag{4-17b}$$

$$Z_{mc} \approx \frac{\sum (A_{j-1} + A_j) H_{j-1-j} (z_{j-1} + z_j)/4}{V_i} \tag{4-17c}$$

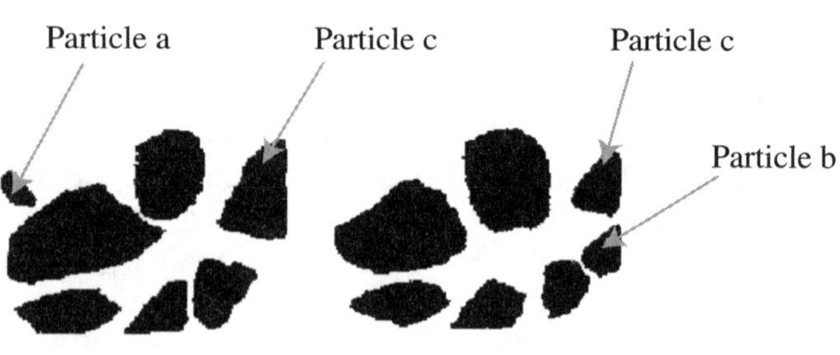

FIGURE 4.15 Special cases for particle identification.

Where x_j, y_j, z_j are the center coordinates of the jth cross-section; A_j is the cross sectional areas of particle i on the jth slice; and H_{j-1-j} is the distance between two adjacent slices ($H_{j-1-j} = z_j - z_{j-1}$).

4.3.6 Computation of Macro Strain

Once the matching particles before and after testing are identified, the translation of the particles can be calculated from the mass center difference before and after testing, which was shown in the following equation:

$$u = X_{mc}{}^a - X_{mc}{}^b \qquad v = Y_{mc}{}^a - Y_{mc}{}^b \qquad w = Z_{mc}{}^a - Z_{mc}{}^b \tag{4-18}$$

Where b represents before testing; a represents after testing. For each of the four closest particles (i, j, k, l), the displacements u_i, u_j, u_k, u_l are in the x direction; v_i, v_j, v_k, v_l are in the y direction; and w_i, w_j, w_k, w_l are in the z direction. If the displacement field is linearly interpolated in a tetrahedron, the displacement can be expressed as the following equation.

$$u = a_0 + a_1 x + a_2 y + a_3 z$$
$$v = b_0 + b_1 x + b_2 y + b_3 z \tag{4-19}$$
$$w = c_0 + c_1 x + c_2 y + c_3 z$$

Where x, y, and z are any coordinates in the tetrahedron before testing; a_0, a_1, a_2, a_3, b_0, b_1, b_2, b_3 and c_0, c_1, c_2, c_3 are constants but vary from tetrahedron to tetrahedron. a_0, a_1, a_2, a_3, b_0, b_1, b_2, b_3 and c_0, c_1, c_2, c_3 can be obtained from the following matrices.

$$\begin{bmatrix} 1,x_i,y_i,z_i \\ 1,x_j,y_j,z_j \\ 1,x_k,y_k,z_k \\ 1,x_l,y_l,z_l \end{bmatrix}\begin{Bmatrix} a_0 \\ a_1 \\ a_2 \\ a_3 \end{Bmatrix} = \begin{Bmatrix} u_i \\ u_j \\ u_k \\ u_l \end{Bmatrix} \quad \begin{bmatrix} 1,x_i,y_i,z_i \\ 1,x_j,y_j,z_j \\ 1,x_k,y_k,z_k \\ 1,x_l,y_l,z_l \end{bmatrix}\begin{Bmatrix} b_0 \\ b_1 \\ b_2 \\ b_3 \end{Bmatrix} = \begin{Bmatrix} v_i \\ v_j \\ v_k \\ v_l \end{Bmatrix} \quad \begin{bmatrix} 1,x_i,y_i,z_i \\ 1,x_j,y_j,z_j \\ 1,x_k,y_k,z_k \\ 1,x_l,y_l,z_l \end{bmatrix}\begin{Bmatrix} c_0 \\ c_1 \\ c_2 \\ c_3 \end{Bmatrix} = \begin{Bmatrix} w_i \\ w_j \\ w_k \\ w_l \end{Bmatrix} \tag{4-20}$$

The macro strain can be calculated as:

$$\varepsilon_x = \frac{\partial u}{\partial x} = a_1 \quad \varepsilon_y = \frac{\partial v}{\partial y} = b_2 \quad \varepsilon_z = \frac{\partial w}{\partial z} = c_3$$

$$\varepsilon_{xy} = \frac{1}{2}\left(\frac{\partial v}{\partial x} + \frac{\partial u}{\partial y}\right) = \frac{b_1 + a_2}{2} \quad \varepsilon_{xz} = \frac{1}{2}\left(\frac{\partial u}{\partial z} + \frac{\partial w}{\partial x}\right) = \frac{a_3 + c_1}{2} \quad \varepsilon_{yz} = \frac{1}{2}\left(\frac{\partial v}{\partial z} + \frac{\partial w}{\partial y}\right) = \frac{b_3 + c_2}{2} \tag{4-21}$$

These strains were named as macro-strains because they represent the average strains in the tetrahedron. The algorithm of strain measurement was shown in Figure 4.16.

Using the above equations, a FORTRAN code was developed to calculate the strain components in the tetrahedron. Results of x, y, z strain and shear strain on one plane are shown in Figure 4.17. Strains in these calculations follow the soil mechanics convention: positive for compressive strain. From Figure 4.17, Y direct strains (in the load direction) are larger than other direct strains. Meanwhile, strains in the area close to the load application zone are apparently larger than those in the zones farther from this zone. Strains in one tetrahedron were also shown in Table 4.5.

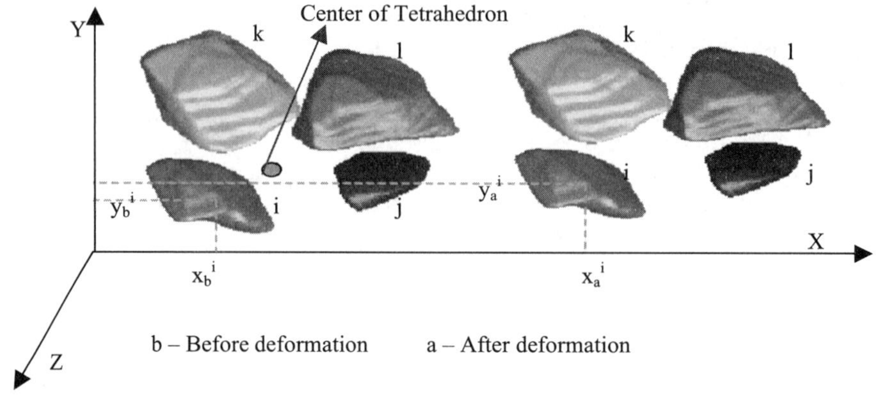

Figure 4.16   Algorithm of macro-strain measurement.

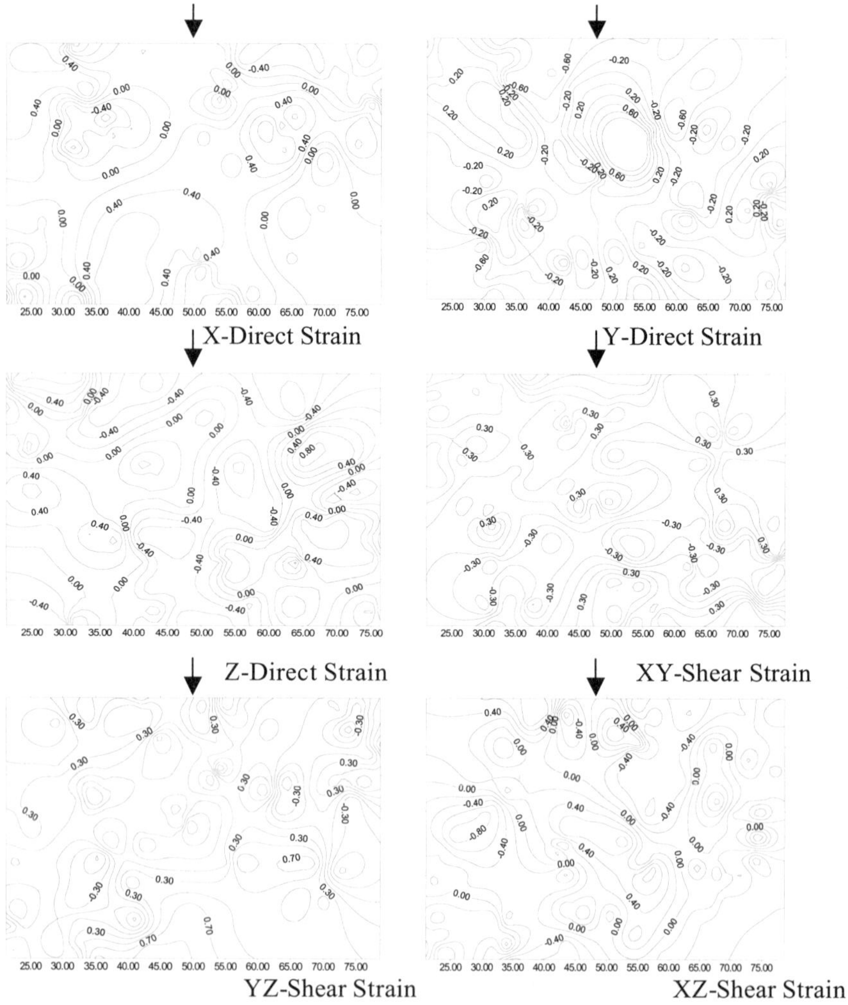

X-Direct Strain

Y-Direct Strain

Z-Direct Strain

XY-Shear Strain

YZ-Shear Strain

XZ-Shear Strain

Figure 4.17   Macro-strain contour of the specimen.

ε_x	ε_y	ε_z	ε_{xy}	ε_{xz}	ε_{yz}
0.004	0.012	0.0038	0.0034	0.0109	0.023

TABLE 4.5   Macro-strain for one tetrahedron.

Figure 4.18 shows the Y direction strain contour of the center section of the rutted area. In Figure 4.18, along the Y-axis of the specimen from the top to the bottom, obtain the macro strain value ε_y in every 10 mm. Then the total deformation of Y direction can be calculated as $\Delta Y = \sum \varepsilon_y dY = 6.1$ mm, which is close to the macro-deformation result obtained from APA test, 6.38 mm. The difference might be caused by the image resolution and image processing. Figure 4.19 shows the Y direction strain contour of the edge section of the rutted area. Using the same deformation measurement method as for the center-rutted area, the deformation for the edge-rutted area is 0.02 mm. The rutted profile, obtained from experiment and measurement, is compared in Figure 4.20.

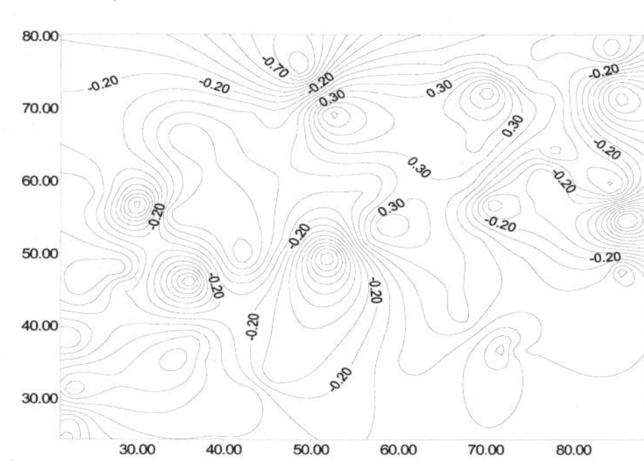

FIGURE 4.18 Y-direction strain of center-rutted area.

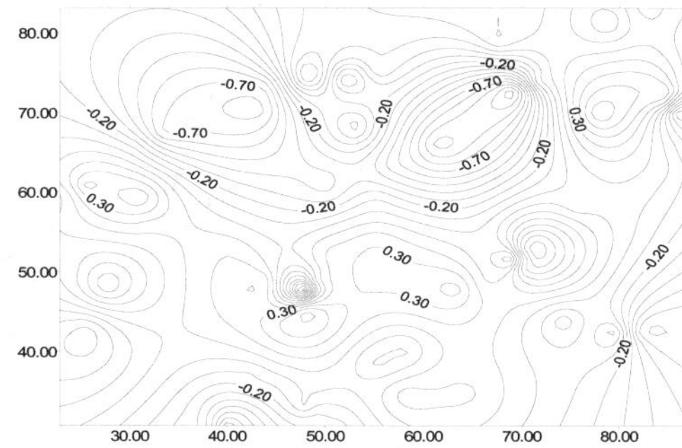

FIGURE 4.19 Y-direction strain contour of edge-rutted area.

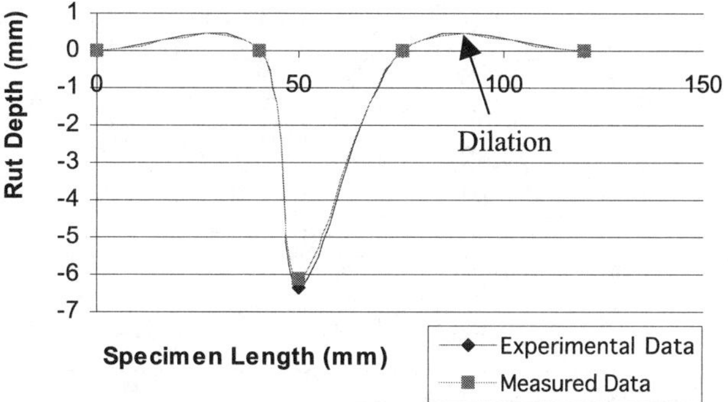

FIGURE 4.20   Comparison of experimental and measured rutted data.

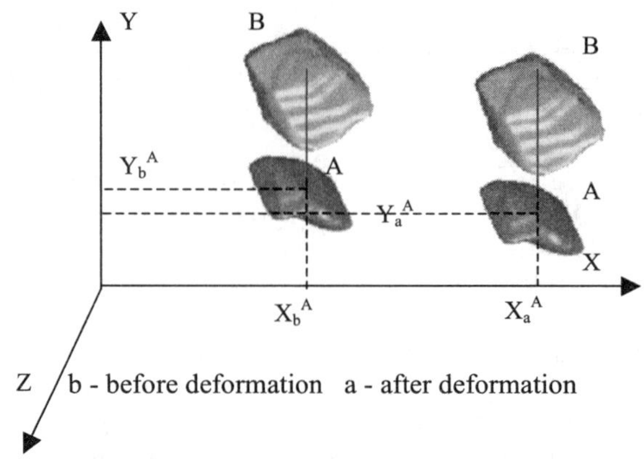

FIGURE 4.21   Algorithm of micro-strain measurement.

4.3.7   Computation of Permanent Strain in the Mastic

Macro-strain described in the previous step was the average strain that could be used to evaluate the overall properties of the mixture. However, it could not reflect the micro-behavior of the mixture. As previously noted, aggregates in the mixture were treated as rigid bodies. The permanent deformation mainly resulted from deformation in the mastics. Therefore, the strains in the mastics are important parameters to study the rutting mechanism in the mixture. It was calculated by considering the relative displacements between Particles A and B in Figure 4.21, which were obtained through the translation of the two particles' mass centers as shown in Equation 4-22.

$$\Delta u_n = (u_A - u_B)\cos\alpha + (v_A - v_B)\cos\beta + (w_A - w_B)\cos\gamma \tag{4-22}$$

Where u_A, u_B, v_A, v_B, w_A, w_B represent the mass center displacements of the two particles, A and B, in the x, y, and z directions. α, β, γ are the angles between the x, y, and z axis and the line connecting the two mass centers (branch vector). Particle rotation effect was not neglected in the above calculation.

Assuming that there is no discontinuity between the particles and the mastics, the normal strain in the mastic can be obtained by the following equation:

$$\varepsilon_n = \frac{\Delta u_n}{L} \tag{4-23}$$

Node No.	X	Y	Z	U	V	W	Strain
1 (B)	53.96	41.24	3.48	0.81	−2.01	0.26	$\varepsilon_n = 0.186$
2 (B)	56.69	33.09	3.82	1.98	−2.35	0.04	$\varepsilon_n = 0.148$
3 (B)	64.38	44.40	3.73	1.19	−2.31	−0.07	$\varepsilon_n = 0.04$
4 (B)	72.39	34.73	3.27	0.90	−1.99	−0.20	$\varepsilon_n = 0.091$
1 (A)	53.10	43.24	3.22				$\varepsilon_n = 0.167$
2 (A)	54.80	35.44	3.78				$\varepsilon_n = 0.171$
3 (A)	63.14	46.71	3.80				
4 (A)	71.49	36.72	3.47				

TABLE 4.6   Strain in the mastics.

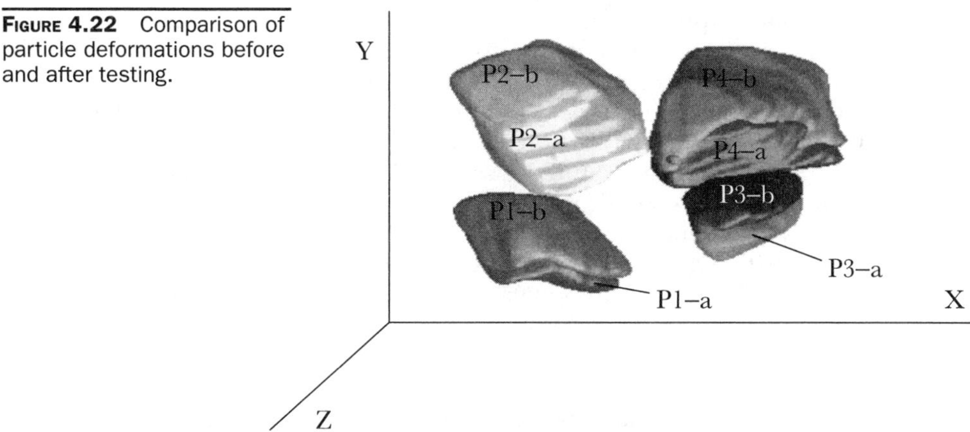

FIGURE 4.22 Comparison of particle deformations before and after testing.

Where L was expressed as: $L = H - r_A - r_B$. r_A and r_B are the radii of the two particles in the direction of their center connection line. Due to the difficulty in measuring the radius of the irregular shapes in the 3D space, the particles were assumed as spheres in the above calculation. Therefore, the radii can be calculated from their volumes. H is the distance between two particles' mass centers as defined in Equation 4-24.

$$H = \sqrt{(x_A - x_B)^2 + (y_A - y_B)^2 + (z_A - z_B)^2} \tag{4-24}$$

The micro-strain calculated for the same tetrahedron in the previous step is illustrated in Table 4.6. Comparing the results in Table 4.5 with those in Table 4.6, it is found that the strain in the mastic was much larger. Tensile strain exists in the zone that is close to the wheel load, which means it occurs in the compression zones. This result is consistent with the properties of AC. The dilation is due to the aggregate skeleton. In Figure 4.16, the configurations of the particles in one tetrahedron were compared before and after testing. The displacements were not very apparent. However, in Figure 4.22, when the particles were overlapped before and after testing, the displacements were evident in that P*-b represents the particles before the testing, while P*-a represents the particles after the testing.

From the results it was found that permanent strain in the mastics was very much localized. The strains in the mastics were generally much larger than macro-strains. The measurement has important implications for evaluating the fatigue and rutting resistance of AC since fatigue and rutting generally result from deformations in the mastics. The experimental results indicate that a larger magnitude of strains must be used in evaluating the mastic properties.

4.4 Digital Image Correlation Method

Digital image correlation (DIC) in general refers to a group of non-contact technologies that acquire and analyze images to extract full-field strain, deformation, or motion measurements. The state of practice in DIC is limited to grayscale images, and DIC involves the measurement of grayscale values (intensity) of individual pixels in the interested domain. The intensity values of deformed specimen images are compared with the initial un-deformed image to determine the movement and displacement of the pixels, and eventually the deformation and strain field of the object being analyzed. Various

DIC matching approaches have been developed involving different types of object-based patterns such as lines, grids, dots, and random arrays. Common use of DIC involves random patterns and compares sub-regions throughout the image to obtain a full field of measurements. As a promising tool for accurate experimental measurement, the DIC method and its application in general mechanical experimental testing, particularly the asphalt mixture testing, are briefly presented as follows. The DIC is superior to conventional displacement and strain measurement methods, such as the linear variable differential transformer (LVDT), in that it has the ability to capture full-field deformation of the specimen and does not entail contact mounting on the specimen surface. The general information on DIC in the following sections is abstracted from an excellent book by Sutton et al. (2009).

4.4.1   History of DIC

Pioneering work in image correlation dates back to the early 1950s. Gilbert Hobrough compared analog representations of photographs to register features from various views. With the emergence of digitized images in the 1960s, vision-based algorithms were developed and applied in artificial intelligence and robotics. In the 1970s, image correlation application centered on character recognition, microscopy, medical radiology, and photogrammetry photography. Extensive applications in engineering shape and deformation measurements were performed (Rosenfeld, 2001). The 1970s also saw the use of laser technology in image correlation, including mainly the laser speckle (Dainty, 1975), laser speckle photography (Archbold et al., 1970), and laser speckle interferometry (Mallik and Roblin, 1972). Significant progress has been made since the 1970s in 2D and 3D DIC development for surface characterization, and in 3D DIC for volumetric characterization. The following sections detail the developments in these regards.

4.4.2   Use of DIC in 2D Surface Analysis

As the experimental testing of materials and mechanical systems gets more complex, researchers have developed methods for digitally recording images of measurement data, algorithms for analyzing images and extracting measurement data, and the approaches for automating the entire process. One of the early works that used computer-based image acquisition and deformation measurements in material testing was conducted by Peters and Ranson (1982). Sutton et al. (1983) developed numerical algorithms known as 2D digital image correlation (2D-DIC). Davidson and Lankford (1983) extended 2D-DIC concepts to the micro-scale for measuring large deformation near fatigue flaws. Han et al. (1994) developed a high magnification optical system to measure deformations around stationary and growing crack tips under nominal Mode I loading. In addition to the work in fracture mechanics, investigators used 2D-DIC to understand the deformation behavior of materials including metals, plastics, wood, ceramics, and tensile loading of papers. In the late 1990s and 2000, investigators applied 2D-DIC to study damage in composites and concrete. The 2D-DIC was applied to measure surface deformations in planar components in the 1990s. Most of the studies prior to 2000 used direct image correlation principles for matching subsets and extracting full-field displacements.

The use of 2D-DIC has gained rapid growth since 2000. 2D-DIC measurement procedures have received various modifications, including the search procedure, correlation approach, and registration method. As to the registration methods, Cheng et al. (2002)

proposed a method for full-field deformation measurement through pixel-by-pixel mapping using a B-spline functional form. Schreier et al. (2000) emphasized the importance of image reconstruction in improving the accuracy of the matching process. 2D-DIC was also combined with the scanning electron microscope (SEM) for large deformations. Inverse methods emerged recently for characterizing elastic properties, properties in heterogeneous materials, hyperelastic properties, micromechanics, and composites. Other applications in material characterization include 2D-DIC measurements in films, polymers, metals, heterogeneous composites, wood, bio-materials, alloys, asphalt, ceramics, concrete, and geomaterials such as clays and sands.

4.4.3   Use of DIC in 3D Surface Analysis

The development of more complicated 3D-DIC surface analysis was intended to overcome the inability of 2D-DIC to handle the out-of-plane motion that changes the magnification and introduces errors in the measured in-plane displacement. With a 3D analysis, the error can be removed by incorporating the displacement in the third direction. Luo et al. (1994) combined stereovision principles with 2D-DIC concepts in single camera imaging and developed a two-camera stereo vision system for the measurement of 3D crack tip deformations. Synnergren and Sjödahl (1999) successfully employed stereovision to make deformation measurements, where the investigators obtained flash X-ray images of a specimen from two directions before and during high-rate loading. The investigators extracted 3D estimates for specimen deformations by comparing features in both sets of X-rays. Andresen (1999) analyzed images of a grating undergoing large deformation using a stereo system. Using SEM imaging, Lockwood and Reynolds (1999) obtained surface stereo images of an in-situ specimen from two orientations.

Entering the new century, 3D digital image correlation has seen a wide range of applications on both large and small structures. Basic studies of typical high-speed imaging systems were performed by Tiwari, et al. (2007). Stereovision system applications have included measurements on flexible wings undergoing aerodynamic loading (Albertani et al., 2007), as well as measurements of shape and deformation on cylindrical surfaces (Luo et al., 1993). Using 3D-DIC, fracture studies were performed under mixed mode I/II (Luo and Huang, 2000) and mixed mode I/III loading conditions (Sutton et al., 2007). Material characterization has also been an active area including microscale studies in engineered materials (Florando et al., 2007) and biomaterials (Nicolella et al., 2001), foam, ceramics, composites, polymers, and high-rate events.

4.4.4   Use of DIC in 3D Volume Analysis

Different from the 3D-DIC stereo surface analysis, 3D volumetric DIC also characterizes the internal changes in deformation and strain of an object, and has gained attention as an active area of research recently, resulting in matured imaging capabilities. 3D volumetric DIC requires technologies for internal characterization. Micro- and macro-computer-aided X-ray tomography (X-ray CT) and magnetic resonance imaging (MRI) are examples of such technologies that are available for general image analysis.

Complementary to development in imaging technology, a wide range of software is now available for image analysis to extract useful information. Veress et al. (2003) developed a method for comparing images to extract full-field estimates for deformations. Bay et al. (1999) extended 2D-DIC concepts to match small sub-volumes before and after undergoing loading to obtain a full volumetric field of 3D motions. Germaneau et al.

(2006) used internally scattered light to obtain a volumetric pattern for imaging and analysis. 3D volumetric DIC has also been applied to analyze on sandwich structures, biomaterials, steel powders, metallic alloys, and rock.

4.4.5   Fundamentals of DIC

Using the 2D-DIC as an example, DIC is performed by maximizing a correlation coefficient that is determined by examining pixel intensity array subsets on two or more corresponding images and extracting the deformation mapping relationship that relates the images. The 2D cross-correlation coefficient r_{ij} is defined as Equation 4-25, where, $F(x_i, y_j)$ is the pixel intensity or the grayscale value at point (x_i, y_j) in the un-deformed image. $G(x^*_i, y^*_j)$ is the grayscale value at point (x^*_i, y^*_j) in the deformed image. $\bar{F}$ and $\bar{G}$ are mean values of the intensity matrices of F and G, respectively.

$$r_{ij} = (u, v, \frac{\partial u}{\partial x}, \frac{\partial u}{\partial y}, \frac{\partial v}{\partial x}, \frac{\partial v}{\partial y}) = 1 - \frac{\sum_i \sum_j \left[F(x_i, y_j) - \bar{F} \right]\left[G(x^*_i, y^*_i) - \bar{G} \right]}{\sqrt{\sum_i \sum_j \left[F(x_i, y_j) - \bar{F} \right]^2 \left[G(x^*_i, y^*_i) - \bar{G} \right]^2}} \tag{4-25}$$

The coordinates (x_i, y_j) and (x^*_i, y^*_j) are related by the deformation that occurs between the two images. If the displacement is perpendicular to the optical axis of the camera, then the relation between (x_i, y_j) and (x^*_i, y^*_j) can be approximated by a 2D transformation according to Equation 4-26, where u and v are translations of the center of the sub-image in the X and Y directions, respectively. The distances from the center of the sub-image to the point (x, y) are denoted by Δx and Δy. The correlation coefficient r_{ij} is a function of displacement components (u, v) and displacement gradients $\frac{\partial u}{\partial x}, \frac{\partial u}{\partial y}, \frac{\partial v}{\partial x}$, and $\frac{\partial v}{\partial y}$.

$$\begin{cases} x^* = x + u + \dfrac{\partial u}{\partial x}\Delta x + \dfrac{\partial u}{\partial y}\Delta y \\[2mm] y^* = y + v + \dfrac{\partial v}{\partial x}\Delta x + \dfrac{\partial v}{\partial y}\Delta y \end{cases} \tag{4-26}$$

4.4.6   DIC in Asphalt Testing and Modeling

Although used widely in the testing of other infrastructure materials such as concrete, metals, wood, and geomaterials, DIC as a non-contact, full-field, surface displacement/strain measurement technique has seldom been used in characterizing asphalt materials. The very early use of DIC in asphalt material testing was conducted by Yue and Morin (1996), where the technology was referred to as digital image processing (DIP). The authors studied the rutting behavior of AC by comparing the orientations of aggregate particles in asphalt mixtures before and after mechanical loading tests. Yue et al. (2003) also used DIP in characterizing other construction materials. Similar work on AC was done by Chen et al. (2005) in studying experimentally the internal structure changes in hot mix asphalt (HMA) under wheel load testing. They found that the rotation of flat and elongated particles contributed significantly to the rutting of asphalt mixes.

Seo et al. (2002) used DIC in mechanical testing (deformation, strain, and fracture) of asphalt mixture to determine the accuracy and convenience in comparison to LVDT.

Vertical displacements by DIC for the middle and bottom sections of a specimen subjected to monotonic tension were compared to those measured by LVDT. A series of DIC images captured during the monotonic and cyclic tests were used to study the evolution of the fracture process zone at the crack tip. The DIC examined deformation to a random speckle pattern placed on the area of interest of the specimen, which yielded an easier and faster specimen preparation method. Through careful manipulation of the spray nozzle with proper pressure from a certain distance, a uniform-sized speckle pattern was applied onto the surface randomly. The black paint was lightly sprayed several times until the desired speckle pattern density was achieved. The optimum speckle pattern density was an even mix of dark and white areas, producing a pattern that was biased toward neither the white nor the black end of the gray scale. DIC demonstrated a more accurate determination of the stress-strain behavior and the fracture process zone. The applicability of DIC to a cylindrical specimen with a curved surface was found to be more complicated than on prismatic specimens with a planar surface. The DIC method was found to provide a more accurate constitutive relationship of the fracture process zone than conventional LVDTs because of its full-field measurement and post-processing nature.

Chehab et al. (2007) adopted DIC in developing a viscoelastoplasticity continuum damage model (VEPCD) of asphalt mixtures. The formation of strains in the asphalt mixtures was found to be highly localized as microcracks got densified, coalesced, and further grew into macrocracks. Conventional LVDTs were found unable to capture the localized process zone strain in the fracture process zone and ceased to accurately predict the performance of asphalt-aggregate mixtures after strain localization. DIC was found to perform well in measuring the fracture process zone strains and could extend the validity of the VEPCD model beyond localization. Methodologies that required transfer from LVDT strains to DIC strains after strain localization for model calibration and validation were presented for a range of loading and temperature conditions.

Romeo et al. (2006) developed a digital image correlation device to more accurately capture localized or non-uniform stress distributions in asphalt mixtures and as a tool to detect first fracture. The experimental analysis of asphalt mixture cracking behavior was based on the HMA fracture mechanics visco-elastic crack growth law. Asphalt mixture cracking mechanism and fundamental tensile failure limits were investigated using multiple laboratory test configurations, namely the Superpave Indirect Tensile (IDT) test, the Semi-Circular Bending (SCB) test and the Three-Point Bending Beam (3PB) test. Montepara et al. (2006) and Martin et al. (2006) extended Romeo's work by measuring more generally the elastic and plastic strains using the same device.

Birgisson et al. (2006, 2007, and 2008) used the DIC to verify fracture energy density as a fundamental fracture threshold in HMA. Fracture energy density was evaluated with the SCB test. The DIC method was based on the careful application of a sophisticated image-matching technique (the least squares matching algorithm) with a proper laboratory test configuration (specimen surface treatment, illumination, etc.). An image sequence of the loaded specimen was acquired with a digital camera; a set of black patterns artificially generated on the specimen surface was accurately tracked by the algorithm along the loading sequence. The essence of the method was to compare the gray-scale values of images of the deformed specimen with those of an initially un-deformed specimen. The DIC was able to provide a dense and accurate displacement-strain field of asphalt mixtures at the microstructural level. The resulting fracture behavior in the SCB was predicted with a displacement discontinuity method to explicitly model the micro-structure of asphalt mixtures and to predict their fracture energy density. The predicted crack initiation and crack propagation patterns were found to be consistent with observed cracking behavior. The results presented lead to the promising statement that fracture energy density for mixtures is a fundamental fracture threshold and can be both consistently measured and predicted for different laboratory test configurations.

In closing, the use of DIC in asphalt testing has been limited exclusively to the 2D, planar, and surface measurement, although the technology provides the advantage of non-contact and full-field displacement/strain measurement. More efforts therefore are needed to evaluate the 3D stereo (surface) DIC measurement and the volumetric DIC measurement to obtain more useful information of a specimen being tested. Technologies and algorithms for these two prospective applications of DIC are ready to be evaluated.

References

Albertani, R., Stanford, B., Hubner, J.P. and Ifju, P.G. (2007). Aerodynamic coefficients and deformation measurements on flexible micro air vehicle wings. *Experimental Mechanics*, Vol.47, No.5, pp.625–635.

Andresen, K. (1999). Strain tensor for large three-dimensional surface deformation of sheet metal from an object grating. *Experimental Mechanics*, Vol.39, No.1, pp.30–35.

Anon. (2003). Obituary-Gilbert Louis Hobrough. *The Photogrammetric Record*, Vol.18, No.104, pp.337–340.

Archbold, E.J., Burch, M. and Ennos, A.E. (1970). Recording of in-plane surface displacements by double exposure speckle photography. *Optica Acta*, Vol.17, pp.883–898.

Bay, B.K., Smith, T.S., Fyhrie, D.P. and Saad, M. (1999). Digital volume correlation: three dimensional strain mapping using X-ray tomography. *Experimental Mechanics*, Vol.39, No.3, pp.217–226.

Birgisson, B., Montepara, A., Napier, J., Romeo, E., Roncella, R. and Tebaldi, G. (2006). Measurement and prediction of HMA fracture energy using micromechanical analyses. *Transportation Research Record*, No.1970, pp.186–195.

Birgisson, B., Montepara, A., Romeo, E., Roncella, R., Napier, J.A.L. and Tebaldi, G. (2007). Determination and prediction of crack patterns in hot mix asphalt (HMA) mixtures. *International Journal of Engineering Fracture Mechanics*, Vol.75, pp.664–673.

Birgisson, B., Montepara, A., Romeo, E., Roque, R., Roncella, R. and Tebaldi, G. (2008). The use of digital image correlation for accurate determination of fracture energy density in hot mix asphalt (HMA). *The 6th RILEM International Conference on Cracking in Pavements*, Chicago, U.S.A., pp.811–820.

Brown, S.F. and Cooper, K.E. (1980). A fundamental study of the stress-strain characteristics of a bituminous material. *Journal of the Association of Asphalt Paving Technologists*, Vol.49, pp.476–498.

Chehab, G.R., Seo, Y. and Kim, Y.R. (2007). Viscoelastoplastic damage characterization of asphalt-aggregate mixtures using digital image correlation. *International Journal of Geomechanics*. Vol.7, No.2, pp.111–118.

Chen, J.S., Wong, S.Y. and Lin, K.Y. (2005). Quantification of movements of flat and elongated particles in hot mix asphalt subject to wheel load test. *Materials and Structures*, Vol.38, pp.395–402.

Cheng, P., Sutton, M.A. Schreier, H.W. and McNeill, S.R. (2002). Full-field speckle pattern image correlation with B-Spline deformation function. *Experimental Mechanics*, Vol.42, No.3, pp.344–352.

Collins, R., Shami, H. and Lai, J.S. (1996). Use of a Georgia loaded wheel tester to evaluate rutting of asphalt concrete samples. *Transportation Research Record*, No.1545, pp.161–168.

Dainty, J.C. (Ed.) (1975). Laser Speckle and Related Phenomena. *Springer*, Berlin.

Davidson, D.L. and Lankford, J. (1983). ASTM STP 811, Fatigue crack tip strains in 7075-T6 aluminum alloy using stereo-imaging and their use in fatigue crack growth models. *Fatigue Mechanisms: Advances in Quantitative Measurement of Physical Damage*, pp.371–399.

Florando, J.N., LeBlanc, M.M. and Lassila, D.H. (2007). Multiple slip in copper single crystals deformed in compression under uniaxial stress. *Scripta Materialia*, Vol.57, No.6, pp.537–540.

Germaneau, A., Doumalin, P. and Dupré, J.C. (2006). Improvement of accuracy of strain measurement by digital volume correlation for transparent materials. *Proceedings of Photomechanics 2006*, Clermont-Ferrand, France (CDROM).

Han, G., Sutton, M.A. and Chao, Y.J. (1994). A study of stationary crack tip deformation fields in thin sheets by computer vision. *Experimental Mechanics*, Vol.34, No.2, pp.751–761.

Hassan, Y.A., Canaan, R.E., Blanchat, T.K. and Seeley, C.H. (1991). Simultaneous velocity measurements of both components of a two phase flow using particle image velocimetry. *American Society of Mechanical Engineers, Fluids Engineering Division*, Vol.128, pp.85–99.

IAEA. (1995). Tomography in nuclear medicine. *Proceedings of International Symposium on Tomography in Nuclear Medicine*, Vienna.

Kandhal, P.S., and Cooley, L.A. (2003). *Accelerated laboratory rutting tests: evaluation of the asphalt pavement analyzer*. National Cooperative Highway Research Program, NCHRP Report 508.

Lockwood, W.D. and Reynolds, A.P. (1999). Use and verification of digital image correlation for automated 3-D surface characterization in the scanning electron microscope. *Materials Characterization*, Vol.42, No.2, pp.123–134.

Luo, P.F. and Huang, F.C. (2000). Application of stereo vision to the study of mixed-mode crack-tip deformations. *Optics and Lasers in Engineering*, Vol.33, No.5, pp.349–368.

Luo, P.F., Chao, Y.J. and Sutton, M.A. (1994). Application of stereo vision to three-dimensional deformation analyses in fracture experiments. *Optical Engineering*, Vol.33, No.3, pp.981–990.

Luo, P.F. Chao, Y.J. Sutton, M.A. and Peters, W.H. (1993). Accurate measurement of three dimensional deformations in deformable and rigid bodies using computer vision. *Experimental Mechanics*, Vol.33, No.2, pp.123–132.

Mallik, S. and Roblin, M.L. (1972). Speckle pattern interferometry applied to the study of phase objects. *Optics Communications*, Vol.6, pp.45–49.

Marin, T., Nicoletto, G., Romeo, E. and Roncella, R. (2006). Determination of elastic-plastic strains by digital image correlation. *Proceeding of Photomechanics 2006*, Luglio 10–12, Clermont-Ferramd, France (CDROM).

Montepara, A., Romeo, E., Roncella, R. and Tebaldi, G. (2006). Local strain measurement in hot mix asphalt by digital image correlation. *Proceeding of Photomechanics 2006*, Luglio 10–12, Clermont-Ferramd. France (CDROM).

Nicolella, D.P., Nicholls, A.E., Lankford, J. and Davy D.T. (2001). Machine vision photogrammetry: a technique for measurement of microstructural strain in cortical bone. *Journal of Biomechanics*, Vol.34, No.1, pp.135–139.

Peters, W.H. and Ranson, W.F. (1982). Digital imaging techniques in experimental stress analysis. *Optical Engineering*, Vol.21, No.3, pp.427–431.

Romeo, E., Roncella, R., Tebaldi, G. and Nicoletto, G. (2006). Crack monitoring and strain measurement in asphalt mixture by digital image correlation. *Crack Paths 2006*, Settembre 14-16, Parma, Italy, pp.87–92.

Rosenfeld, A. (2001). From image analysis to computer vision: An annotated bibliography, 1955–1979. *Computer Vision and Image Understanding*, Vol.84, pp.298–324.

Schreier, H.W., Braasch, J. and Sutton, M.A. (2000). Systematic errors in digital image correlation caused by intensity interpolation. *Optical Engineering*, Vol.39, No.11, pp.2915–2921.

Seo, Y., Kim, Y.R., Witczak, M.W. and Bonaquist, R. (2002). Application of digital image correlation method to mechanical testing of asphalt-aggregate mixtures. *Transportation Research Record*, Vol.1789, pp.162–172.

Sutton, M. A. Orteu, J.J. and Schreier, H.W. (2009). Image Correlation for Shape, Motion and Deformation Measurements, Basic Concepts, Theory and Applications. *Springer.*

Sutton, M.A., Wolters, W.J., Peters, W.H., Ranson, W.F. and McNeill, S.R. (1983). Determination of displacements using an improved digital correlation method. *Image and Vision Computing*, Vol.1, No.3, pp.133–139.

Sutton, M.A., Yan, J., Deng, X., Cheng, C.S. and Zavattieri, P. (2007). Three-dimensional digital image correlation to quantify deformation and crack-opening displacement in ductile aluminum under mixed-mode I/III loading. *Optical Engineering*, Vol.46, No.5, pp.1–17.

Synnergren, P. and Sjödahl, M.A. (1999). Stereoscopic digital speckle photography system for 3-D displacement field measurements. *Optics and Lasers in Engineering*, Vol.31, No.6, pp.425–443.

Tiwari, V., Sutton, M.A. and McNeill, S.R. (2007). Assessment of high speed imaging systems for 2D and 3D deformation measurements: methodology development and validation. *Experimental Mechanics*, Vol.47, No.4, pp.561–579.

Veress, A.I., Weiss, J.A., Gullberg, G.T., Vince, D.G. and Rabbitt, R.D. (2003). Strain measurement in coronary arteries using intravascular ultrasound and deformable images. *Journal of Biomechanical Engineering*, Vol.124, pp.734–741.

Warr, S., Jacques, G.T.H. and Huntley, J.M. (1994). Tracking the translational and rotational motion of granular particles: use of high-speed photography and image processing. *Powder Technology*, Vol.81, No.10, pp.41–56.

Wang, L.B. (1998). *An Evaluation of the Microstructures and the Macro-Behavior of Unbonded and Bonded Granular Materials. Ph.D. Thesis,* Georgia Institute of Technology, Atlanta.

Wang, L.B., Lai, J.S. and Frost, J.D. (1997). Fourier morphological descriptors of aggregate particle Profiles. *Proceedings of the Second International Conference on Imaging Technologies: Techniques and Applications in Civil Engineering*, ASCE, Switzerland, pp.76–87.

Wang, L.B., Frost, J.D. and Lai. J.S. (1999). Non-invasive measurement of permanent strain field resulting from rutting in asphalt concrete. *Transportation Research Record*, No.1687, pp.85–94.

Wang, L.B. (2004). Three-dimensional digital representation of granular material micro structure from X-ray tomography. *Journal of Computing Engineering*, Vol.18, No.1, pp.28–35.

Yue, Z., Chen, S. and Tham, L. (2003). Finite-element modeling of geomaterials using digital image processing. *Computers and Geotechniques*, Vol.30, pp.375–397.

Yue, Z. and Morin, I. (1996). Digital image processing for aggregate orientations in asphalt concrete. *Canadian Journal of Civil Engineering*, Vol.23, No.2, pp.480–489.

CHAPTER 5

Mixture Theory and Micromechanics Applications

5.1 Mixture Theory and Its Application

Asphalt concrete (AC) is a heterogeneous mixture of three constituents of asphalt binder, aggregate, and air voids. The local volume fractions of these constituents vary spatially and therefore result in the spatial gradients of the local volume fractions. The local volume fractions and their spatial gradients are important field variables in the mixture theory whose objective is to predict the mixture behavior out of the structure of the mixture and the properties of the constituents. In this section (based on Wang et al., 2003), the fundamentals of mixture theory and a general method for solving boundary value problems using mixture theories are presented. A simplified mixture theory for two-constituent mixtures of solids and air voids is presented to model the initial stress distribution of AC under static loading. The analytical solutions of simple two-dimensional (2D) and one-dimensional (1D) cases using the simplified theory are also presented to illustrate how this theory predicts the effective stress distribution of a heterogeneous mixture. The final part of this section will present a study on using mixture theory to model the air void reduction (Krishnan and Rao, 2000, 2001).

5.1.1 A Brief Introduction

Mixture theory (Truesdell, 1957, 1969; Green and Naghdi, 1967; Bowen, 1976; Eringen and Ingram, 1976) was initially proposed for modeling the mixture behavior of fluids, and the granular flow properties (Goodman and Cowin, 1971, 1972; Passman, 1977; Kanatani, 1979). Two books (Bear, 1972; Coussy, 1995) offered excellent literature on mixture theory applications in porous media. In recent years, mixture theory has also been used in modeling the properties of AC (Krishnan and Rao, 2000, 2001). For solid state and granular materials, volume-fraction-based mixture theories have been shown to be more suited than the traditional mixture theories. For this type of mixture theory, important field variables include the local volume fraction and the spatial gradients of the local volume fractions of the constituents (i.e., Goodman and Cowin, 1971, 1972). However, due to the lack of an efficient method to characterize these field variables, application of mixture theory to general boundary value problems is limited.

In recent years, X-ray computed tomography (XCT) has become a reliable tool in obtaining the microstructure of AC and other construction materials (Desruses, 1996; Shashidhar, 1999; Braz et al., 1999; Rogasik et al., 1999; Shi 1999; Wang et al., 2001;

Masad et al., 2002). This technique can be used to quantify not only the local volume fractions of the constituents but also the spatial gradients of the local volume fractions (Wang et al., 2002; Chapter 3 of this book). With this technique, it is possible to perform verification of mixture theory predictions by incorporating the required field variables, and to apply it to model the behavior of AC and/or other mixture materials.

Many mixture theories (i.e., Muller, 1968; Twiss and Eringen, 1971; Twiss, 1972; Druheller and Bedford, 1980; Dobran, 1985) have been proposed in this century. Most of these theories follow the general philosophy that was originally proposed by Truesdell (1957). These theories defined the single mixture variables based on the momentum equivalency, which results in some erroneous formulations (Hansen, 1989). Hansen's formulations (volume-fraction-based) and Bowen's formulations (Bowen, 1976) for the general balance equations for the mixture will be followed in this presentation.

5.1.2 General Framework of Mixture Theory

For volume-fraction-based mixture theories, important field variables include local volume fractions of the constituents and their gradients (Chapter 3). The local volume fractions are usually assumed to be absolutely continuous spatially and differentiable to the second order to their independent variables. For each of the constituents, $x_\alpha = \chi(X_\alpha, t)$ represents the motion of the particle belonging to the αth constituent ($\alpha = 1, N$).

$$x_\alpha = \chi(X_\alpha, t) \tag{5-1}$$

Where x_α represents the coordinates of the deformed configuration while X_α represents the coordinates in the reference configuration. The velocity, acceleration, and velocity gradient of the particle can be represented as:

$$\dot{x}_\alpha = \frac{\partial \chi(X_\alpha, t)}{\partial t} \tag{5-2}$$

$$\ddot{x}_\alpha = \frac{\partial^2 \chi(X_\alpha, t)}{\partial t^2} \tag{5-3}$$

$$L_\alpha = \frac{\partial \dot{x}^i_\alpha}{\partial x^j} \tag{5-4}$$

For each constituent, the following balance equations should be satisfied.

Balance of mass

$$\dot{\rho}_\alpha + (\nabla \bullet \rho_\alpha \dot{x}_\alpha) = c_\alpha \tag{5-5}$$

c_α = the mass supply to the αth constituent

ρ_α = dispersed density, related to the actual local density of the constituent by

$\qquad \rho_\alpha = \phi_\alpha \gamma_\alpha$

ϕ_α = local volume fraction of the αth constituent and has a restriction of $\sum\limits_{\alpha=1}^{N} \phi_\alpha = 1$.

Balance of linear momentum

$$\nabla \bullet \sigma_\alpha + \rho_\alpha b_\alpha + p_\alpha = \rho_\alpha \ddot{x}_\alpha \tag{5-6}$$

p_α = the linear momentum supply to the αth constituent

σ_α = the partial stress of the αth constituent

b_α = the body force on the αth constituent

Balance of angular momentum

$$m_{\alpha^1} = \sigma_{\alpha^{23}} - \sigma_{\alpha^{32}}^T \tag{5-7a}$$

$$m_{\alpha^2} = \sigma_{\alpha^{31}} - \sigma_{\alpha^{13}}^T \tag{5-7b}$$

$$m_{\alpha^3} = \sigma_{\alpha^{12}} - \sigma_{\alpha^{21}}^T \tag{5-7c}$$

$m_{\alpha^1}, m_{\alpha^2}, m_{\alpha^3}$ = components of the angular momentum in the x, y, and z directions.

Balance of energy

$$\rho_\alpha e_\alpha = tr(t_\alpha^T L_\alpha) - \nabla \bullet q_\alpha + \rho_\alpha r_\alpha + \hat{e}_\alpha \tag{5-8}$$

e_α = internal energy

q_α = heat flux

r_α = heat supply

$\hat{e}_\alpha$ = energy supply

Unlike many of the traditional mixture theories, no entropy inequality is proposed for each of the constituents. The second law of thermodynamics is proposed for the mixture only.

By defining the corresponding mixture quantities as follows, balance equations for the mixture can be obtained.

$$\rho = \sum_{\alpha=1}^N \phi_\alpha \gamma_\alpha = \sum_{\alpha=1}^N \rho_\alpha \tag{5-9}$$

$$\dot{x} = \sum_{\alpha=1}^N \phi_\alpha \dot{x}_\alpha \tag{5-10}$$

$$M = \sum_{\alpha=1}^N \phi_\alpha \gamma_\alpha \dot{x}_\alpha = \sum_{\alpha=1}^N \rho_\alpha \dot{x}_\alpha \tag{5-11}$$

$$\sigma = \sum_{\alpha=1}^N \sigma_\alpha \tag{5-12}$$

$$\varepsilon = \sum_{\alpha=1}^N \phi_\alpha \varepsilon_\alpha \tag{5-13}$$

$$q = \sum_{\alpha=1}^{N} q_\alpha \tag{5-14}$$

$$\eta = \sum_{\alpha=1}^{N} \phi_\alpha \eta_\alpha \tag{5-15}$$

where ρ = density of the mixture
$\dot{x}$ = velocity of the mixture
M = linear momentum of the mixture
σ = stress tensor of the mixture
ε = strain tensor of the mixture
q = heat supply to the mixture
η = entropy of the mixture
η_α = entropy of the constituent

Balance of mass for the mixture

$$\frac{\partial \rho}{\partial t} + \nabla \bullet [\sum_{\alpha=1}^{N} \rho_\alpha \dot{x}_\alpha] = 0 \tag{5-16}$$

$\sum_{\alpha=1}^{N} c_\alpha = 0$ means that no net mass should disappear or produce.

By the definitions of the mixture quantities, the above equation can be rewritten as:

$$\frac{\partial \rho}{\partial t} + \nabla \bullet (\rho \dot{x}) + \nabla \bullet [\sum_{\alpha=1}^{N} \rho_\alpha \dot{u}_\alpha] = 0 \tag{5-17}$$

Where $\dot{u}_\alpha = \dot{x}_\alpha - \dot{x}$ is the diffusion speed.

It is clear only when $\nabla \bullet [\sum_{\alpha=1}^{N} \rho_\alpha \dot{u}_\alpha] = 0$, will the mixture formula have the same form for a single constituent.

Balance of linear momentum for the mixture

By summing the linear momentum equations of the constituents, the following equation is produced.

$$\nabla \bullet \sum_{\alpha=1}^{N} \sigma_\alpha + \sum_{\alpha=1}^{N} \rho_\alpha b_\alpha = \sum_{\alpha=1}^{N} \rho_\alpha \ddot{x}_\alpha \tag{5-18}$$

The net momentum supply vanishes.

Balance of angular momentum for the mixture

$$\sigma = \sigma^T \tag{5-19}$$

It should be noted that the partial stress may not be symmetric, but the stress of the mixture is symmetric.

Balance of energy

$$\sum_{\alpha=1}^{N} \rho_\alpha e_\alpha = \sum_{\alpha=1}^{N} tr(t_\alpha^T L_\alpha) - \nabla \bullet \sum_{\alpha=1}^{N} q_\alpha + \sum_{\alpha=1}^{N} \rho_\alpha r_\alpha \tag{5-20}$$

The second law of thermodynamics

$$\sum_{\alpha=1}^{N} [\frac{\partial}{\partial t}(\rho_\alpha \eta_\alpha) + \nabla \bullet (\rho_\alpha \eta_\alpha \dot{x}_\alpha) + \frac{q_\alpha}{\theta_\alpha} + \frac{\rho_\alpha r_\alpha}{\theta_\alpha}] \geq 0 \tag{5-21}$$

Linear momentum interaction

The linear momentum supply p_α in Equation 5-6 represents the interaction between any two of the constituents. This interaction may also be affected by other constituents in the mixture to a secondary degree (Truesdell, 1969). Following Bowen (1976), this supply can be represented as:

$$p_\alpha = -\sum_{\beta=1}^{N} \xi_{\alpha\beta} grad\rho_\beta - \sum_{\beta=1}^{N} \zeta_{\alpha\beta} \dot{x}_\beta \tag{5-22}$$

with $\sum_{\beta=1}^{N} \xi_{\alpha\beta} = 0$, $\sum_{\beta=1}^{N} \zeta_{\alpha\beta} = 0$, $\xi_{\alpha\beta}$ and $\zeta_{\alpha\beta}$ are functions of dispersed densities of the constituents and the temperature.

For a homogeneous material, it can be proved that: $grad\rho_\beta = \gamma_\beta grad\phi_\beta$.

5.1.3   Two-constituent Case

The following refers to the two-constituent mixture of a solid (i.e., aggregate coated with asphalt binder) and air voids. The local solid volume fraction is ϕ. The symbols used follow those in the previous sections and will not be repetitively explained. The objective of this section is to illustrate how mixture theory predicts the mechanical properties. Due to this consideration, the main focus will be on the balance equation of linear momentum.

Balance of linear momentum

For the solids:

$$\nabla \bullet \sigma_s + \rho_s b_s + p_s = \rho_s \ddot{x}_s \tag{5-23}$$

For the air:

$$\nabla \bullet \sigma_a + \rho_a b_a + p_a = \rho_a \ddot{x}_a \tag{5-24}$$

Constitutive relations

Assume that the solids are the aggregates coated with asphalt binder (with density equal to γ-the maximum density). The mixture therefore consists of only two constituents: the coated aggregate (effective aggregate) and air void. This application is to evaluate the mechanical properties, especially the initial stress distribution (not the creeping behavior). It is more applicable to the situation of low-temperature conditions.

By these considerations, it is assumed that the effective aggregate is a linearly elastic material observing the following constitutive equation:

$$\tau_{sij} = \lambda \varepsilon_{skk} + 2\mu \varepsilon_{sij} \tag{5-25}$$

Where λ and μ are the *Lame's* constants; τ_{sij} and ε_{sij} are the stress and the strain tensor in the real material, and the subscript "s" represents solids.

It is assumed that the air voids behave as an ideal compressible gas observing the following constitutive equation:

$$\tau_{aij} = -p\delta_{ij} \tag{5-26}$$

$$p = \gamma_a RT \tag{5-27}$$

Where R = the universal gas constant, T = the absolute temperature, γ_a = the density of air, and τ_{aij} = the stress in the air; the subscript a represents air.

The stress calculated from the above two equations is the actual stress in the material. It is related to the partial stress by the following equation:

$$\sigma_{sij} = \phi\tau_{sij} \tag{5-28}$$

$$\sigma_{aij} = (1-\phi)\tau_{aij} \tag{5-29}$$

Momentum interaction

With the consideration of static loading, Equation 5-22 can be simplified as:

$$p_s = -\xi_{11}\gamma_s grad\phi + \xi_{12}\gamma_a grad\phi \tag{5-30}$$

$$p_a = -\xi_{21}\gamma_s grad\phi + \xi_{22}\gamma_a grad\phi \tag{5-31}$$

Boundary conditions

There are many methods proposed to handle the boundary condition problem (Rajagopal, 1996). Krishnan and Rao (2000) also followed Rajagopal's method. The basic idea is to use the relation that the area fraction is equal to the volume fraction and distribute the mixture surface traction to the partial surface traction of constituents according to their area fraction.

$$\sigma_\alpha n_s = t_\alpha \tag{5-32}$$

$$t_\alpha = \phi_\alpha t \tag{5-33}$$

Where σ_α is the partial stress; n_s is the normal to the surface, and t_α is the partial traction.

By the distribution proportions expressed in Equation 5-33, it is easy to show that $\sum_{\alpha=1}^{N} t_\alpha = \sum_{\alpha=1}^{N} \phi_\alpha t = t$. Formally, it satisfies the requirements. However, it should be noted that Equation 5-33 might not hold universally. For example, the shear components of the surface traction cannot be allocated to the air constituent at all. In general, boundary conditions might be case dependent. In the two-constituent situation (solids and air voids), the stress in the air voids may be neglected. The boundary conditions may be written as:

$$\sigma_s n_s = t_s \tag{5-34}$$

$$t_s = t / \phi \tag{5-35}$$

5.1.4 Analytic Solution of Two Simple Cases

Equations 5-24 through 5-31 and 5-34 through 5-35 plus the strain-compatible conditions constitute a whole set of equations for solving mechanical boundary value problems. Obviously, unlike the classic continuum theory, field variables ϕ and $grad\,\phi$ are needed for the solution of the set of equations. It should be noted, if $grad\,\phi = 0$, the set of the equations becomes the same as those for a single continuum. Therefore, it may be deduced that the spatial gradients of the local volume fractions are important in affecting the mixture properties. The following will look into two simple cases: a 2D case and a 1-D case. In both cases, a static problem is assumed and the momentum supply from the air and the body force are considered negligible. With these simplifications, only the stress component in the solid exists.

Two-dimensional case

Equation 5-23 becomes:

$$\frac{\partial \sigma_{11}}{\partial x_1} + \frac{\partial \sigma_{12}}{\partial x_2} = 0 \tag{5-36}$$

$$\frac{\partial \sigma_{12}}{\partial x_1} + \frac{\partial \sigma_{22}}{\partial x_2} = 0 \tag{5-37}$$

or

$$\frac{\partial (\phi\tau_{11})}{\partial x_1} + \frac{\partial (\phi\tau_{12})}{\partial x_2} = 0 \tag{5-38}$$

$$\frac{\partial (\phi\tau_{12})}{\partial x_1} + \frac{\partial (\phi\tau_{22})}{\partial x_2} = 0 \tag{5-39}$$

A simple way to solve the problem is to solve the partial stress Equations 5-36 and 5-37. This way, the physical implication is not so explicit. However, the solution to the material Equations 5-38 and 5-39 will be interesting. The above two equations can be rewritten as:

$$\phi\frac{\partial (\tau_{11})}{\partial x_1} + \tau_{11}\frac{\partial (\phi)}{\partial x_1} + \phi\frac{\partial (\tau_{12})}{\partial x_2} + \tau_{12}\frac{\partial \phi}{\partial x_2} = 0 \tag{5-40}$$

$$\phi\frac{\partial (\tau_{12})}{\partial x_1} + \tau_{12}\frac{\partial (\phi)}{\partial x_1} + \phi\frac{\partial (\tau_{22})}{\partial x_2} + \tau_{22}\frac{\partial \phi}{\partial x_2} = 0 \tag{5-41}$$

The terms $\tau_{11}\dfrac{\partial (\phi)}{\partial x_1} + \tau_{12}\dfrac{\partial \phi}{\partial x_2}$ and $\tau_{12}\dfrac{\partial (\phi)}{\partial x_1} + \tau_{22}\dfrac{\partial \phi}{\partial x_2}$ play the roles of driving forces in the above two equations. These terms are similar to the materials forces (Maugin, 1993) and will be called inhomogeneity-induced material forces.

In the case of a rectangular area subjected to a uniform boundary pressure (Figure 5.1a), if it is further assumed that $\phi = \phi_0 + ax_2$, then:

$$\phi\frac{\partial(\tau_{12})}{\partial x_2} + \tau_{12}a = 0 \tag{5-42}$$

$$\phi\frac{\partial(\tau_{22})}{\partial x_2} + \tau_{22}a = 0 \tag{5-43}$$

In conjunction with the strain compatibility equation, one has $\tau_{11} = C_1(\phi_0 + ax_2)$, $\tau_{22} = C_2(\phi_0 + ax_2)$, $\tau_{12} = C_3(\phi_0 + ax_2)$. Obviously, if the gradient of ϕ is equal to zero, the equation is the same as the partial stress equation. The normalized partial stress $\dfrac{\sigma_{22}}{\sigma_{22}\big|_{x_2=0}}$ and the normalized stress in the material $\dfrac{\tau_{22}}{\tau_{22}\big|_{x_2=0}}$ are plotted in Figure 5.1b to illustrate how the two stresses vary spatially when the gradient ($a = \phi_0 / L$) is not equal to zero. For general cases, numerical methods such as the Finite Element Method (FEM, [Arduino, 1996]) must be used.

One-dimension case

The governing equation for the 1D case is very simple:

$$\frac{\partial(\phi\tau_{11})}{\partial x_1} = 0 \tag{5-44}$$

Therefore $\phi\tau_{11} = C$. This basically demonstrates that the stress in the material is varying with the local volume fraction.

The two simple analytical solutions have the following implication: stress in the material is very much related to the void structure. Even though the total volume of voids in the specimen is the same in each, the distribution of the voids will cause different stress

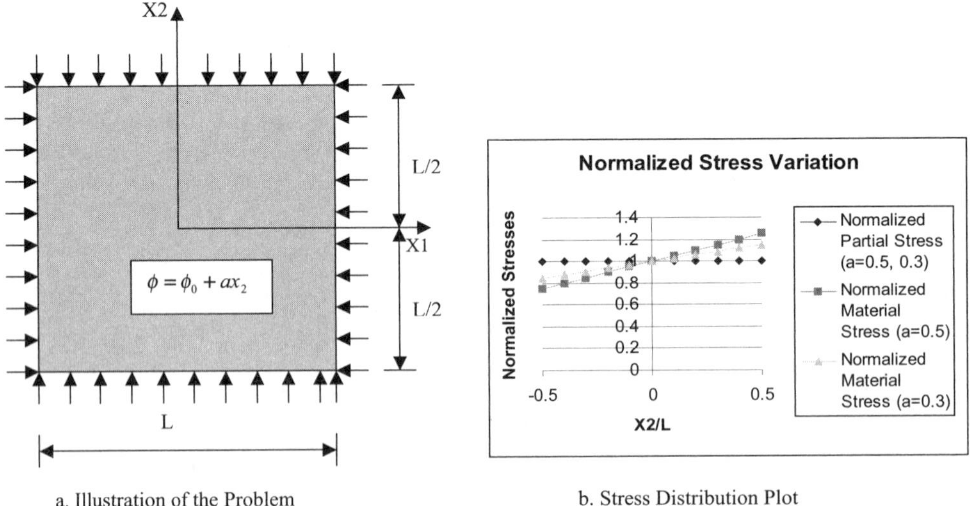

a. Illustration of the Problem   b. Stress Distribution Plot

FIGURE 5.1   Illustration of a 2D case, isotropic pressure on a structural heterogeneous material.

distributions. As deformations take place in the materials (solids in this case), the stress in the material actually counts. Both permanent deformation and fatigue cracking should be related to the stresses in the material rather than the nominal stress.

5.1.5 Characterization of the Filed Variable ϕ and $grad\phi$

A procedure is presented in Chapter 3. There are other methods to define the local volume fractions such as the material cell and the influence cell concepts.

5.1.6 Air-void Reduction Simulation Using Mixture Theory

Krishnan and Rao (2000) developed a mixture theory and applied it to the modeling of the air void reduction process. His theory is briefly summarized as follows.

Partial tractions and partial stresses

$$t = \sum_{\alpha=1}^{3} t_\alpha \tag{5-45}$$

$$\sigma = \sum_{\alpha=1}^{3} \sigma_\alpha \tag{5-46}$$

$$t = \sigma^T n_s \tag{5-47}$$

Conservation of angular momentum

$$\sum_{\alpha=1}^{3} M_\alpha = 0 \tag{5-48}$$

Momentum interaction terms

Asphalt-aggregate interaction

$$f_{(asp--agg)} = \tau_{ij}^{(1)} grad\phi_1 - \frac{(\phi_1)^2 \gamma_1 u_{(agg--asp)}}{\kappa} \tag{5-49}$$

$$f_{(agg--asp)} = \tau_{ij}^{(2)} grad\phi_2 - \frac{(\phi_2)^2 \gamma_2 u_{(agg--asp)}}{\kappa} \tag{5-50}$$

$u_{(agg--asp)} = v_2 - v_1$ is relative seepage speed.

Asphalt-air voids interaction

$$f_{(asp-air)} = \tau_{ij}^{(3)} grad\phi_3 \tag{5-51}$$

$$f_{(air-asp)} = \tau_{ij}^{(2)} grad\phi_2 \tag{5-52}$$

Aggregate-air voids interaction

$$f_{(agg--air)} = \tau_{ij}^{(3)} grad\phi_3 \tag{5-53}$$

$$f_{(air--agg)} = \tau_{ij}^{(1)} grad\phi_1 \tag{5-54}$$

Here 1, 2, and 3 represent asphalt binder, aggregates, and voids, respectively.
Total force of interaction

$$f_{(asp--agg)} + f_{(agg--asp)} = f_{(air--agg)} + f_{(agg--air)} = f_{(air--asp)} + f_{(asp--air)} = 0 \tag{5-55}$$

In his model, aggregates are isotropic and linear elastic. Binder is modeled using the Burger's model.

By introducing the interaction forces into the governing equations of motion, and mass conservation equation for each constituent, Krishnan established a complete set of equations for the three-constituent asphalt mixture. Krishnan also applied his theory to model the air reduction processes at different pressures and achieved rational results.

5.2   Micromechanics and Its Application

Micromechanics has many branches and it is difficult to give a rigorous and scientific definition. The author would like to define micromechanics as a branch of mechanics that predicts the mechanical behavior of a material by incorporating any distribution and properties of constituents. The distribution information could include the total volume fractions of a representative volume, the local volume fractions of constituents and their gradients, interfaces, and particle arrangements. It could be in the form of analytical solutions or through computational simulations. This section will mainly focus on the introduction of Eshelby Mechanics. Computational techniques including the use of the FEM, Boundary Element Method (BEM), and Discrete Element Method (DEM) will be described in other chapters.

5.2.1   Eshelby Mechanics

5.2.1.1   Green Function

Eshelby (1952) developed a solution to predict the stress and strain field of a medium that has an ellipsoid inclusion. Through the use of Green function (see Chapter 1), he obtained a solution for the disturbed strain field. The Green function in an unbounded elastic domain can be represented in Equation 5-56 (a concise derivation of the Green function is given in Qu and Cherkaoui, 2006).

$$G_{ij}^{\infty}(x,y) = \frac{1}{(2\pi)^3} \int_{-\infty}^{\infty} N_{ij}(\xi)D^{-1}(\xi)e^{i\xi\bullet(x-y)}d\xi \tag{5-56}$$

Where N and D are functions of the stiffness tensor.

By placing $G_{ij}^{\infty}(x,y)$ in the equilibrium Equation 5-57, one can prove that the Green function is actually the displacement field created by a unit force (for an infinite domain). It is also the basis for the BEM to be discussed in Chapter 8.

$$L_{ijkl}\frac{\partial^2 G_{km}^{\infty}(x,y)}{\partial x_i \partial x_j} + \delta_{lm}\delta(x-y) = 0 \tag{5-57}$$

There are a few interesting features for the Green function, for example, symmetry for locations x, y, that is the displacement at point y produced by a unit force at point x will be equal to the displacement at point x produced by a unit force at point y. For isotropic materials, the Green function can be further expressed as:

$$G_{ij}^{\infty}(x,y) = \frac{1}{16\pi\mu(1-v)|x-y|}\{(3-4v)\delta_{ij} + \frac{(x_i-y_i)(x_j-y_j)}{|x-y|^2}\} \tag{5-58}$$

The above solution refers to the unbound (infinite) domain problem. It can include both the stress and displacement boundary conditions.

5.2.1.2 Eigenstrains

Eigenstrains refer to inelastic strains such as thermal strains, plastic strains, etc. It is typically represented as ε_{ij}^{*} while ε_{ij} and e_{ij} often represent total strain and elastic strain. Hooke's law can be presented as:

$$\sigma_{ij} = L_{ijkl}(\varepsilon_{kl} - \varepsilon_{kl}^{*}) \text{ or } e_{ij} = M_{ijkl}\sigma_{kl} \tag{5-59a, b}$$

By replacing $\sigma_{ij} = L_{ijkl}(\varepsilon_{kl} - \varepsilon_{kl}^{*})$ into the equilibrium equation, it can be proven that the eigenstrain problem is equivalent to the problem where the body force is equal to $f_i = -L_{ijkl}\varepsilon_{kl,j}^{*}$. Therefore, the eigenstrain problem can be related to the Green function. It can be shown that the resulted displacement field will be:

$$u_i(x) = \int_{-\infty}^{\infty} L_{mjkl}\varepsilon_{kl}^{*}(y)\frac{\partial G_{mi}^{\infty}(x,y)}{\partial y_j}dy \tag{5-60}$$

From this displacement, the strain field and stress field can be obtained.

A typical problem involves the strain and stress field due to an ellipsoid inclusion with uniform eigenstrain in the inclusion. The solution to this problem is the Eshelby solution, which has significant meaning in micromechanics. The derivation can be found in several references such as Qu and Cherkaoui (2006). The problem can be stated as follows:

For an unbounded domain with an ellipsoid inclusion of size a_1, a_2, and a_3, embedded at location (0,0,0) [domain, $(\frac{x_1}{a_1})^2 + (\frac{x_2}{a_2})^2 + (\frac{x_3}{a_3})^2 \leq 1$], if there is a uniform eigenstrain ε_{ij}^{*} within the inclusion, what is the total strain and the stress in the inclusion? Through the use of the Green's function, the solution can be obtained as:

$$\varepsilon_{ij}(x) = S_{ijkl}\varepsilon_{kl}^{*} \tag{5-61}$$

Where S_{ijkl} is the Eshelby inclusion tensor. S_{ijkl} poses certain symmetry such that $S_{ijkl} = S_{jikl} = S_{ijlk}$ but $S_{ijkl} \neq S_{klij}$. It is nonsingular and independent of the eigenstrain. With the Eshelby inclusion tensor, the stress can be calculated in the inclusion, $\sigma_{ij} = L_{ijkl}(S_{ijkl}\varepsilon_{kl}^{*} - \varepsilon_{kl}^{*})$. The traction continuity will result in surface tractions. The elements for the Eshelby inclusion tensor of inclusions of different shapes are different. It is related to the shape of the inclusions. The elements of the Eshelby tensor for special cases such as ellipsoids and spheres are documented in Appendix B. These Eshelby tensor elements for special cases can be used for computing the effective modulus presented in the following sections.

5.3   Effective Properties of Mixture

The Eshelby solution is very important in various topics in micromechanics. One major application of the solution is in the estimation of the effective properties, mainly the effective modulus of elasticity of the mixture from the properties, of components, and their volume fractions. The effective modulus is defined on the average stress and strain relationship.

5.3.1   Average Stress and Average Strain Theorem

Average stress and stress rate (Nemat-Nasser and Hori, 1999)
It is understood that even though the surface traction follows the distribution resulted from a uniform stress condition (when the material is homogenous), due to the inhomogeneities, the stress distribution is not uniform. It can be proved that as long as the prescribed surface traction is self-equilibrating on ∂V, the volume average of the stress field $\sigma(x)$ is related only to the prescribed boundary tractions (regardless of how the inhomogeneities are distributed in the volume).

If the volume average of a field integrable variable is denoted as:

$$<T> = \frac{1}{V}\int_V T(x)dV \tag{5-62}$$

The unweighted volume average stress is:

$$\bar{\sigma} \equiv <\sigma> \tag{5-63}$$

The average stress is:

$$\bar{\sigma}_{ij} = \frac{1}{V}\int_V \sigma_{ij}dV = \frac{1}{V}\int_{\partial V} x_i t_j dS \tag{5-64}$$

$$\bar{\sigma}_{ij} = \frac{1}{V}\int_{\partial V} x_i n_k \sigma_{kj} dS$$

Using the divergence theorem, the above surface integral can be converted into a volume integral.

$$\frac{1}{V}\int_V \frac{\partial(x_i\sigma_{kj})}{\partial x_k}dV = \frac{1}{V}\int_V (\delta_{ik}\sigma_{kj} + x_i\frac{\partial\sigma_{kj}}{\partial x_k})dV = \frac{1}{V}\int_V (\sigma_{ij} + x_i\frac{\partial\sigma_{kj}}{\partial x_k})dV \tag{5-65}$$

Considering the equilibrium condition, $\dfrac{\partial\sigma_{kj}}{\partial x_k} = 0$ the theorem is proved.
It can also be proven that:

$$\bar{\sigma}_{ij} \equiv <\sigma_{ij}> = \frac{1}{V}\int_{\partial V} x_i t_j dS \tag{5-66}$$

Average strain and strain rate (Nemat-Nasser and Hori, 1999)
It can also be proven that the average strain is completely determined by the surface displacements regardless of the distribution and volume fractions of the heterogeneities. The theorems on average stress, strain, and stress rate and strain rate serve as the foundation of the representative volume element (RVE) methods.

$$\overline{u_{j,i}} = <u_{j,i}> = \frac{1}{V}\int_{\partial V} n_i u_j dS \tag{5-67}$$

Average strain and average strain rate:

$$\overline{\varepsilon}_{ij} =< \varepsilon_{ij} >= \frac{1}{V}\int_V \varepsilon_{ij}dV = \frac{1}{V}\int_{\partial V}\frac{1}{2}(n_i u_j + u_i n_j)dS \tag{5-68}$$

$$\overline{\dot{\varepsilon}}_{ij} =< \dot{\varepsilon}_{ij} >= \frac{1}{V}\int_{\partial V}\frac{1}{2}(n_i \dot{u}_j + \dot{u}_i n_j)dS \tag{5-69}$$

An important relationship among the averages (and also the strain energy) is the Hill's Lemma. A concise proof is documented in Qu and Cherkaoui (2006).

$$\overline{\sigma_{ij}\varepsilon_{ij}} - \overline{\sigma}_{ij}\overline{\varepsilon}_{ij} = \frac{1}{V}\int_S (u_i - x_j\overline{\varepsilon}_{ij})(\sigma_{ik}n_k - \overline{\sigma}_{ik}n_k)dS \tag{5-70}$$

Where $\overline{\sigma_{ij}\varepsilon_{ij}} = \dfrac{1}{V}\int_V \sigma_{ij}\varepsilon_{ij}dv$

Therefore, effective properties are defined as:

$$\overline{\sigma}_{ij} = \overline{L}_{ijkl}\,\overline{\varepsilon}_{kl}$$

$$\overline{\varepsilon}_{ij} = \overline{M}_{ijkl}\,\overline{\sigma}_{kl}$$

Therefore:

$$\overline{LM} = \overline{ML} = I \text{ or } \overline{L}^{-1} = \overline{M} \tag{5-71}$$

5.3.2 General Philosophy of Estimating Effective Properties

For a finite homogeneous linearly elastic solid with elasticity tensor L and compliance tensor M, containing a linearly elastic and homogeneous inclusion Ω of arbitrary geometry, with elasticity and compliance tensors L^Ω and M^Ω.

$$u^o = x \bullet \varepsilon^o \text{ on } \partial V$$

$$\sigma^o = L : \varepsilon^o \tag{5-72}$$

Hooke's law:

$$\sigma = \begin{cases} L : (\varepsilon^o + \varepsilon^d(x)) & \text{in } M = V - \Omega \\ L^\Omega : (\varepsilon^o + \varepsilon^d(x)) & \text{in } \Omega \end{cases} \tag{5-73}$$

$$\varepsilon = \begin{cases} M : (\sigma^o + \sigma^d(x)) & \text{in } M = V - \Omega \\ M^\Omega : (\sigma^o + \sigma^d(x)) & \text{in } \Omega \end{cases} \tag{5-74}$$

$\varepsilon^d(x)$ and $\sigma^d(x)$ are perturbed strain and stress.

The Eigenstrain theorem states that the heterogeneity can be treated by finding a suitable strain field $\varepsilon^*(x)$ (eigenstrain) in Ω, such that the equivalent homogeneous solid has the same strain and stress fields as the actual heterogeneous solid has under the applied tractions or displacements.

$$\varepsilon^*(x) = \begin{cases} 0 & \text{in } M \\ \varepsilon^* & \text{in } \Omega \end{cases} \tag{5-75}$$

The corresponding stress and strain fields:

$$\sigma(x) = L:(\varepsilon(x) - \varepsilon^*(x)) = \begin{cases} L:(\varepsilon^o + \varepsilon^d(x)) \text{ in } M & \text{in } M \\ L:(\varepsilon^o + \varepsilon^d(x) - \varepsilon^*(x)) \text{ in } \Omega \end{cases} \tag{5-76}$$

Integral operator:

$$\varepsilon_{ij}^d(x) \equiv S_{ij}(x;\varepsilon^*) \tag{5-77}$$

Similarly, the concept of eigenstress can be introduced:

$$\sigma^*(x) = \begin{cases} 0 & \text{in } M \\ \sigma^* & \text{in } \Omega \end{cases} \tag{5-78}$$

So that elasticity tensor is uniform everywhere, including Ω, so that the strains and stress can be represented in the following equations.

$$\varepsilon(x) = \varepsilon^o + \varepsilon^d(x)$$

$$\sigma(x) = L:\varepsilon(x) + \sigma^*(x) = \begin{cases} L:(\varepsilon^o + \varepsilon^d(x)) & \text{in } M \\ L:(\varepsilon^o + \varepsilon^d(x)) + \sigma^*(x) \text{ in } \Omega \end{cases} \tag{5-79}$$

$$\overline{\varepsilon}_{ij} = \frac{1}{V}\int_V \varepsilon_{ij} dV = \frac{1}{V}\sum_{i=0}^N \int_{\Omega_i} \varepsilon_{ij} dV = \sum_{i=0}^N c_i \overline{\varepsilon}_i = \overline{\varepsilon} \tag{5-80}$$

Where c_i is the volume fraction of the ith heterogeineity and c_0 is the volume fraction of the matrix. Strain concentration tensor is so defined that (Qu and Cherkaoui, 2006):

$$\overline{\varepsilon}_i = A_i \overline{\varepsilon} \tag{5-81}$$

$\overline{\varepsilon}$ is the average strain for the entire composite material. Equation 5-80 can be further written in the following format:

$$c_0 \overline{\varepsilon}_0 = \overline{\varepsilon} - \sum_{i=1}^N c_i \overline{\varepsilon}_i = \overline{\varepsilon} - \sum_{i=1}^N c_i A_i \overline{\varepsilon} \tag{5-82}$$

For stresses, the following similar representations can be used:

$$\overline{\sigma}_{ij} = \frac{1}{V}\int_V \sigma_{ij} dV = \overline{\sigma} = \sum_{i=0}^N c_i \overline{\sigma}_i \tag{5-83}$$

$$\overline{\sigma}_i = L_i \overline{\varepsilon}_{i\,(i=0,N)}$$

Making use of Equation 5-82:

$$\overline{\sigma} = c_0 L_0 \overline{\varepsilon}_0 + \sum_{i=1}^N c_i L_i \overline{\varepsilon}_i \tag{5-84}$$

$$\overline{\sigma} = [L_0 + \sum_{i=1}^N c_i (L_i - L_0) A_i] \overline{\varepsilon} \tag{5-85}$$

In other words, the effective modulus is:

$$\bar{L} = L_0 + \sum_{i=1}^{N} c_i (L_i - L_0) A_i \tag{5-86}$$

Please note that the above formulation does not make assumptions regarding the shape of the inclusions. It does not address the misfit of the strain fields between the matrix and the inclusions. Following the similar philosophy, equation for the compliance tensor can be derived:

$$\bar{M} = M_0 + \sum_{i=1}^{N} c_i (M_i - M_0) B_i \tag{5-87}$$

$$\bar{\sigma}_i = B_i \bar{\sigma} \tag{5-88}$$

B_i is the stress concentration tensor.

It should be noted that the above formulations are based on relationships with the average strain or stress of the entire composite. Formulations can be derived using the relationship between strains/stresses of inhomogeneities with the average strains/stresses in the matrix.

$$\varepsilon_i = G_i \overline{\varepsilon}_0 \qquad \bar{\sigma}_i = H_i \bar{\sigma}_0 \tag{5-89a, b}$$

The above formulations and the Eshelby tensor form the basis of effective moduli computations of the Eshelby method, the Mori-Tanaka method, the Self-consistent method and the Differential Area method. These methods vary with the choice of the reference matrix. The general philosophy can be illustrated in Figure 5.2 and Table 5.1. It is also illustrated in Equation 5-90. Equation 5-90 expresses how to calculate the effective modulus when the ith inhomogeneity is included. It should be noted that when the ith inclusion is embedded, it is surrounded by the matrix and the inhomogeneities (1, i-1).

$$L_i(\varepsilon_r + \varepsilon_i^d) = L_r(\varepsilon_r + \varepsilon_i^d - \varepsilon_i^*) \tag{5-90}$$

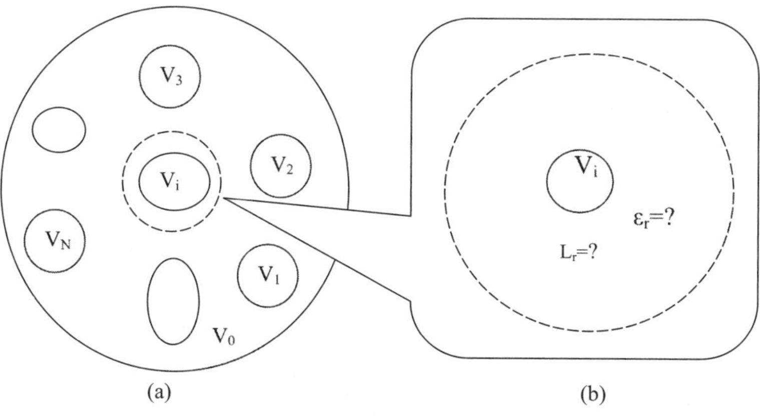

FIGURE 5.2   (a) The ith inhomogeneity in the composite. (b) The ith inhomogeneity background (Qu and Cherkaoui, 2006).

5.3.3 Eshelby Dilute Solution

Therefore the Eshelby method (Qu and Cherkaoui, 2006) (see Figure 5.2) will have:

$$L_r = L_0 \text{ and } \varepsilon_r = \varepsilon_0 \tag{5-91}$$

$$\varepsilon_i^* = [(L_i - L_0)S_i + L_0]^{-1}(L_i - L_0)\varepsilon_0 \tag{5-92}$$

$$T_i = [I + S_i L_0^{-1}(L_i - L_0)]^{-1} \tag{5-93}$$

$$\bar{L} = L_0 + \sum_{i=1}^{N} c_i(L_i - L_0)T_i = L_0 + \sum_{r=1}^{N} c_i(L_i - L_0)[I + S_i(M_0 L_i - I)]^{-1} \tag{5-94}$$

$$\bar{M} = M_0 + \sum_{i=1}^{N} c_i(M_i - L_0)L_i T_i M_0 \tag{5-95}$$

It should be noted that:

$$\overline{ML} \neq I.$$

For spherical inclusions, the formulations are as follows:

$$\bar{K} = K_0 + \frac{c_1(K_1 - K_0)(3K_0 + 4\mu_0)}{3K_1 + 4\mu_0} \tag{5-96}$$

$$\bar{\mu} = \mu_0 + \frac{5c_1\mu_0(\mu_1 - \mu_0)(3K_0 + 4\mu_0)}{3K_0(3\mu_0 + 2\mu_1) + 4\mu_0(2\mu_0 + 3\mu_1)} \tag{5-97}$$

5.3.4 Mori-Tanaka Method

The Mori-Tanaka Method (Qu and Cherkaoui, 2006):

$$L_r = L_0 \text{ and } \varepsilon_r = \bar{\varepsilon}_0 \tag{5-98}$$

$$\hat{S}_i = S_i \tag{5-99}$$

$$\bar{L} = L_0 + \sum_{i=1}^{N} c_i(L_i - L_0)A_i \tag{5-100}$$

$$\bar{L} = \sum_{i=0}^{N} c_i L_i T_i \left[\sum_{k=0}^{N} c_k T_k \right]^{-1} \tag{5-101}$$

$$\bar{M} = \sum_{i=0}^{N} c_i T_i \left[\sum_{k=0}^{N} c_k L_k T_k \right]^{-1} \tag{5-102}$$

$$\overline{ML} = I$$

For one inclusion:

$$\bar{L} = (c_0 L_0 T_0 + c_1 L_1 T_1)(c_0 T_0 + c_1 T_1)^{-1} = (c_0 L_0 + c_1 L_1 T_1)(c_0 I + c_1 T_1)^{-1} \tag{5-103}$$

For spherical particles the predictions are:

$$\bar{K} = K_0 + \frac{c_1(K_1 - K_0)(3K_0 + 4\mu_0)}{3K_0 + 4\mu_0 + 3(1 - c_1)(K_1 - K_0)} \tag{5-104}$$

$$\bar{\mu} = \mu_0 + \frac{5c_1\mu_0(\mu_1 - \mu_0)(3K_0 + 4\mu_0)}{5\mu_0(3K_0 + 4\mu_0) + 6(1 - c_1)(\mu_1 - \mu_0)(K_0 + 2\mu_0)} \tag{5-105}$$

Method	L_r	ε_r
Eshelby method	L_0 (matrix modulus)	ε_0 Prescribed uniform strain in matrix
Mori-Tanaka	L_0 (matrix modulus)	$\bar{\varepsilon}_0$ Average strain in matrix
Self-consistent	$\bar{L}$ (effective composite modulus)	$\bar{\varepsilon}$ Average strain of the composite

TABLE 5.1   Reference conditions of the three methods.

5.3.5   Self-Consistent Method

For the Self-consistent method (Qu and Cherkaoui, 2006):

$$L_r = \bar{L} \text{ and } \varepsilon_r = \bar{\varepsilon} \tag{5-106}$$

$$\varepsilon_i = \bar{\varepsilon} + \varepsilon_i^{pt} = \bar{\varepsilon} + \bar{S}\varepsilon_i^* = \bar{T}_i\bar{\varepsilon} \tag{5-107}$$

$$\bar{T}_i = \bar{S}_i\bar{L} \; / (L_i - \bar{L}) \tag{5-108}$$

$$\sigma_i = L_i\varepsilon_i = L_i\bar{T}_i\bar{\varepsilon} = L_i\bar{T}_i\bar{M}\bar{\sigma} \tag{5-109}$$

$$A_i = \bar{T}_i, \text{ and } B_i = L_i\bar{T}_i\bar{M} \tag{5-110}$$

$$\bar{L} = L_0 + \sum_{i=1}^{N} c_i(L_i - L_0)\bar{T}_i \tag{5-111}$$

$$\bar{M} = \bar{L}^{-1} + \sum_{i=1}^{N} c_i(M_0 - M_i)L_i\bar{T}_i\bar{L}^{-1} \tag{5-112}$$

$$\overline{ML} = I$$

For spherical particles, there are the following predictions:

$$\bar{K} = K_0 + \frac{c_1\bar{K}(K_1 - K_0)}{\bar{K} + 3\bar{\gamma}(K_1 - \bar{K})} \tag{5-113}$$

$$\bar{\mu} = \mu_0 + \frac{c_1\bar{\mu}(\mu_1 - \mu_0)}{\bar{\mu} + 2\bar{\delta}(\mu_1 - \bar{\mu})} \tag{5-114}$$

5.3.6   Differential Schemes

The differential scheme (Qu and Ckerkaoui, 2006) follows the philosophy of adding the inclusions into the matrix with each time following the dilute solution but updating the matrix.

$$\bar{L}(c_1 + \Delta c_1) = \bar{L}(c_1) + \frac{\Delta\Omega_1}{\Omega_0 + \Omega_1 + \Delta\Omega_1}(L_1 - \bar{L}(c_1)) : A_1(\bar{L}(c_1)) \tag{5-115}$$

$$\frac{\bar{L}(c_1 + \Delta c_1) - \bar{L}(c_1)}{\Delta c_1} = \frac{1}{(1 - c_1)}(L_1 - \bar{L}(c_1)) : A_1(\bar{L}(c_1)) \tag{5-116}$$

$$\frac{d\bar{L}(c_1)}{dc_1} = \frac{1}{(1 - c_1)}(L_1 - \bar{L}(c_1)) : A_1(\bar{L}(c_1)) \tag{5-117}$$

For spherical particles:

$$\frac{d\bar{K}}{dc} + \frac{(\bar{K} - K_1)(3\bar{K} + 4\bar{\mu})}{(1-c)(3K_1 + 4\bar{\mu})} = 0 \tag{5-118}$$

$$\frac{d\bar{\mu}}{dc} + \frac{5\bar{\mu}(\bar{\mu} - \mu_1)(3\bar{K} + 4\bar{\mu})}{(1-c)[3\bar{K}(3\bar{\mu} + 2\mu_1) + 4\bar{\mu}(2\bar{\mu} + 3\mu_1)]} = 0 \tag{5-119}$$

5.3.7   Relationships Among the Four Methods (Qu and Cherkaoui, 2006)

Low-concentration case

For the spherical particle cases, in the case of low concentrations, it can be proven that the Mori-Tanaka method will reduce to the Elsheby solution.

$$\bar{K} = K_0 + \frac{c_1(K_1 - K_0)(3K_0 + 4\mu_0)}{3K_1 + 4\mu_0} \tag{5-120}$$

$$\bar{\mu} = \mu_0 + \frac{5c_1\mu_0(\mu_1 - \mu_0)(3K_0 + 4\mu_0)}{3K_0(3\mu_0 + 2\mu_1) + 4\mu_0(2\mu_0 + 3\mu_1)} \tag{5-121}$$

By expanding the self-consistent formulation around zero concentration and neglecting the higher order terms, the following is produced (Qu and Cherkaoui, 2006):

$$\bar{K} = K_0 + c_1 K_0^{(1)} + c_1^2 K_0^{(1)} + \ldots \tag{5-122}$$

$$\bar{\mu} = \mu_0 + c_1 \mu_0^{(1)} + c_1^2 \mu_0^{(1)} + \ldots \tag{5-123}$$

It also reduces to the Eshelby solution.

High-concentration case

By allowing the concentration to Approach 1, the following predictions by the Eschelby method are produced:

$$\bar{K} = \frac{K_0(2K_1 - K_0) + 4\mu_0 K_1}{3K_1 + 4\mu_0} \tag{5-124}$$

$$\bar{\mu} = \frac{\mu_0[3K_0(7\mu_1 - 2\mu_0) + 4\mu_0(8\mu_1 - 3\mu_0)]}{3K_0(3\mu_0 + 2\mu_1) + 4\mu_0(2\mu_0 + 3\mu_1)} \tag{5-125}$$

Obviously, it does not predict accurately. Both the Mori-Tanaka and the Self-consistent methods predict it correctly.

Rigid inclusions

For rigid inclusions, the following can be assumed:

$$\frac{\mu_0}{\mu_1} \to 0, \quad \frac{K_0}{K_1} \to 0$$

$$\frac{\bar{\mu}}{\mu_1} \to 0, \quad \frac{\bar{K}}{K_1} \to 0$$

The Eshelby method yields:

$$\bar{K} = K_0 + \frac{c_1(3K_0 + 4\mu_0)}{3}, \quad \bar{\mu} = \mu_0 + \frac{5c_1\mu_0\ (3K_0 + 4\mu_0)}{6\ (K_0 + 2\mu_0)} \qquad (5\text{-}126a, b)$$

The Mori-Tanaka method gives:

$$\bar{K} = K_0 + \frac{c_1(3K_0 + 4\mu_0)}{3(1 - c_1)}, \quad \bar{\mu} = \mu_0 + \frac{5c_1\mu_0\ (3K_0 + 4\mu_0)}{6\ (K_0 + 2\mu_0)(1 - c_1)} \qquad (5\text{-}127a, b)$$

And the Self-consistent method yields:

$$\bar{K} = K_0 + \frac{c_1(3\bar{K} + \bar{\mu})}{3}, \quad \bar{\mu} = \mu_0 + \frac{5c_1\bar{\mu}\ (3\bar{K} + 4\bar{\mu})}{6\ (\bar{K} + 2\bar{\mu})} \qquad (5\text{-}128a, b)$$

These methods predict complex patterns and may not be correct.

Void inclusions

The Elshelby method yields:

$$\bar{K} = K_0 - \frac{c_1 K_0(3K_0 + 4\mu_0)}{4\mu_0}, \quad \bar{\mu} = \mu_0 - \frac{5c_1\mu_0\ (3K_0 + 4\mu_0)}{9K_0 + 8\mu_0} \qquad (5\text{-}129a, b)$$

The Mori-Tanaka method gives:

$$\bar{K} = K_0 - \frac{c_1 K_0(3K_0 + 4\mu_0)}{3K_0 + 4\mu_0 - 3(1 - c_1)K_0}, \quad \bar{\mu} = \mu_0 - \frac{5c_1\mu_0\ (3K_0 + 4\mu_0)}{5\ (3K_0 + 4\mu_0) - 6(1 - c_1)(K_0 + 2\mu_0)} \qquad (5\text{-}130a, b)$$

The Self-consistent method gives:

$$\bar{K} = K_0 - \frac{c_1 K_0(3\bar{K} + 4\bar{\mu})}{4\bar{\mu}}, \quad \bar{\mu} = \mu_0 - \frac{5c_1\mu_0(3\bar{K} + 4\bar{\mu})}{9\bar{K} + 8\bar{\mu}} \qquad (5\text{-}131a, b)$$

5.3.8 The Composite Sphere Method (Qu and Cherkaoui, 2006)

The general philosophy of the Composite Sphere method is illustrated in Figure 5.3.

The two-phase, three-phase, and four-phase model predictions are presented respectively as follows.

$$\bar{K} = \bar{K}_r = K_0 + \frac{c_1(K_1 - K_0)(3K_0 + 4\mu_0)}{3K_0 + 4\mu_0 + (1 - c_1)(K_1 - K_0)} \qquad (5\text{-}132)$$

$$\bar{K} = \bar{K}_r = K_0 + \frac{c_1(K_1 - K_0)(3K_0 + 4\mu_0)}{3K_0 + 4\mu_0 + 3(1 - c_1)(K_1 - K_0)} \qquad (5\text{-}133)$$

$$\bar{K} = K_0 + \frac{(K_e - K_0)(3K_0 + 4\mu_0)(b^3 / c^3)}{3K_0 + 4\mu_0 + 3(K_e - K_0)(1 - b^3 / c^3)} \qquad (5\text{-}134)$$

In the case of a multicoated inclusion problem, the prediction is as follows:

$$\bar{K}_n = K_n + \frac{(R_{n-1}^3 / R_n^3)(\bar{K}_{n-1} - K_n)(3K_n + 4\mu_n)}{3K_n + 4\mu_n + 3(1 - R_{n-1}^3 / R_n^3)(\bar{K}_{n-1} - K_n)}, \quad n \geq 2 \qquad (5\text{-}135)$$

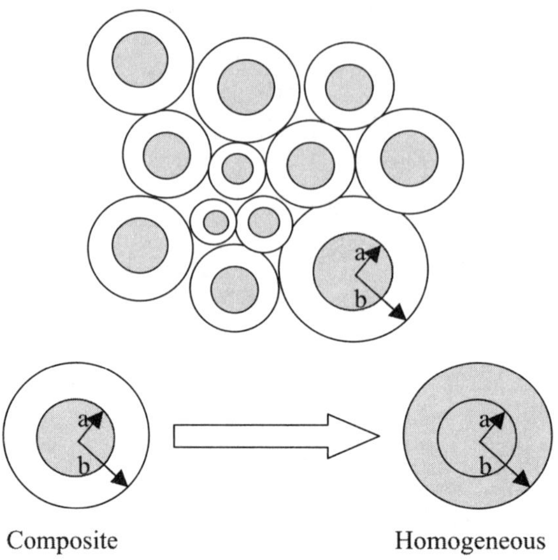

FIGURE 5.3   Composite-sphere model (Qu and Cherkaoui, 2006).

5.3.9   Lower and Upper Bounds (Qu and Cherkaoui, 2006)

The Minimum Potential Energy theorem and Minimum Complementary Energy theorem are used to predict the bounds of the effective moduli.

$$\Pi[u_i] = \frac{1}{2}\int_V L_{ijkl} u_{k,l} u_{i,j}\, dv - \int_{S\sigma} p_i^0 u_i\, ds \tag{5-136}$$

$$\Pi_c[\sigma_{ij}] = \frac{1}{2}\int_V M_{ijkl}\sigma_{kl}\sigma_{ij}\, dv - \int_{Su} u_i^0 \sigma_{ij} n_j\, ds \tag{5-137}$$

Where V is the total volume and S_u is the portion of the boundary of V where displacement u_i^0 is prescribed.

The Voigt Upper Bounder and Reuss Lower Bound can be represented as follows:

$$2\bar{\varepsilon}\hat{\sigma} - \hat{\sigma}\bar{M}^R\hat{\sigma} \leq 2\bar{U} \leq \bar{\varepsilon}\bar{L}'\bar{\varepsilon} \tag{5-138}$$

$$\bar{L}^R \leq \bar{L} \leq \bar{L}^V \tag{5-139}$$

For spherical particles, it results in the following bounds:

$$\frac{K_0 K_1}{K_1 - c_1(K_1 - K_0)} \leq \bar{K} \leq K_0 + c_1(K_1 - K_0) \tag{5-140}$$

$$\frac{\mu_0 \mu_1}{\mu_1 - c_1(\mu_1 - \mu_0)} \leq \bar{\mu} \leq \mu_0 + c_1(\mu_1 - \mu_0) \tag{5-141}$$

The Hashin-Shtrikman Variational Principle states that when L^p is negative semi-definite, among all the symmetric second-order tensors, the true solution to the stress polarization tensor $\hat{\tau}$ renders the following minimum:

$$I_h\left[(\hat{\mathrm{T}})\right]=\bar{U}^h-\frac{1}{2V}H\left[\hat{\mathrm{T}}\right]\tag{5-142}$$

$$\min\left\{I_h\left[\hat{\mathrm{T}}\right]\right\}=\frac{1}{2}\bar{\varepsilon}\bar{L}\bar{\varepsilon}$$

$$I^c\left[\hat{\mathrm{T}}\right]=\bar{U}^h-\frac{1}{2V}H\left[\hat{\mathrm{T}}\right]-\frac{1}{2V}\int_V L^h\eta\left[\hat{\mathrm{T}}\right]M^p L^h\eta\left[\hat{\mathrm{T}}\right]dv\tag{5-143}$$

The famous Hashin-Shtrikman Bounds are then represented as the following:

$$L=L^h+\left[\sum_{r=0}^{N}c_r(I+\bar{L}_r^p P)^{-1}\right]^{-1}\sum_{r=0}^{N}c_r(I+\bar{L}_r^p P)^{-1}\bar{L}_r^p\tag{5-144}$$

For spherical particles, the bounds are:

$$\bar{K}\le\frac{K_0(3K_0+4\mu_0)+c_1(K_1-K_0)[4\mu_0-3(K_1-K_0)]}{3K_0+4\mu_0-3c_1(K_1-K_0)}\tag{5-145}$$

$$\bar{\mu}\le\frac{5\mu_0^2(3K_0+4\mu_0)+c_1(\mu_1-\mu_0)[5\mu_0(3K_0+4\mu_0)-6\mu_1(K_0+2\mu_0)]}{5\mu_0(3K_0+4\mu_0)-6c_1(K_0+2\mu_0)(\mu_1-\mu_0)}\tag{5-146}$$

5.3.10 Hirsch Model

The Hirsch model is based on the rule mixtures.

In parallel:

$$E_c=v_1E_1+v_2E_2\tag{5-147}$$

In series:

$$1/E_c=v_1/E_1+v_2/E_2\tag{5-148}$$

The Hirsch model combines parallel and series arrangement (Figure 5.4), and assumes the relative proportions of Phase 1 and 2 are the same in the series.

$$1/E_c=v_{1s}/E_1+v_{2s}/E_2+(v_{1p}+v_{2p})^2/(v_{1p}E_2++v_{2p}E_2)^2\tag{5-149}$$

If the proportion between the parallel and the serial components is variable, it will result in the following formulation:

$$\frac{1}{E_c}=(1-x)\left(\frac{v_1}{E_1}+\frac{v_2}{E_2}\right)+x\left(\frac{1}{v_1E_1+v_2E_2}\right)\tag{5-150}$$

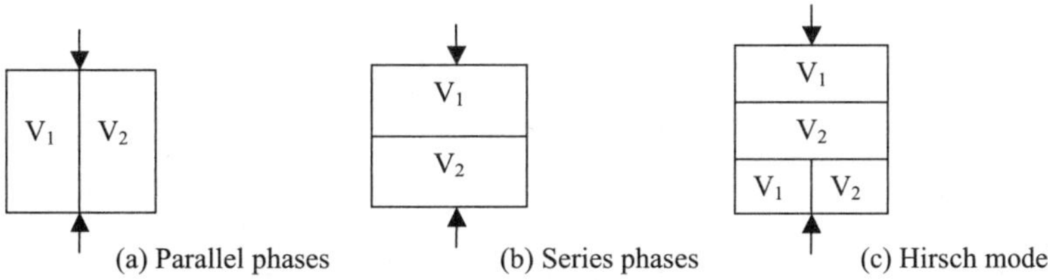

Figure 5.4 Schematic representation of composite models for parallel, series, and Hirsch (combination) arrangement of phases (Christensen et al., 2003).

5.4   Micromechanics Application to Mastics and Asphalt Concrete

When fillers or aggregates are added to binder or mastics, aggregate or filler particles will reinforce the binder or mastics. For some aggregate/filler binder combinations, there may be physiochemical reactions at their interfaces, forming a thin layer of asphalt binder that becomes stronger due to absorption, adsorption, and selective sorption. When aggregate/filler volume fractions are large enough so that filler particles are forming contact and skeleton, the particle interaction effects will take place. The rest can be attributed to the dilute inclusion situations that can be well addressed by the generalized self-consistent scheme (GSCS).

One of the applications of micromechnics that was widely used dated back to the 70s (Shell, 1978). The Shell nomography actually utilized micromechanics principles to obtain the stiffness modulus of asphalt mixture from binder stiffness, the volume fraction of mineral aggregates, and the volume fraction of binder. The binder stiffness is obtained from the penetration index (Equation 5-101), the loading frequency, and temperature in service. More recent systematic studies using micromechanics started from Buttlar's group.

$$PI = \frac{20 - 500A}{1 + 50A} \tag{5-151}$$

Where $A = \dfrac{\log(penatT_1) - \log(penatT_2)}{T_1 - T_2}$ represents the temperature sensitivity of penetration of asphalt binder.

Two major publications of Butllar's research present evaluations of accuracy of various micromechanics formulations for predicting mastics and mixture modulus, as well as use of the micromechanics approach to assess the volume fraction of recovered AC.

Buttlar and Dave (2005) applied a) Paul's Rule of Mixture (Paul, 1960); b) Arbitrary Phase Geometry (APG) model (Hashin and Shtrikman, 1963); c) GSCS Model (Christensen and Lo, 1979); d) Mori-Tanaka method (Mori and Tanaka, 1973); and e) Hirsch Model (Hirsch, 1962). The predictions of the mixture modulus using various models were compared with experimental results. The Hashin's APG model was identified as offering the best potentials in accuracy.

Christensen, et al. (2003) evaluated four alternatives under the general philosophy of the Hirsch model to combine the parallel and series models differently (Figure 5.5). The fundamental idea is based on the consideration that AC is a thermal sensitive material and the load transfer proportion will vary with temperature; the real distribution mechanisms are more complicated than the simple combination of the original Hirsch model.

The formulations corresponding to the four different alternatives are as follows:

(a) Serial version

$$Ec = \left[\frac{Va'}{Ea} + \frac{(1 - Va')^2}{PcEa + VmEm} \right]^{-1} \tag{5-152}$$

(b) Parallel version

$$Ec = VcEa + (1 - Vc)^2 \left(\frac{Va'}{Ea} + \frac{Vm}{Em} \right)^{-1} \tag{5-153}$$

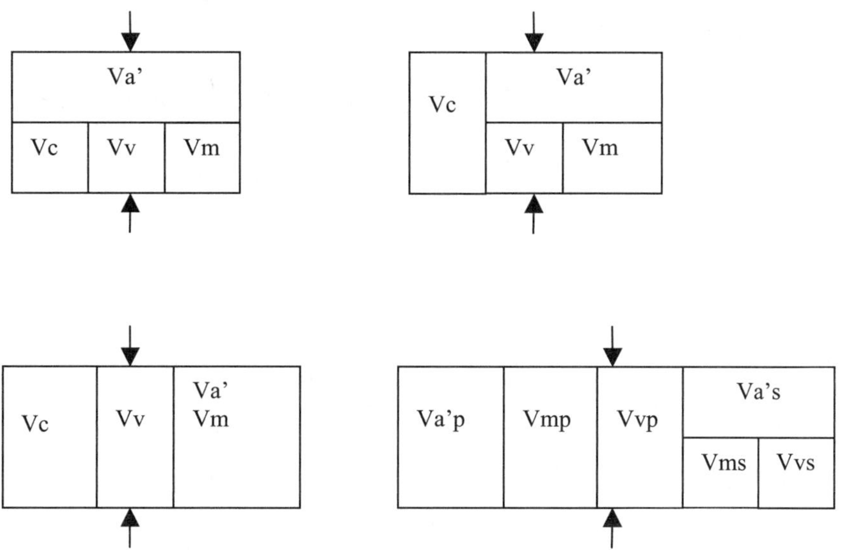

Figure 5.5   Schematic representation of four alternate versions of modified Hirsch model (Christensen, 2003).

(c) Dispersed version

$$E_c = VcEa + (1 - Vc)^{1-1/m}\left(Va'Ea^m + VmEm^m\right)^{1/m} \qquad (5\text{-}154)$$

Einstein equation by Shashidhar and his associates:

$$Em = \left[\frac{1 + AVf'}{1 - (1 + CVf')Vf'}\right]Eb \qquad (5\text{-}155)$$

where $A = K_E - 1$

$\qquad K_E$ = generalized Einstein coefficient

$\qquad Vf'$ = volume fraction filler in mastic = Vf/Vm

$\qquad C = (1 - Vf'_{max})/Vf'^2_{max}$

Vf'_{max} = maximum volume fraction of filler in mastic

Generalized law of mixtures:

$$Em = \left(Vf'Ea^n + Vb'Eb^n\right)^{1/n} \qquad (5\text{-}156)$$

where Vb' = volume fraction binder in mastic = Vb/Vm

$\qquad Eb$ = modulus of binder

$\qquad n$ = exponent with values from -1 to $+1$

(d) Alternative version

$$Ec = Va'pEa + VmpEm + \left(Va's + Vms + Vvs\right)^2 \left[\frac{Va's}{Ea} + \frac{(Vms + Vvs)^2}{VmsEm}\right]^{-1}$$

$$Ec = Pc(Va'Ea + VmEm) + \left(1 - Pc\right)\left[\frac{Va'}{Ea} + \frac{(Vm + Vv)^2}{VmEm}\right]^{-1} \tag{5-157}$$

$$Pc = \frac{\left(P_0 + \dfrac{VFM \times Em}{VMA'}\right)^{P_1}}{P_2 + \left(\dfrac{VFM \times Em}{VMA'}\right)^{P_1}}$$

Simple three-phase system of aggregate, asphalt binder, and air voids:

$$Ec = Pc(VaEa + VbEb) + \left(1 - Pc\right)\left[\frac{Va}{Ea} + \frac{(Vb + Vv)^2}{VbEb}\right]^{-1} \tag{5-158}$$

Incorporate film thickness:

$$Ec = Pc(VaEa + VbEb') + \left(1 - Pc\right)\left[\frac{Va}{Ea} + \frac{(Vb + Vv)^2}{VbEb'}\right]^{-1} \tag{5-159}$$

Effective binder modulus:

$$Eb' = \frac{t_F EbE_g}{(t_F - t_T)E_g + t_T Eb} \tag{5-160}$$

Comparison between model predictions and experimental results indicates that the parallel version of model AC properties is better.

Luo and Lytton (2009) made use of the Hashin and Shtrikman lower bound and extended it into general coupled self-consistent formulations.

$$\sum_{i=1}^{n} c_i \frac{K_i - K^*}{3K_i + G^*} = 0 \tag{5-161}$$

$$\sum_{i=1}^{n} \frac{c_i(G_i - G^*)}{G^* \dfrac{9K^* + 8G^*}{K^* + 2G^*} + 6G_i} = 0 \tag{5-162}$$

For the three-constituent asphalt mixture, the above formulations become:

$$\frac{c_1(K_1 - K^*)}{3K_1 + G^*} + \frac{c_2(K_2 - K^*)}{3K_2 + G^*} + \frac{c_3(K_3 - K^*)}{3K_3 + G^*} = 0 \tag{5-163}$$

$$\frac{c_1(G_1 - G^*)}{G^* \dfrac{9K^* + 8G^*}{K^* + 2G^*} + 6G_1} + \frac{c_2(G_2 - G^*)}{G^* \dfrac{9K^* + 8G^*}{K^* + 2G^*} + 6G_2} + \frac{c_3(G_3 - G^*)}{G^* \dfrac{9K^* + 8G^*}{K^* + 2G^*} + 6G_3} = 0 \tag{5-164}$$

K_1 = bulk modulus of aggregate

K_2 = bulk modulus of asphalt binder

K_3 = bulk modulus of air

K^* = bulk modulus of AC

G_1 = bulk modulus of aggregate

G_2 = bulk modulus of asphalt binder

G_3 = shear modulus of air

G^* = shear modulus of AC

c_1 = volumetric concentration of aggregate

c_2 = volumetric concentration of asphalt binder

c_3 = volumetric concentration of air

Luo compared the model predictions with experimental observations. The results indicated that the model was reasonably accurate.

Shu and Huang (2008) applied the concept of equivalent medium (Eshelby, 1957; Christensen and Lo, 1979) to first obtain the effective elastic modulus of a two-constituent (asphalt binder and aggregates) solid. Through the Laplace transform they obtained the modulus in the transformed domain:

$$E_0(s,a) = \frac{E_1(s)(1-2v_0)(1-n)}{x_1 - \dfrac{9E_2 n(1-v_1)^2}{4(1-2v_2)(1-n)E_1(s)+4E_2 x_2}} \tag{5-165}$$

The equivalent complex modulus at a certain frequency for a fixed size of particle (graduation) is then obtained.

$$E_0(\omega,a) = \frac{E_1^*(\omega)(1-2v_0)(1-n)}{x_1 - \dfrac{9E_2 n(1-v_1)^2}{4(1-2v_2)(1-n)E_1^*(\omega)+4E_2 x_2}} \tag{5-166}$$

The complex modulus of HMA mixtures that includes gradation information $(P(a_i) - P(a_i + 1))$ is expressed as:

$$E_0^*(\omega) = \frac{1}{2}\sum_{i=1}^{N+1}[E_0^*(\omega,a_i)+E_0^*(\omega,a_{i+1})][P(a_i)-P(a_{i+1})] \tag{5-167}$$

Following the Prony series expression of the relaxation modulus, they obtained both the real and imaginary part of the modulus and the phase angle:

$$E(t) = E_e + \sum_{i=1}^{m} E_i e^{-(1/\rho_i)} \tag{5-168}$$

Using a two-step procedure with Step 1 to obtain the effective modulus of asphalt binder and aggregates, and Step 2 to treat air voids as a medium of zero modulus, they then obtained the effective modulus of mixtures in the following equation, where air void distribution has been considered:

$$E_0(a) = \frac{2E_1(1-n)(1-2v_0)}{n(1+v_1)+2(1-2v_1)} \tag{5-169}$$

Lytton in Song et al., 2005, following the Energy Equivalency theorem, developed equations to relate the modulus in a damaged material to the modulus of an undamaged material. The formations are related to whether the state is stress-controlled or strain-controlled. For the strain controlled state, the following equations are obtained.

Cohesive:

$$\frac{G'}{G} = \left[1 - 2\pi^2\left(\frac{m}{A}\right)\frac{\bar{r}^3}{\bar{t}}\left(1 - \frac{G_f \Delta G_f^c}{\pi \tau^2 \bar{r}}\right)\right] = \left[1 - 2\pi\bar{\xi}\frac{\bar{r}}{\bar{t}}\left(1 - \frac{G_f \Delta G_f^c}{\pi \tau^2 \bar{r}}\right)\right] \tag{5-170}$$

Adhesive:

$$\frac{G'}{G} = \left[1 - \pi^2\left(\frac{m}{A}\right)\frac{\bar{r}^3}{\bar{t}}\left[\left(1 + \frac{G_f}{G_s}\right) - \frac{4G_f \Delta G^a}{\pi \tau^2 \bar{r}}\right]\right] = \left[1 - \pi\bar{\xi}\frac{\bar{r}}{\bar{t}}\left[\left(1 + \frac{G_f}{G_s}\right) - \frac{4G_f \Delta G^a}{\pi \tau^2 \bar{r}}\right]\right] \tag{5-171}$$

Other Studies

Abbas, et al. (2000) compared micromechanics predictions using the Hashin and Shtrikman lower bound, the GSCS, and the inverse rule of mixture with their DEM simulations. Li, et al. (2005) compared micromechanics model predictions using the Hirsch model, the Hashin Composite Sphere model, and the Christensen and Lo model with experimental observations. Chen and Peng (1998) applied the Hirsch model to study the mastic modulus. Druta, et al. (2007a, b) also applied the model to predict the modulus of mastic and mixture.

5.5   Doublet Mechanics

There have been significant developments in the area of micromechanics of granular materials in the last two decades. Two different approaches have been generally followed in micromechanics. One follows the routine of describing particle contact and configuration (Christoffersen et al., 1981; Chang and Ma, 1991); the other, doublet mechanics (Granik and Ferrari, 1993; Ferrari et al., 1997) assumes a Bravis Lattice as the microstructure and considers the doublet deformation and particle interaction by taking different orders of the deformation field expressed as a series of the coordinates. When the particle size approaches zero, it becomes a continuum. Both methods could explain the phenomenon of tensile stress induced under compressive forces. The following discussion presents a doublet mechanics explanation.

Granik et al. (1993) and Ferrari et al. (1997) assumed a face-centered cubic packing of spheres as the microstructure model of granular materials (with bond among the contacts) to solve the classical Flamant problem: a semi-infinite plate subjected to a point force normal to the boundary (Todhunter and Pearson, 1960). The analytical solution by Granik and Ferrari (based on the microstructure model illustrated in Figure 5.6)

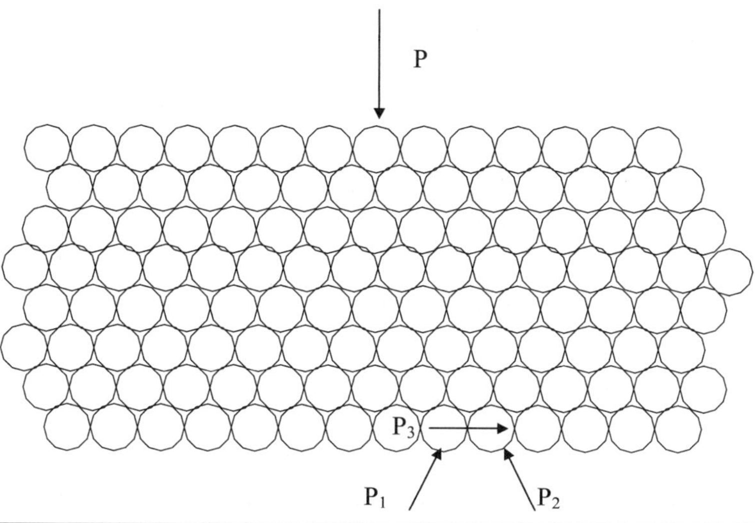

FIGURE 5.6 Microstress distribution of bonded granular material in semi-space.

showed that the macro stress distributions were the same as those of the Flamant solution, based on the Linear Elastic Continuum model, as presented in Equations 5-172*a*, *b*, and *c*:

$$\sigma_x = -2Px^2y \; [\pi(x^2 + y^2)^2]^{-1} \tag{5-172a}$$

$$\sigma_{xy} = 2Pxy^2[\pi(x^2 + y^2)^2]^{-1} \tag{5-172b}$$

$$\sigma_y = -2Py^3[3\pi(x^2 + y^2)^2]^{-1} \tag{5-172c}$$

However, the micro-stress of P_1, P_2, P_3, illustrated in Figure 5-6, represented in Equations 5-173*a*, *b*, and *c*, are significantly different:

$$p_1 = -4Py^2(\sqrt{3}x + y)[3\pi(x^2 + y^2)^2]^{-1} \tag{5-173a}$$

$$p_2 = 4Py^2(\sqrt{3}x - y)[3\pi(x^2 + y^2)^2]^{-1} \tag{5-173b}$$

$$p_3 = -2Py(3x^2 - y^2)[3\pi(x^2 + y^2)^2]^{-1} \tag{5-173c}$$

Where p_1, p_2, and p_3 are the micro-stresses along the directions of valences where particles are in contact (Figure 5.6). This solution shows that the micro-stress of granular materials can be in tension in certain zones under the compressive force (positive indicates compression). For example, p_3 would be in tension if $|y| > \sqrt{3}|x|$. However, the macro-stresses are the same as those in a continuum (Equations 5-172*a*, *b*, *c*). It was also shown (Granik, 1993) that micro-stresses can be altered with microstructure (e.g., the orientation of valence), while the macro-stresses remain the same as in Equations 5-172*a*, *b*, and *c*.

For a distributed load ranging from $-a$ to a, the micro-stresses at location (x, y) can be obtained by integration as in Equations 5-174*a, b,* and *c.* Figure 5.7 graphically presents the stress distributions.

$$F_1 = \int_{-a}^{a} p_1 du = \frac{2qy}{3\pi}\left[\frac{\sqrt{3}y - x - a}{(x+a)^2 + y^2} - \frac{\sqrt{3}y - x + a}{(x-a)^2 + y^2} - \frac{1}{y}arctg\left(\frac{x+a}{y}\right) + \frac{1}{y}arctg\left(\frac{x-a}{y}\right)\right] \quad (5\text{-}174a)$$

$$F_2 = \int_{-a}^{a} p_2 du = \frac{2qy}{3\pi}\left[\frac{\sqrt{3}y + x + a}{(x+a)^2 + y^2} - \frac{\sqrt{3}y + x - a}{(x-a)^2 + y^2} + \frac{1}{y}arctg\left(\frac{x+a}{y}\right) - \frac{1}{y}arctg\left(\frac{x-a}{y}\right)\right] \quad (5\text{-}174b)$$

$$F_3 = \int_{-a}^{a} p_3 du = \frac{2qy}{3\pi}\left[\frac{2x - 2a}{(x+a)^2 + y^2} - \frac{2x - 2a}{(x-a)^2 + y^2} - \frac{1}{y}arctg\left(\frac{x+a}{y}\right) + \frac{1}{y}arctg\left(\frac{x-a}{y}\right)\right] \quad (5\text{-}174c)$$

The micro-stresses are related to the configurations of the particles. This indicates that the microstructure of the material must be considered in the micromechanics analysis.

FIGURE 5.7 Stress distribution under a strip of loading.

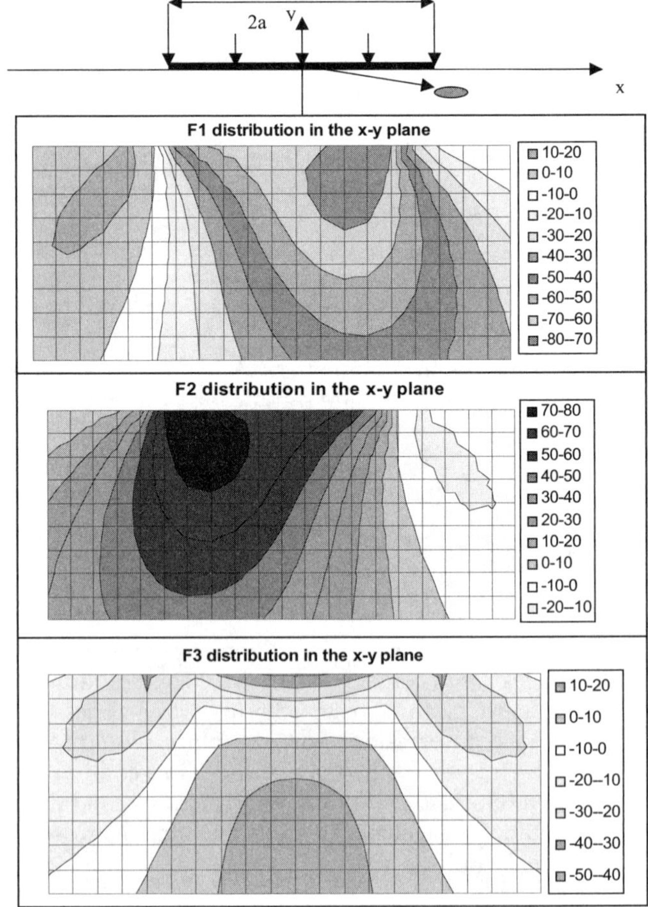

5.6   Micromechanics Applications for Pavement Analysis

5.6.1   Top-down Cracking (Wang et al, 2003)

Top-down cracking has been recognized as a severe cracking type that significantly reduces the quality service life of pavements (Gerritsen et al., 1987; Dauzats and Rampal, 1987). However, the causes of top-down cracking are very complicated. As a result, different causes, including tire-pavement contacting properties, thermal gradient, type of mix, and construction factors have been identified (Svasdisant et al., 2002; Matsuno and Nishizawa, 1992; Uhlmeyer et al., 2000). So far, research in this field has mainly focused on stress analysis using the FEM to identify causes that induce tensile stress at the surface (Myers et al., 1998, 1999; Roque et al., 2000; Myers et al., 2001, 2002). Material inhomogeneity was qualitatively recognized as a cause; however, no analytical and experimental methods have been developed to investigate how material inhomogeneity would cause tensile stress and how significant the tensile stress would be. In addition, this type of top-down cracking is currently assumed to be Type I cracking. Whether there exists Type II or III cracking has not been verified. Due to the lack of sufficient experimental investigations into the causes of top-down cracking, and especially from the point of view of materials structure, no effective preventive measures can be developed.

Investigation of top-down cracking causes in the view of materials properties is important. For example, why are fewer top-down cracking cases reported in cement concrete even when loading and tire configuration are the same? A most intuitive academic guess would be straightforward: the binding strength of cement paste is much stronger than that of asphalt binder. Nevertheless, this straightforward answer involves the micromechanics and microstructures of the material. From a materials standpoint, AC is different from cement concrete in its binder. The weak and thermal sensitive binder in asphalt concrete causes most of the distresses such as rutting, fatigue cracking, low-temperature cracking, and moisture damage. Most of the current research uses a continuum mechanics approach. This approach cannot account for the material structure and is not convenient for use in identifying the internal causes that lead to top-down cracking.

Figure 5.8a presents a simple illustration for understanding the top-down cracking from a materials standpoint. The spheres represent aggregate particles. It is apparent that under vertical loading, the binder between the two bottom particles may be subjected to tensile stress. The maximum tensile force between the two particles can be calculated as large as:

$$T = \sqrt{3}P / 6 \tag{5-175}$$

However, due to friction and restraint from the surrounding medium, this maximum force may not be reached.

Figure 5.8b presents another algorithm that will introduce tensile strains (tensile stress) by shearing the materials, or the so-called dilatancy mechanism. In Figure 5.8b, a densely packed granular system is subjected to a tensile force. When Particle C moves away from Particle A, the mastic between the two particles will be in tension, and produces volume increase or dilatancy.

These two simple cases have important implications in better understanding the causes of the top-down cracking. The aggregate particle skeleton structure and strength

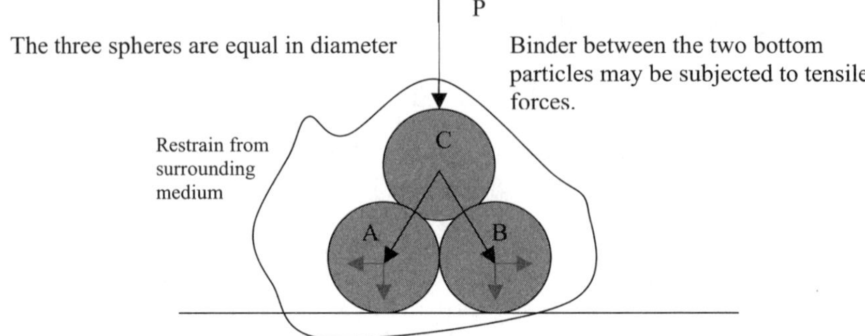

a) Tensile Stress Induced by a Compressive Loading

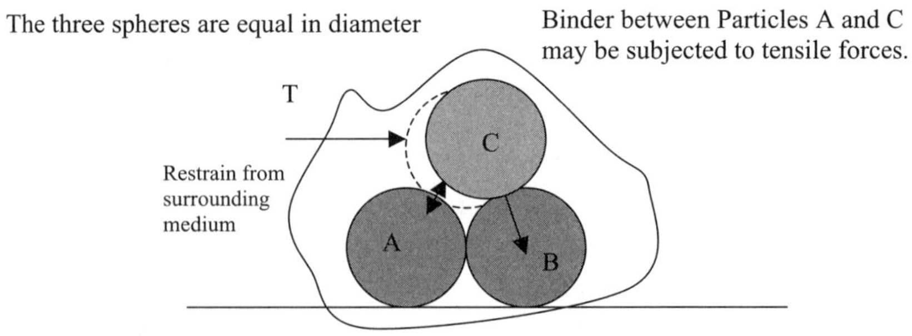

b) Tensile Stress Induced by a Shear Loading

FIGURE 5.8   Illustration of potential tensile stress induction by compressive or shear loading.

of the mastics are two important factors that may affect the location where the top-down cracking initiates and the number of cyclic loads needed to initiate and propagate the cracking.

Figure 5.9 presents an illustration of the location of the maximum tensile forces, where the top-down cracking most likely takes place. Due to the symmetry of loading, there would be no tensile forces between Particles 2 and 3, 3 and 4, and 4 and 5. The

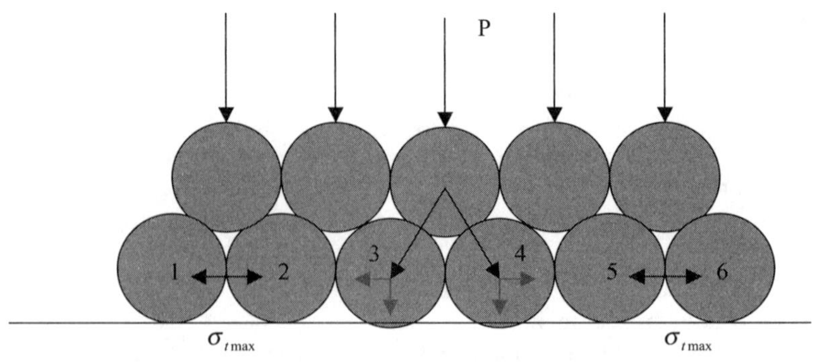

FIGURE 5.9   Illustration of locations where large tensile stress exists.

maximum tensile forces may take place between Particles 1 and 2 and 5 and 6. In other words, the top-down cracking may likely occur around the edges of the tire. This is coincident with the conclusions drawn in Myers,- et al., (2001). However, the interpretation methods are so different: one uses a linear elastic fracture mechanics approach; the other uses a micromechanics approach. The advantage of the micromechanics approach lies in its capability to naturally incorporate the materials' microstructure and the relative stiffness and strength of the constituents.

5.6.2 Doublet Mechanics Interpretation

From these formulations, it can be deduced that tensile micro-stress will generate in certain zones, representing the causes of top-down cracking. Nevertheless, the analytical solution assumes a periodic materials structure, which is quite different from the real structure of the materials. This microstructure difference implies that the real stress distribution may be more complicated than that of theoretical predictions.

5.6.3 DEM Interpretation

To account for the more complicated structure of the material, the DEM, initially developed by Cundall (1979) to simulate the properties of granular soil, was applied to analyze the stress distribution. This method has also been applied to simulate the behavior of AC in recent years (Chang and Meegoda, 1997, 1999). By allowing the particles to be bonded together, this method can be used to simulate actual behavior of AC.

Figure 5.10 presents the stress distribution under a linear compressive load simulating traffic loading. A random skeleton structure of aggregates is assumed. A parallel-bond is used among the particles to represent the binding effects from the binder. Figure 5.10 presents the force distribution in the material, where the red line represents tensile force distribution; and the black line represents the compressive force distribution.

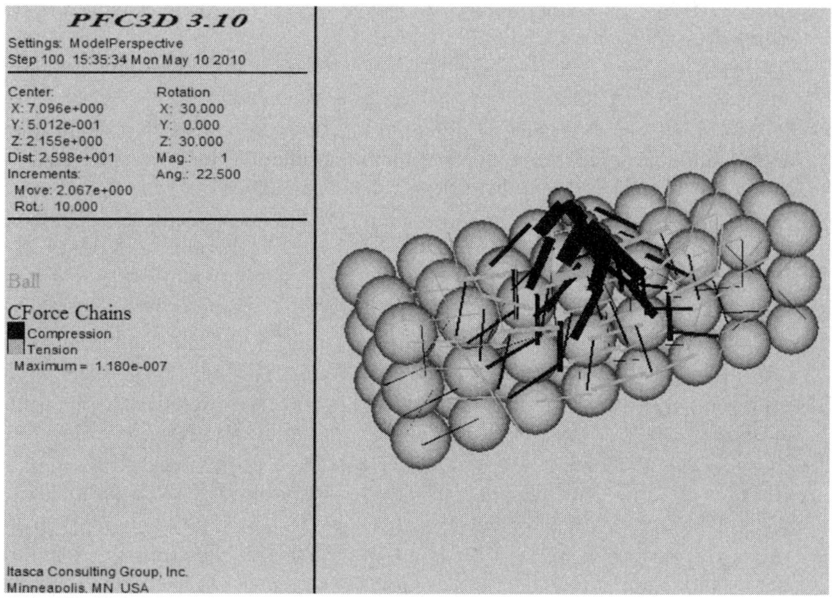

FIGURE 5.10   Stress distribution under a line load.

The thickness of the lines shows the magnitude of the forces. Clearly, tensile forces are distributed in a very complicated pattern between the contacts.

5.6.4 Anisotropy

From Section 5.3.9, it is clear that even if the two constituents are isotropic, the thus formed mixture is no longer isotropic. The anisotropic property may have significant implications to pavement analysis and design. Chapter 12 will focus on the anisotropic properties of AC.

Suggested Readings

Sections 5.2 and 5.3 are based on the teaching notes of the author. These notes are prepared following the presentations from two excellent books by Nemat-Nasser and Hori (1999) and Qu and Cherkaoui (2006). If readers need more backgrounds, please read these books. For convenience and connections, the symbols adopted in these two sections are consistent with those used in these books. The author would like to take this opportunity to express his appreciations of Dr. Qu for his inspiring lectures that attracted him into the research and study of micromechanics.

References

Abbas, A., Masad, E., Papagiannakis, T. and Shenoy, A. (2000). Modeling asphalt mastic stiffness using discrete element analysis and micromechanics-based models. *The International Journal of Pavement engineering*, Vol.6, No.2, pp.137–146.

Arduino, P. (1996). *Multiphase description of deforming porous media by the finite element method.* Ph.D. Thesis, Georgia Institute of Technology, Atlanta.

Bear, J. (1972). Dynamics of Fluids in Porous Media. *Dover*, New York.

Bowen, R.M. (1976). Theory of mixtures. In Eringen, A.C. (Ed.). *Continuum Physics*, Vol.3, Academic Press, New York, USA.

Braz, D., Da Motta, L.M.G. and Lopes, R.T. (1999). Computed tomography in the fatigue test analysis of an asphaltic mixture. *Applied Radiation and Isotopes*, Vol.50, No.4, pp.661–671.

Buttlar, W.G. and Dave, E.V. (2005) A micromechanics-based approach for determining presence and amount of recycled asphalt pavement material in asphalt concrete. *Journal of the Association of Asphalt Paving Technologist*, Vol.74, pp.829–884.

Chang, C.S. and Ma, L. (1991). A micromechanical-based micropolar theory for deformation of granular solids. *International Journal of Solids and Structures*, Vol.28, No.1, pp.67–86.

Chang, G.K. and Meegoda, J.N. (1997). Micromechanical simulation of hot mix asphalt. *Journal of Engineering Mechanics*, Vol.123, No.5, pp.495–503.

Chang, G.K. and Meegoda, J.N. (1999). Micromechanical model for temperature effects of hot-mix asphalt concrete. *Transportation Research Record*, No.1687. pp.95–103.

Chen, J.S. and Peng, C.H. (1998). Analysis of tensile failure properties of asphalt-mineral filler mastics. *Journal of Materials in Civil Engineering*, Vol.10, No.4, pp.256–262.

Christensen, D.W., Pellinen, T. and Bonaquist, R.F. (2003). Hirsch model for estimating the modulus of asphalt concrete. *Journal of the Association of Asphalt Paving Technologists*, Vol.72, pp.97–121.

Christensen, R.M. and Lo, K.H. (1979). Solution for effective shear properties in three phase sphere and cylinder models. *Journal of the Mechanics and Physics of Solids*, Vol.27, pp.315–330.

Christoffersen, J., Mehrabadi, M.M. and Nemat-Nasser, S. (1981). A micromechanical description of granular material behavior. *Journal of Applied Mechanics*, Vol.48, No.2, pp.339–344.

Coussy, O. (1995). Mechanics of Porous Continua. *John Wiley and Sons,* New York.

Cundall, P.A. and Strack. O.D.L. (1979). A discrete numerical model for granular assemblies. *Geotechnique*, Vol.29, No.1, pp.47–65.

Dauzats, M. and Rampal. A. (1987). Mechanism of surface cracking in wearing courses. *Proceeding of 6th International Conference on the Structural Design of Asphalt Pavements*, Ann Arbor, Michigan, pp.232–247.

DeHoff, R.T. and Rhines, F.R. (1968). Quantitative Microscopy. *McGraw-Hill*, New York.

Desruses, J., Chambon, R., Mokni, M. and Mazerolle, F. (1996). Void ratio evolution inside shear bands in triaxial sand specimen studied by computed tomography. *Geotechnique*, Vol.46, No.3, pp.529–546.

Dobran, F. (1985). Theory of multiphase mixtures—A thermodynamic formulation. *International Journal of Multiphase Flow*, Vol.11, No.1, pp.1–30.

Druheller, D.S. and Bedford, A. (1980). A thermodynamical theory of reacting immiscible mixtures. *Archive for Rational Mechanics and Analysis*, Vol.73, No.3, pp.257–284.

Druta, C., Wang, L.B., Voyiadjis, G.Z. and Abadie, C. (2007a). Estimation of the stiffness of asphalt mastics using hirsch model. *Geotechnical Special Publication*, No.176, pp.33–44.

Druta, C., Wang, L.B., Voyiadjis, G.Z. and Abadie, C. (2007b). Estimation of the stiffness of asphalt mixture using a micromechanics approach. *Geotechnical Special Publication*, No.182, pp.155–163.

Eringen, A.C. and Ingram, J.D. (1967). A continuum theory of chemically reacting media-II Constitutive equations of reacting fluid mixtures. *International Journal of Engineering Science*, Vol.5, No.4, pp.289–322.

Eshelby, J.D. (1957). The determination of the elastic field of an ellipsoidal inclusion and related problems. *Proceedings of the Royal Society of London. Series A, Mathematical and Physical Sciences*, Vol.241, No.1226, pp.376–396.

Ferrari, M., Granik, V.T., Imam, A. and Nadeau, J.C. (Eds.) (1997). Advances in Doublet Mechanics. *Springer*, New York.

Gerritsen, A.H., Gurp, C.A.P.M., Heide, J.P.J., Molenaar, A.A.A. and Pronk, A.C. (1987). Prediction and prevention of surface cracking in asphaltic pavements. *Proceeding of 6th International Conference on the Structural Design of Asphalt Pavements*, Ann Arbor, Vol.1, pp.378–391.

Goodman, M.A. and Cowin, S.C. (1971). Two problems in the gravity flow of granular materials. *Journal of Fluid Mechanics*, Vol.45, pp.321–339.

Goodman, M.A. and Cowin, S.C. (1972). A continuum theory for granular materials. *Archive for Rational Mechanics Analysis*, Vol.44, No.4, pp.249–266.

Granik, V.T. and Ferrari, M. (1993). Microstructural mechanics of granular media. *Mechanics of Materials*, Vol.15, No.4, pp.301–332.

Green, A.E. and Naghdi, P.M. (1967). A theory of mixtures. *Archive for Rational Mechanics Analysis*, Vol.24, No.4, pp.243–263.

Hansen, A.C. (1989). Reexamining some basic definitions of modern mixture theory. *International Journal of Engineering Science*, Vol.27, No.12, pp.1531–1544.

Hashin, Z. and Shtrikman, S. (1963). A variational approach to the theory of the elastic behaviour of multiphase materials. *Journal of the Mechanics and Physics of Solids*, Vol.11, No.2, pp.127–140.

Hirsch, T.J. (1962). Modulus of Elasticity of Concrete Affected by Elastic Moduli of Cement Paste Matrix and Aggregate, *Journal Proceedings of the American Concrete Institute*, Vol.59, No.3, pp.427–452.

Kanatani, K.I. (1979). A micropolar continuum theory for the flow of granular materials. *International Journal of Engineering Science*, Vol.17, No.4, pp.419–432.

Krishnan, J.M. and Rao, C.L. (2000). Mechanics of air voids reduction of asphalt concrete using mixture theory. *International Journal of Engineering Science*, Vol.38, No.12, pp.1331–1354.

Krishnan, J.M. and Rao, C.L. (2001). Permeability and bleeding of asphalt concrete using mixture theory. *International Journal of Engineering Science*, Vol.39, No.6, pp.611–627.

Li, Y.Q. and Metcalf, J.B. (2005). Two-step approach to prediction of asphalt concrete modulus from two-phase micromechanical models. *Journal of Materials in Civil Engineering*, Vol.17, No.4, pp.407–415.

Luo, R. and Lytton, R.L. (2009). Self-consistent micromechanics models of an asphalt mixture. *Transportation Research Board 88th Annual Meeting*, Washington, D.C. (DVD).

Masad, E., Jandhyala, V.K., Dasgupta, N., Somadevan, N. and Shashidhar, N. (2002). Characterization of air void distribution in asphalt mixes using X-ray CT. *Journal of Materials in Civil Engineering*, Vol.14, No.2, pp.122–129.

Matsuno, S. and Nishizawa, T. (1992). Mechanism of Longitudinal Surface Cracking in Asphalt Pavement. *Proc. 7th International Conference on the Structural Design of Asphalt Pavements*, Ann Arbor, Vol.2, pp.277–291.

Morin, T. and Tanaka, K. (1973). Average stress in matrix and average elastic energy of materials with misfitting inclusions. *Acta Metallurgica*, Vol.21, No.5, pp.571–574.

Muller, I. (1968). A thermodynamic theory of mixtures of fluids. *Archive for Rational Mechanics and Analysis*, Vol.28, No.1, pp.1–39.

Myers, L., Roque, R. and Ruth, B.E. (1998). Mechanisms of surface-initiated longitudinal wheel path cracks in high-type bituminous pavements. *Journal of the Association of Asphalt Paving Technologists*, Vol.67, pp.401–432.

Myers, L., Roque, R., Ruth, B. and Drakos, C. (1999). Measurement of contact stresses for different truck tire types to evaluate their influence on near-surface cracking and rutting. *Transportation Research Record*, No.1655, pp.175–184.

Myers, L., Roque, R. and Birgisson, B. (2001). Propagation mechanisms for surface-initiated longitudinal wheel path cracks. *Transportation Research Record*, No.1778, pp.113–122.

Myers, L., Roque, R. and Birgisson, B. (2002). Evaluation of two- and three-dimensional finite element analysis of asphalt pavement structures. Accepted to International Journal of Pavement Engineering.

Nemat-Nasser, S. and Hori, M. (1999). Micromechanics: Overall Properties of Heterogeneous Materials. 2nd Edition. *Elsevier*, Amsterdam.

Passman, S.L. (1977). Mixtures of granular materials. *International Journal of Engineering Science*, Vol.15, pp.117–129.

Paul, B. (1960). Prediction of elastic constants of multiphase materials. *Transactions of the Metallurgical Society of AIME*, Vol.218, pp.36–41.

Qu, J.M. and Cherkaoui, M. (2006). Fundamentals of Micromechanics of Solids. *John Wiley and Sons*, New York.

Rajagopal, K.R. and Rao, L. (1996). Mechanics of Mixtures. *World Scientific*, Singapore.

Rogasik, H., Crawford, J.W., Wendroth, O., Young, I.M., Joschko, M. and Ritz, K. (1999). Discrimination of soil phases by dual energy X-ray tomography. *Soil Science Society of America Journal*, Vol.63, No.4, pp.741–751.

Roque, R., Myers, L.A. and Birgisson, B. (2000). Evaluating measured tire contact stresses to predict pavement response and performance. *Transportation Research Record*, No.1716, pp.73–81.

Shashidhar, N. (1999). X-ray tomography of asphalt concrete. *Transportation Research Board*, No.1681, pp.186–192.

Shell International Petroleum Company Limited (1978). *Shell Pavement Design Manual-Asphalt Pavements and Overlays for Road Traffic*. London.

Shi, B.M., Wu, Y., Chen, Z. and Inyang, J.H. (1999). Monitoring of internal failure evolution in soils using computerization X-ray tomography. *Engineering Geology*, Vol.54, No.3–4, pp.321–328.

Shu, X. and Huang, B.S. (2008). Dynamic modulus prediction of HMA mixtures based on the viscoelastic micromechnical model. *Journal of Materials in Civil Engineering*, Vol.20, No.8, pp.530–538.

Song, I.J., Little, D.N., Masad, E.A., Lytton, R. (2005). Comprehensive evaluation of damage in asphalt mastics using x-ray ct, continuum mechanics, and micromechanics. *Journal of the Association of Asphalt Paving Technologists*, Vol.74, pp.885-920

Svasdisant, T., Schorsch, M., Baladi, G.Y. (2002). Mechanistic analysis of top-down cracks in asphalt pavements. *Transportation Research Record*, No.1809, pp.126–136.

Todhunter, I. and Pearson, K. (1960). A History of the Theory of Elasticity and Strength of Materials. *Dover Publications, Inc.*, New York.

Truesdell, C. (1957). Sulle basi della termomeccanica. *Accademia Nazionale dei Lincei, Rendiconti della Classe di Scienze Fisiche, Matematiche e Naturali*, Vol.22, No.8, pp.33–88, 158–166.

Truesdell, C. (1969). Rational Thermodynamics. *McGraw Hill*, New York.

Twiss, R.J. and Eringen, A.C. (1971). Theory of mixtures for micromorphic materials-I: balance laws. *International Journal of Engineering Science*, Vol.9, No.10, pp.1019–1044.

Twiss, R.J. (1972) Theory of mixtures for micromorphic materials-II: elastic constitutive equations. *International Journal of Engineering Science*, Vol.10, pp.437–465.

Uhlmeyer, J.S., Willoughby, K., Pierce, L.M. and Mahoney, J.P. (2000). Top-down cracking in Washington State asphalt concrete wearing courses. *Transportation Research Record*, No.1730, pp.110–116.

Wang, L.B., Frost, J.D. and Shashidhar, N. (2001). Microstructure study of Westrack mixes from X-ray tomography images. *Transportation Research Record*, No.1767, pp.85–94.

Wang, L.B., Myers, L.A., Mohammad, L.N. and Fu, Y.R. (2003). A micromechanics study on top-down cracking. *Transportation Research Record* (TRR), No.1853, pp.121–133.

Wang, L.B., Wang, Y.P. Mohammad, L. and Harman, T. (2002). Voids distribution and performance of asphalt concrete. *International Journal of Pavement*, Vol.1, No.3, pp.22–33.

Wang, L.B., Wang, Y.P. and Mohammad, L.N. (2003). Application of mixture theory in the evaluation of mechanical properties of asphalt concrete. *ASCE Journal of Materials in Civil Engineering*, Vol.16, No.2, pp.167–174.

CHAPTER 6

Fundamentals of Phenomenological Models

The fundamental concepts of elasticity, viscosity, plasticity, viscoelasticity, and viscoplasticity are mathematical descriptions of the experimentally observed results on stress-strain or stress-strain rate relationships. There are research efforts and results on more fundamental mechanisms of the macroscopic deformation mechanism behind these phenomenological models; however, they are beyond the scope of this chapter. Continuum damage mechanics and fracture mechanics are a little different. They involve some physical models regarding micro-cracking and macro-cracking. Nevertheless, if possible, a concise description of the mechanism at lower scales (lower than the macroscopic scale) will be presented. The guiding philosophy for this chapter is to present minimum backgrounds of these topics so readers can be prepared for reading more complicated topics and understanding Chapter 7 on the typical models used in asphalt concrete.

Figure 6.1 presents a typical one-dimensional (1D) stress-strain relationship for a metal specimen in tension (well defined behavior). The proportional limit, elasticity limit (or yield stress), the hardening, the peak strength, and the residual or fracture strength are the thresholds where different mechanics theories can be utilized or more accurately utilized as different mechanisms control the deformation or strength characteristics in these zones.

Within the proportional limit (Zone I), the material can be considered linear elastic. The corresponding model may include linear elasticity. Between the proportional limit and the yield stress, or Zone II, non-linear elasticity such as hyperelasticity, may be a valid model. For most models, Zone I and Zone II are often modeled as either linear elasticity or non-linear elasticity for simplicity. Once the yield stress is reached, plastic deformation will be produced. The yield stress may increase due to strain hardening or work hardening. The small zone of perfect plasticity (where the yield stress is constant) might be very small and is often neglected. This zone (Zone III) is often modeled as plasticity. For plasticity, the components may include the initial yield stress or yielding criteria for complicated stress states; the hardening rule to describe how the magnitude and center of the yield surface increase or move with plastic strains or plastic work; the flow rule regarding the direction of the plastic deformation; and finally the peak strength or the ultimate yield surface. From the peak strength point on, the material

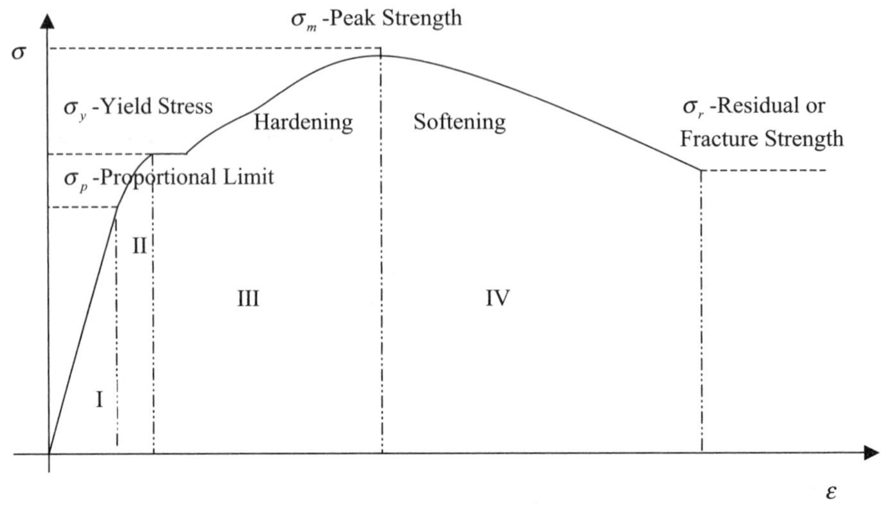

FIGURE 6.1   Typical stress-strain curve for metals and other materials.

starts softening. This zone (Zone IV) might be better described by continuum damage mechanics (CDM) or fracture mechanics.

Many materials may not follow the well defined behavior. In some cases certain zones may not exist. For different stress modes, the behavior of the material may become very complicated.

Elasticity comes from resilient deformation of perfect atom lattice or crystals. Weiner (1983) presents excellent explanation on elasticity and its atomistic scale reasons. Due to the existence and movements of such as a point, line, surface, and volume defects, and breaking up of bonds or fracture, plastic deformation, or irrecoverable deformation occurs. The traditional plasticity considers dislocations as the resources of plastic deformation. The limit forces exist that drive the dislocation to move. This explains the yield strength of metals. For AC, scientifically valid and verified microscopic mechanisms regarding yielding are needed.

6.1   Elasticity

6.1.1   Linear Elasticity

Generally speaking, the phenomenological modeling of the stress-strain relationship using mathematic functions can be represented as a Taylor series:

$$\varepsilon_{ij} = \varepsilon_{ij}^0 + \frac{\partial \varepsilon_{ij}}{\partial \sigma_{kl}}\sigma_{kl} + \frac{1}{2}\frac{\partial^2 \varepsilon_{ij}}{\partial \sigma_{kl}\partial \sigma_{pq}}\sigma_{kl}\sigma_{pq} + \ldots \tag{6-1}$$

If there exists a state where zero stresses corresponds to zero strains, $\varepsilon_{ij}^0 = 0$, which means no residual strains (plastic strains, eigenstrains), then the material is elastic. If the high order terms are neglected, then the following relationship holds:

$$\varepsilon_{ij} = \frac{\partial \varepsilon_{ij}}{\partial \sigma_{kl}}\sigma_{kl} \tag{6-2}$$

Denoting $C_{ijkl} = \dfrac{\partial \varepsilon_{ij}}{\partial \sigma_{kl}}$, the above equation can be represented as:

$$\varepsilon_{ij} = C_{ijkl}\sigma_{kl} \tag{6-3}$$

Generally, $C_{ijkl} = \dfrac{\partial \varepsilon_{ij}}{\partial \sigma_{kl}}$ is a function of σ_{kl}, which may include a constant term, a linear term, and a high order term. The simplest case is that it has only the constant term. In that case, it represents the linear elasticity in the most well established elasticity, plasticity, viscoelasticity, and viscoplasticity. In theory, C_{ijkl} should have $3^4 = 81$ constants. Due to the symmetry of the stress tensor and the strain tensor, $C_{ijkl} = C_{jikl} = C_{ijlk}$, therefor this will reduce the number of combinations to $6 \times 6 = 36$ (the nine combinations of ij or kl are reduced to six independent combinations and therefore will have 36 independent coefficients). If the strain energy exists, then $u = \dfrac{1}{2}\sigma_{ij}\varepsilon_{ij} = \dfrac{1}{2}\sigma_{ij}C_{ijkl}\sigma_{kl}$, which means if one replaces σ_{ij} with σ_{kl}, u will not change and therefore, $C_{ijkl} = C_{klij}$. This symmetry will reduce the inde-pendent constants to $\left(\dfrac{36-6}{2}+6\right) = 21$. This can be represented as a matrix C_{PQ}, where P and Q have six independent variations. So if C_{PQ} is symmetric, it has only 21 independent variables. Through the transformation of stress and strain tensors and making use of the symmetry properties of the tensors, it can be proved that there are only five independent parameters for the cross-anisotropic case and there will be only two independent parameters for the isotropic case. A very good textbook on elasticity is by Green and Zerna (1954).

For isotropic elasticity, there are several methods to express the stress-strain relationship in terms of the two independent parameters. Typical ones include Young's modulus (E), shear modulus (G), Lame's constant (λ), Poisson's ratio (v), and bulk modulus (K). Their interrelationship is presented in Table 6.1.

The Hooke's Law represented by these constants is respectively:

	Lame's Modulus, λ	Shear Modulus, G	Young's Modulus, E	Poisson's Ratio, v	Bulk Modulus, K
λ,G			$\dfrac{G(3\lambda+2G)}{\lambda+G}$	$\dfrac{\lambda}{2(\lambda+G)}$	$\dfrac{3\lambda+2G}{3}$
λ,v		$\dfrac{\lambda(1-2v)}{2v}$	$\dfrac{\lambda(1+v)(1-2v)}{v}$		$\dfrac{\lambda(1+v)}{3v}$
λ,K		$\dfrac{3(k-\lambda)}{2}$	$\dfrac{9K(K-\lambda)}{3K-\lambda}$	$\dfrac{\lambda}{3K-\lambda}$	
G,v	$\dfrac{2Gv}{1-2v}$		$2G(1+v)$		$\dfrac{2G(1+v)}{3(1-2v)}$
G,K	$\dfrac{3K-2G}{3}$		$\dfrac{9KG}{3K+G}$	$\dfrac{3K-2G}{2(3K+G)}$	
E,v	$\dfrac{vE}{(1+v)(1-2v)}$	$\dfrac{E}{2(1+v)}$			$\dfrac{E}{3(1-2v)}$
E,K	$\dfrac{3K(3K-E)}{(9K-E)}$	$\dfrac{3EK}{9K-E}$		$\dfrac{3K-E}{6K}$	

TABLE 6.1 Interrelationship among elastic constants.

$$\sigma_{ij} = \lambda\varepsilon_{kk}\delta_{ij} + 2\mu\varepsilon_{ij} \tag{6-4}$$

$$\varepsilon_{ij} = \frac{1}{E}[(1+v)\sigma_{ij} - v\sigma_{kk}\delta_{ij}] \tag{6-5}$$

$$\sigma_{kk} = 3K\varepsilon_{kk}, \; s_{ij} = 2\mu e_{ij} \tag{6-6}$$

6.1.2 Hyperelasticity (Bower, 2010)

Hyperelasticity is used to describe the constitutive relationship of elastic materials at large strains. It is often non-linear. The non-linearity may result in directly the large-strain definition and strain energy dependency on strains of higher order terms.

If $W(F)$ is the specific strain-energy, the first Piola-Kirchoff (PK) stress tensor is conjugated to the deformation gradient F and therefore:

$$P = \frac{\partial W}{\partial F} \; \text{or} \; P_{iK} = \frac{\partial W}{\partial F_{iK}} \tag{6-7}$$

If the strain energy is represented as a function of the Lagrangian Green strain E:

$$P = F \cdot \frac{\partial W}{\partial E} \; \text{or} \; P_{iK} = F_{iL}\frac{\partial W}{\partial E_{LK}} \tag{6-8}$$

In the case that the strain energy is a function of the right Cauchy-Green deformation tensor (C), then:

$$P = 2F \cdot \frac{\partial W}{\partial C} \; \text{or} \; P_{iK} = 2F_{iL}\frac{\partial W}{\partial C_{LK}} \tag{6-9}$$

Considering the definition of the second PK tensor S and its relationship with the first PK tensor, the following relationships will be valid:

$$S = F^{-1} \cdot \frac{\partial W}{\partial F} \; \text{or} \; S_{IJ} = F_{IK}^{-1}\frac{\partial W}{\partial F_{kJ}} \tag{6-10}$$

$$S = \frac{\partial W}{\partial E} \; \text{or} \; S_{IJ} = \frac{\partial W}{\partial E_{IJ}} \tag{6-11}$$

$$S = 2\frac{\partial W}{\partial C} \; \text{or} \; S_{IJ} = 2\frac{\partial W}{\partial C_{IJ}} \tag{6-12}$$

It can be proved (see also Chapter 1), for small strains, one can use Cauchy stress.

6.1.2.1 Hyperelastic Models

Saint Venant-Kirchhoff Model

The Saint Venant-Kirchhoff model is the simplest hyperelastic material model. It is an extension of the linear elastic material model to the nonlinear regime. This model has the form:

$$S = \lambda tr(E)I + 2\mu E \tag{6-13}$$

Where S is the second PK stress and E is the Lagrangian Green strain, and λ and μ are the Lame constants.

The specific strain-energy for the St. Venant-Kirchhoff model is:

$$W(E) = \frac{\lambda}{2}[tr(E)]^2 + \mu tr(E^2) \tag{6-14}$$

And the second PK stress can be derived from the relation:

$$S = \frac{\partial W}{\partial E} \tag{6-15}$$

Mooney-Rivlin Model

In continuum mechanics, a Mooney-Rivlin solid is a generalization of the Neo-Hookean solid model, where the strain energy W is a linear combination of two invariants of Finger tensor $\{B\}$:

$$W = C_1(\bar{I}_1 - 3) + C_2(\bar{I}_2 - 3) \tag{6-16}$$

Where C_1 and C_2 are constants; $\bar{I}_1$ and $\bar{I}_2$ and are the first and second invariant of the deviatoric component of the Finger tensor:

$$I_1 = \lambda_1^2 + \lambda_2^2 + \lambda_3^2$$
$$I_2 = \lambda_1^2\lambda_2^2 + \lambda_2^2\lambda_3^2 + \lambda_3^2\lambda_1^2 \tag{6-17}$$
$$I_3 = \lambda_1^2\lambda_2^2\lambda_3^2$$

Where $\lambda_1 = \sqrt{\omega_1}$, $\lambda_2 = \sqrt{\omega_2}$, $\lambda_3 = \sqrt{\omega_3}$, and ω_1, ω_2 and ω_3 and the eigenvalues of $\mathbf{B}$.

If $C_1 = \frac{1}{2}G$ (where G is the shear modulus) and $C_2 = 0$, we obtain a Neo-Hookean solid, a special case of a Mooney-Rivlin solid.

The stress tensor T depends upon the left Cauchy-Green deformation tensor and is related to B by the following equation:

$$T = -pI + 2C_1 B + 2C_2 B^{-1} \tag{6-18}$$

Uniaxial Tension

For the case of uniaxial elongation, true stress can be calculated as:

$$T_{11} = (2C_1 + \frac{2C_2}{\alpha_1})(\alpha_1^2 - \alpha_1^{-1}) \tag{6-19}$$

And the engineering stress can be calculated as:

$$T_{11eng} = (2C_1 + \frac{2C_2}{\alpha_1})(\alpha_1 - \alpha_1^{-2}) \tag{6-20}$$

The Mooney-Rivlin solid model usually fits experimental data better than the Neo-Hookean solid model does, but requires an additional empirical constant.

Ogden Material Model

In the Ogden Material model, the strain energy density is expressed in terms of the principal stretches $\lambda_j, j = 1,2,3$ as:

$$W(\lambda_1, \lambda_2, \lambda_3) = \sum_{p=1}^{N} \frac{\mu_p}{\alpha_p}(\lambda_1^{\alpha_p} + \lambda_2^{\alpha_p} + \lambda_3^{\alpha_p} - 3) \tag{6-21}$$

Where N, μ_p, and α_p are material constants. Under the assumption of incompressibility one can rewrite as:

$$W(\lambda_1, \lambda_2) = \sum_{p=1}^{N} \frac{\mu_p}{\alpha_p} (\lambda_1^{\alpha_p} + \lambda_2^{\alpha_p} + \lambda_1^{-\alpha_p} \lambda_2^{-\alpha_p} - 3) \tag{6-22}$$

In general, the shear modulus results from:

$$2\mu = \sum_{p=1}^{N} \mu_p \alpha_p \tag{6-23}$$

With $N = 3$ and by fitting the material parameters, the material behavior of rubbers can be described very accurately. For particular values of material constants, the Ogden model will reduce to either the Neo-Hookean solid ($N = 1$, $\alpha = 2$) or the Mooney-Rivlin material ($N = 2$, $\alpha_1 = 2$, $\alpha_2 = -2$).

Using the Ogden material model, the three principal values of the Cauchy stresses can now be computed as:

$$\sigma_\alpha = p + \lambda_\alpha \frac{\partial W}{\partial \lambda_\alpha} \tag{6-24}$$

Where use is made of $\sigma_\alpha = \lambda_\alpha P_\alpha$.

Uniaxial Tension
An incompressible material under uniaxial tension is now considered, with the stretch ratio given as $\lambda = \frac{l}{l_0}$. The principal stresses are given by:

$$\sigma_\alpha = p + \sum_{p=1}^{N} \mu_p \lambda_p^{\alpha_p} \tag{6-25}$$

The pressure p is determined from incompressibility and boundary condition $\sigma_2 = \sigma_3 = 0$, yielding:

$$\sigma_\alpha = \sum_{p=1}^{N} (\mu_p \lambda_p^{\alpha_p} - \mu_p \lambda_p^{-\frac{1}{2}\alpha_p}) \tag{6-26}$$

Yeoh Model
The original model proposed by Yeoh had a cubic form with only I_1 dependence and is applicable to purely incompressible materials. The strain energy density for this model is written as:

$$W = \sum_{i=1}^{3} C_i (I_1 - 3)^i \tag{6-27}$$

Where C_i are material constants. The quantity $2C_1$ can be interpreted as the initial shear modulus.

Today, a slightly more generalized version of the Yeoh model is used. This model includes n terms and is written as:

$$W = \sum_{i=1}^{n} C_i (I_1 - 3)^i \tag{6-28}$$

When $n = 1$ the Yeoh model reduces to the neo-Hookean model for incompressible materials.

The Cauchy stress for the incompressible Yeoh model is given by:

$$\sigma = -p\mathbf{1} + 2\frac{\partial W}{\partial I_1}B; \quad \frac{\partial W}{\partial I_1} = \sum_{i=1}^{n} iC_i(I_1 - 3)^{i-1} \tag{6-29}$$

Uniaxial Extension

For uniaxial extension in the n_1 direction, the principal stretches are $\lambda_1 = \lambda, \lambda_2 = \lambda_3$. From incompressibility $\lambda_1 \cdot \lambda_2 \cdot \lambda_3 = 1$. Hence, $\lambda_2^2 = \lambda_3^2 = 1/\lambda$. Therefore,

$$I_1 = \lambda_1^2 + \lambda_2^2 + \lambda_3^2 = \lambda^2 + \frac{2}{\lambda} \tag{6-30}$$

The left Cauchy-Green deformation tensor can then be expressed as:

$$B = \lambda^2 n_1 \otimes n_1 + \frac{1}{\lambda}(n_2 \otimes n_2 + n_3 \otimes n_3) \tag{6-31}$$

If the directions of the principal stretches are oriented with the coordinate basis vectors, it follows:

$$\sigma_{11} = -p + 2\lambda^2 \frac{\partial W}{\partial I_1}; \quad \sigma_{22} = -p + \frac{2}{\lambda}\frac{\partial W}{\partial I_1} = \sigma_{33} \tag{6-32}$$

Since $\sigma_{22} = \sigma_{33} = 0$, the resulting situation is:

$$p = \frac{2}{\lambda}\frac{\partial W}{\partial I_1} \tag{6-33}$$

Therefore,

$$\sigma_{11} = 2(\lambda^2 - \frac{1}{\lambda})\frac{\partial W}{\partial I_1} \tag{6-34}$$

The engineering strain is $\lambda - 1$. The engineering stress is:

$$T_{11} = \sigma_{11}/\lambda = 2(\lambda - \frac{1}{\lambda^2})\frac{\partial W}{\partial I_1} \tag{6-35}$$

Equibiaxial Extension

For equibiaxial extension in the n_1 and n_2 directions, the principal stretches are $\lambda_1 = \lambda_2 = \lambda$. From incompressibility $\lambda_1 \cdot \lambda_2 \cdot \lambda_3 = 1$. Hence, $\lambda_3 = 1/\lambda^2$. Therefore,

$$I_1 = \lambda_1^2 + \lambda_2^2 + \lambda_3^2 = 2\lambda^2 + \frac{1}{\lambda^4} \tag{6-36}$$

The left Cauchy-Green deformation tensor can then be expressed as:

$$B = \lambda^2 n_1 \otimes n_1 + \lambda^2 n_2 \otimes n_2 + \frac{1}{\lambda^4}n_3 \otimes n_3 \tag{6-37}$$

If the directions of the principal stretches are oriented with the coordinate basis vectors, then:

$$\sigma_{11} = -p + 2\lambda^2 \frac{\partial W}{\partial I_1} = \sigma_{22};\ \sigma_{33} = -p + \frac{2}{\lambda^4}\frac{\partial W}{\partial I_1} \tag{6-38}$$

Since $\sigma^{33} = 0$, the following results:

$$p = \frac{2}{\lambda^4}\frac{\partial W}{\partial I_1} \tag{6-39}$$

Therefore,

$$\sigma_{11} = 2(\lambda^2 - \frac{1}{\lambda^4})\frac{\partial W}{\partial I_1} = \sigma_{22} \tag{6-40}$$

The engineering strain is $\lambda-1$. The engineering stress is:

$$T_{11} = \sigma_{11}/\lambda = 2(\lambda - \frac{1}{\lambda^5})\frac{\partial W}{\partial I_1} = T_{22} \tag{6-41}$$

Planar Extension

Planar extension tests are carried out on thin specimens that are constrained from deforming in one direction. For planar extension in the n_1 directions with the n_3 direction constrained, the principal stretches are $\lambda_1 = \lambda$, $\lambda_3 = 1$. From incompressibility $\lambda_1 \cdot \lambda_2 \cdot \lambda_3 = 1$. Hence, $\lambda_2 = 1/\lambda$. Therefore,

$$I_1 = \lambda_1^2 + \lambda_2^2 + \lambda_3^2 = \lambda^2 + \frac{1}{\lambda^2} + 1 \tag{6-42}$$

The left Cauchy-Green deformation tensor can then be expressed as:

$$B = \lambda^2 n_1 \otimes n_1 + \frac{1}{\lambda^2} n_2 \otimes n_2 + n_3 \otimes n_3 \tag{6-43}$$

If the directions of the principal stretches are oriented with the coordinate basis vectors, it follows:

$$\sigma_{11} = -p + 2\lambda^2 \frac{\partial W}{\partial I_1},\ \sigma_{22} = -p + \frac{2}{\lambda^2}\frac{\partial W}{\partial I_1},\ \sigma_{33} = -p + 2\frac{\partial W}{\partial I_1} \tag{6-44}$$

Since $\sigma^{22} = 0$, then:

$$p = \frac{2}{\lambda^2}\frac{\partial W}{\partial I_1}$$

Therefore,

$$\sigma_{11} = 2(\lambda^2 - \frac{1}{\lambda^2})\frac{\partial W}{\partial I_1}$$

$$\sigma_{22} = 0$$

$$\sigma_{33} = 2(1 - \frac{1}{\lambda^2})\frac{\partial W}{\partial I_1}$$

The engineering stress is:

$$T_{11} = \sigma_{11} / \lambda = 2(\lambda - \frac{1}{\lambda^3})\frac{\partial W}{\partial I_1}$$

Neo-Hookean Solid Model

The model of neo-Hookean solid assumes that the extra stresses due to deformation are proportional to the Left Cauchy-Green deformation tensor:

$$T = -pI + GB \tag{6-45}$$

Where T = stress tensor, p = pressure, I = the unity tensor, G = a constant equal to shear modulus, B = the Left Cauchy-Green deformation tensor (sometimes referred as Finger tensor).

The strain energy for this model is:

$$W = \frac{1}{2}GI_B \tag{6-46}$$

Where W is potential energy and $I_B = tr(B)$ is the trace (or first invariant) of the Left Cauchy-Green deformation tensor B.

Usually the model is used for incompressible media.

Uniaxial Tension

Under uniaxial extension from the definition of the Left Cauchy-Green deformation tensor:

$$T_{11} = -p + G\alpha_1^2, \; T_{22} = T_{33} = -p + \frac{G}{\alpha_1} \tag{6-47}$$

Where α_1 is the stretch ratio in the 1 direction.

Assuming no traction on the sides, T22 = T33 = 0, so:

$$T_{11} = G(\alpha_1^2 - \alpha_1^{-1}) = G\frac{3\varepsilon + 3\varepsilon^2 + \varepsilon^3}{1 + \varepsilon}$$

Where $\varepsilon = \alpha_1$ is the strain.

The equation above is for the true stress (ratio of the elongation force to deformed cross-section), for engineering stress the equation is:

$$T_{11eng} = G(\alpha_1 - \alpha_1^{-2})$$

For small deformations $\varepsilon << 1$ we will have:

$$T_{11} = 3G\varepsilon$$

Thus, the equivalent Young's modulus of a neo-Hookean solid in uniaxial extension is 3G (shear modulus). For simple shear:

$$T_{11} = G\gamma$$

$$T_{11} - T_{22} = G\gamma^2 \tag{6-48}$$

$$T_{22} - T_{33} = 0$$

Where γ is shear deformation. Thus, neo-Hookean solid shows linear dependence of shear stresses upon shear deformation.

These models are not well utilized in AC. They may be well suited for modeling the large strain elastic behavior of modified asphalt binder.

6.1.3 Crystal Elasticity

Assume there is a collection of N atoms and the existence of potential energy representing the many-body interactions. The force acting on the ith atom due to the other N-1 atoms can be expressed as:

$$f_i = -\frac{\partial}{\partial r_i}\Phi(r_1,..,r_N)$$ (6-49)

Where $r_1,..,r_N$ represent the position vectors of N atoms and f_i is the force vector. The simplest case involves the interaction of two atoms. In this situation the popular Lenard-Jones potential may be used, although it may not be correct for most of the metals.

$$\Phi_{LJ}(r) = 4\varepsilon(\frac{\sigma^{12}}{r^{12}} - \frac{\sigma^6}{r^6})$$ (6-50)

Where ε, σ are material constants. One of the atoms is located at the origin of the frame. The force acting on the two atoms is the same and equal to:

$$f = -\frac{\partial}{\partial r}\Phi_{LJ}(r) = 24\varepsilon(2\frac{\sigma^{12}}{r^{13}} - \frac{\sigma^6}{r^7})$$ (6-51)

The equilibrium position is where the force is zero or the repulsive and attractive forces are equal to each other, $d_e = 2^{1/6}\sigma$. In other words, this corresponds to the zero-stress states in the material.

Now considering the force induced due to either pulling apart or contracting together by a distance dr from the equilibrium position, the force would be equal to df and df.

$$df = -24\varepsilon(26\frac{\sigma^{12}}{r^{14}} - 7\frac{\sigma^6}{r^8})\Big|_{r=d_e} dr$$ (6-52)

It may be stated that at equilibrium position, for very small changes (distance or strains), the force is linearly proportional to the distance. This is the basis for the linear elasticity.

For many-body problems, a simple case is that the overall potential is a sum of all the two body potentials.

$$\Phi_{LJ}(r_1,..,r_N) = \sum_{i=1}^{N}\sum_{j\neq i}^{N-1}\Phi_{LJ}(r_i,r_j)$$ (6-53)

6.2   Plasticity (Bower, 2010)

6.2.1 General Concept

From Figure 6.1, one can see the effect of yielding for the 1D case. When the total strain increases, the material cannot take up additional load or stress (or some more loads for the hardening materials). One can no longer use the elastic constitutive model to calculate the stress. However, as the unloading and reloading (up to the stress where unloading starts) are elastic, one can calculate the stress using the following equation:

$$\sigma = E\varepsilon^e = E(\varepsilon - \varepsilon^p) \tag{6-54}$$

Where σ is the stress, ε is the total strain, while ε^e and ε^p are the elastic and plastic strains respectively.

Clearly from Figure 6.1, to describe the plasticity behavior, one needs to determine the initial yield stress (could be zero) where yielding starts, the magnitude of the plastic strain and its directions, and the increase of the yielding stress with plastic strain or plastic work.

The uniaxial strain decomposition $\varepsilon = \varepsilon^e + \varepsilon^p$ can be extended to the 3D cases $\varepsilon_{ij} = \varepsilon^e_{ij} + \varepsilon^p_{ij}$ and could be in an incremental format. The incremental format is more useful in that the plastic strains are usually quite non-linear with stresses.

$$d\varepsilon_{ij} = d\varepsilon^e_{ij} + d\varepsilon^p_{ij} \tag{6-55}$$

If it is linear elastic, the following equation will calculate the incremental stress.

$$d\sigma_{ij} = C_{ijkl}d\varepsilon^e_{kl} \tag{6-56}$$

6.2.2 Yielding and Hardening

For 1-D cases, the initial yield stress can be conveniently expressed as $\sigma = \sigma_Y$. For 3D problems, it is much more complicated. In other words, one may have to express the yielding as a yielding condition such as the von Mises condition, the Tresca condition, and the Druck Pluger condition. For geomaterials or granular materials, the yielding stress is very much affected by the effective confining stress I, the larger the I, the larger the yielding stress.

The yielding conditions can then be generalized into a yielding function (including a hardening component).

For example, $f = f(\sigma_{ij}, \overline{\varepsilon}^p, \theta) = f(I_1, I_2, I_3, \overline{\varepsilon}^p, \theta)$.

For objectivity considerations, the yielding function is often represented as a function of the stress invariants, which means that the function will not change during co-ordinate transformation. Overall, the typical criteria can be classified into three types: maximum tensile/compression stress, maximum shear stress, and shear energy. More complicated yielding criteria may include the anisotropic characteristics (Hill, 1948; Gotoh, 1977a,b).

6.2.2.1 General Tresca Yield Criterion

$$f(\sigma_{ij}, \overline{\varepsilon}^p) = \max\left\{|\sigma_1 - \sigma_2|, |\sigma_1 - \sigma_3|, |\sigma_2 - \sigma_3|\right\} - Y(\overline{\varepsilon}^p) = 0 \tag{6-57}$$

Where $Y(\overline{\varepsilon}^p)$ also carries a hardening component.

The effective plastic strain magnitude is defined as:

$$\overline{\varepsilon}^p = \sqrt{\frac{2}{3}\varepsilon^p_{ij}\varepsilon^p_{ij}} \qquad \overline{\varepsilon}^p = \int \sqrt{\frac{2}{3}d\varepsilon^p_{ij}d\varepsilon^p_{ij}} \tag{6-58}$$

The Tresca criterion actually states that the material will yield when the maximum shear stress reaches a certain value.

Perfectly plastic: $Y = \text{constant}$

Linear strain hardening: $Y(\bar{\varepsilon}^p) = Y_0 + h\bar{\varepsilon}^p$

Power-law hardening: $Y = Y_0 + h(\bar{\varepsilon}^p)^{1/m}$

One can have various combinations for the yielding conditions and the hardening laws.

6.2.2.2   General von Mises Yield Function

$$f(\sigma_{ij}, \bar{\varepsilon}^p) = \sigma_e - Y(\bar{\varepsilon}^p) \tag{6-59}$$

$$\sigma_e = \sqrt{\frac{3}{2} S_{ij} S_{ij}} \tag{6-60}$$

$$S_{ij} = \sigma_{ij} - \frac{1}{3}\sigma_{kk}\delta_{ij} \tag{6-61}$$

The von Mises yielding criterion that includes a kinematic component (the back stress) can be expressed as:

$$f(\sigma_{ij}, \alpha_{ij}, \bar{\varepsilon}^p) = \sqrt{\frac{3}{2}(S_{ij} - \alpha_{ij})(S_{ij} - \alpha_{ij})} - Y(\bar{\varepsilon}^p) = 0 \tag{6-62}$$

The physical meaning of the kinematic terms is the center of the yielding surface is also moving in the stress space. The direct implication of the back stress is the change of the reference state (an offset of the zero stress state). There are different rules that govern the update of the back stress. The simplest approach is:

$$d\alpha_{ij} = \frac{2}{3} c d\varepsilon_{ij}^p \tag{6-63}$$

A more sophisticated approach is:

$$d\alpha_{ij} = \frac{2}{3} c d\varepsilon_{ij}^p - \gamma\alpha_{ij}d\bar{\varepsilon}^p \tag{6-64}$$

6.2.2.3   Mohr-Coulomb Criterion

Neither the Tresca criterion nor the von Mises criterion is dependent on the mean stress or the confining stress (the first invariant). The Mohr-Coulomb criterion presents one that is dependent on the normal stress (therefore, the mean stress). For a specific plane with an applied normal stress σ, when its shear stress reaches:

$$f(\sigma) = c - \sigma\tan(\varphi) \tag{6-65}$$

The material will yield. Where c and φ are the cohesion and the friction angle correspondingly.

In 3D cases, the above criterion transfers to:

$$f(\sigma) = (\sigma_1 - \sigma_3) + (\sigma_1 + \sigma_3)\sin\varphi - 2c\cos\varphi \tag{6-66}$$

6.2.2.4 Druker Prager Criterion

Druker and Prager (1952) proposed a criterion that combines the Mises criterion with the Mohr-Coulomb criterion, which has been known as the Druker-Prager criterion.

$$f(s, I_1) = \sqrt{J_2} + I_1 \tan\varphi / \sqrt{6} - Y(\bar{\varepsilon}^p) \tag{6-67}$$

It may be interpreted as the Mohr-Coulomb criterion applied to the octahedral plane.

6.2.3 Plastic Flow Directions

Another function that governs the magnitude and directions of the plastic flow is named the plasticity potential function, typically represented as $g = g(\sigma_{ij}, \theta)$ and:

$$d\varepsilon_{ij}^p = \lambda \frac{\partial g}{\partial \sigma_{ij}} \tag{6-68}$$

Where λ is a scalar called plastic multiplier. λ can be combined into g so that $d\varepsilon_{ij}^p = \frac{\partial g}{\partial \sigma_{ij}}$. While this presents a simpler format it requires more criteria in selecting the potential function. If g and f are the same function, it is named associated flow; otherwise it is non-associated flow.

6.2.3.1 von Mises Criteria and Associated Flow

The following makes use of the von Mises criteria and associated flow as an example to illustrate how one can calculate the plastic strain in increments.

Isotropic Hardening

$$d\varepsilon_{ij}^p = d\bar{\varepsilon}^p \frac{\partial f}{\partial \sigma_{ij}} = d\bar{\varepsilon}^p \frac{3}{2}\frac{S_{ij}}{Y} \tag{6-69}$$

where

$$f(\sigma_{ij}, \bar{\varepsilon}^p) = \sqrt{\frac{3}{2}S_{ij}S_{ij}} - Y(\bar{\varepsilon}^p)$$

$$S_{ij} = \sigma_{ij} - \frac{1}{3}\sigma_{kk}\delta_{ij}$$

$$f(\sigma_{ij} + d\sigma_{ij}, \bar{\varepsilon}^p + d\bar{\varepsilon}^p) = f(\sigma_{ij}, \bar{\varepsilon}^p) + \frac{\partial f}{\partial \sigma_{ij}}d\sigma_{ij} + \frac{\partial f}{\partial \bar{\varepsilon}^p}d\bar{\varepsilon}^p = 0 \tag{6-70}$$

$$\Rightarrow \frac{\partial f}{\partial \sigma_{ij}}d\sigma_{ij} - \frac{\partial Y}{\partial \bar{\varepsilon}^p}d\bar{\varepsilon}^p = 0 \tag{6-71}$$

$$\Rightarrow d\bar{\varepsilon}^p = \frac{1}{h}\frac{\partial f}{\partial \sigma_{ij}}d\sigma_{ij} = \frac{1}{h}\frac{3}{2}\frac{S_{ij}d\sigma_{ij}}{Y} \tag{6-72}$$

$$h = \frac{dY}{d\bar{\varepsilon}^p}$$

Linear Kinematic Hardening

$$d\varepsilon_{ij}^{p} = d\bar{\varepsilon}^{p}\,\frac{\partial f}{\partial \sigma_{ij}} = d\bar{\varepsilon}^{p}\,\frac{3}{2}\frac{\left(S_{ij}-\alpha_{ij}\right)}{Y}$$

(6-73)

Where the yield criterion is:

$$f(\sigma_{ij},\alpha_{ij}) = \sqrt{\frac{3}{2}\left(S_{ij}-\alpha_{ij}\right)\left(S_{ij}-\alpha_{ij}\right)} - Y$$

$$S_{ij} = \sigma_{ij} - \frac{1}{3}\sigma_{kk}\delta_{ij}$$

Where Y = constant:

$$f(\sigma_{ij}+d\sigma_{ij},\alpha_{ij}+d\alpha_{ij}) = f(\sigma_{ij},\alpha_{ij}) + \frac{\partial f}{\partial \sigma_{ij}}d\sigma_{ij} + \frac{\partial f}{\partial \alpha_{ij}}d\alpha_{ij} = 0 \qquad (6\text{-}74)$$

If the linear kinematic hardening law is followed, one has:

$$d\alpha_{ij} = \frac{2}{3}cd\varepsilon_{ij}^{p} = cd\bar{\varepsilon}^{p}\,\frac{\left(S_{ij}-\alpha_{ij}\right)}{Y} \qquad (6\text{-}75)$$

And therefore:

$$d\bar{\varepsilon}^{p} = \frac{1}{c}\frac{3}{2}\frac{\left(S_{ij}-\alpha_{ij}\right)d\sigma_{ij}}{Y} \qquad (6\text{-}76)$$

Following the similar philosophy, one can obtain the solution for combined isotropic hardening and kinematic hardening.

6.2.4   General Laws on Plastic Deformations

There are several general principles regarding plastic deformations. These include: the loading, unloading conditions; normality and regularity; and stability conditions.

6.2.4.1   Unloading Condition

For 1-D situations, unloading is easy to judge. For multi-dimensions, it is much more difficult.

For isotropic hardening, the unloading condition is:

$$S_{ij}d\sigma_{ij} < 0 \qquad (6\text{-}77)$$

For kinematic hardening, it is:

$$\left(S_{ij}-\alpha_{ij}\right)d\sigma_{ij} < 0 \qquad (6\text{-}78)$$

The physical meaning is that the incremental stress will point to the inside of the yielding surface.

6.2.4.2   The Kuhn-Tucker Conditions

It can be observed that $d\bar{\varepsilon}^{p}$ and f obey the Kuhn-Tucker conditions:

$$d\bar{\varepsilon}^{p}f = 0,\, d\bar{\varepsilon}^{p} \geq 0,\, f \leq 0 \qquad (6\text{-}79)$$

6.2.4.3 Principle of Maximum Plastic Resistance

For the actual stress (σ_{ij}) on the yield surface and any other stress (σ^*_{ij}) within the yield surface (the stress state outside the yield surface is impossible), the Principle of Maximum Plastic Resistance states:

$$\left(\sigma_{ij} - \sigma^*_{ij}\right)d\varepsilon^p_{ij} \geq 0 \text{ or } \sigma_{ij}d\varepsilon^p_{ij} \geq \sigma^*_{ij}d\varepsilon^p_{ij} \tag{6-80}$$

That is the actual stress state will offer the maximum plastic resistance.

6.2.4.4 Drucker's Postulate ΔT_i

Druker proposed the following conditions for:

$$\dot{\sigma}\dot{\varepsilon}^p \begin{cases} \geq \ 0, \text{ hardening material} \\ = \ 0, \text{ perfectly plastic material} \\ \leq \ 0, \text{ softening material} \end{cases} \tag{6-81}$$

For a stable plastic material, if the surface traction has an increase of ΔT_i and its corresponding displacement increase is Δu_i, the following inequality will hold.

$$\Delta W = \int \left\{ \int_A \Delta T_i \frac{\Delta u_i}{dt} \right\} dt \geq 0 \tag{6-82}$$

The implications of the inequality are:

- The yield surface f(σ_{ij}) must be convex;
- The plastic strain rate must be normal to the yield surface $d\varepsilon^p_{ij} = d\bar{\varepsilon}^p \dfrac{\partial f}{\partial \sigma_{ij}}$;
- The rate of strain hardening must be positive or zero $\dfrac{dY}{d\bar{\varepsilon}^p} \geq 0$.

6.3 Viscoelasticity (Findley et al., 1989)

6.3.1 Creep, Recovery, and Relaxation

For a material of linear, isotropic elasticity, Hook's law applies. The 1-D stress-strain relation follows the linear relation presented in Equation 6-83.

$$\varepsilon = \frac{\sigma}{E} \tag{6-83}$$

Where ε and σ are correspondingly the uniaxial strain and stress, E is the Young's modulus. Obviously, when the stress is a constant, the strain is also a constant and vice versa. Where the stress is removed, the strain will return to zero immediately. Figure 6.2 illustrates this relationship.

However, for a material of viscoelasticity, the straining process is quite different under the same stressing process. First, the strain is not a constant under the same constant stress σ_0; it increases with time. Second, when the stress is removed, the strain does not return to zero immediately; instead, it gradually returns to zero (Figure 6.3).

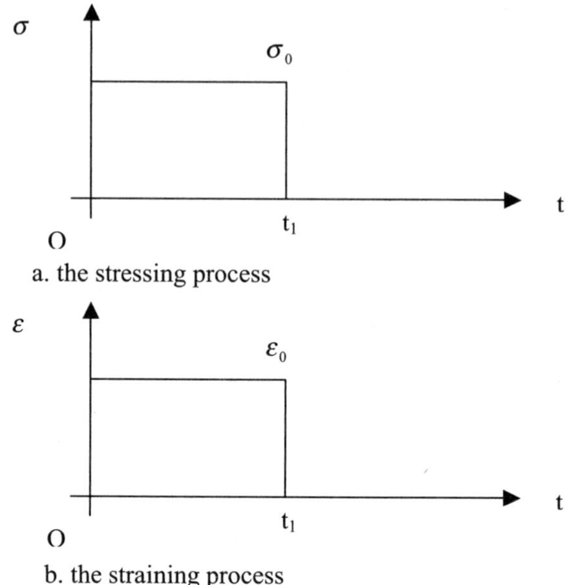

a. the stressing process

b. the straining process

FIGURE 6.2 Illustration of stressing and straining of an elastic process.

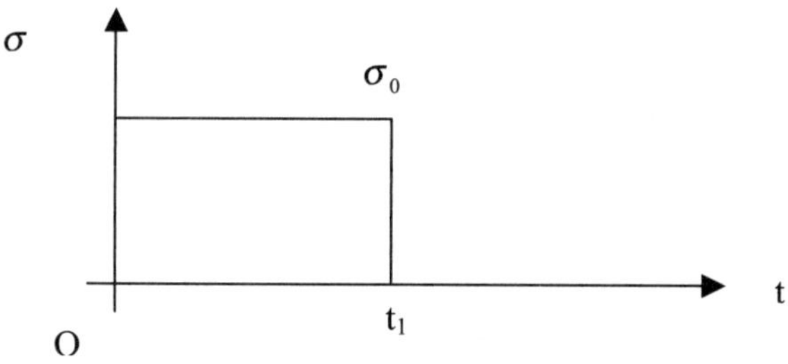

a. the stressing process

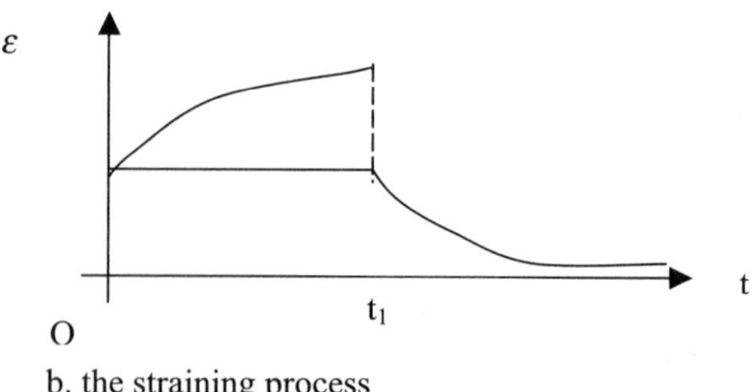

b. the straining process

FIGURE 6.3 Illustration of stressing and straining of a viscoelastic process.

6.3.1.2 Creep

Under a constant stress, the phenomenon that the strain gradually increases is called creep.

Equations 6-84*a* and *b* describe this phenomenon.

$$\sigma(t) = \sigma_0 \tag{6-84a}$$

$$\varepsilon(t) = \varepsilon_e + \varepsilon_c(t) \tag{6-84b}$$

Where σ_0 represents a constant stress, ε_e represents the elastic strain that does not vary with time, and $\varepsilon_c(t)$ represents the creep strain that varies with time. Equation 6.85 represents the strain rate. Depending on this rate, the creep process can be divided into three stages—the primary creep, the secondary creep, and the tertiary creep. Figure 6.4 presents the three stages qualitatively. At the primary stage, the strain rate decreases; it remains constant at the secondary stage and increases at the tertiary stage:

$$\dot{\varepsilon} = \frac{d\varepsilon(t)}{dt} \tag{6-85}$$

And the following ratio is called creep compliance:

$$C(t) = \frac{\varepsilon(t)}{\sigma_0} \tag{6-86}$$

6.3.2 Recovery

Like elastic recovery, for viscoelastic material, the elastic part ε_e is immediately recovered when the load is removed. However, creep part $\varepsilon_c(t)$ will be recovered gradually (Figure 6.5). For metals, only a small part of $\varepsilon_c(t)$ will be recovered, while it may be recovered completely over a long period of time for some plastics and polymers.

6.3.3 Relaxation

For linear elastic materials, a constant stress yields a constant strain and vice versa. For a viscoelastic material, a constant strain will not yield a constant stress, but instead it will yield a stress $\sigma(t)$ that decreases with time (Figure 6.6). This phenomenon is called stress relaxation.

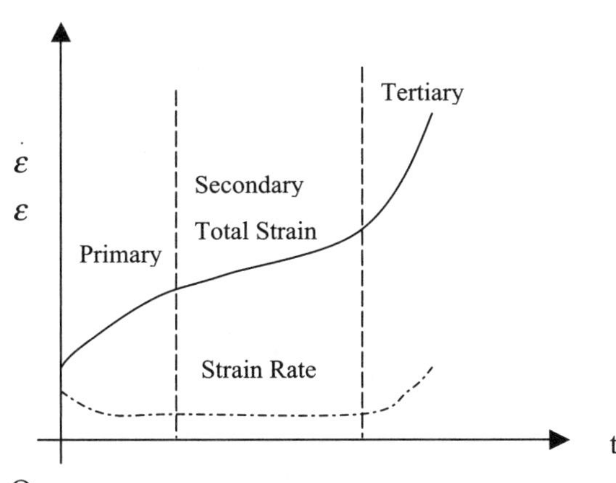

FIGURE 6.4 Illustration of the creep stages.

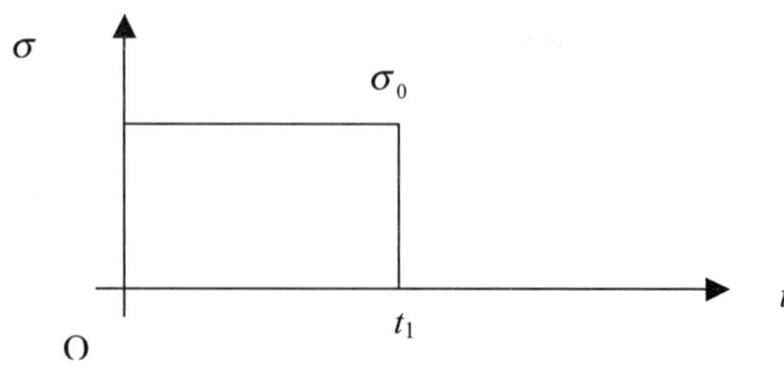

Figure 6.5 Illustration of recovery process.

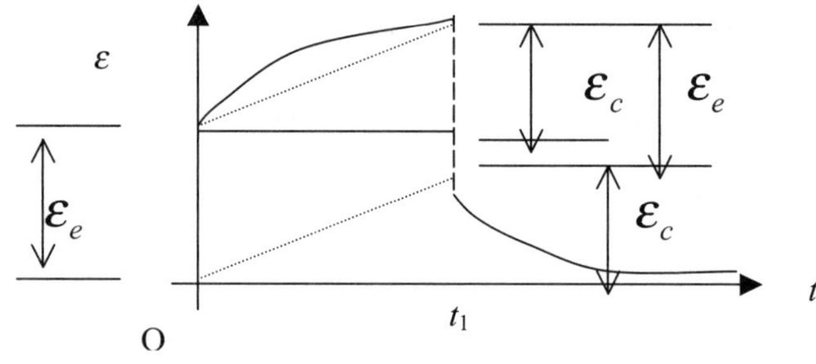

a. the stressing process

b. the straining process

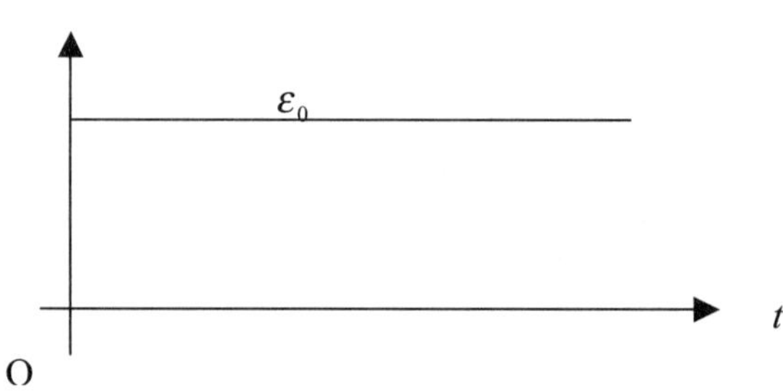

Figure 6.6 Illustration of stressing and straining of an elastic process.

a. the straining process

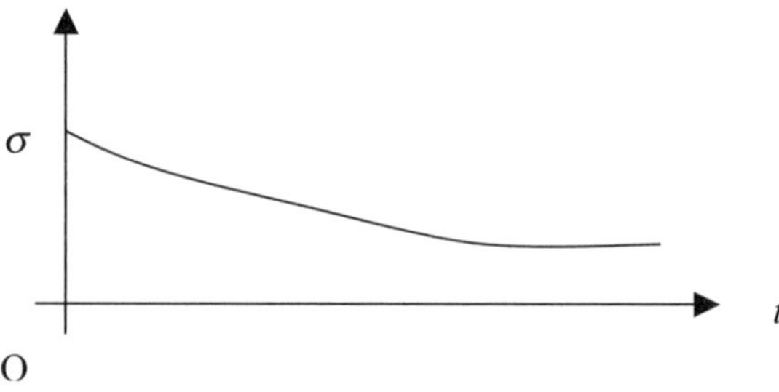

b. the stress relaxation

The following ratio is called relaxation modulus:

$$E(t) = \frac{\sigma(t)}{\varepsilon_0} \tag{6-87}$$

The above phenomena are the topics of viscoelasticity. For a mixture of AC, among the three constituents, asphalt binder is a viscoelastic (viscoplastic) material, aggregate is approximately an elastic material under normal loading. Due to its complex structure, in an AC specimen, creep, recovery, and relaxation may happen simultaneously, adding the complexity of characterization and modeling.

Generally, these phenomena are a non-linear process. In future sections, the relationship between creep and relaxation will be established.

6.3.4   Linearity and Boltzmann's Superposition Principle

If a material is considered as linearly viscoelastic, its stress is proportional to strain and vice versa. In linear viscoelasticity, the linear superposition principle is valid. Mathematically, this principle can be written in the following two equations:

$$\varepsilon[\alpha\sigma(t)] = \alpha\varepsilon[\sigma(t)] \tag{6-88}$$

$$\varepsilon[\sigma_1(t) + \sigma_2(t - t_1)] = \varepsilon[\sigma_1(t)] + \varepsilon[\sigma_2(t - t_1)] \tag{6-89}$$

6.3.5   Complex Loading

From Boltzmann's superposition principle, one can derive the strain response of a complex loading process as illustrated in Figure 6.7.

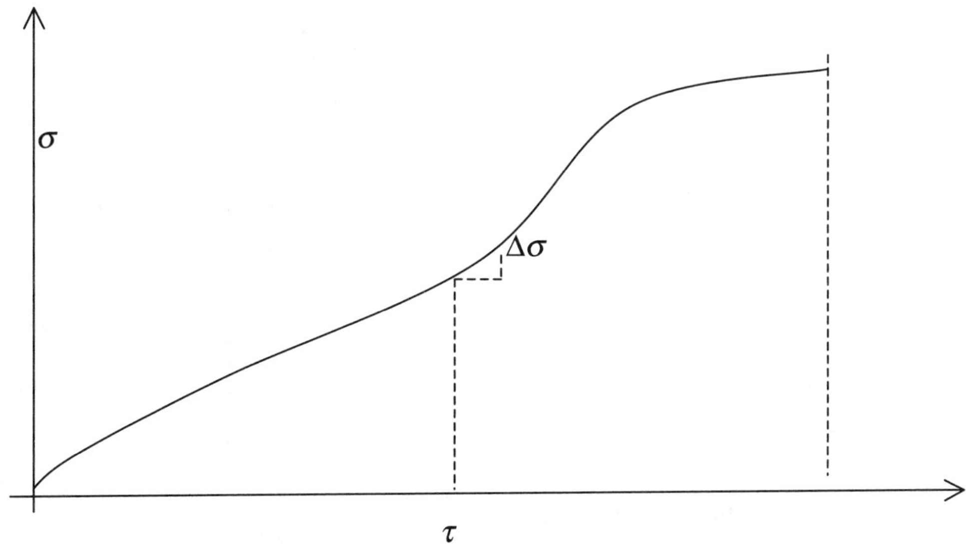

FIGURE 6.7   Complex loading.

If the creep compliance is $C(t) = \dfrac{\varepsilon(t)}{\sigma_0}$, the strain response (measured at time t) due to the load increment $\Delta\sigma$ at time instant τ should be $\Delta\varepsilon = \Delta\sigma C(t-\tau)H(t-\tau)$.

Where $H(t-\tau)$ is the step function.

$$H(t-\tau) = \begin{matrix} 1 & \text{for } t \geq \tau \\ 0 & \text{for } t < \tau \end{matrix} \tag{6-90}$$

Decompose the complex loading into a series of incremental loading, and the overall response can be represented as:

$$\varepsilon(t) \approx \sum_{i=1}^{N} \Delta\varepsilon(i) \approx \sum_{i=1}^{N} \Delta\sigma_i C(t-\tau_i)H(t-\tau_i) \tag{6-91}$$

When $\Delta\sigma_i \to 0$, the above summation can be represented as an integral.

$$\varepsilon(t) = \int_0^t d\sigma C(t-\tau)H(t-\tau) = \int_0^t C(t-\tau)H(t-\tau)d\sigma \tag{6-92}$$

Since $t \geq \tau$, the above integral can be simplified as $\varepsilon(t) = \int_0^t C(t-\tau)d\sigma$.

$$\varepsilon(t) = [C(t-\tau)\sigma(\tau)]\big|_0^t + \int_0^t \sigma(\tau)dC(t-\tau) \tag{6-93}$$

6.3.6 Simple Linear Models

Many of the linear viscoelastic constitutive models can be formed though a linear combination of the springs (elastic) and dashpots (viscous) (Figure 6.8).

Typical models include the Maxwell model, the Kelvin model, the Burgers model, and the generalized models.

6.3.6.1 Maxwell Model

The Maxwell model can be represented as the spring in series with the dashpot (Figure 6.9).

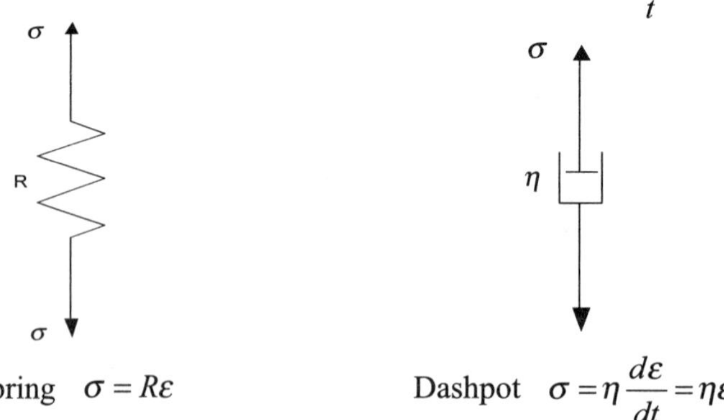

Figure 6.8   Illustration of the spring and dashpot.

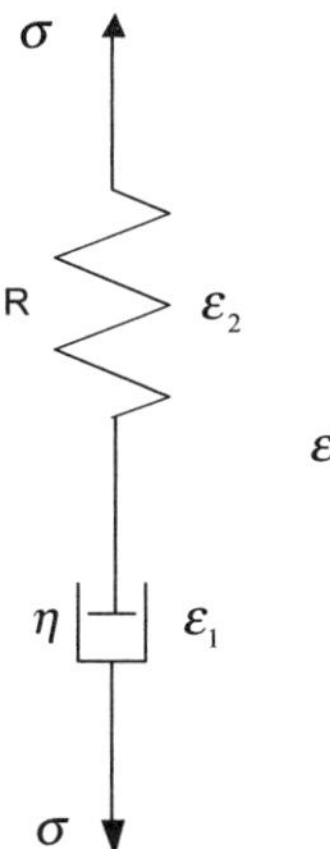

FIGURE 6.9   Illustration of the Maxwell model.

The governing equations include:

$$\varepsilon = \varepsilon_1 + \varepsilon_2, \ \dot{\varepsilon} = \dot{\varepsilon}_1 + \dot{\varepsilon}_2, \ \dot{\varepsilon} = \frac{\dot{\sigma}}{R} + \frac{\sigma}{\eta} \tag{6-94}$$

For constant stress (creep effect, $\dot{\sigma} = 0$), the following solution can be obtained:

$$\varepsilon(t) = \frac{\sigma_0}{R} + \frac{\sigma_0}{\eta} t \tag{6-95}$$

For constant strain situations ($\dot{\varepsilon} = 0$ relaxation process) the following equation can be obtained:

$$\sigma = \sigma_0 e^{-Rt/\eta} \tag{6-96}$$

Clearly, the stress will decrease with time. The rate of decreasing is non-linear and equal to:

$$\dot{\sigma} = -(\sigma_0 R / \eta) e^{-Rt/\eta} \tag{6-97}$$

The initial decreasing rate is $\dot{\sigma}_{t=0} = -(\sigma_0 R/\eta)$, if the stress is decreasing at this rate constantly (following a straight line $\sigma = -(\sigma_0 R/\eta)t + \sigma_0$), the stress will reduce to zero at time $t_R = \eta/R$. This is the relaxation time. For the linear system, the relaxation time is independent of the magnitude of the load.

6.3.6.2   Kelvin Model

The Kelvin model represents the parallel combination of a spring with a dashpot (Figure 6-10).

The governing equations include:

$$\begin{array}{c} \sigma_1 = R\varepsilon \\ \sigma_2 = \eta\dot{\varepsilon} \end{array}, \ \sigma = \sigma_1 + \sigma_2 \tag{6-98}$$

$$\eta\dot{\varepsilon} + R\varepsilon = \sigma \ \text{or} \ \dot{\varepsilon} + \frac{R}{\eta}\varepsilon = \frac{\sigma}{\eta} \tag{6-99}$$

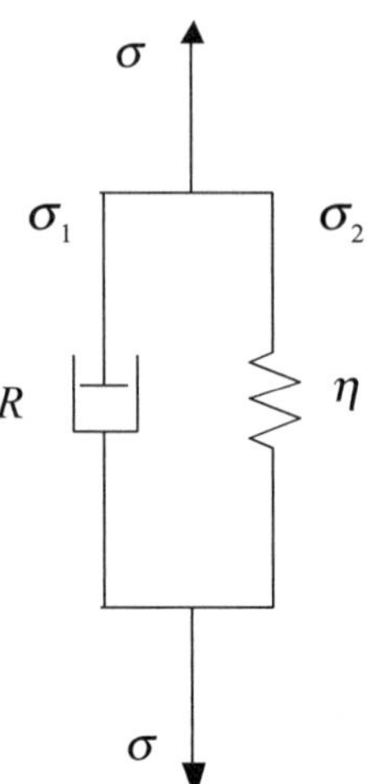

FIGURE 6.10 Illustration of the Kelvin model.

For constant stress (creep effect, $\dot{\sigma} = 0$), the following solution can be obtained:

$$\varepsilon = \frac{\sigma_0}{R}(1 - e^{-Rt/\eta}) \tag{6-100}$$

It should be noted that the Kelvin model cannot describe the relaxation phenomenon.

6.3.6.3 Burgers Model

The Burgers model can be graphically expressed as in Figure 6.11.

The governing equations include:

$$\varepsilon = \varepsilon_1 + \varepsilon_2 + \varepsilon_3 \tag{6-101}$$

$$\varepsilon_1 = \frac{\sigma}{R_1}$$

$$\dot{\varepsilon}_2 = \frac{\sigma}{\eta_1}$$

$$\dot{\varepsilon}_3 + \frac{R_2}{\eta_2}\varepsilon_3 = \frac{\sigma}{\eta_2} \tag{6-102}$$

The above equations lead to the following governing equation:

$$\sigma + (\frac{\eta_1}{R_1} + \frac{\eta_1}{R_2} + \frac{\eta_2}{R_2})\dot{\sigma} + \frac{\eta_1\eta_2}{R_1R_2}\ddot{\sigma} = \eta_1\dot{\varepsilon} + \frac{\eta_1\eta_2}{R_2}\ddot{\varepsilon} \tag{6-103}$$

At constant stress, the above equation leads to the following solution (creep effect):

$$\varepsilon(t) = \frac{\sigma_0}{R_1} + \frac{\sigma_0}{\eta_1}t + \frac{\sigma_0}{R_2}(1 - e^{-R_2t/\eta_2}) \tag{6-104}$$

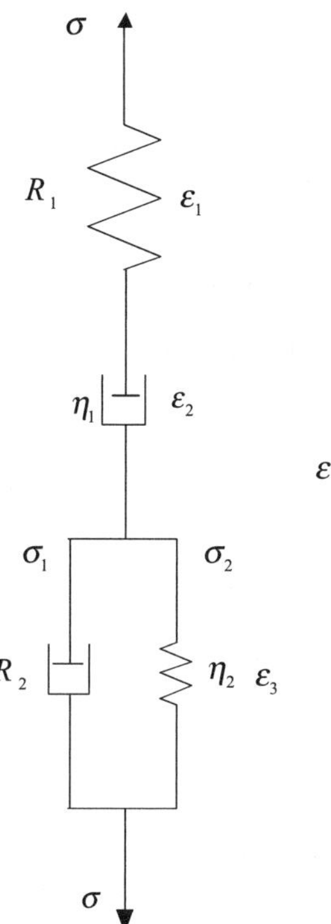

FIGURE 6.11 Illustration of the Burger model.

For the constant-strain process, the generic governing equation can be expressed as:

$$\sigma + p_1\dot{\sigma} + p_2\ddot{\sigma} = q_1\varepsilon_0\delta(t) + q_2\varepsilon_0\frac{d\delta(t)}{dt} \tag{6-105}$$

where

$$p_1 = \frac{\eta_1}{R_1} + \frac{\eta_1}{R_2} + \frac{\eta_2}{R_2}, p_2 = \frac{\eta_1\eta_2}{R_1R_2}, q_1 = \eta_1, q_2 = \frac{\eta_1\eta_2}{R_2} \tag{6-106}$$

The Laplace transform of the above equation yields:

$$\hat{\sigma} + p_1 s\hat{\sigma} + p_1 s^2\hat{\sigma} = q_1\varepsilon_0 + q_2\varepsilon_0 s \tag{6-107}$$

$$\hat{\sigma} = \frac{\varepsilon_0(q_1 + q_2 s)}{1 + p_1 s + p_2 s^2} \tag{6-108}$$

The inverse transform of the above equation yields:

$$\sigma(t) = \frac{\varepsilon_0}{A}[(q_1 - q_2 r_1)e^{-r_1 t} - (q_1 - q_2 r_2)e^{-r_2 t}] \tag{6-109}$$

where

$$r_1 = (p_1 - A)/2p_2, r_2 = (p_1 + A)/2p_2, A = \sqrt{p_1^2 - 4p_2}$$

6.3.6.4 Generalized Models

Generalized Maxwell and Kelvin models are comprised of a number of the basic elements in parallel or series. The solutions to these generalized models can be derived following the Laplace transform approach. In some cases, they can be obtained through simple linear addition. Generally, the governing equations of these models can be expressed as a generic format:

$$P(\sigma,\dot{\sigma},\ddot{\sigma},...) = Q(\varepsilon,\dot{\varepsilon},\ddot{\varepsilon},...)$$

In operator format, it is expressed as:

$$P\sigma = Q\varepsilon$$

$$P\sigma = p_0\sigma + p_1\dot{\sigma} + p_2\ddot{\sigma} + ... + p_a\frac{\partial^a\sigma}{\partial t^a} = q_0\varepsilon + q_1\dot{\varepsilon} + q_2\ddot{\varepsilon} + ... + q_b\frac{\partial^b\sigma}{\partial t^b}Q\varepsilon \qquad (6\text{-}110)$$

Through Laplace transform, one can have:

$$\hat{P}(s)\hat{\sigma}(s) = (p_0 + p_1 s + p_2 s^2 + ... + p_a s^a)\hat{\sigma}(s) = \hat{Q}(s)\hat{\varepsilon}(s) = (q_0 + q_1 s + q_2 s^2 + ... + q_b s^b)\hat{\varepsilon}(s)$$

$$\frac{\hat{Q}(s)}{\hat{P}(s)} = \frac{\hat{\sigma}}{\hat{\varepsilon}} \qquad (6\text{-}111)$$

Through an inverse Laplace transform one can obtain time-domain solutions from following equations:

$$\hat{\sigma}(s) = [\frac{q_0 + q_1 s + q_2 s^2 + ... + q_b s^b}{p_0 + p_1 s + p_2 s^2 + ... + p_a s^a}]\hat{\varepsilon}(s) \qquad (6\text{-}112)$$

or

$$\hat{\sigma}(s) = [\frac{q_0 + q_1 s + q_2 s^2 + ... + q_b s^b}{p_0 + p_1 s + p_2 s^2 + ... + p_a s^a}]\hat{\varepsilon}(s) \qquad (6\text{-}113)$$

The above methods can be extended to the stress tensors and strain tensors in 3D cases:

$$P_1\sigma_{ij}(t) = Q_1\varepsilon_{ij}(t) \qquad (6\text{-}114)$$

It can be decomposed into the deviator stress-strain relationship and the volumetric stress-strain relationship:

$$[p_0' + p_1'\frac{\partial}{\partial t} + p_2'\frac{\partial^2}{\partial t^2} + ... + p_a'\frac{\partial^a}{\partial t^a}]s_{ij}(t) = [q_0' + q_1'\frac{\partial}{\partial t} + q_2'\frac{\partial^2}{\partial t^2} + ... + q_b'\frac{\partial^b}{\partial t^b}]d_{ij}(t) \qquad (6\text{-}115)$$

$$P_2\sigma_{ij}(t) = Q_2\varepsilon_{ij}(t)$$

And the bulk stress and volumetric strain relationship:

$$[p_0'' + p_1''\frac{\partial}{\partial t} + p_2''\frac{\partial^2}{\partial t^2} + ... + p_a''\frac{\partial^a}{\partial t^a}]\sigma_{ii}(t) = [q_0'' + q_1''\frac{\partial}{\partial t} + q_2''\frac{\partial^2}{\partial t^2} + ... + q_b''\frac{\partial^b}{\partial t^b}]\varepsilon_{ii}(t)$$

$$P_3\sigma_{ii}(t) = Q_3\varepsilon_{ii}(t) \qquad (6\text{-}116)$$

This actually represents that there is no coupling between volumetric strain and deviatory stresses:

$$s_{ij} = 2Gd_{ij}$$
$$\sigma_{ii} = 3K\varepsilon_{ii}$$

(6-117)

G is shear modulus and K is the bulk modulus:

$$G = \frac{1}{2}\frac{Q_2}{P_2}, \; K = \frac{1}{3}\frac{Q_3}{P_3}, \; E = \frac{9KG}{3K+G}, \; v = \frac{3K-2G}{6K+2G}$$

6.3.7 Complex Modulus and Compliance

It would be interesting to see the responses of a viscoelastic material to a cyclic loading, for example the material modeled with the Maxwell model $\dot{\varepsilon} = \dfrac{\dot{\sigma}}{R} + \dfrac{\sigma}{\eta}$, subjected a stress $\sigma = \sigma_0\cos\omega t$. By solving the governing equation, one can obtain the following strain response: $\varepsilon = \varepsilon_0\cos(\omega t - \delta)$:

Where $\tan\delta = \dfrac{R}{\eta}$ and $\varepsilon_0 = \dfrac{\sigma_0}{\sqrt{\dfrac{1}{R^2} + \dfrac{1}{\omega^2\eta^2}}}$.

(6-118)

For a more general case, when the load is $\sigma = \sigma_0 e^{i\omega t}$

Where $e^{i\omega t} = \cos\omega t + i\sin\omega t$

The strain response is:

$$\varepsilon = \varepsilon_0\cos(\omega t - \delta)$$

Or $\varepsilon = \varepsilon_0 e^{i(\omega t - \delta)}$

Or $\varepsilon = (\varepsilon_0 e^{-\delta i})e^{i\omega t} = \varepsilon^* e^{i\omega t}$

Denoting $\varepsilon^* = \varepsilon_0 e^{-i\delta} = \varepsilon_0(\cos\delta - i\sin\delta)$ and placing both the σ and ε in the general governing equation, one can have the following equation:

$$[p_0 + i\omega p_1 + (i\omega)^2 p_2 + \ldots]\sigma_0 e^{i\omega t} = [q_0 + i\omega q_1 + (i\omega)^2 q_2 + \ldots]\varepsilon^* e^{i\omega t}$$

(6-119)

or

$$E^* = \frac{\sigma_0}{\varepsilon_0}e^{i\delta} = \frac{\sigma_0}{\varepsilon_0}[\cos\delta + i\sin\delta] = E_1 + iE_2 = |E^*|e^{i\delta}$$

(6-120)

6.3.8 Energy Dissipation

It would be interesting to see how energy is dissipated in one loading cycle (T is the period):

$$\Delta W = \int_0^T \sigma\frac{d\varepsilon}{dt}dt$$

(6-121)

$$\Delta W = \int_0^T \varepsilon_0\sigma_0\omega\sin\omega t\cos(\omega t - \delta)dt$$

$$\Delta W = \pi\sigma_0\varepsilon_0\sin\delta$$

$$\Delta W = \pi\varepsilon_0^2 E_2$$

If the strain is in pace with stress, the total strain energy will be stored (the entire loop will involve unloading) and the dissipated energy would be equal to zero:

$$W = \int_0^{T/4} \sigma \frac{d\varepsilon}{dt} dt = \int_0^{T/4} \sigma_0 \varepsilon_0 \sin \omega t \cos \omega t\, dt = \sigma_0 \varepsilon_0 / 2 \tag{6-122}$$

Therefore, the energy dissipation rate is:

$$\frac{\Delta W}{W} = 2\pi \sin \delta \tag{6-123}$$

6.3.9 Temperature Effect and Time-Temperature Superposition Principle

For thermo-elastic materials, the elastic modulus is a function of temperature and time $E = E(T,t)$. The relaxation modulus described previously shows a function of time at constant temperature. The modulus-temperature relationship deserves study. Based on experimental studies, the Time-Temperature Superposition principle was proposed. The principle actually states that the relaxation modulus E at temperature T is equal to the modulus at the reference temperature T_0 at a scaled time by a factor of $a_T(T)$ and this factor is a function of the temperature difference only. Mathematically, this relationship can be expressed as:

$$E(T,t) = E(T_0,\zeta)$$
$$\zeta = t / a_T(T) \tag{6-124}$$

Where t is actual time, T is the temperature, and ζ is the scaled time. A well-known relationship is the Williams, Landel, and Ferry (Williams, et al., 1955) relationship:

$$\log_{10} a_T(T) \equiv \log \frac{t}{\zeta} = \frac{-k_1(T - T_0)}{k_2 + (T - T_0)} \tag{6-125}$$

Where k_1 and k_2 are material constants. Experimental results indicate that k_1 and k_2 are usually not "true" constants. They also vary with temperature and time, showing the empirical nature of the above relationship. For transient temperature conditions, Morland and Lee (1960), proposed the following relationship:

$$\zeta(t) = \int_0^t \frac{dt'}{a_T[T(t')]} \tag{6-126}$$

6.3.10 The Correspondence Principle

For stress analysis of quasi-static viscoelastic problems, the Correspondence Principle will allow one to obtain viscoelatic solution if the corresponding elastic solution is known, or directly solving a problem in the transformed space with transformed elasticity components.

For a boundary value problem in viscoelasticity, mathematically it is equivalent to solving the following set of equations.

A) The equilibrium equations:

$$\frac{\partial \sigma_{ij}(x,t)}{\partial x_i} + F_j(x,t) = 0 \qquad (6\text{-}127)$$

B) The time-dependent constitutive equations:

$$s_{ij}(t) = 2\int_0^t G(t-\xi)\dot{d}_{ij}(\xi)d\xi$$

$$\sigma_{ii}(t) = 3\int_0^t K(t-\xi)\dot{\varepsilon}_{ii}(\xi)d\xi \qquad (6\text{-}128)$$

C) The strain-displacement relations (small strain):

$$\varepsilon_{ij}(t) = \frac{1}{2}[\frac{\partial u_i(t)}{\partial x_j} + \frac{\partial u_j(t)}{\partial x_i}] \qquad (6\text{-}129)$$

D) The compatibility relations (using 2D case as simplification):

$$2\frac{\partial^2 \varepsilon_{xy}(t)}{\partial x \partial y} = \frac{\partial^2 \varepsilon_{xx}(t)}{\partial y^2} + \frac{\partial^2 \varepsilon_{yy}(t)}{\partial x^2} \qquad (6\text{-}130)$$

E) Boundary conditions:

$$T_j(x,t) = \sigma_{ij}(x,t)n_i$$

$$U_j(x,t) = u_j(x,t) \qquad (6\text{-}131)$$

A solution using the elastic-viscoelastic correspondence principle applies the Laplace transform to the above equations and obtains a set of equations in the transformed space.

A) $\dfrac{\partial \hat{\sigma}_{ij}(x,s)}{\partial x_i} + \hat{F}_j(x,s) = 0 \qquad (6\text{-}132)$

B) $\hat{S}_{ij}(s) = 2s\hat{G}(s)\hat{d}_{ij}(s)$

$\hat{\sigma}_{ii}(s) = 3s\hat{K}\hat{\varepsilon}_{ii}(s) \qquad (6\text{-}133)$

C) $\hat{\varepsilon}_{ij}(s) = \dfrac{1}{2}[\dfrac{\partial \hat{u}_i(s)}{\partial x_j} + \dfrac{\partial \hat{u}_j(s)}{\partial x_i}] \qquad (6\text{-}134)$

D) $2\dfrac{\partial^2 \hat{\varepsilon}_{xy}(s)}{\partial x \partial y} = \dfrac{\partial^2 \hat{\varepsilon}_{xx}(s)}{\partial y^2} + \dfrac{\partial^2 \hat{\varepsilon}_{yy}(s)}{\partial x^2} \qquad (6\text{-}135)$

E) $\hat{T}_j(x,s) = \hat{\sigma}_{ij}(x,s)n_i$

$\hat{U}_j(x,s) = \hat{u}_{ij}(x,s) \qquad (6\text{-}136)$

By comparing this set of equations with the equations in small-strain linear elasticity, one can notice that if G is replaced with $s\hat{G}(s)$, K with $3s\hat{K}$ the set of equations is the same as the set of equations in elasticity. This will allow one to solve the viscoelasticity problems by using the elasticity theory. Quite a few examples were presented in Findley et al. (1989).

Nevertheless, the above principle is not valid if the boundary conditions such as the contact areas are changing with time.

6.3.11   Relationship between Creep Compliance and Relaxation Modulus

Apply the Laplace transform to the viscoelastic stress-strain relationships:

$$\varepsilon(t) = \int_0^t C(t-\tau)\frac{\partial\sigma}{\partial\tau}d\tau \qquad \sigma(t) = \int_0^t E(t-\tau)\frac{\partial\varepsilon}{\partial\tau}d\tau \tag{6-137}$$

$$\hat{\varepsilon}(s) = s\hat{C}(s)\hat{\sigma}(s)$$
$$\hat{\sigma}(s) = s\hat{E}(s)\hat{\varepsilon}(s) \tag{6-138}$$

$$\frac{\hat{\sigma}(s)}{\hat{\varepsilon}(s)} = s\hat{E}(s) = \frac{1}{s\hat{C}(s)} \tag{6-139}$$

or

$$\hat{C}(s)\hat{E}(s) = \frac{1}{s^2} \tag{6-140}$$

Apply the inverse Laplace transform to 6-140:

$$\int_0^t C(t-\xi)E(\xi)d\xi = t \tag{6-141}$$

or

$$\int_0^t E(t-\xi)C(\xi)d\xi = t$$

$$E(0)C(0) = 1$$
$$E(\infty)C(\infty) = 1 \tag{6-142}$$

Generalization of the integral representation to 3D:

$$\varepsilon_{ij}(t) = \int C_{ijkl}(t-\xi)\dot{\sigma}_{kl}(\xi)d\xi \tag{6-143}$$

Or in the deviatoric stress-strain relationship, and the bulk stress relationship:

$$s_{ij}(t) = 2\int_0^t G(t-\xi)\frac{\partial d_{ij}(\xi)}{\partial\xi}d\xi \tag{6-144}$$

$$\sigma_{ii}(t) = 3\int_0^t K(t-\xi)\frac{\partial\varepsilon_{ii}(\xi)}{\partial\xi}d\xi \tag{6-145}$$

$G(t)$ is the stress relaxation modulus in shear.

The linear stress-strain relationship in elasticity $\sigma_{ij} = \lambda\varepsilon_{kk}\delta_{ij} + 2G\varepsilon_{ij}$ can be extended to the viscoelasticity as:

$$\sigma_{ij}(t) = \delta_{ij}[\lambda\varepsilon_{kk}(t) - \int_0^t \psi_1(t-\xi)\frac{\partial\varepsilon_{kk}(\xi)}{\partial\xi}d\xi] + 2G\varepsilon_{ij}(t) - \int_0^t \psi_2(t-\xi)\frac{\partial\varepsilon_{ij}(\xi)}{\partial\xi}d\xi \tag{6-146}$$

Where λ and G are constants describing the linear time-independent stress-strain relationships, Ψ_1 and Ψ_2 are the stress relaxation functions for ε_{kk} and ε_{ij}.

$$\varepsilon_{ij} = -\frac{\lambda}{2G(2G+3\lambda)}\sigma_{kk}\delta_{ij} + \frac{1}{2G}\sigma_{ij}$$

or

$$\varepsilon_{ij}(t) = \delta_{ij}[a_0\sigma_{kk}(t) + \int_0^t \varphi_1(t-\xi)\frac{\partial\sigma_{kk}(\xi)}{\partial\xi}] + b_0\sigma_{ij}(t) + \int_0^t \varphi_2(t-\xi)\frac{\partial\sigma_{ij}(\xi)}{\partial\xi}d\xi \qquad (6\text{-}147)$$

Where a_0 and b_0 are constants describing the linear time-independent strain-stress relations, φ_1 and φ_2 are creep functions for σ_{kk} and σ_{ij}.

There are non-linear viscoelasticity theories. These theories may not be necessary and can be replaced with viscoplastcity and therefore will not be discussed. Interested readers may refer to Findley et al. (1989).

6.4   Viscoplasticity (Perzyna, 1966)

6.4.1   General Concept

For a set of tests at different strain rates, one may obtain a set of curves as illustrated in Figure 6.12. In other words, the Young's modulus, the initial yield stress, the hardening parameters, the peak strength, and the residual strength are not only a function of the strain and plastic strain, but also a function of the strain rate. This phenomenon can then be modeled as viscoplastcity. There are many viscoplasticity theories. Typical ones used in AC followed the Perzyna model (isotropic hardening) and the Chaboche model (isotropic and kinematic hardening). For example, the SHRP permanent deformation model has both isotropic and kinematic hardening.

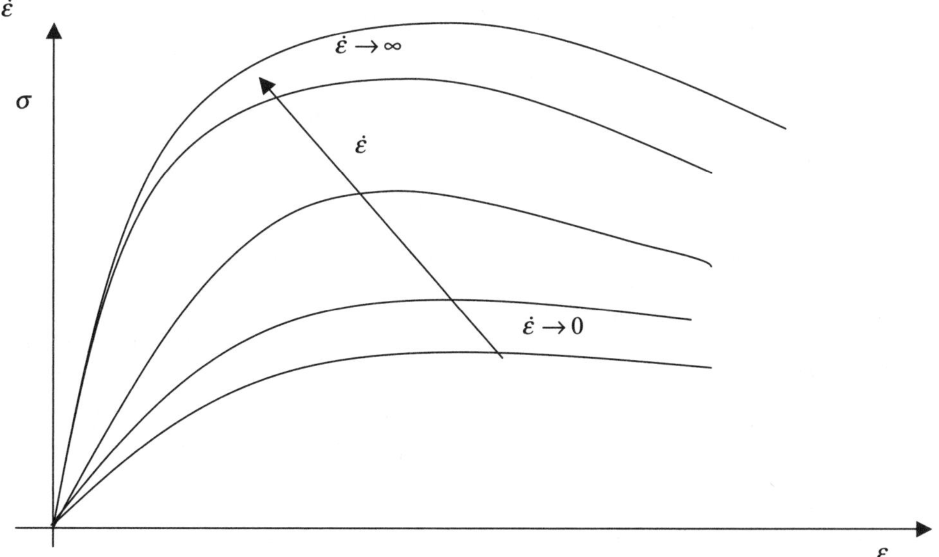

Figure 6.12   Typical stress-strain curves for metals and other materials.

In theory, according to the objectivity principle, the modulus, initial yielding stress, the hardening laws, the ultimate yielding stress, the rate of softening, and the residual strength are all related to strain rate and should carry it as an independent variable. A review indicates that most of the viscoplasticity models used in AC do not include strain rate in all the equations. However, many of them include the strain rate effect in the yielding criteria.

6.4.2 Perzyna Models

Under the title of viscoplasticity, many models developed in AC do not have adequate consistency. Most of the viscoplascticity models are based on Perzyna (1966). There is no ambiguity in viscoelasticity and elastoplastcity. However, there is no unique definition for viscoplasticity. The two major sub-branch definitions, according to Perzyna, are elastic-viscoplastic and elastic/viscoplastic; the former demonstrates viscous properties in both elastic and plastic regions; the latter shows viscous properties in the plastic region only. Considering the following decomposition:

$$\varepsilon_{ij} = \varepsilon_{ij}^{e} + \varepsilon_{ij}^{v} + \varepsilon_{ij}^{p} \tag{6-148}$$

Therefore the elastic-viscoplastic model can be considered as:

$$\varepsilon_{ij} = \varepsilon_{ij}^{ve} + \varepsilon_{ij}^{vp} \tag{6-149}$$

The elastic/viscoplastic relationship can be considered as:

$$\varepsilon_{ij} = \varepsilon_{ij}^{e} + \varepsilon_{ij}^{vp} \tag{6-150}$$

The rate (time) equation is:

$$\dot{\varepsilon}_{ij} = \dot{\varepsilon}_{ij}^{e} + \dot{\varepsilon}_{ij}^{v} + \dot{\varepsilon}_{ij}^{p} \tag{6-151}$$

Therefore for the elastic-viscoplastic model, the rate equation can be represented as:

$$\dot{\varepsilon}_{ij} = \dot{\varepsilon}_{ij}^{ve} + \dot{\varepsilon}_{ij}^{vp} \tag{6-152}$$

The elastic/viscoplastic relationship can be represented as:

$$\dot{\varepsilon}_{ij} = \dot{\varepsilon}_{ij}^{e} + \dot{\varepsilon}_{ij}^{vp} \text{ (time rate should be valid, please verify it)} \tag{6-153}$$

Where the elastic component is not time-dependent.

6.4.2.1 Elastic-viscoplastic Model

One example may involve the use of the general visecoelastic constitutive relation to represent:

$$\varepsilon_{ij}^{ve}(t) = \delta_{ij}[a_0\sigma_{kk}(t) + \int_0^t \varphi_1(t-\xi)\frac{\partial\sigma_{kk}(\xi)}{\partial\xi}] + b_0\sigma_{ij}(t) + \int_0^t \varphi_2(t-\xi)\frac{\partial\sigma_{ij}(\xi)}{\partial\xi}d\xi \tag{6-154}$$

For the plastic part, the viscoplastic yielding will be dependent not only on the stress state and the plastic deformation but also on the viscoelastic strain, which is time dependent. Therefore, even within the elastic region, depending on the stress path it may reach the yield surface at different locations. Naghdi and Murch (1963) proposed the so-called flow surface concept.

$$f = f(\sigma_{ij}, \varepsilon_{ij}^p, \kappa, \beta) \tag{6-155}$$

In addition to the stress state σ_{ij}, plastic strain ε_{ij}^p and the strain hardening parameter κ, a parameter β, a function of the viscoelastic strain $\beta = \beta(\varepsilon_{kl}^v)$ was introduced. Introduction of β makes the yield surface "flow" with time. An examination of the time rate of the yield function will allow one to define loading and unloading.

$$\dot{f} = \frac{\partial f}{\partial \sigma_{ij}} \dot{\sigma}_{ij} + \frac{\partial f}{\partial \varepsilon_{kl}^p} \dot{\varepsilon}_{ij}^p + \frac{\partial f}{\partial \kappa} \dot{\kappa} + \frac{\partial f}{\partial \beta} \dot{\beta} \tag{6-156}$$

If $\dot{f} < 0$ then, $f + \dot{f}dt < 0$, leading to the viscoelastic state, and therefore the rate of the plastic strain (no incremental plastic strain), or $\dot{\varepsilon}_{ij} = 0$. This will also lead to $\dot{\kappa} = 0$. Under these conditions, one can have:

$$\dot{f} = \frac{\partial f}{\partial \sigma_{ij}} \dot{\sigma}_{ij} + \frac{\partial f}{\partial \beta} \dot{\beta} = \zeta(\dot{\sigma}_{ij}) \tag{6-157}$$

$f = 0 \ \zeta(\dot{\sigma}_{ij} < 0)$, unloading

$f = 0 \ \zeta(\dot{\sigma}_{ij} = 0)$, neutral process

$f = 0 \ \zeta(\dot{\sigma}_{ij} > 0)$, process yielding plastic strains

Unlike the traditional yielding, where neutral loading will be tangent to the yield surface, for the viscoplastic situation $\dfrac{\partial f}{\partial \sigma_{ij}} \dot{\sigma}_{ij} + \dfrac{\partial f}{\partial \beta} \dot{\beta} = 0$ does not require $\dfrac{\partial f}{\partial \sigma_{ij}} \dot{\sigma}_{ij} = 0$.
It should be noted that the above formulations indicate that loading, unloading, and neutral loading conditions are related to strain rates of both $\dot{\sigma}_{ij}$ and $\dot{\beta}$.

Like plasticity theories, the viscoplasticity theory also requires the satisfaction of the conditions for stable materials, and convexity conditions.

Naghdi and Murch (1963) proposed for the following relationship (please note its similarity to the incremental plastic strain formulation, but here it is time rate of the plastic strain):

$$\dot{\varepsilon}_{ij}^p = \Lambda \frac{\partial f}{\partial \sigma_{ij}} \tag{6-158}$$

Through the use of the condition $\dot{f} = 0$, one can obtain:

$$\Lambda = -\left(\frac{\partial f}{\partial \sigma_{kl}} \dot{\sigma}_{kl} + \frac{\partial f}{\partial \beta} \dot{\beta}\right)\left[\frac{\partial f}{\partial \varepsilon_{mn}^p} \frac{\partial f}{\partial \sigma_{mn}} + \frac{\partial f}{\partial \kappa} \tilde{\kappa}\left(\frac{\partial f}{\partial \sigma_{pq}}\right)\right]^{-1} \tag{6-159}$$

Where $\dot{\kappa}(\dot{\varepsilon}_{kl}^p) = \Lambda \tilde{\kappa}\left(\dfrac{\partial f}{\partial \sigma_{pq}}\right).$

By introducing the viscoplastic potential function h, Perzyna arrived at the following general formulation for the plastic deformation rate:

$$\dot{\varepsilon}_{ij}^p = N \frac{\partial h}{\partial \sigma_{ij}} \left\langle \frac{\partial f}{\partial \sigma_{ij}} \dot{\sigma}_{ij} + \frac{\partial f}{\partial \beta} \dot{\beta} \right\rangle = N \frac{\partial h}{\partial \sigma_{ij}} \left\langle \zeta(\dot{\sigma}_{ij}) \right\rangle \tag{6-160}$$

6.4.2.2 The Elastic/Viscoplastic Model

$$\varepsilon_{ij} = \varepsilon_{ij}^{e} + \varepsilon_{ij}^{vp}$$

The rate equation is:

$$\dot{\varepsilon}_{ij} = \dot{\varepsilon}_{ij}^{e} + \dot{\varepsilon}_{ij}^{vp}$$

Perzyna introduced the static yield function (initial yield function):

$$F(\sigma_{ij},\varepsilon_{kl}^{p}) = f(\sigma_{ij},\varepsilon_{kl}^{p}) / \kappa - 1 \qquad (6\text{-}161)$$

$$\kappa = \kappa(W_{p})$$

$$W_{p} = \int_{0}^{\varepsilon_{kl}^{p}} \sigma_{ij} d\varepsilon_{ij}^{p}$$

W_{p} is the accumulative plastic work.

The static yield function is assumed to be regular and convex. The formulation for the rate of the viscoplastic strain is similar to that for the plastic strain:

$$\dot{\varepsilon}_{ij}^{vp} = \gamma < \Phi(F) > \frac{\partial f}{\partial \sigma_{ij}} \qquad (6\text{-}162)$$

γ and $\Phi(F)$ are a viscosity parameter of the material, and a function showing the dynamic behavior of the material.

$$\dot{\varepsilon}_{ij} = \frac{1}{2\mu}\dot{s}_{ij} + \frac{1-2v}{E}\dot{s}\delta_{ij} + \gamma < \Phi(F) > \frac{\partial f}{\partial \sigma_{ij}} \qquad (6\text{-}163)$$

6.5 Continuum Damage Mechanics (Lemaitre, 1996)

6.5.1 General Concepts of Continuum Damage Mechanics

Elasticity is for material structures with no defects: bonds among atoms are perfect and atom configurations follow regular patterns. Viscoelasticity is for material structure with relatively weak bonds. Plasticity is for materials with dislocations. Continuum damage mechanics is valid for materials when their bonds start to break, or start to sustain damage. Damage for different materials may be demonstrated as breakage of bonds of long chain molecules of polymers; debonding between fibers and matrix in composite materials; microdecohensions between inclusions and matrix; pore coalescing in the interfacial transition zones in concrete. In general, damage mechanics may be interpreted as a mechanics that take into consideration existing defects and damage, but smooths the local damage effects with continuum mechanics by using one or a few damage parameters. It is unlike micromechanics where the details of microstructure change must be considered. The two concise books that are recommended are Lemaitre (1996) and Voyiadjis and Kattan (1999).

In AC, damage may be demonstrated as pore initiation in the binder, micro-pore coalescence in the binder, mastic, and aggregate-binder or filler-binder interfaces. CDM is valid when the continuum types of governing equations are valid through the concept of effective stresses. There are several types of damage phenomena including brittle damage, ductile damage, creep damage, low-cycle fatigue damage, and high-cycle fatigue damage. Brittle damage refers to the cases where plastic strain is smaller than elastic strain. Ductile damage refers to the cases where plastic strains can be equal or even larger than elastic strains. Creep damage refers to the cases when significant creep strains occur (for example at an elevated temperature). Cyclic loading also causes damage. At relatively higher stresses, the number of loading repetitions to rupture could be smaller, for example, smaller than 10,000. This process is referred to as low-cycle fatigue damage. If the number of loading repetitions to rupture is larger than 10,000, it is called high-cycle fatigue damage.

All materials have inherent defects or induced damage and microstructure. The representative material properties should be defined on a representative volume element (RVE). Depending on the material structure, the distribution of inherent defects, and induced damage, the sizes of the representative volume elements are different. Typically this size is $(0.1 \text{ mm})^3$ for metals and ceramics; $(1 \text{ mm})^3$ for polymers and most composites; $(10 \text{ mm})^3$ for wood; and $(100 \text{ mm})^3$ for concrete. Damage parameters and the governing equations are valid for materials of sizes larger that the RVE.

6.5.2   Damage Parameter

Considering a section through an RVE at location M and in the n direction, the damage is defined as:

$$D(M,n,x) = \frac{\delta S_{Dx}}{\delta S} \tag{6-164}$$

Where δS_{Dx} and δS are the areas that cannot take stresses and the overall cross section area; x is the location along the n direction. Typically, damage thus defined will vary with x. The maximum value of the damage variable is then defined as the damage value for location M in the orientation n.

$$D_{(M,n)} = \underset{(x)}{Max}[D_{(M,n,x)}]$$

$$D_{(M,n)} = \frac{\delta S_D}{\delta S} \tag{6-165}$$

Clearly, this damage variable is dependent on orientations as well. It can be usually represented as a damage tensor or a damage effect tensor.

In the 1-D and homogeneous case, it reduces to a scalar $D = \frac{S_D}{S}$. Clearly, this D may vary between 0 and 1.

$$0 \le D \le 1$$

With $D = 0$, no damage; $D = 1$, complete damage. In reality, material will fail at D much smaller than 1.

6.5.3   Effective Stress

In the same 1-D and homogeneous case for defining the damage parameter, the effective stress can be defined. In the simplest case where the force on the surface is in the normal direction, the normal stress is defined as:

$$\sigma = \frac{F}{S} \tag{6-166}$$

In the case where micro voids exist on the surface, which cannot take the loads, the effective stress is defined as:

$$\tilde{\sigma} = \frac{F}{S - S_D}$$

$$(6\text{-}167)$$

$$\tilde{\sigma} = \frac{F}{S(1 - \frac{S_D}{S})} \quad \text{or} \quad \tilde{\sigma} = \frac{\sigma}{1 - D}$$

$$(6\text{-}168)$$

Obviously, if the force is tangent to the normal, the above definition is also valid. Nevertheless, in either case, the definitions provide only nominal measurements. For example, when the pore size is very small, it may be able to take load.

6.5.4   Effective Modulus

If there is no damage $D = 0$, Hooke's law will give the strain:

$$\varepsilon_e = \frac{\sigma}{E}$$

In the case with damage $0 < D < 1$, one may have the effective strain as:

$$\varepsilon_{11}^e = \frac{\sigma}{E(1 - D)}$$

$$(6\text{-}169)$$

$$\varepsilon_{22}^e = \varepsilon_{33}^e = -\nu\varepsilon_{11}^e$$

$$(6\text{-}170)$$

It should be noted that the above equations are approximate, as the modulus and the Poisson's ratio is also a function of the damage parameter. Obviously, the effective strain thus defined may not be the same as the true effective strain measured. The degree of deviation is a measure of the validity of CMD. By using the effective modulus concept, one can define the effective modulus as:

$$\tilde{E} = E(1 - D)$$

$$(6\text{-}171)$$

And the effective strain is thus defined as:

$$\varepsilon_{11}^e = \frac{\sigma}{\tilde{E}}$$

$$(6\text{-}172)$$

6.5.5   Strain Equivalency Principle

Lamaitre (1971) postulated a strain equivalency principle. It states that "the constitutive equation of a damage material can be derived using the same way as for virgin material except that the usual stress is replaced by the effective stress."

In the elasticity case the effective strains for virgin and damaged materials are correspondingly defined as:

$$\varepsilon_e = \frac{\sigma}{E} \qquad \varepsilon_e = \frac{\tilde{\sigma}}{E}$$

$$(6\text{-}173)$$

One may assume that the strain decompositions are still valid:

$$\varepsilon = \varepsilon_e + \varepsilon_p$$

$$(6\text{-}174)$$

In terms of plasticity, the elasticity part may utilize the equivalent strain principle. The yielding criterion may be revised accordingly.

Considering the yielding criterion:

$$\sigma = \sigma_y + R + X \tag{6-175}$$

or

$$f = |\sigma - X| - R - \sigma_y = 0 \tag{6-176}$$

Where σ_y is the initial yielding stress; R is the hardening part; and X is the back stress or the kinematics hardening part.

It can be rewritten as:

$$f = \left| \frac{\sigma}{1-D} - X \right| - R - \sigma_y = 0 \tag{6-177}$$

or

$$\sigma = (\sigma_y + R + X)(1 - D) \tag{6-178}$$

Which actually means the yield stress is proportionally reduced by a factor of (1-D). In other words, it could also be written as:

$$\tilde{\sigma} = (\sigma_y + R + X)(1 - D) \tag{6-179}$$

6.5.6   Assessments of the Damage Parameter

6.5.6.1   Direct Measurement

A direct measurement of the damage parameter is to use the RVE and following the direct definitions $D = \dfrac{\delta S_D}{\delta S}$. Chapter 3 presents methods for such measurements. This approach requires the direct measurements of the defects on the surfaces. More practical methods are those of the indirect methods. The following presents a brief summary of these methods.

6.5.6.2   Young's Modulus Degradation

Considering $\tilde{E} = E(1 - D)$, one may obtain:

$$D = 1 - \frac{\tilde{E}}{E} \tag{6-180}$$

Through measuring the degrading of the Young's modulus, the damage parameter can be obtained.

6.5.6.3   Mechanical Wave Velocity

The longitudinal wave velocity for the virgin material and the damaged materials are respectively $V_L^2 = \dfrac{E}{\rho} \dfrac{1-v}{(1+v)(1-2v)}$ and $\tilde{V}_L^2 = \dfrac{\tilde{E}}{\rho} \dfrac{1-\tilde{v}}{(1+\tilde{v})(1-2\tilde{v})}$, if the change in Poisson's ratio and density is negligible, one can obtain:

$$D = 1 - \frac{\tilde{E}}{E} \approx 1 - \frac{\tilde{V}_L^2}{V_L^2} \tag{6-181}$$

It can be conveniently proved that the above equation is also valid if the longitudinal wave velocity is replaced with shear wave velocity.

6.5.6.4 Micro Hardness Test

Hardness is a measurement of the stress defined as the ratio between indentation force and the indentation area, $H = \dfrac{F_i}{S_i}$. Both theory analysis and experimental data indicate this measurement is proportional to the yield stress or $\tilde{H} = \alpha(\sigma_y + R + X)(1 - D)$.

Therefore one may obtain: $D = 1 - \dfrac{\tilde{H}}{H^*}$ $\hspace{4em}$ (6-182)

Where $\tilde{H}$ is measured hardness for damaged materials, H^* corresponds to the hardness measured on the undamaged material for low-cycle fatigue (Lamaitre, 1996).

6.5.6.5 Other Methods

There are other approaches that can be used to measure the degree of damage. They include plastic strain variation method due to cyclic stress, electrical resistance method, tertiary creep rate method, and acoustic emission method. Table 6.2 (Lamaitre, 1996) presents a summary of these methods that may be used for AC.

In the AC modeling studies, the major methods rest on the elasticity modulus degradation approach.

6.5.7 Validity of the Above Relationships

For fatigue tests of AC, a criterion to judge whether the material has reached failure is to use the modulus. When the modulus is reduced to 50% of the original modulus, the material is considered being failed. Obviously, at this stage, $D = 0.5$, or the cross-section

Methods	Damage	Brittle	Ductile	Creep	Low-cycle fatigue	High-cycle fatigue
Density	$D = \left(1 - \dfrac{\tilde{\rho}}{\rho}\right)^{2/3}$		**	*	*	
Elasticity modulus	$D = 1 - \dfrac{\tilde{E}}{E}$	**	***	***	***	
Ultrasonic waves	$D = 1 - \dfrac{\tilde{V}_L^2}{V_L^2}$	***	**	**	*	*
Cyclic stress amplitude	$D = 1 - \dfrac{\Delta\sigma}{\Delta\sigma^*}$		*	*	**	*
Tertiary creep	$D = 1 - \left(\dfrac{\dot{\varepsilon}_p^*}{\dot{\varepsilon}_p}\right)^{1/N}$		*	***	*	
Electrical resistance	$D = 1 - \dfrac{V}{\tilde{V}}$	*	**	**	*	*

TABLE 6.2 Damage assessment methods (from Lamaitre, 1996) *Not sure, **Good, ***Very good).

will have 50% pores or voids. This is equivalent to about 45% volume increase (considering the original air void content being around 5%). However, experimental observations do not indicate such big volume changes. Therefore, it may be concluded that damage effects cannot be completely represented by the damage effect thus defined. In other words, the fatigue phenomena cannot be explained by CDM only.

Lamaitre (1996) also presented summaries on the development of plasticity and viscoplasticty theories that include damage and coupling. These theories have not been applied to AC, and are not presented here due to limited space.

6.5.8 Damage Concept Extended to 3D Cases

In 3D, the degree of damage on surfaces of different orientations may vary. The formulations for the 1-D or homogenous damage may not be valid for the 3D cases. Murakami (1988) introduced a symmetric fourth-rank tensor named damage effect tensor M. M will link the effective stress and the stress in the undamaged material by the following formulation:

$$\tilde{\sigma}_{ij} = M_{ijkl}\sigma_{kl} \tag{6-183}$$

The matrix format of the damage effect tensor, according to Murakami (1988), is linked to the second-rank damage tensor $\ddot{o}$ by the following relationship:

$$M = (I - \ddot{o})^{-1} = \det(G^{-1})G^{T} \tag{6-184}$$

Where G is a fictitious deformation gradient given by:

$G_{ij} = \dfrac{\partial \bar{x}_i}{\partial x_j}$ where x and $\bar{x}$ are the coordinates in the damaged state and the fictitious effective undamaged state, respectively. Details regarding the fictitious effective undamaged state can be found in Vojiadjis and Kattan (1999).

6.6 Fracture Mechanics (Anderson, 1995)

6.6.1 General Concept

Fracture mechanics has been widely used in the modeling of fracture of AC, especially fatigue cracking. This section presents only the most fundamental concepts essential to the understanding of the research in binder, mastic, and AC. Fracture due to cyclic loading is an important topic in this section as well. An excellent textbook reference is Anderson (1995).

6.6.2 Stress Concentration-Macroscopic View

Figure 6.13 illustrates an elliptic hole embedded in a large plate (width $w \gg 2a$ and its height $h \gg 2b$). An elastic analysis for the problem with its boundary subjected a uniform stress σ indicates the following relationship:

$$\sigma_A = \sigma\left(1 + \frac{2a}{b}\right) \tag{6-185}$$

In terms of the radius of curvature ρ ($\rho = \dfrac{b^2}{a}$ for the ellipsis) at Point A, the above equation can be represented as:

$$\sigma_A = \sigma\left(1 + 2\sqrt{\frac{a}{\rho}}\right) \tag{6-186}$$

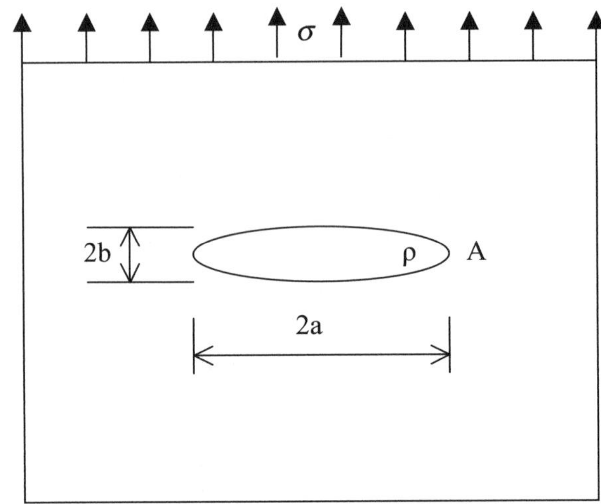

Figure 6.13 Stress analysis in an elliptic voids in a thin plate.

When $2a \gg 2b$

$$\sigma_A \approx 2\sigma\sqrt{\frac{a}{\rho}} \tag{6-187}$$

It was shown that even though the entire void is not elliptical, as long as its shape around the tip (Point A) is elliptical, the above formulation gives a very good approximation for the stress concentration $\left(\dfrac{\sigma_A}{\sigma}\right)$. It should be noted that when the stress is large enough the material close to the tip will yield and is dominated by plastic deformation and will blunt the tip. While mathematically the radius of curvature could reach zero, realistically this radius may not be smaller than the radius of an atom x_0. Therefore, one may conclude that the maximum stress is $\sigma_A = 2\sigma\sqrt{\dfrac{a}{x_o}}$. For actual materials, the stress concentration is much smaller than the above value due to yielding and plastic deformation.

6.6.3   Energy Required to Create a Free Surface, an Atomic View of Fracture

Fracture ultimately goes to the dimension of atoms. Two atoms must be pulled apart so that the attraction between the two atoms will no longer be able to push them back without external forces.

Figure 6.14 illustrates the potential energy between two atoms of the same element. The energy required to pull apart the two atoms x_0 (it is defined as the radius of the atoms) at equilibrium is called bond energy and can be calculated as:

$$E_b = \int_{x_o}^{\infty} P\,dx \tag{6-188}$$

One may approximate the force distribution between the two atoms as: $P = P_c \sin\left(\dfrac{\pi x}{\lambda}\right)$

The meanings of the terms are presented in Figure 6.14, the maximum force P_c is called the cohesive force. For small displacements, the force displacement relationship

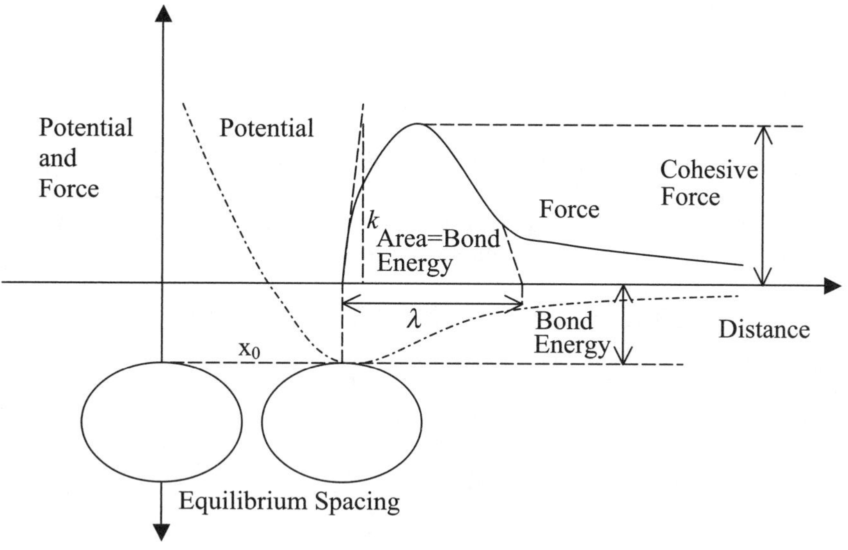

Figure 6.14 Potential, bond energy and cohesive force.

can be approximated as a linear relationship $P = P_c\left(\dfrac{\pi x}{\lambda}\right)$. The bond stiffness can then be obtained as:

$$k = P_c \frac{\pi}{\lambda} \tag{6-189}$$

Multiplying both sides of the above equation by the number of bonds per unit area and the gage length x_0 leads to the following equation:

$$kNx_0 = Nx_0 P_c \frac{\pi}{\lambda} \tag{6-190}$$

One can obtain the following equation:

$$\sigma_c = \frac{E\lambda}{\pi x_0} \tag{6-191}$$

Considering λ and x_0 are at a similar magnitude, the above equation can be approximated as:

$$\sigma_c \approx \frac{E}{\pi}$$

And the surface energy can be calculated as:

$$\gamma_S = \frac{1}{2}\int_0^\lambda \sigma_c \sin\left(\frac{\pi x}{\lambda}\right) dx = \sigma_c \frac{\lambda}{\pi} \tag{6-192}$$

And therefore:

$$\sigma_c = \sqrt{\frac{E\gamma_s}{x_0}} \tag{6-193}$$

And the remote stress at failure will be:

$$\sigma_f = \left(\frac{E\gamma_s}{4a}\right)^{1/2}, \quad \sigma_f = \alpha\left(\frac{E\gamma_s}{a}\right)^{1/2} \tag{6-194}$$

There are quite a few assumptions in the above calculations. There is no possibility that only two atoms are interacting with each other; the yielding process may weaken the bonds significantly when the displacements reach a certain level; other factors such as dislocations and defects are more dominating.

6.6.4 The Griffith Energy Criteria for Crack Opening

Griffith (1920) applied the First Law of Thermodynamics, which states that a system going from a non-equilibrium state to equilibrium will have a net decrease in energy. Therefore, a critical condition is that the total external energy input is equal to the energy consumed in creating new fractured surfaces (cracking opening or propagation) so the total energy in the system is equal to zero.

$$\frac{dE}{dA} = \frac{d\Pi}{dA} + \frac{dW_s}{dA} = 0 \tag{6-195}$$

Where E is the total energy, Π is the potential energy provided by internal strain energy and external forces, and W_s is the work required to create new surfaces. Therefore:

$$-\frac{d\Pi}{dA} = \frac{dW_s}{dA} \tag{6-196}$$

Griffith used the stress analysis by Inglis (1913) and showed that for a crack of width a (b approaches zero) embedded in a thin plate of thickness B under external stress σ, the driving potential energy is:

$$-\frac{d\Pi}{dA} = \frac{\pi\sigma^2 a}{E} \tag{6-197}$$

To create crack surfaces ($2aB$ on each surface, or the $2A$), the energy required:

$$W_s = 4aB\gamma_s \tag{6-198}$$

Therefore,

$$\frac{dW_s}{dA} = 2\gamma_s \tag{6-199}$$

The stress that is required to drive the crack (fracture stress) is:

$$\sigma_f = \left(\frac{2E\gamma_s}{\pi a}\right)^{1/2} \tag{6-200}$$

The fracture stress for a penny-shaped flaw can be shown as:

$$\sigma_f = \left(\frac{\pi E\gamma_s}{2(1-v^2)a}\right)^{\frac{1}{2}} \tag{6-201}$$

Where v is Poisson's ratio.

6.6.5 Modified Griffith Equation

The Griffith criterion is applicable to brittle fracture. Irwin (1948) and Orowan (1948) modified the Griffith criterion to include the energy (γ_p) required to produce plastic flow:

$$\sigma_f = \left(\frac{2E(\gamma_s + \gamma_p)}{\pi a} \right)^{\frac{1}{2}} \tag{6-202}$$

6.6.5.1 Energy Release Rate

Griffith also proved that the energy release rate for the elliptic crack is equal to:

$$G = -\frac{d\Pi}{dA} = \frac{\pi \sigma^2 a}{E} \tag{6-203}$$

The above sections present a general philosophy to link the external stress that will cause crack opening and fracture to the crack size, surface energy, and energy required to produce plastic deformation. It offers an assessment approach of materials in terms of fracture failure. Nevertheless, AC is full of inherent defects of different sizes and shapes. It is difficult to directly use such criteria in AC.

6.6.6 Stress Analysis

Westergaard (1939), Irwin (1957), Sneddon (1946), and Williams (1957) developed solutions for the stress distributions around the crack tips. For 2D cases, the stress state around the tip (polar coordinate system with origin at the tip) is illustrated in Figure 6.15.

Depending on the directions of the external forces, there are three modes of fracturing as illustrated in Figure 6.16.

For Mode I fracturing, the three components of the stresses are:

$$\sigma_{xx} = \frac{K_I}{\sqrt{2\pi r}} \cos\left(\frac{\theta}{2}\right)\left[1 - \sin\left(\frac{\theta}{2}\right)\sin\left(\frac{3\theta}{2}\right)\right]$$

$$\sigma_{yy} = \frac{K_I}{\sqrt{2\pi r}} \cos\left(\frac{\theta}{2}\right)\left[1 + \sin\left(\frac{\theta}{2}\right)\sin\left(\frac{3\theta}{2}\right)\right]$$

$$\tau_{xy} = \frac{K_I}{\sqrt{2\pi r}} \cos\left(\frac{\theta}{2}\right)\sin\left(\frac{\theta}{2}\right)\cos\left(\frac{3\theta}{2}\right) \tag{6-204}$$

Where K_I is the Mode I stress intensity factor.

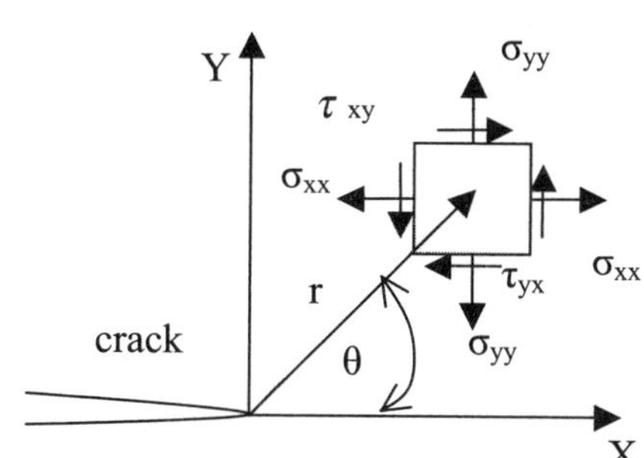

FIGURE 6.15 Stress around the tip of a crack.

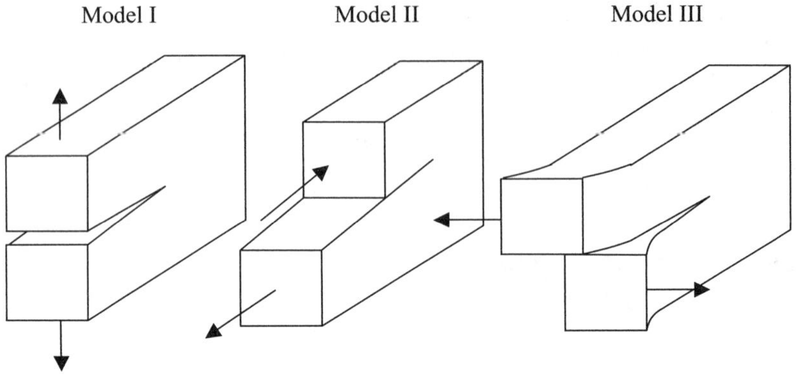

Figure 6.16   The stress modes of cracks.

In general, the stress fields ahead of a crack tip in an isotropic linear elastic material can be expressed as follows:

$$\lim_{r \to 0} \sigma_{ij}^{(I)} = \frac{K_I}{\sqrt{2\pi r}} f_{ij}^{(I)}(\theta)$$

$$\lim_{r \to 0} \sigma_{ij}^{(II)} = \frac{K_{II}}{\sqrt{2\pi r}} f_{ij}^{(II)}(\theta)$$

$$\lim_{r \to 0} \sigma_{ij}^{(III)} = \frac{K_{III}}{\sqrt{2\pi r}} f_{ij}^{(III)}(\theta)$$

$$(6\text{-}205)$$

$$\sigma_{ij}^{(total)} = \sigma_{ij}^{(I)} + \sigma_{ij}^{(II)} + \sigma_{ij}^{(III)}$$

$$(6\text{-}206)$$

Where K_I, K_{II}, K_{III} are the corresponding stress intensity factors; $f_{ij}^{I}(\theta)$, $f_{ij}^{II}(\theta)$ and $f_{ij}^{III}(\theta)$ are functions representing stress distributions. There are methods to calculate the stress intensity of mixed modes.

6.6.7   Crack Tip Plasticity

The elasticity analysis of the stresses around a tip results in infinite stresses when r approaches zero. Real material will yield at a much lower stress. Considering the stress on the plane of crack ($\theta = 0$), one has:

$$\sigma_{xx} = \frac{K_I}{\sqrt{2\pi r}} \qquad \sigma_{yy} = \frac{K_I}{\sqrt{2\pi r}} \qquad (6\text{-}207)$$

By setting $\sigma_{yy} = \sigma_{YS}$ one can have an approximate solution for the plastic zone.

$$r_y = \frac{1}{2\pi}\left(\frac{K_I}{\sigma_{YS}}\right)^2 \qquad (6\text{-}208)$$

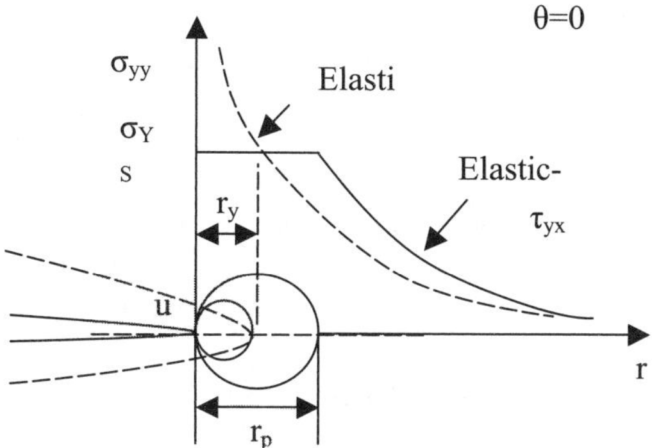

A more accurate estimation will need to account for the stress redistribution:

$$\sigma_{YS} r_p = \int_0^{r_y} \sigma_{yy} dr = \int_0^{r_y} \frac{K_I}{\sqrt{2\pi r}} dr$$

$$r_p = \frac{1}{\pi}\left(\frac{K_I}{\sigma_{YS}}\right)^2 \tag{6-209}$$

The meaning of some of the symbols are illustrated in Figure 6.17. There are more accurate approaches to do the estimation (Irwin, 1961). Additional inform ation will not enhance the understanding of the tip plasticity concept and will not be presented.

6.6.8 Crack Tip Open Distance (CTOD)

Wells (1961) discovered that due to plastic deformation, the sharp crack will become blunt around the tip. He found that the degree of blunting increased proportionally to the toughness of the material. Wells then proposed the opening at the tip as a measure of the fracture toughness. Irwin (1961) showed that the plasticity effect is equivalent to a "longer" tip starting r_y from the actual tip (Figure 6.17).

Therefore,

$$u_y = \frac{\kappa + 1}{2\mu} K_I \sqrt{\frac{r_y}{2\pi}} \tag{6-210}$$

Considering the Irwin's plastic zone correction, CTOD is:

$$\delta = 2u_y = \frac{4K_I^2}{\pi \sigma_{YS} E} \tag{6-211}$$

It is related to the energy release rate G by:

$$\delta = \frac{4}{\pi}\frac{G}{\sigma_{YS}} \tag{6-212}$$

$$\delta = \frac{G}{m\sigma_{YS}} \tag{6-213}$$

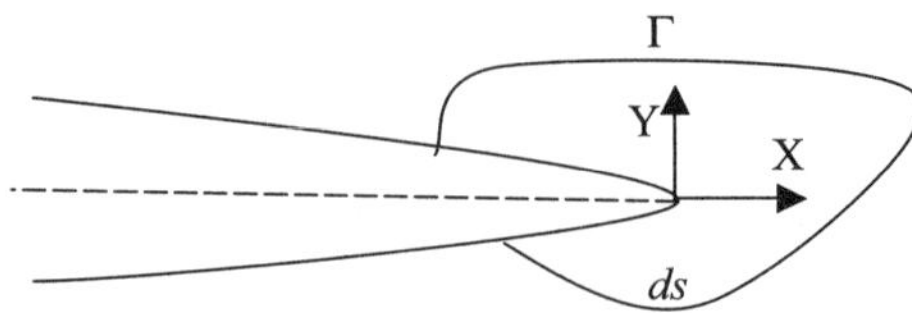

FIGURE 6.18 Illustration of the J integral concept.

Where m is a dimensionless constant and is approximately equal to 1.0 for plane stress and 2.0 for plane strain. Please note that CTOD is related to the toughness of a material and therefore a measure of the material's resistance against fracturing.

6.6.9 The J Contour Integral

For non-linear materials, the J contour integral is a better measurement. Rice (1968) showed that the following integral is path independent and it is equal to the energy release rate:

$$J = \int_{\Gamma} \left(wdy - T_i \frac{\partial u_i}{\partial x} ds \right) = -\frac{d\Pi}{da} \tag{6-214}$$

Where $w = \int_0^{\varepsilon_{ij}} \sigma_{ij} d\varepsilon_{ij}$ is strain energy density; T_i ($T_i = \sigma_{ij} n_j$) and u_i are the components of traction vector and displacement vector, respectively. This integral is independent of the path of integral (see Figure 6.18).

6.6.10 Relations Between J and CTOD

Considering the relationship between energy release rate and the CTOD, one can have the following relation:

$$J = m\sigma_{YS}\delta \tag{6-215}$$

6.6.11 The C* Integral

For materials subjected to steady state creep, Landes and Begley (1976); Ohji et al. (1976) and Nikbin et al. (1976) independently proposed the C* integral method.

$$C^* = \int_{\Gamma} \left(\dot{w}dy - T_i \frac{\partial \dot{u}_i}{\partial x} ds \right) \tag{6-216}$$

where $\dot{w} = \int_0^{\dot{\varepsilon}_{kl}} \sigma_{ij} d\dot{\varepsilon}_{ij}$

It can be proved that C^* is path independent. The dotted parameters in Equation 6-215 represent the time rate of the parameters.

6.6.12 Cracking in Viscoelasticity Materials

Schapery (1984) proposed a generalized J integral that is applicable to a wide range of viscoelastic materials. To better understand his method, the fundamental viscoelastic constitutive relationships are briefly reviewed. The hereditary integrals for viscoelastic materials include the following both 1-D and 3D cases:

$$\varepsilon(t) = \int_0^t C(t - \tau)\frac{d\sigma}{d\tau}d\tau \qquad \sigma(t) = \int_0^t E(t - \tau)\frac{d\varepsilon}{d\tau}d\tau \tag{6-217}$$

$$\varepsilon_{ij}(t) = \int_0^t C_{ijkl}(t - \tau)\frac{d\sigma_{kl}}{d\tau}d\tau \qquad \sigma_{ij}(t) = \int_0^t E_{ijkl}(t - \tau)\frac{d\varepsilon_{kl}}{d\tau}d\tau \tag{6-218}$$

Scharpery (1984) showed that through introducing pseudo-elastic strain, the hereditary integrals can be converted into elastic stress-strain relations following Hooke's Law:

$$\varepsilon^e(t) = \frac{\sigma(t)}{E_R} \tag{6-219}$$

$\varepsilon^e(t)$ and E_R are the pseudo-elastic strain and reference modulus, respectively.

$$\varepsilon(t) = E_R \int_0^t C(t-\tau)\frac{d\varepsilon^e(\tau)}{d\tau}d\tau \tag{6-220}$$

$$\varepsilon^e(t) = E_R^{-1} \int_0^t E(t-\tau)\frac{d\varepsilon(\tau)}{d\tau}d\tau \tag{6-221}$$

$$\varepsilon_{ij}(t) = E_R \int_0^t C_{ijkl}(t-\tau)\frac{d\varepsilon_{kl}^e(\tau)}{d\tau}d\tau \tag{6-222}$$

$$\varepsilon_{ij}^e(t) = E_R^{-1} \int_0^t E_{ijkl}(t-\tau)\frac{d\varepsilon_{kl}(\tau)}{d\tau}d\tau \tag{6-223}$$

$$\varepsilon_{ij}^e(t) = E_R^{-1}[(1+v)\sigma_{ij} - v\sigma_{kk}] \tag{6-224}$$

By defining the following two operators, the above relationships can be concisely presented as:

$$\{Ddf\} = E_R \int_0^t C(t-\tau)\frac{\partial f}{\partial\tau}d\tau \tag{6-225}$$

$$\{Edf\} = E_R^{-1} \int_0^t E(t-\tau)\frac{\partial f}{\partial\tau}d\tau \tag{6-226}$$

The correspondence principle by Schapery (1984) states that the elastic solution and the viscoelastic solution are associated by the following relationships:

$$\sigma_{ij} = \sigma_{ij}^e \quad \varepsilon_{ij} = \{Dd\varepsilon_{ij}^e\} \quad u_i = \{Ddu_i^e\} \quad T_i = \sigma_{ij}n_j = \sigma_{ij}^e n_j \tag{6-227}$$

Therefore, if one has the elastic solution for a boundary value problem, the above correspondence principle will allow one to obtain the corresponding viscoelastic solution.

Through those operations, the generalized J integrals are:

$$J = \int_\Gamma \left(w^e dy - T_i \frac{\partial u_i^e}{\partial x}ds \right) \tag{6-228}$$

where $w^e = \int_0^{\varepsilon_{kl}} \sigma_{ij}d\varepsilon_{ij}^e$

6.6.13   Fatigue Cracking

The fatigue crack growth rate in metals:

$$\frac{da}{dN} = C(\Delta K)^m$$

$$\Delta K = K_{max} - K_{min}$$

(6-229)

6.7   General Considerations of Inelasticity

By extending equation $\sigma = E(\varepsilon - \varepsilon^{in})$ to include various sources of inelastic deformation, the unified theories based on internal state variables have been proposed. A systematic description was presented in Lublinar (1992). It is worthwhile to develop a unified theory to describe plasticity, viscoelasticity, viscoplasticty, and continuum damage mechanics in a unified format.

Suggested Readings

This chapter is mainly based on the teaching notes of the author. These notes are prepared following the presentations from several excellent textbooks. More specifically, hyperelasticity is based on Bower (2010); plasticity is based on Bower (2010) and Lubliner (1990); viscoelasticity is based on Findley et al. (1989); continuum damage mechanics is based on Lemaitre (1992); fracture mechanics is based on Anderson (1995). If readers need more backgrounds, please read these books. For convenience and connections, the symbols adopted in this chapter are consistent with those used in these books. In recent years, quite a few books have been published to address the fundamentals for asphalt mechanics. They include Creep Mechanics (Betten, 2005), Mechanics of Fatigue (Bolotin, 1999), Microcontinuum Field Theories (Eringen, 1999), Fatigue of Materials (Suresh, 1998), and Deformation and Fracture Mechanics of Engineering Materials (Hertzberg, 1996). Interested readers should try to read these books.

References

Anderson, T.L. (1995). Fracture Mechanics: Fundamentals and Applications. 2nd Edition. *CRC Press*, Moscow.

Betten, J. (2005). Creep Mechanics. *Springer*, Berlin.

Bolotin, V.V. (1999). Mechanics of Fatigue. *CRC Press*, Moscow.

Bower, F.A. (2010). Applied Mechanics of Solids. *CRC Press, Taylor & Francis Group*.

Drucker, D.C. and Prager, W. (1952). Soil mechanics and plastic analysis for limit design. *Quarterly of Applied Mathematics*, Vol.10, No.2, pp.157–165.

Eringen, A.C. (1999). Microcontinuum Field Theories. *Springer*, New York.

Findley, W.N., Lai, J. and Onaran, K. (1989). Creep and Relaxation of Nonlinear Viscoelastic Materials, with an Introduction to Linear Viscoelasticity. *Dover Publications, Inc.*, New York.

Gotoh, M., (1977a). A theory of plastic anisotropy based on a yield function of fourth order (plane stress state)-I. *International Journal of Mechanical Sciences*, Vol.19, No.9, pp.505–512.

Gotoh, M., (1977b). A theory of plastic anisotropy based on a yield function of fourth order (plane stress state)-II. *International Journal of Mechanical Sciences*, Vol.19, No.9, pp.513–520.

Green, A.E., and Zerna, W. (1954). Theoretical Elasticity. *Dover Publications, Inc.*, New York.

Griffith, A.A. (1920). The phenomena of rupture and flow in solids. *Philosophical Transactions of the Royal Society of London*, Series A, Vol.221, pp.163–198.

Hertzberg, R.W. (1996). Deformation and Fracture Mechanics of Engineering Materials. 4th Edition. *John Wiley & Sons, Inc.*, Hoboken, NJ.

Hill, R. (1948). A theory of the yielding and plastic flow of anisotropic metals. *Proceedings of the Royal Society of London*, Series A, Vol.193, No.1033, pp.281–297.

Inglis, C.E. (1913). Stresses in a plate due to the presence of cracks and sharp corners. *Transactions of the Institute of Naval Architects*, Vol.55, pp.219–241.

Irwin, G.R. (1948). Fracture dynamics. *Fracturing of Metals, American Society for Metals,* Cleveland, pp.147–166.

Irwin, G.R. (1957). Analysis of stresses and strains near the end of a crack traversing a plate. *Journal of Applied mechanics*, Vol.24, pp.361–364.

Irwin, G.R. (1961). Plastic Zone near a crack and fracture toughness. *Proceedings of Seventh Sagamore Conference*, Vol.4, pp.63–76.

Landes, J.D. and Begley, J.A. (1976). A fracture mechanics approach to creep growth. *ASTM STP 590*. pp.128–148.

Lemaitre, J. (1996). A Course on Damage Mechanics. 2nd Edition. *Springer.*

Lubliner, J. (1990). Plasticity Theory. *MacMillan Publishing Company*, New York.

Morland, L.W. and Lee, E.H. (1960). Stress analysis for linear viscoelastic materials with temperature variation. *Journal of Rheology*, Vol.4, No.1, pp.233–263.

Murakami, S. (1988). Mechanical modeling of material damage. *Journal of Applied Mechanics*, Vol.55, pp.280–286.

Naghdi, P.M. and Murch, S.A. (1963). On the mechanical behaviour of viscoelastic plastic solids. *Journal of Applied Mechanics*, Vol.30, pp.321–328

Nikibin, K.M., Webster, G.A. and Turner, C.E. (1976). ASTM STP 601, *American Society for Testing and Materials*, Philadelphia, pp.47–62.

Ohji, K., Ogura, K. and Kubo, S. (1976). Transactions. *Japanese Society of Mechanical Engineers*, Vol.42, No.354, pp.350–358.

Orowan, E. (1948). Fracture and Strength of Solids. Reports on *Progress in Phsics*, Vol.XII, p.185.

Perzyna, P. (1966). Fundamental problems in viscoplasticity. *Advances in Applied Mechanics*. Vol.9, pp.241–257.

Rice, J.R. (1968). A path independent integral and the approximate analysis of strain concentration by notches and cracks. *Journal of Applied Mechanics*, Vol.35, pp.379–386.

Schapery, R.A. (1984). Correspondence principles and a generalized J integral for large deformation and fracture analysis of viscoelastic media. *International Journal of Fracture*, Vol.25, pp.195–223.

Sneddon, I.N. (1946). The distribution of stress in the neighborhood of a crack in an elastic solid. *Proceedings, Royal Society of London*, Vol.A187, pp.229–260.

Suresh, S. (1998). Fatigue of Materials. 2nd Edition. *Cambridge University Press*, Cambridge.

Voyiadjis, G.Z. and Kattan, P.I. (1999). Advances in Damage Mechanics: Metals and Metal Matrix Composites. *Elsevier Science*, The Netherland.

Weiner, J.H. (1983). Statistical Mechanics of Elasticity, *Wiley Press*, New York.

Wells, A.A. (1961). Unstable crack propagation in metals: cleavage and fast fracture. *Proceedings of the Crack Propagation Symposium*, Cranfield, UK, Vol.1, pp.210–230.

Westergaard, H.M. (1939). Bearing pressures and cracks. *Journal of Applied Mechanics*, Vol.6, pp.49–53.

Williams, M.L. (1957). On the stress distribution at the base of a stationary crack. *Journal of Applied Mechanics*, Vol.24, pp.109–114.

Williams, M.L., Landel, R.F. and Ferry, J.D. (1955). The temperature dependence of relaxation mechanism in amorphorous polymers and other glass-liquids. *J. of Am. Chem. Soc.*, Vol.77, p.370.

CHAPTER 7

Models for Asphalt Concrete

7.1 Introduction

There are many rational models developed in the past 20 years for asphalt concrete (AC). Detailed descriptions of these models are beyond the scope of this book. Due to page limitations, only those complete models are presented. Works on empirical models and/or some components of a model are not summarized. Elasticity models and viscoelasticity models are not reviewed. Models reviewed in this chapter are mainly those viscoplasticity models that consider 1) viscoelasticity; 2) yielding; 3) hardening; and 4) failures. Most of the fatigue models are actually fatigue failure criteria.

The total strains in AC can be decomposed into elastic, viscoelastic, and viscoplastic components. Depending on the ways of the individual components or a combination of several components (viscoelastic, viscoplastic, elastoplastic), a large number of models have been proposed. The interpretation of the mechanism regarding the plastic deformation and the viscoplastic deformation will add more varieties including the damage mechanics mechanism.

7.2 Viscoplasticity with Damage

Richard Kim (Kim et al., 1995, 2004, 2009) and his group (Chehab et al., 2005; Chehab and Kim, 2005; Daniel and Kim, 2002) devoted significant efforts to the modeling of AC using viscoelastoplastic continuum damage (VEPCD) mechanics. Their models followed a series of publications by Schapery (1975, 1981, 1984, 1987a, 1987b, 1990, and 1999). Major features of their models are summarized as follows.

7.2.1 Strain Components

$$\varepsilon_{Total} = \varepsilon_{ve} + \varepsilon_{vp} \tag{7-1}$$

where ε_{Total} = total strain
ε_{ve} = viscoelastic (VE) strain
ε_{vp} = viscoplastic (VP) strain

7.2.2 Viscoelastic Stress-Strain Relationship

The stress-strain equation for linear viscoelastic materials can be represented as:

$$\sigma_{ij} = \int_0^{\xi} E_{ijkl}(\xi - \tau)\frac{\partial \varepsilon_{kl}}{\partial \tau}d\tau \tag{7-2}$$

where σ_{ij}, ε_{kl} = stress and strain tensors

$\quad\quad E_{ijkl}(t)$ = the relaxation modulus tensor

$\quad\quad\quad t$ = physical time

$\quad\quad\quad \xi = t/a_T$ = reduced time

$\quad\quad\quad a_T$ = the time-temperature shift factor

$\quad\quad\quad \tau$ = the integration variable

Through introducing the reference modulus E_R and pseudostrain ε_{kl}^R (Schapery, 1984):

$$\varepsilon_{kl}^R = \frac{1}{E_R}\int_0^\xi E_{ijkl}(\xi - \tau)\frac{\partial\varepsilon_{kl}}{\partial\tau}d\tau \tag{7-3}$$

The stress and pseudostrain relationship can be written in the same format as in elasticity:

$$\sigma_{ij} = E_R\varepsilon_{kl}^R \text{ or } \varepsilon_{kl}^R = \frac{\sigma_{ij}}{E_R} \tag{7-4}$$

Clearly, if the relaxation modulus is constant E_0 and $E_R = E_0$, the above relationship reduces to those of linear elasticity.

In the above formulations, the reduced time is calculated as follows for transient and steady phenomena respectively:

$$\xi = \int_0^t \frac{dt}{a_T} \quad\quad \xi = \frac{t}{a_T} \tag{7-5a, b}$$

The pseudostrain can be obtained through a piecewise integration:

$$\varepsilon^R = \frac{1}{E_R}\left[\int_0^{t_1} E(t - \tau)\frac{d\varepsilon_1}{d\tau}d\tau + \int_{t_1}^{t_2} E(t - \tau)\frac{d\varepsilon_2}{d\tau}d\tau + ... + \int_{t_{n-1}}^{t_n} E(t - \tau)\frac{d\varepsilon_n}{d\tau}d\tau\right] \tag{7-6}$$

Through the Prony representation of the relaxation modulus:

$$E(t) = E_\infty + \sum_{i=1}^m E_i e^{-t/\rho_i} \tag{7-7}$$

The pseudostrain can be obtained as:

$$\varepsilon^{R(n+1)} = \frac{1}{E_R}\left[\eta_0^{n+1} + \sum_{i=1}^m \eta_i^{n+1}\right] \tag{7-8}$$

Where η_0 and η_i are the internal state variables for the elastic responses and for the specific Maxwell element I at time step $n + 1$.

For the general case, that is viscoelastic body with damage, the stress-pseudostrain relationship can be written as:

$$\sigma = C(S_m)E_R\varepsilon^R \tag{7-9}$$

Where $C(S_m)$ is a function of damage parameters S_m.

By introducing the strain energy density function to include terms related to damage $W = (\varepsilon_{ij}, S_m)$ and the dissipated energy due to damage growth $W_s = W_s(S_m)$. They expressed the stress-strain relationships as $\sigma_{ij} = \dfrac{\partial W}{\partial \varepsilon_{ij}}$ and the damage evolution laws as $-\dfrac{\partial W}{\partial S_m} = \dfrac{\partial W_s}{\partial S_m}$

where σ_{ij} = stresses

$\qquad \varepsilon_{ij}$ = strains

$\qquad S_m$ = internal state variables (or damage parameters)

In analogy to the elasticity situations, the pseudostrain energy density function $W^R = W^R(\varepsilon^R, S_m)$ was introduced. Correspondingly, the stress-strain relationship and the damage evolution law becomes:

$$\sigma = \frac{\partial W^R}{\partial \varepsilon^R} \qquad \dot{S}_m = \left(-\frac{\partial W^R}{\partial S_m} \right)^{\alpha_m} \qquad\qquad (7\text{-}10a, b)$$

where $\dot{S}_m$ = damage evolution rate

$\qquad \alpha_m$ = material-dependent constants related to the viscoelasticity
$\qquad\qquad$ of the material

Through these manipulations, the viescoelastic stress-strain relationship with damage becomes:

$$\varepsilon_{ve} = E_R \int_0^{\xi} D(\xi - \tau) \frac{d\left(\dfrac{\sigma}{C(S)} \right)}{d\tau} d\tau \qquad\qquad (7\text{-}11)$$

Following the strain hardening viscoplastic model proposed by (Uzan, 1996; Seibi et al., 2001),

$$\dot{\varepsilon}_{VP} = \frac{g(\sigma)}{\eta_{vp}} \qquad\qquad (7\text{-}12)$$

Where $\dot{\varepsilon}_{VP}$ is the viscoplastic strain rate at $g(0) = 0$; η_{vp} is viscosity. By introducing the power law of viscosity, $\dot{\varepsilon}_{VP} = \dfrac{g(\sigma)}{A\varepsilon_{vp}^p}$. Integration of this equation results in:

$$\varepsilon_{vp}^{p+1} = \frac{p+1}{A} \int_0^t g(\sigma) dt \qquad\qquad (7\text{-}13)$$

or

$$\varepsilon_{vp} = \left(\frac{p+1}{A} \right)^{1/(p+1)} \left(\int_0^t g(\sigma) dt \right)^{1/(p+1)} \qquad\qquad (7\text{-}14)$$

Through adopting $g(\sigma) = B\sigma^q$ following Uzan (1996), Perl et al. (1983), and Kim et al. (1997), and coupling the A and B into a coefficient Y, they obtained the viscoplastic strain expression as:

$$\varepsilon_{vp} = \left(\frac{p+1}{Y} \right)^{1/(p+1)} \left(\int_0^{\xi} \sigma^q d\xi \right)^{1/(p+1)} \qquad\qquad (7\text{-}15)$$

And the viscoelastoplastic continuum damage model became (the total strain):

$$\varepsilon_T = E_R \int_0^\xi D(\xi - \xi') \frac{d\left(\dfrac{\sigma}{C(S)}\right)}{d\xi'} d\xi' + \left(\frac{p+1}{Y}\right)^{1/(p+1)} \left(\int_0^\xi \sigma^q d\xi\right)^{1/(p+1)} \tag{7-16}$$

Logically, the major step for calibrating the above VECD or VEPCD models is to characterize the damage function and damage parameter $C(S_m)$ and S_m. Through more than 20 years continuing efforts, Kim's group has established the models, the corresponding methods to calibrate models, and numerical tools through FEM implementation.

7.3   Disturbed State Models

The concept of disturbed state is based on the assumption that the response of a damaged material can be interpolated between two extreme conditions, the relative intact (RI) condition and the fully adjusted (FA) condition through a scalar damage parameter D (could be a tensor) (Desai, 2001, 2007, 2009). In the case of the stress-strain relationship, it can be represented as:

$$d\underset{\sim}{\sigma}^a = (1-D)d\underset{\sim}{\sigma}^i + Dd\underset{\sim}{\sigma}^c + dD(\underset{\sim}{\sigma}^c - \underset{\sim}{\sigma}^i) \tag{7-17}$$

The above equation can be considered as the incremental format of the following static formulation with the consideration that D is an evolving state parameter.

$$\underset{\sim}{\sigma}^a = (1-D)\underset{\sim}{\sigma}^i + D\underset{\sim}{\sigma}^c \tag{7-18}$$

$$d\underset{\sim}{\sigma}^a = (1-D)\underset{\sim}{C}^i d\underset{\sim}{\varepsilon}^i + D\underset{\sim}{C}^c d\underset{\sim}{\varepsilon}^c + dD(\underset{\sim}{\sigma}^c - \underset{\sim}{\sigma}^i) \tag{7-19}$$

$$d\underset{\sim}{\sigma}^a = \underset{\sim}{C}^{DSC} d\underset{\sim}{\varepsilon} \tag{7-20}$$

where a represents any states observed

$\quad i$ and c represent the corresponding parameters at RI and FA states, respectively

$\quad \sigma$ and ε = stress and strain vectors, respectively

$\quad \underset{\sim}{C}^i, \underset{\sim}{C}^c$ = compliance tensor or stress-strain matrix at corresponding RI and FA states

$\quad D$ = disturbance, a similar concept to damage parameter

$\quad dD$ = increment or rate of D

$\quad \underset{\sim}{C}^{DSC}$ = compliance tensor for the disturbed state

Taking the plasticity model as an example, the format of the yield function could be modified as:

$$F = \overline{J}_{2D} - (-\alpha \overline{J}_1^n + \gamma J_1^2)(1 - \beta S_r)^{-0.5} = 0 \tag{7-21}$$

where $\overline{J}_{2D} = J_{2D}/p_a^2$ = dimensionless second invariant (J_{2D}) of the deviatoric stress tensor S_{ij}

$\quad p_a$ = atmospheric pressure constant

$\quad \overline{J}_1 = (J_1 + 3R)/p_a$ = dimensionless first invariant of the total stress tensor σ_{ij}

$\quad R$ = parameter proportional to cohesion

$\quad S_r$ = stress ratio $J_{3D} \cdot J_{2D}^{-\frac{3}{2}}$

J_{3D} = third invariant of the deviatoric stress tensor

n = parameter associated with the phase change from contractive to dilative response

γ and β are associated with the characteristics of the ultimate yield surface

$\alpha = \dfrac{a_1}{\xi^{\eta_1}}$ is the hardening or growth function

a_1 and η_1 are parameters to describe the hardening features.

Where:

$$\xi = \xi_v + \xi_D \tag{7-22}$$

$$\xi_v = \frac{1}{\sqrt{3}}\varepsilon_{ii}^{p} \tag{7-23}$$

$$\xi_D = \int (dE_{ij}^{p} \bullet dE_{ij}^{p})^{1/2} \tag{7-24}$$

ε_{ij}^{p}, E_{ij}^{p} and ε_{ii}^{p} are the total, devitoric, and volumetric strains.

The disturbance D follows the following evolution law:

$$D = D_u(1 - e^{-A\xi_D^{z}}) \tag{7-25}$$

Where D_u, A, and Z are parameters to characterize the disturbance (could be healing) evolution. While the definition of D is quite clear, the association of D with the microscopic damage or other macroscopic observable quantity is not an easy task. For the rate independent plasticity, by plotting the plastic strain-loading process, one can calculate ξ_D and then fit the disturbance evolution law.

In the case of cyclic viscoplasticity, one may assume a power law to associate ξ_D at different loading cycles.

$$\xi_D(N) = \xi_D(N_r)\left(\frac{N}{N_r}\right)^{b} \tag{7-26}$$

$$N = N_r\left[\frac{1}{\xi_D(N_r)}\left\{\frac{1}{A}\ln\left(\frac{D_u}{D_u - D}\right)\right\}^{1/Z}\right]^{1/b} \tag{7-27}$$

Through plotting the accumulative plastic strain-N relationship, one may obtain the parameter b and therefore define the damage evolution law. The DSC model has a rational core. More applications in using it to model the behavior of AC are necessary.

7.4   The Benedetto Model

Another model that targets at the general deformation and failure of AC in a unified format is the Di Benedetto and Neifar (DBN) Model. In a series of publications (Benedetto et al., 2004a, b, 2007a, b, 2009), Benedetto and his colleagues presented a general model of viscoplasticity. The major features of the model lie in its general format to

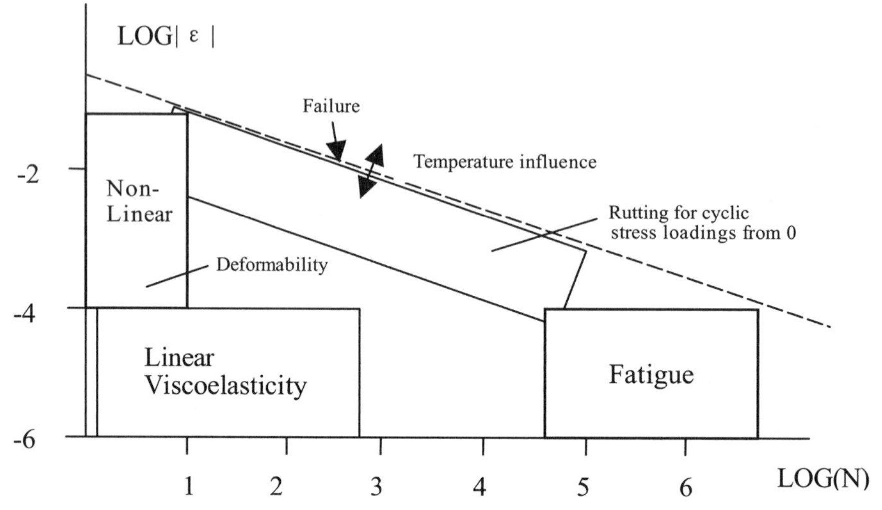

FIGURE 7.1 Typical bituminous mix behavior domains (Benedetto).

describe the behavior of AC at different strain magnitudes and number of loading repetitions (Figure 7.1). The general concept can be illustrated in Figures 7.2 and 7.3. In Figures 7.2 and 7.3, elements Ei , EPi and Vi are correspondingly the elastic, elasto-plastic, and viscous components.

Figure 7.4 illustrates several major concepts in forming the DBN model. In the 1-D case, the stress increment ($\Delta\sigma$) and strain increment ($\Delta\varepsilon$) can be generically represented as a function:

$$\Delta\sigma = f(\Delta\varepsilon) \tag{7-28}$$

For cyclic loading, it has a hyperbolic loading curve and a hyperbolic unloading curve bounded by two ultimate strengths in tension and compression (anisotropic behavior can be considered) correspondingly. While each element may be different, the ultimate strengths are the same.

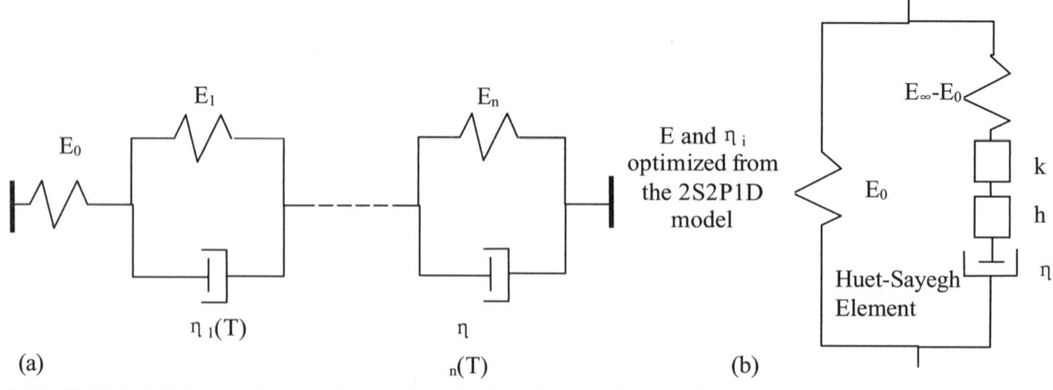

FIGURE 7.2 (a) analogical asymptotic form of the law in the linear viscoelastic (LIVE) domain, (b) 2S2P1D model (Olard and Di Benedetto, 2003).

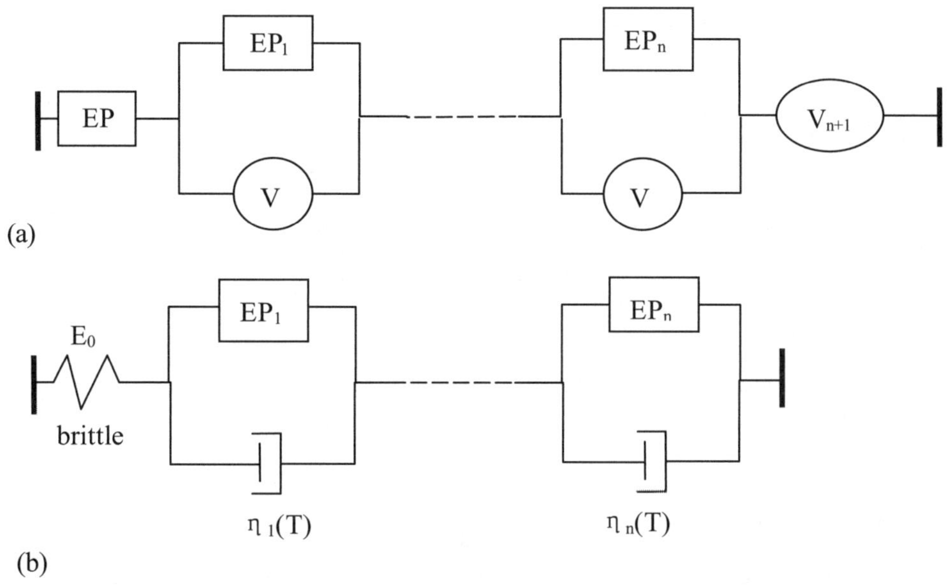

FIGURE 7.3 (a) generalized analogical body, (b) viscoplastic DBN model for bituminous mixtures (Di Benedetto, Ph.D. Dissertation, in French, 1987).

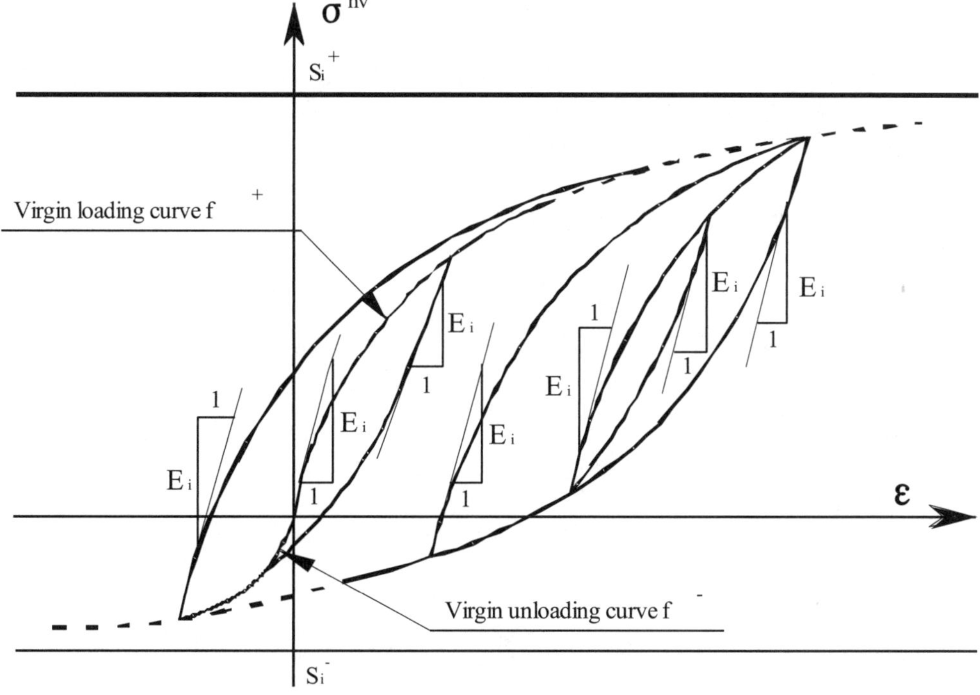

FIGURE 7.4 Major concepts in the DBN model (Benedetto).

The virgin extension curve is:

$$f_i^+(\varepsilon_i) = \sigma_i = \frac{E_i\varepsilon_i}{1 + \dfrac{E_i\varepsilon_i}{s_i^+}}$$

(7-29)

Replacing S_i^+ with S_i^- one can obtain the unloading curve:

$$f_i^-(\varepsilon_i) = \sigma_i = \frac{E_i\varepsilon_i}{1 + \dfrac{E_i\varepsilon_i}{s_i^-}}$$

(7-30)

Denoting $k = \dfrac{S^-}{S^+}$ one can obtain the following relation (this applies to the magnitude of the strains):

$$f_i^-(\varepsilon_i) = kf_i^+(\frac{\varepsilon_i}{k}) = \sigma_i$$

(7-31)

For small strains, the linear viscoelasticity applies and the modulus can be represented as:

$$E^{*DBN}(\omega,T) = \left(\frac{1}{E_0} + \sum_{i=1}^{n}\frac{1}{E_i + j\omega\eta_i(T)}\right)^{-1}$$

(7-32)

where j = complex number, $j^2 = -1$
$\quad\quad \omega$ = radian frequency ($\omega = 2\pi f, f$ = frequency)
$\quad\quad T$ = temperature

E_i, η_i can be obtained through a minimization process with a curve fitting the experimental results. By introudcing a function $\tau = \tau_0\alpha_T(T)$, where τ_0 is a material constant and $\alpha_T(T)$ is the shift function, a complex modulus expression can be obtained.

Assuming that the WLF (William, Landel and Ferry) equation holds:

$$\tau = \tau_0 a_T(T) = \tau_0 10^{-\frac{C_1(T-T_s)}{C_2+T-T_s}}$$

(7-33)

The model can consider large strains, brittle failure at low temperatures, and be extended to modeling damage, thixotropy, and healing (Benedetto and Olard, 2009).

Damage

$$d(D_{iN}) / dN = function_E(\varepsilon_{iN}, D_{iN})$$

(7-34)

$$d(D\eta_{iN}) / dN = function_\eta(\varepsilon_{iN}, \eta_{iN})$$

(7-35)

Thixotropy

$$D\eta_{iNthixo} = function_{thixo} \text{ (dissipated energy, dissipated energy }_N, \varepsilon_{iN})$$

(7-36)

Healing

$$d(D\eta_{iN})/dt = function_{healing} \text{ (loading parameters, } D_{iN}, D\eta_{iN})$$

(7-37)

7.5 Viscoplastic Model with Microstructural Characteristics

Masad et al. (2005, 2007) and Tashman et al. (2005) developed a viscoplasticity model for AC at relatively high temperatures. The unique features of this model include the incorporation of microstructure and the particle orientation information measured from 2D and 3D images. The general principle follows the Perzyna model. For relatively high temperatures, the viscoelastic component was considered as negligible and only the viscoplastic component is considered.

$$\varepsilon_{Total} = \varepsilon^{ve} + \varepsilon^{vp} \tag{7-38}$$

They adopted a non-associated flow rule:

$$\dot{\varepsilon}_{ij}^{vp} = \Gamma \cdot \left\langle \phi(\bar{f}) \right\rangle \cdot \frac{\partial \bar{g}}{\partial \sigma_{ij}} \tag{7-39}$$

Where $\dot{\varepsilon}_{ij}^{vp}$ is the viscoplastic strain rate tensor; Γ is the fluidity parameter, which establishes the relative rate of viscoplastic straining; $\bar{g}$ is the viscoplastic potential function; $\bar{f}$ is the viscoplastic yield function; and <> are the Macauley brackets to ensure the non-positive $\phi(\bar{f})$ leads to $\dot{\varepsilon}_{vp} = 0$. These follow the same format as that by Perzyna (1966).

Yield Function
Assuming a power law function for the yield surface, the yield function in the above equation becomes:

$$\left\langle \phi(\bar{f}) \right\rangle = \begin{cases} 0 & \phi(\bar{f}) < 0 \\ \phi(\bar{f}) = \bar{f}^N & \phi(\bar{f}) > 0 \end{cases} \tag{7-40}$$

Where N is a constant to be determined experimentally.

Hardening
The hardening parameter is defined as:

$$\kappa = \kappa_0 + \frac{H}{C_r + \dfrac{C_s}{\dot{\bar{\varepsilon}}_{vp}}} \left[1 - \exp\left(-(C_r + \frac{C_s}{\dot{\bar{\varepsilon}}_{vp}})\bar{\varepsilon}_{vp}\right) \right] \tag{7-41}$$

H is an isotropic hardening coefficient; C_r is a dynamic recovery coefficient; C_s is a static recovery coefficient; and $\dot{\bar{\varepsilon}}_{vp}$ is the effective viscoplastic strain rate. κ_0 defines the initial yield surface and $\bar{\varepsilon}_{vp}$ is the effective viscoplastic strain.

Damage

$$\xi = \frac{A_v}{S}$$

$$\sigma_e = \frac{\sigma}{1-\xi} \quad \text{where } 0 \le \xi \le 1 \tag{7-42}$$

ξ is the damage parameter; A_v is the area of cracks and air voids; S is total cross-sectional area of the specimen; σ is the stress applied to the damaged specimen; and σ_e is the effective stress applied to the fictitious undamaged specimen.

Microstructure Effects

Microstructure directional distribution in HMA is developed based on the modified Drucker-Prager yield function:

$$\bar{f} = \sqrt{\bar{J}_2^e} - \bar{\alpha}\bar{I}_1^e - \bar{\kappa}$$

$$\bar{I}_1^e = \frac{\bar{I}_1}{1-\xi^*}$$

$$\bar{J}_2^e = \frac{\bar{J}_2}{(1-\xi^*)^2}$$

$$\bar{I}_1 = (a_1\delta_{ij} + a_2 F_{ij})\sigma_{ij}$$

$$\bar{J}_2 = (2b_1\delta_{ik}\delta_{jl} + 4b_2 F_{ik}\delta_{lj})S_{ij}S_{kl} \tag{7-43}$$

σ_{ij} and S_{ij} are the stress tensor and the corresponding deviatoric stress tensor, respectively, and they are related as $S_{ij} = \sigma_{ij} - \frac{1}{3}\sigma_{kk}\delta_{ij}$. $\bar{\alpha}$ is a parameter that reflects the material internal function, $\bar{\kappa}$ is an isotropic hardening parameter that reflects the cohesive properties, and a_1, a_2, b_1 and b_1 are functions of the second invariant (D_2) of the deviatoric microstructure tensor (F_{ij}) as follows:

$$a_1 = 1 - \lambda(2D_2)^{1/2}$$

$$a_2 = 3\lambda(2D_2)^{1/2}$$

$$b_1 = \frac{1}{4} - \frac{\mu}{2}(2D_2)^{1/2}$$

$$b_2 = \frac{3}{4}\mu(2D_2)^{1/2} \tag{7-44}$$

Where λ and μ are material coefficients that account for the effect of anisotropy on the confining and shear stresses, respectively.

The microstructure tensor adopted follows Tobita (1989):

$$F_{ij} = \begin{pmatrix} (1-\Delta)/(3+\Delta) & 0 & 0 \\ 0 & (1+\Delta)/(3+\Delta) & 0 \\ 0 & 0 & (1-\Delta)/(3+\Delta) \end{pmatrix} \tag{7-45}$$

Where Δ is a microstructure parameter that quantifies the average anisotropy of the aggregate orientation distribution measured on 2D axial cut sections of the material as:

$$\Delta = \frac{1}{M}[(\sum_{k=1}^{M}\cos 2\theta^k)^2 + (\sum_{k=1}^{M}\sin 2\theta^k)^2]^{\frac{1}{2}} \tag{7-46}$$

The modified yield function reduces to the classical Drucker–Prager function for the special case of isotropic and intact (no damage) material. Assuming a power law for the viscoplastic yield function, it becomes:

$$\langle \phi(f) \rangle = \begin{cases} 0, & \text{No Viscoplastic Strain,} & \bar{J}_2^e - \alpha \bar{I}_1^e - \kappa \leq 0 \\ [\sqrt{\bar{J}_2^e} - \alpha \bar{I}_1^e - \kappa] & \text{Viscoplastic Strain Occurs} & \bar{J}_2^e - \alpha \bar{I}_1^e - \kappa > 0 \end{cases} \tag{7-47}$$

7.6 Temperature Dependent Viscoplastic Hierarchical Single Surface (HiSS) Model

Huang et al. (2002) extended the HiSS model by Desai et al. (1986) to include temperature dependence of the model parameters. The major structure follows the Perzyna model. Similar work in this category includes the Delft model (Erkens et al., 2003) and the University of Maryland (UOM) model (Gibson et al., 2003; Gibson, 2006). The UOM model is more inclusive but similar to the VEPCD model presented in Section 7.2. For the purpose of conciseness, neither the Delft model nor the UOM model is presented in detail.

Strain Components (rate equation)

$$e_{ij} = e^{e\theta}{}_{ij} + e_{ij}^{vp\theta} \tag{7-48}$$

Elastic Component

The elastic component follows the linear elasticity.

$$d\sigma_{ij} = C_{ijkl}^{e\theta} de_{kl}^{e\theta} = C_{ijkl}^{e\theta}(de_{kl} - de_{kl}^{p\theta} - \alpha_T d\theta \delta_{kl}) \tag{7-49}$$

Viscoplastic Component

Yield function:

$$F = \left(\frac{J_{2D}}{p_a^2}\right) - \left\{ \gamma(\theta) \cdot \left(\frac{J_1}{p_a}\right)^2 - \alpha(\theta) \cdot \left(\frac{J_1}{p_a}\right)^n \right\} \cdot [1 - \beta(\theta) \cdot S_r]^{-0.5} \tag{7-50}$$

J_1 is the first invariant of the stress tensor; J_{2D} is the second invariant of the deviatoric stress tensor; P_a is atmospheric pressure; $\alpha(\theta)$ is the hardening function; and $\beta(\theta)$ is material parameters. S_r is stress ratio and θ is temperature.

For associated isotropic hardening material, the thermoplastic strain increment flow rule is:

$$de_{ij}^{p\theta} = \lambda \frac{\partial F}{\partial \sigma_{ij}} \tag{7-51}$$

Where λ is the proportionality factor.

Considering the time dependency, the total strain rate is:

$$\dot{e}_{ij} = \dot{e}_{ij}^{e\theta} + \dot{e}_{ij}^{vp\theta} \tag{7-52}$$

The thermo-viscoplastic strain rate is written as:

$$\dot{e}_{ij} = \dot{e}_{ij}^{e\theta} + \Gamma(\theta)\left\langle \phi(\frac{F}{F_0}) \right\rangle \frac{\partial F}{\partial \sigma_{ij}} + \alpha_T \dot{\theta}\delta_{ij} \qquad (7\text{-}53)$$

Where F is yield function; $\Gamma(\theta)$ is temperature dependent fluidity parameter; ϕ is the function that plays the role of a flow coefficient depending on the distance from the present stress point to the yield surface F; and <> are the Macauley brackets.

The model did not include the damage effect and the microstructure effect. Through the use of triaxial tests including isotropic compression, conventional triaxial compression, reduced triaxial extension, and creep tests at various temperatures, the model parameters are calibrated.

7.7   Stress-Dependent Elastoviscoplastic Constitutive Model with Damage

Collop et al. (2003) developed an elastoviscoplastic model with damage considerations. The following presents its 1-D formulations. The model followed the Perzyna decomposition of the total strain into three components.

$$\varepsilon_{Total} = \varepsilon_{el} + \varepsilon_{ve} + \varepsilon_{vp} \qquad (7\text{-}54)$$

The Elastic Component

$$\varepsilon_{el}(t) = \sigma(t) / E_0 \qquad (7\text{-}55)$$

Where σ is the stress and E_0 is the modulus of elasticity of the elastic element.

The Viscoelastic Component

$$\varepsilon_{ve}(t) = J_{ve}(0)\sigma(t) + \int_0^t \frac{dJ_{ve}(t-t')}{d(t-t')}\sigma(t')dt' \qquad (7\text{-}56)$$

J_{ve} is viscoelastic creep compliance
t' is dummy integration variable

The Viscoplastic Component

$$\varepsilon_{vp}(t) = J_{vp}(0)\sigma(t) + \int_0^t \frac{dJ_{vp}(t-t')}{d(t-t')}\sigma(t')dt' \qquad (7\text{-}57)$$

J_{vp} is viscoplastic creep compliance

Stress Dependency

$$\frac{d\varepsilon}{dt} = K\sigma^n = K\sigma^{n-1}\sigma = \frac{\sigma}{\lambda_m} \qquad (7\text{-}58)$$

Where K and n are materials constants. The stress dependency and non-linearity are represented.

Damage

The strain is assumed to be a function of both stress and damage (D, same definition as 7-42).

$$\frac{d\varepsilon}{dt} = \frac{C_1 \sigma^n}{(1-D)^m}$$

$$\frac{dD}{dt} = \frac{C_2 \sigma^v}{(1-D)^\mu} \tag{7-59}$$

Where C_1, C_2, n, m, μ are material constants depending on temperature.

Collop et al. (2006) extended the above formulas into 3D cases and developed numerical implementations. The simplicity of the model structure is advantageous.

7.8   3D Constitutive Model for Asphalt Pavements

Oeser and Moller (2004) developed a generalized 3D viscoplastic constitutive model for AC. Its 1-D presentation is as follows:

Strain Components

$$\dot{\varepsilon}_{ges}(t,\sigma,T) = \dot{\varepsilon}_{el}(\sigma,T) + \dot{\varepsilon}_{ve}(t,\sigma,T) + \dot{\varepsilon}_{vp}(t,\sigma,T) + \dot{\varepsilon}_{th}(t,T) \tag{7-60}$$

In which σ is stress; T is temperature; and $\dot{\varepsilon}_{ges}(t,\,\sigma,\,T)$ is overall strain rate, dependent on σ, T;

The Elastic Component

$$\dot{\varepsilon}_{el}(\sigma,T) = \frac{\sigma}{E_H(\sigma,T)} \tag{7-61}$$

The Viscoelastic Component

$$\dot{\varepsilon}_{ve}(t,\sigma,T) = \frac{\sigma - E_k(\sigma,T) \cdot \varepsilon_{ve}(t,\sigma,T)}{\eta_K(\sigma,T)} \tag{7-62}$$

This actually indicates that the elastic component and the viscoelastic component are in parallel.

The Viscoplastic Component

$$\dot{\varepsilon}_{vp}(t,\sigma,T) = \frac{\sigma}{\eta_N(\sigma,T)} \tag{7-63}$$

This actually indicates that the viscoplastic component is in series with the viscoelastic part.

The Thermal Component

$$\dot{\varepsilon}_{th}(t,T) = \dot{T} \cdot \alpha_T \tag{7-64}$$

The material parameters E_H, E_k, η_K, η_N are dependent on both temperature and stress.

The intergranular tensile stress σ_H develops as a result of the lateral strain of the inhomogeneous material.

Extended Model

$$\dot{\varepsilon}_{ges}(t,\sigma,T) = \dot{\varepsilon}_{el}(\sigma,T) + \dot{\varepsilon}_{ve}(t,\sigma,T) + \dot{\varepsilon}_{vp}(t,\sigma,T) + \dot{\varepsilon}_{th}(t,T) + \dot{\varepsilon}_{tr}(t,\sigma,T) \qquad (7\text{-}65)$$

$\dot{\varepsilon}_{tr}(t,\,\sigma,\,T)$ is tertiary strain rate.

The above formulations can be extended to 3D cases.

Elastic

$$\sigma_{ij} = \frac{E_H}{1-\mu_H} \cdot [\varepsilon_{ij}^{el} + \frac{\mu_H}{1-2\cdot\mu_H} \cdot \bar{\varepsilon}^{el} \cdot \delta_{ij}] \qquad (7\text{-}66)$$

Viscoelastic

$$\sigma = \sigma_{KH} + \sigma_{KN} \qquad (7\text{-}67)$$

$$\sigma_{KH} = E_K(\sigma,T) \cdot \varepsilon_{ve}(t,\sigma,T) \qquad (7\text{-}68)$$

$$\sigma_{KN} = \eta_K(\sigma,T) \cdot \dot{\varepsilon}_{ve}(t,\sigma,T) \qquad (7\text{-}69)$$

The stress σ_{KH} is the part of the overall stress σ giving rise to the elastic strain at time t. In contrast, the stress σ_{KN} constitutes the part of the overall stress σ responsible for the increase in the viscoelastic strain at time t.

Viscoplastic

For viscoplastic strain the corresponding evolution equation is in the form of Equation 7-70 below.

$$\dot{\varepsilon}_{vp}(t,\sigma,T) = \frac{\sigma}{\eta_N(\sigma,T)} \qquad (7\text{-}70)$$

Tertiary Part

$$\sigma = \sigma_{DH} + \sigma_{DN}$$

$$\sigma_{DH} = -\frac{D_1(\sigma,T)}{D_2(\sigma,T)} \cdot \varepsilon_{tr}(t,\sigma,T)$$

$$\sigma_{DN} = \frac{1}{D_2(\sigma,T)} \cdot \dot{\varepsilon}_{tr}(t,\sigma,T) \qquad (7\text{-}71)$$

$$\sigma = \sigma_{HH} + \sigma_{HN}$$

$$\sigma_{HH} = -\frac{H_1(\sigma,T)}{H_2(\sigma,T)} \cdot \varepsilon_{tr}(t,\sigma,T)$$

$$\sigma_{HN} = -\frac{1}{H_2(\sigma,T)} \cdot \dot{\varepsilon}_{tr}(t,\sigma,T) \qquad (7\text{-}72)$$

The stresses σ_{DH} and σ_{HH} are coupled to the tertiary strain via the corresponding ratios of the parameters D_1 and D_2 or H_1 and H_2.

More general models should also include the kinematic hardening. The SHRP permanent deformation is a good example (see Chapter 8). A good review paper was presented by Krempl (1987). Other examples in the general models include the ones by Chaboche (1986) and Krempl and Khan (2003). There are other models such as (Nguyen et al., 2007). A good literature resource can be found from Kim et al. (2009).

7.9 Generalization of the Models

The elastic strain energy can be stored and recovered. Depending on the dissipation mechanism, one can assume various dissipation mechanisms to cover plastic deformation, viscoplastic deformation, and damage (fracture) dissipation. The possibilities of unifying these models exist. A potential approach is the internal variable method. A detailed description can be found in Lubliner (1990).

7.10 Fatigue Modeling

7.10.1 Overview

The fundamental mechanisms of fatigue of AC are complicated. In general, fatigue is the accumulation of damage (in broad sense) in materials under the effect of repeated loading. Fatigue damage accumulation in AC mixtures results in cracking, which is one of the main distresses in flexible pavements. A thorough description of damage caused by fatigue therefore becomes essential if mechanistic-based pavement design principles are to be applied accurately.

The fatigue properties of AC are usually obtained by repeated-load laboratory testing. The early fatigue models of asphalt mixtures are simple phenomenological formulas of fatigue under cyclic loading. Paris's law plays an important role in linking the rate of crack growth to tensile strain developed in asphalt mixture. In addition, fatigue models based on continuum damage or fracture mechanics have been developed. Paris's law was also used to link the rate of crack growth to the degradation of fracture toughness indicators such as the stress intensity factor when LEFM is used. Models were also reviewed involving the concepts of dissipated energy or surface energy that were proposed taking the format of the phenomenological formula or incorporated in the Paris's law. Numerical methods using the cohesive zone approach will be briefly reviewed in this section.

7.10.2 Empirical Phenomenological Models

Models falling into this category were developed based on experimental data to link fatigue life (maximum allowable number of repetitions N_f) to tensile strain ε_t and the dynamic modulus E^* (sometimes) of AC. A typical formulation can be represented as:

$$N_f = k_1 (\varepsilon_t)^{-k_2} \left| E^* \right|^{-k_3} \tag{7-73}$$

In which k_1, k_2, k_3 are regression coefficients. A typical failure criterion for fatigue life at a specific strain is the loading repetition at which AC loses its modulus by 50%. If $k_2 = k_3$ and for direct tension tests, $N_f = k_1 (\varepsilon_t E^*)^{-k_2} = k_1 (\sigma_t)^{-k_2}$.

There are many models in this category. Following are two examples.

The current Mechanistic-Empirical Pavement Design Guide (MEPDG) uses a model shown below:

$$N_f = 0.00432 \times \beta_{f1} \times C \times \left(\frac{1}{\varepsilon_t}\right)^{3.291\beta_{f2}} \left(\frac{1}{E}\right)^{0.854\beta_{f3}} \tag{7-74}$$

$$C = 10^{4.84*[V_b/(V_a+V_b)-0.69]} \tag{7-75}$$

$\beta_{f1}, \beta_{f2}, \beta_{f3}$ = calibration factors
$\quad C$ = laboratory to field adjustment factor
$\quad \varepsilon_t$ = critical tensile strain
$\quad E$ = stiffness of the AC surface layer
$\quad V_a$ = air voids (%)
$\quad V_b$ = effective binder content (%)

The model is similar to that by El-Basyouny et al. (2005).

$$N_f = 0.00432C\left(\frac{1}{\varepsilon_t}\right)^{3.291}\left(\frac{1}{E}\right)^{0.854}$$

$$C = 10^M$$

$$M = 4.84\left(\frac{V_b}{V_a + V_b} - 0.69\right)$$

where N_f = number of repetitions to fatigue cracking,
$\quad t$ = tensile strain at the critical location,
$\quad C$ = correction factor,
$\quad E$ = stiffness of the material (psi),
$\quad M$ = power factor,
$\quad V_b$ = effective binder content (%), and
$\quad V_a$ = air voids (%)

Sousa et al. (1998) evaluated a series of phenomenological fatigue models under an SHRP contract. These models were calibrated by applying shift factors based on field observations to provide reasonable estimates of the in-service life of a pavement. The shift factors, of the order 10 and more, are necessary to correct deficiencies in the approach. One such deficiency is the neglect of the crack propagation phase, which is usually not properly represented or simulated in the conventional laboratory fatigue tests, and thus the phenomenological models. Several other models in this category are listed as follows.

The Monismith Model

Monismith et al. (1969) proposed a model as the following format:

$$N = K\left(\frac{1}{\varepsilon_t}\right)^a \left(\frac{1}{S_{mix}}\right)^b \tag{7-76}$$

Where S_{mix} is the flexural stiffness; K is a factor that recognizes the influence of asphalt content and degree of compaction; ε_t = tensile strain applied; and a, b are the experimentally determined coefficients.

The Shell Model

$$N_f = \left[\frac{\varepsilon_t}{(0.856V_b + 1.08)S_{mix}^{-0.36}} \right]^{-5} \tag{7-77}$$

Where N_f = fatigue life; ε_t = tensile strain; S_{mix} = mixture flexural stiffness; and V_b = asphalt content by volume.

The Asphalt Institute Model

$$N_f = S_f * 10^{\left[4.84(VFB-0.69)\right]} * 0.004325 * \varepsilon_t^{-3.291} * S_{mix}^{-0.845} \tag{7-78}$$

Where N_f = fatigue life; S_f = shift factor to convert laboratory test results to field expected results (the recommended factor is 18.4 for a 10% cracked area); ε_t = tensile strain applied; S_{mix} = flexural stiffness of a mix (psi); and VFB = voids filled with bitumen.

The Tayebali (1996) Model (SHRP Project A-003A)

$$N_f = S_f * 2.738 \times 10^5 * e^{0.077 VFB} * \varepsilon_0^{-3.6224} S_0^{"-2.720} \tag{7-79}$$

Where S_f = shift factor to convert laboratory results to field expected results (the recommended factor is 10 for 10% cracked area and 14.0 for 45% cracked area), e = base of natural logarithm, VFB = percentage of voids filled by bitumen, ε_0 = strain level, and $S_0^{"}$ = loss of stiffness as measured in flexure.

The Medani and Molenaar Model (2000)

$$N_f = k_1 \left(\frac{1}{\varepsilon_t} \right)^n \tag{7-80}$$

$$n = \frac{2}{m\left(0.541 + \dfrac{0.346}{m} - 0.03524V_a\right)}$$

$$\log k_1 = 6.589 - 3.762n + \frac{3209}{S_m} + 2.332\log V_b + 0.149\frac{V_b}{V_a} + 0.928PI - 0.0721T_{R\&B} \tag{7-81}$$

Where k_1 = coefficient; ε_t = initial tensile strain; m = slope of the mix stiffness master curve; $T_{R\&B}$ = softening point for binder (determined by ring and ball test) (°C); S_m = mixture stiffness (MPa); n = fracture parameter; V_a = air void content (%); and V_b = volume of binder (as a percentage).

As it is understood that AC mixtures are typically assumed to exhibit linear-viscoelastic behavior, their response is dependent on time of loading and test temperature. This behavior is represented by the following relationship where the stiffness is a function of temperature and time:

$$S_{mix} = \frac{\sigma}{\varepsilon}(t, T) \tag{7-82}$$

Where S_{mix} = mixture stiffness; σ = stress level; ε = strain level; t = time of loading; and T = test temperature.

7.10.3 Fracture Mechanics Models

The fracture models made use of the cracking propagation law for both linear elastic fracture mechanics and the non-linear fracture mechanics. For LEFM the Paris law is often used. For the non-linear elastic mechanics, the J integral is often used.

The Uzan Model

Uzan (2007) modeled the fatigue cracking as a two-stage process consisting of crack initiation and crack propagation. The crack initiation stage is characterized by conventional laboratory fatigue tests; while the crack-propagation stage is described using the Paris-Erdogan law.

Uzan adopted the model developed by Tayebali et al. in SHRP Project A-003A as the crack-initiation model (Equation 7-79).

The Paris-Erdogan law $\dfrac{dc}{dN} = A(\Delta K)^n$ was used as the crack-propagation model for evaluating the number of load repetitions needed to propagate the crack. Where, c = crack length; N = number of load repetitions; ΔK = difference between maximum and minimum stress intensity factor K; and A, n = Paris law fracture parameters for AC.

In the case of $\Delta K = K$, it is given by the following equation:

$$N_p = \frac{1}{A} \cdot \int_{c0}^{h} \frac{dc}{K^n} = \frac{1}{A} \cdot I_K \tag{7-83}$$

Where N_p is number of load repetitions to propagate a crack of initial length c_0 to the surface; h is layer thickness; c_0 is initial crack length; K is stress-intensity factor (K_I for Mode I and K_{II} for Mode II); n, A, are material properties; and $I_K = N_p A$.

Owusu-Antwi Model

Owusu-Antwi et al. (1998) used the principles of fracture mechanics and developed a mechanistic-based performance model for predicting the amount of reflective cracks in composite AC/PCC pavements. It is illustrated that a mechanistic-based model can be developed to closely model the real-life behavior of composite pavements and predict the amount of reflective cracks. Because of the mechanistic nature of the model, it is particularly effective for performance prediction for design checks and pavement management. Also, the model has great potential for application in cost allocation since it can take into account the relative damaging effect of the actual axle loads in any traffic distribution.

For a 1-kilometer long section, the number of reflective cracks can be determined from the following equation:

$$RCRACKS = \frac{DAMTOT^{1.9}}{DAMTOT^{1.9} + 1} * \frac{1000 + S}{S} \tag{7-84}$$

Where S is the joint spacing; the total damage (DAMTOT) from both temperature and traffic loading can be calculated using the following equation:

$$DAMTOT = 0.0132 \sum \frac{n_1}{N_i} + \frac{AGE}{N_{temp}} (8.79 + 0.000795 * FI * AGE) \tag{7-85}$$

Where AGE = the age of the pavement after overlay, in years; FI = the freezing index; N_{temp} = the number of temperature load applications to failure; n_i = the actual number of axle passes for axle weight; i, and N_i = the allowable number of axle passes according to the following equation. The N_i was developed based on Paris's crack growth law that was used to model the relationship between crack propagation in AC and their stress intensity factors and other material fatigue properties.

$$N_i = \frac{h_{OL}}{10^{-12} K_C^{2.4}}$$ (7-86)

Where K_C = the stress intensity factor; h_{OL} = the thickness of the AC overlay.

Non-Linear Fracture Mechanics

For non-linear fracture mechanics, plastic deformation is considered. A commonly used parameter for characterization of non-linear behavior is the J contour path independent integral, which can be used as both an energy parameter and a stress intensity parameter.

$$J = \int_{\Gamma} (W dy - \sigma_{ij} n_j \frac{\partial u_i}{\partial x} ds)$$ (7-87)

Where W = the strain energy density, Γ = any contour, and ds = a length increment along the contour (see Chapter 6). Like the stress intensity factor K in linear elastic mechanics, the growth rate of crack is a function of J instead of K in non-linear fracture mechanics. If the material is viscoelastic, strain rates, stress rates, displacement rates and work rates should be used in the above formulas (see Chapter 6).

Smith and Hesp Model

Smith and Hesp (2000) also modeled the crack propagation phase using Paris's law, which relates the rate of crack growth to fundamental properties of the material and experimental conditions and stress intensity factor K.

7.10.4   Damage-Based Models

The models falling in this category rest on the accumulative damage concept.

Castro and Sanchez Model

Castro and Sanchez (2008) proposed a phenomenological model based on the continuum damage theory. The three point bending fatigue test was used to determine the parameters in the equation shown below:

$$N = a \cdot \varepsilon_0^b \cdot D^c$$ (7-88)

N is the number of loading cycles and ε_0 is the initial strain; a, b and c are the parameters of asphalt concrete determined experimentally; D is the damage parameter,

$D = \dfrac{\left||E_0^*| - |E^*|\right|}{|E_0^*|}$, defined as the loss of the norm of the complex modulus that takes place in a specimen during a test.

Bodin Model

Bodin, et al. (2004) proposed a non-local damage model to predict pavement fatigue cracking, which was implemented in a finite-element code along with a self-adaptive jump-in-cycle procedure for high-cycle fatigue computations.

The mathematical model used to describe mechanical damage is an elasticity-based damage model for fatigue.

$$N_{crit} = \frac{F(d_{crit})(\beta+1)}{\overline{\varepsilon}_a^{\beta+1}} \quad \text{with } F(d_{crit}) = \text{constant}, \ F(d) = \sum_{cycle 1}^{N} \frac{\overline{\varepsilon}_a^{\beta+1}}{\beta+1} \tag{7-89}$$

Where d is the damage variable; $\overline{\varepsilon}_a$ is the amplitude of the equivalent strain over one cycle; $f(d)$ is the function of damage and $F(d)$ is the scalar function of damage; and β is a model parameter.

$$f(d) = Cd^{\alpha}, \ \text{and} \ F(d) = \alpha_1 \left\{ 1 - \exp\left[\frac{-d}{\alpha_2}\right]^{\alpha_3} \right\} \tag{7-90}$$

Where C is the secant stiffness of the material; α_1, α_2 and α_3 are the three model parameters.

Lee Model

Based on the elastic-viscoelastic correspondence principle and continuum damage mechanics through mathematical simplifications, a fatigue performance prediction model of AC was developed from a uniaxial constitutive model by Lee et al. (2000). The fatigue model is similar to the phenomenological tensile strain-based fatigue model, a comparison between the two models yields the regression coefficients in the phenomenological model as functions of viscoelastic properties of the materials, loading conditions, and damage characteristics. The fatigue model in terms of the total number of loading cycles to failure N_f, is given as:

$$N_{f,Total} = \frac{f(S_{1,f})^{p_1}}{p_1(0.125IC_{11}C_{12})^{\alpha_1}} |E*|^{-2\alpha_1} \varepsilon_0^{-2\alpha_1} + \sum_{i=1}^{M} \frac{f(S_{3e})^{p_3}}{p_3\left[0.125I(C_2 + S_B^R)C_{31}C_{32}\right]^{\alpha_3}} |E*|^{-2\alpha_3} \varepsilon_0^{-2\alpha_3} \tag{7-91}$$

Where f is the loading frequency; S^R is the secant pseudostiffness; I is the initial pseudostiffness; S_m = internal state variables (or damage parameters) that account for the effects of damage; S_{1f} = value of damage parameter S_1 at failure; $p_1 = 1 + (1 - C_{12})\,C_1$; $p_3 = 1 + (1 - C_{32})\,C_3$; C_{ij} = regression coefficients; C_i = material constants ($i = 1, 2, 3$); ε_0 is the initial strain; and E^* is the complex modulus.

7.10.5 Micromechanics-Based Model

Guddati et al. (2002) adopted a random truss lattice model to simulate the linear elastic and viscoelastic deformation of homogeneous materials in axial compression and shearing experiments. The linear elastic deformation and the stress field in heterogeneous materials in an axial compression experiment, and the damage evolution in elastic solids under an indirect tensile test were simulated. The simulation results agree well with the theoretical solutions and show excellent promise in predicting cracking patterns in the indirect tensile test.

7.10.6 Surface Energy-Related Model

Based on Schapery's fundamental law of fracture mechanics and Lytton et al.'s healing model, Cheng et al. (2002) derived the following cohesive fracture law for cyclic fatigue testing, which is the equivalent of Paris's law.

$$\frac{dc}{dN} = \int_0^{\Delta t_f} \frac{K_a \alpha (D_{1f} E_R J_v)^{\frac{1}{m_f}}}{(2\Gamma_f - D_{0f} E_R J_v)^{\frac{1}{m_f}}} dt - \frac{dh}{dN} \tag{7-92}$$

Where dh/dN = the rate of crack healing per load cycle; dc/dN = the rate of cracking per load cycle; Γ_f = fracture surface energy density of a crack surface (FL^{-1}); α = the length of the fracture process zone before the crack surface; E_R = reference modulus used to represent a nonlinear viscoelastic material as equivalent nonlinear elastic (FL^{-2}); D_{0f}, D_{1f}, m_f = creep compliance coefficients of the power law; $D_f(t) = D_{0f} + D_f t^{mf}$; K_f = constants that depend on the value of m, the slope of the log creep compliance versus log time curve (a common value is one-third); and J_v = strain energy (this is the rate of change of dissipated energy per unit of crack growth area from one tensile load cycle to the next) (FL^{-1}).

Cohesive fracture and healing were treated in this model as a special case of adhesive fracture and healing. Asphalt pavement fatigue failure occurs at the interface between asphalt and aggregate, which is caused by adhesive bonding failure. For adhesive fracture, the following, more general, equations hold:

$$\Gamma_1 + \Gamma_2 - \Gamma_{12} = E_R D_f(t_\alpha) J_V$$

$$\Gamma_1 + \Gamma_2 - \Gamma_{12} = E_R D_h(t_\alpha) H_V$$

$$\frac{dc}{dN} = \int_0^{\Delta t_f} \frac{K_f \alpha (D_{1f} E_R J_v)^{\frac{1}{m_f}}}{(\Gamma_1 + \Gamma_2 - \Gamma_{12} - D_{0f} E_R J_v)^{\frac{1}{m_f}}} dt - \frac{dh}{dN} \tag{7-93}$$

Where Γ_1 = surface energy of the asphalt, Γ_2 = surface energy of the aggregate, and Γ_{12} = energy of interaction between the asphalt and the aggregate.

7.10.7 Dissipated Energy-Based Model

Based on the relationship between total dissipated energy and the number of cycles up to fatigue or fracture, the following energy model was developed by van Dijk (1975):

$$N = \left(\frac{\pi S_{fat} Sin\phi}{A_\varphi}\right)^{\frac{1}{z-1}} \varepsilon_0^{\frac{2}{z-1}} \tag{7-94}$$

Where N is the number of load applications to fatigue; S_{fat} is initial stiffness modulus; ϕ-phase angle between stress and strain; and z and A are material constants.

Ghuzlan et al. (2000) proposed a dissipated energy model to model the fatigue of AC. The following equation is used to calculate dissipated energy in the flexural fatigue test:

$$W_i = \pi\sigma_i\varepsilon_i \sin\varphi_i \tag{7-95}$$

Where W_i = dissipated energy at load cycle I; σ_i = stress amplitude at load cycle I; ε_i = strain amplitude at load cycle i; and ϕ_i = phase angle between stress and strain.

Then the total (cumulative) dissipated energy at failure will be as follows:

$$W_{fat} = \sum_{i=1}^{n} W_i \tag{7-96}$$

Conventionally, fatigue life was related to the total dissipated energy in the fatigue test as follows, where N is the number of cycles to failure and A, z are experimentally determined coefficients:

$$W_{fat} = A \cdot (N)^z \tag{7-97}$$

The conventional failure criterion in fatigue testing, 50% reduction in modulus, however, was found not to provide a consistent indicator of the onset of failure when different modes of loading are used. Dissipated energy, when examined as a change between two load cycles, provides a more fundamentally correct indication of damage from one load cycle to the next than does cumulative dissipated energy.

A new failure criterion has been presented for fatigue characterization based on the premise that the change in dissipated energy from one load cycle to the next is the appropriate indicator of the damage done to the material by that load cycle. The new failure criterion was defined as the change in dissipated energy (ΔDE) between cycles a and $a + 1$ (representative of the actual damage to the sample) divided by the total dissipated energy (DE) to load cycle a. Because of equipment readout limitations, this change was usually calculated approximately every 100 load cycles. The new failure criterion is given as follows, where N is the number of cycles to failure and A, z are experimentally determined coefficients.

$$\frac{\Delta DE}{DE} = A(N)^z \tag{7-98}$$

The damage accumulation ratio ($\Delta DE/DE$) provides a consistent failure indicator that appears to be independent of the mode of loading.

Bonnetti et al. (2002) performed fatigue tests on a set of unmodified and modified binders and analyzed the test results using the dissipated energy ratio concept. The number of cycles to crack propagation, N_p, was used as the fatigue criterion for the analysis. Using the initial dissipated energy per cycle (W_i) as the main independent variable for modeling fatigue of binders appears to be a promising technique to normalize some of the testing conditions. The parameter N_{p20}, defined as the number of cycles at which the dissipated energy ratio shows 20% deviation from the no-damage ratio, appears to be a promising parameter to define failure.

It was found that the most suitable means of evaluating the effect of modifiers in the fatigue response of the binders is using the cumulative dissipated energy ratio (DER):

$$DER = \frac{\displaystyle\sum_{i=1}^{n} W_i}{W_n} \tag{7-99}$$

Where W_i = dissipated energy per cycle; W_n = dissipated energy at cycle n; and $\sum_{i=1}^{n} W_i$ = total sum of dissipated energy up to cycle n. Then fatigue life, based on the DER criterion, can be represented as:

$$N_f = K_2 \left(\frac{1}{W_i}\right)^{K_1} \tag{7-100}$$

Where K_1 and K_2 are the slope and the Y intercept, respectively, of the fitted fatigue curves (W_i versus N_p) for a given asphalt binder. The slope and the Y intercept from fatigue curves obtained from different types of fatigue testing were used in the equation to determine the number of cycles to failure (N_f).

The major limitations for all the fatigue models include the criteria for fatigue life; the testing boundary conditions; and the lack of fundamental mechanism at microscopic levels.

7.11   Cohesive Zone Models for Numerical Simulations

7.11.1   General Ideas

The idea to describe fracture as a material separation across a surface was started by Barenblatt (1962). Generally, it is called a cohesive zone model. The cohesive zone is a surface in a bulk material where displacement discontinuities occur. Continuum is extended with discontinuities in the form of displacement jump which requires additional constitutive description. Equations relating normal and tangential displacement jumps across the cohesive surfaces with the proper tractions define a specific cohesive zone model.

Based on the elementary functions used in cohesive zone models, they can be classified as (1) multilinear, (2) polynomial, (3) trigonometric, and (4) exponential. Almost all these models are constructed as follows: tractions increase, reach a maximum, and then approach zero with increasing separation. Needleman (1987) introduced the cohesive zone models in computational practice. Since then, cohesive zone models are used widely in finite element simulations of crack tip plasticity and creep; crazing in polymers; adhesively bonded joints; interface cracks in bimaterials; delamination in composites and multilayered materials; fast crack propagation in polymers, and so on.

Four typical cohesive zone models are briefly introduced here. They present relationships between surface tractions T and displacement jumps Δ across the cohesive zone. Maximum surface traction T_{max} is the cohesive strength. The corresponding displacement jump is Δ_{max}. Dimensionless parameters used are:

$$\sigma = \frac{T}{T_{max}} \tag{7-101}$$

$$\delta = \frac{\Delta}{\Delta_{max}} \tag{7-102}$$

The work of separation is:

$$J = \int T d\Delta \tag{7-103}$$

Or dimensionless:

$$\phi = \frac{J}{T_{max}\Delta_{max}} \tag{7-104}$$

Bilinear Model

Bilinear form in a cohesive zone model is described by:

$$\sigma = \begin{cases} \delta & 0 \le \delta \le 1 \\ 2-\delta & 1 \le \delta \le 2 \\ 0 & 2 \le \delta \end{cases} \qquad (7\text{-}105)$$

The separation work is $\phi = 1$.

Parabolic Model

The parabolic cohesive zone model is described by:

$$\sigma = \begin{cases} 2\delta - \delta^2 & 0 \le \delta \le 2 \\ 0 & 2 \le \delta \end{cases} \qquad (7\text{-}106)$$

The separation work is $\phi = \dfrac{4}{3} = 1.33$.

Sinusoidal Model

The sinusoidal cohesive zone model is described by:

$$\sigma = \begin{cases} \sin(\dfrac{\pi\delta}{2}) & 0 \le \delta \le 2 \\ 0 & 2 \le \delta \end{cases} \qquad (7\text{-}107)$$

The separation work is $\phi = \dfrac{4}{\pi} = 1.273$.

The Exponential Model

The exponential cohesive zone model is described by:

$$\sigma = \delta e^{1-\delta} \qquad 0 \le \delta \qquad (7\text{-}108)$$

The separation work is $\phi = e = \exp(1) \cong 2.718$.

7.11.2 Cohesive Zone Models used in Modeling Fracture of AC

As mentioned before, cohesive zone models are widely used as a tool to investigate the fracture of materials. They have been used in AC simulations in recent years. Kim (2008) used a clustered discrete element method (DEM) as a tool to investigate fracture mechanisms in AC at low temperatures. The bilinear cohesive zone model was implemented into the DEM to enable simulation of crack initiation and propagation in AC. Song et al. (2005, 2006) used both the bilinear and the exponential cohesive zone models to describe cracking in AC. Cohesive elements are generated by means of a user subroutine of the ABAQUS software and calibrated by simulation of the double cantilever beam test. Song et al. (2008) conducted another study to investigate an important parameter of the cohesive zone model, softening shape, which becomes very important as the relative size of the fracture process zone compared to the structure size increases. The parabolic cohesive zone model was used in that study. Liu (2008) incorporated the digital image processing techniques and finite element method to simulate the indirect tension test in the lab. The fracture mechanism is described by a cohesive zone model.

The initiation and propagation of the crack in the cylindrical specimen under conventional indirect tensile test (IDT) are modeled. This study is unique in its consideration of material heterogeneity. It allows researchers to link the micro-scale damage with the real pavement failure on the global scale.

7.11.3 Cohesive Zone Model Used in Modeling Fatigue of AC

Few studies (Song et al., 2005, 2006, 2008; Kim et al, 2006) have been conducted to investigate the fatigue of AC. Kim et al. (2006), Kim and Lutif (2008) started to incorporate the cohesive zone model into a damage model to explain the mechanism of the AC fatigue. Kim et al. (2006) developed a computational damage model which is able to account for material viscoelasticity, heterogeneity, and path-and rate-dependent energy dissipation as cracks initiate and propagate. The model incorporates elastic behavior of the aggregate particles, viscoelastic behavior of the asphalt matrix, and time-dependent fracture both within the asphalt matrix and along boundaries between matrix and aggregate particles. Rate-dependent progressive cracking up to failure was implemented by incorporation of a cohesive zone fracture model. In the model, the tractions are described in terms of the displacement differences across the cohesive zone. In Kim's study, a nonlinear cohesive zone model developed by Allen and Searcy (2001) was used because that model can reflect nonlinear viscoelastic damage growth in the asphalt mixtures and predict damage evolution, microcracking, post peak material softening, and eventual fracture failure. The equation is shown below:

$$T_i(t) = \frac{u_i(t)}{\lambda(t)\delta_i}[1 - \alpha(t)][\int_0^t C^{CZ}(t - \pi)\frac{\partial\lambda(t)}{\partial\tau}d\tau] \qquad (7\text{-}109)$$

where $i = n$ (normal direction), t (tangential direction), or r (radial direction),
$\quad T_i(t)$ = cohesive zone traction,
$\quad u_i(t)$ = cohesive zone displacement,
$\quad \lambda(t)$ = cohesive zone strain,
$\quad \delta_i$ = cohesive zone material length parameter,
$\quad \alpha(t)$ = cohesive zone damage evolution function, and
$\quad C^{CZ}(t)$ = linear viscoelastic stress relaxation modulus of cohesive zone

The damage evolution function $\alpha(t)$ can be experimentally determined by performing small-scale fracture tests. The cohesive zones, which are subjected to damage and fracture, are modeled by employing interface elements along the aggregate surface. The elements were randomly specified in the sample to maximize the power of the model to predict random crack initiation and propagation. The laboratory dynamic frequency sweep test and dynamic shear rheometer test are used to determine the viscoelastic property parameters in the model. The damage and fracture properties of the cohesive zone interfaces were determined by utilizing a tensile fracture test. Damage growth was captured by using an optical microscope and analyzed by video.

7.12 Other Fatigue Studies

Al-Qadi et al. (2005) discussed thermal fatigue cracking in flexible pavements based on experimental results from the Virginia Smart Road. Thermal fatigue is caused by repeated thermal cycles; a finite element simulation model was developed to simulate the heat transfer and thermal stress. Ullidtz (2005) proposed a damage model based on Miner's law with a nonlinear relation between damage and the number of load applications. Damage was defined as a decrease in the modulus of material or as an increase in permanent deformation. The model was based on experiment data. Ker et al. (2008) used an LTPP database to develop an improved fatigue cracking prediction model. The generalized additive model (GAM) extended from the generalized linear model along with assumption of the Poisson distribution and adoption of the quasi-likelihood estimation method. The proposed model included variables such as KESALs, pavement age, annual precipitation, annual temperature, critical tensile strain under the AC surface layer, and freeze-thaw cycle for the prediction of fatigue cracking.

7.13   Non-Local Theory

Traditional elasticity relates the stress of a point to the strain at that point only. This type of theory is usually named local theory. A more complicated theory could associate the stress at a point with the average strain surrounding that point. In a simple case, one may assume that the elastic strain at r (vector) surrounding a position x (vector) depends linearly on the relative distance between r and x, which occurs in a small but finite material volume V surrounding x. The average strain $\bar{\varepsilon}$ can be obtained by the volume average of the local strain distribution within the representative volume element RVE as follows:

$$\bar{\varepsilon} = \frac{1}{V} \int_{V_{Ref}} \varepsilon(x-r)dV \tag{7-110}$$

Taylor's series expansion limited to the second-order term of the function $(x\text{-}r)$ around x gives:

$$\bar{\varepsilon} = \varepsilon(x) + l_c^2 \nabla^2 \varepsilon \tag{7-111}$$

Where l_c = characteristic length scale of the material microstructure. $\nabla^2 \varepsilon$ is the strain gradient around point x. The following gradient elasticity model is obtained when the above average strain is substituted in Hooke's law:

$$\sigma = E\varepsilon + l_c^2 E \nabla^2 \varepsilon \tag{7-112}$$

Where σ and ε = stress and strain tensors, respectively, and E = fourth-order elasticity tensor.

Dessouky et al. (2006) applied the gradient elasticity theory and implemented an FEM code for modeling the AC behavior. The microstructure of AC is also incorporated into their modeling work.

References

Allen, D.H. and Searcy, C.R. (2001). A micromechanically-based model for predicting dynamic damage evolution in ductile polymers. *Mechanics of Materials*, Vol.33, pp.177–184.

Al-Qadi, I.L., Hassan, M.M. and Elseifi, M.A. (2005). Field and theoretical evaluation of thermal fatigue cracking in flexible pavements. *Transportation Research Record*, No.1919, pp.87–95.

Barenblatt, G. (1962). The mathematical theory of equilibrium cracks in brittle fracture. *Advances in Applied Mechanics*, Vol.7, pp.55–129.

Benedetto, H.D. and Olard, F. (2009). DBN law for thermo-visco-elasto-plastic behavior of asphalt concrete. *Modeling of Asphalt Concrete*, Kim, Y.R (Editor), McGraw-Hill, ASCE.

Bodin, D., Pijaudier-Cabot, G., de La Roche, C., Piau, J.M. and Chabot, A. (2004). Continuum damage approach to asphalt concrete fatigue modeling. *Journal of Engineering Mechanics*, Vol.130, No.6, pp.700–708.

Bonnetti, K.S., Nam, K. and Bahia, H.U. (2002). Measuring and defining fatigue behavior of asphalt binders. *Transportation Research Record*, No.1810, pp.33–43.

Castro, M. and Sanchez, J. A. (2008). Estimation of asphalt concrete fatigue curves – A damage theory approach. *Construction and Building Materials*, Vol.22, pp.1232–1238.

Chehab, G.R., Kim, Y.R., Schapery, R.A., Witczak, M.W. and Bonaquist, R. (2002). Time-temperature superposition principle for asphalt concrete mixtures with growing damage in tension state. *Journal of the Association of Asphalt Paving Technologists*, Vol.71, pp.559–593.

Chehab, G.R., and Kim, Y.R. (2005). Viscoelastoplastic continuum damage model application to thermal cracking of asphalt concrete. *Journal of Materials in Civil Engineering*, Vol.17, No.4, pp.384–392.

Chaboche, J. L. (1986). Time-independent constitutive theories for cyclic plasticity. *International Journal of Plasticity*, Vol.2, No. 249, pp. 149–188.

Cheng, D., Little, D. N., Lytton, R. L., and Holste, J. C. (2002) Surface energy measurement of asphalt and its application to predicting fatigue and healing in asphalt mixtures. *Transportation Research Record*. No.1810, pp.44–53.

Collop, A. C., Scarpas, A.T., Kasbergen, C. and Bondt, A.D. (2003). Development and finite element implementation of a stress dependent elasto-visco-plastic constitutive model with damage for asphalt. *Transportation Research Record*. No. 1832, pp.96–104.

Daniel, J.S., and Kim, Y.R. (2002). Development of a simplified fatigue test and analysis procedure using a viscoelastic continuum damage model. *Journal of the Association of Asphalt Paving Technologists*, Vol.71, pp.619–650.

Desai, C.S. (2001). Mechanics of Materials and Interfaces: the Disturbed State Concept. *CRC Press*, Boca Raton.

Desai, C.S. (2007). Unified DSC constitutive model for pavement materials with numerical implementation. *International Journal of Geomechanics*, Vol.7, No.2, pp.83–101.

Desai, C. (2009). Unified disturbed state consitutive modeling of asphalt concrete. *Modeling of Asphalt Concrete*, Kim, Y.R (Editor), McGraw-Hill, ASCE.

Desai, C.S., Somasundaram, S. and Frantziskonis, G. (1986). A hierarchical approach for constitutive modeling of geologic materials. *International Journal of Numerical Analytical Methods in Geomechanics*, Vol. 10, pp. 225–257.

Dessouky, S., Masad; E., Little, D. and Zbib, H. (2006). Finite-element analysis of hot mix asphalt microstructure using effective local material properties and strain gradient elasticity. *Journal of Engineering Mechanics*, Vol.132, No.2, pp.158–171.

Di Benedetto, H., Roche, C., Baaj, H., Pronk, A. and Lundstrom, R. (2004a). Fatigue of bituminous mixes. *Journal of Materials and Structures*, Vol.37, pp.202–216.

Di Benedetto, H., Olard, F., Sauzeat, C. and Delaporte, B. (2004b). Linear viscoelastic behavior of bituminous materials: from binders to mixes. Special Issue of the *International Journal of Road Materials and Pavement Design*, 1st EATA, Vol.4, pp.163–202.

Di Benedetto, H., Delaporte, B. and Sauzeat, C. (2007a). Three-dimensional linear behavior of bituminous materials: experiments and modeling. *International Journal of Geomechanics*, Vol.7, No.2, pp.149–157.

Di Benedetto, H., Neifar, M., Sauzeat, C. and Olard, F. (2007b). Three-dimensional thermoviscoplastic behaviour of bituminous materials: the DBN model. The *International Journal of Road Materials and Pavement Design*, Vol.8, No.2, pp.285–316.

El-Basyouny, M. and Witczak,M.W. (2005). Part 2: flexible pavements: calibration of alligator fatigue cracking model for 2002 design guide. Transportation Research Record, No. 1919, pp.76–86

Erkens, S., Molenaar, A. and Scarpas, A. (2003). A better understanding of asphalt concrete response, *16th ASCE Engineering Mechanics Conference*, University of Washington, Seattle.

Erkens, S.M., Liu, X., Scarpas, T., Molenaar, A.A.A. and Blaauwendraad, J. (2003). *Modeling of asphalt concrete–numerical and experimental aspects. Geotechnical Special Publication*, No. 123, pp. 160–177.

Ghuzlan, K.A. and Samuel, H.T. (2006). Fatigue damage analysis in asphalt concrete mixtures using the dissipated energy approach. *Canada Journal of Civil Engineering*, Vol.33, pp.890–901.

Gibson, N., Schwartz, C.W., Schapery, R. and Witczak, M.W. (2003). Viscoelastic, viscoplastic, and damage modeling of asphalt concrete in unconfined compression. *Trasnportation Research Record*, No.1860, pp.3–15.

Gibson, N.H. (2006). *A viscoelastoplastic continuum damage model for the compressivebehavior of asphalt concrete. Ph.D. Dissertation*, University of Maryland.

Guddati, M.N., Feng, Z., and Kim, Y.R. (2002). Toward a micromechanics-based procedure to characterize fatigue performance of asphalt concrete. *Transportation Research Record* , No.1789, pp.121–128.

Huang, B., Mohammad, L., and Wathugala, W. (2002). Development of a thermo-viscoplastic constitutive model for HMA mixtures. *Journal of the Association of Asphalt Paving Technologists*, Vol. 71, pp.594–618.

Ker, H.W., Lee, Y.H. and Wu, P.H. (2008). Development of fatigue cracking prediction models using long-term pavement performance database. *Journal of Transportation Engineering*, Vol.134, No.11, pp.477–482.

Kim, Y.R. (Editor). Modeling of Asphalt Concrete. McGraw-Hill, ASCE.

Kim, Y. R., Guddati, M.N, Underwood, B.S., Yun, T.Y., Subramanian,V. and Savadatti, S. (2009). Development of a multiaxial viscoelastoplastic continuum damage model for asphalt mixtures, FHWA-HRT-08-073.

Kim, Y.R. and Lee, H.J. (1997). Healing of microcracks in asphalt and asphalt concrete. *Report to Western Research Institute and Federal Highway Administration.*

Kim, Y.R., Lee, Y.C. and Lee, H.J. (1995). Correspondence principle for characterization of asphalt concrete. *Journal of Materials in Civil Engineering*, Vol.7, No.1, pp.59–68.

Kim, Y.R. and Chehab, G.R. (2004). Development of a viscoelastoplastic continuum damage model for asphalt-aggregate mixtures: Final Report as Part of Tasks F and G in the NCHRP 9-19 Project. *National Cooperative Highway Research Program*, National Research Council, Washington, D.C.

Kim,Y.R., Allen, D.H. and Little, D.N. (2006). Computational model to predict fatigue damage behavior of asphalt mixtures under cyclic loading. *Transportation Research Record*, No.1970, pp.196–206.

Kim,Y.R. and Lutif, J.S. (2008). Computational micromechanics modeling for damage-induced behavior of asphalt mixtures considering viscoelasticity and cohesive zone fracture. *Geotechnical Special Publication*, No.184, pp.17–25.

Kim, H., Wagoner, M.P. and Buttlar, W.G. (2008). Simulation of fracture behavior in asphalt concrete using a heterogeneous cohesive zone discrete element model. *Journal of Materials in Civil Engineering*, Vol.20, No.8, pp.552–563.

Krempl, E. (1987). Models of viscoplasticity: some comments on equilibrium (back) stress and drag stress. *Acta Mechanica*, Vol.69, pp. 25–42.

Krempl, E. and Khan, F., (2003). Rate (time)-dependent deformation behavior: an overview of some properties of metals and solid polymers. *International Journal of Plasticity*, Vol.19, pp.1069–1095.

Lee, H.J. and Kim, Y.R. (1998b). A viscoelastic continuum damage model of asphalt concrete with healing. *Journal of Engineering Mechanics*, Vol.124, No.11, pp.1224–1232.

Lee, H.J., Daniel, J.S. and Kim, Y.R. (2000). Continuum damage mechanics-based fatigue model of asphalt concrete. *Journal of Materials in Civil Engineering*. Vol.12, No.2, pp.105–112.

Liu, J.H. and Wang, D.Y. (2008). Numerical simulation of a crack in the cement stabilized stone using cohesive zone models. *Proceeding* of *International Conference on Experimental Mechanics 2008*.

Lubliner, J. (1990). Plasticity Theory. *Macmillan*, New York.

Masad, E., Dessouky, S. and Little, D. (2007). Development of an elastoviscoplastic microstructural-based continuum model to predict permanent deformation in hot mix asphalt. *Interntional Journal of Geomechanics*, Vol.7, No.2, pp. 119–130.

Masad, E., Tashman, L., Little, D. and Zbib, H. (2005). Viscoplastic modeling of asphalt mixes with the effects of anisotropy, damage and aggregate characteristics. *Journal of Mechanics of Materials*, Vol. 37, No.12, pp.1242–1256.

Medani, T.O. and Molenaar, A.A.A. (2000). Estimation of fatigue characteristics of asphaltic mixes using simple tests. Heron, Vol.45, No. 3, pp.155–165.

Monismith, C.L. and Deacon, J.A. (1969). Fatigue of asphalt paving mixtures. Transportation Engineering Journal, Vol.95, No.2, pp.317–346.

Needleman, A. (1987). A continuum model for void nucleation by inclusion debonding. *Journal of Applied Mechanics*, Vol.54, pp.525–531.

Nguyen, D.T., Nedjar, B. and Tamagny, P. (2007). Cyclic elasto-viscoplastic model for asphalt concrete materials. *International Journal of Road Materials and Pavement Design*, Vol.8, No.2, pp. 239–255.

Oeser, M. and Moller, B. (2004). 3D constitutive model for asphalt pavements. *International Journal of Pavement Engineering*, Vol.5, pp.153–161.

Olard, F. and Benedetto, H.D. (2003). General 2S2P1D model and relation between the linear viscoelastic behaviors of bituminous binders and mixes. *International Journal of Road Materials and Pavement Design*, Vol.4, No.2, pp.185–224.

Owusu-Antwi, E. B., Khazanovich, L., and Titus-Glover, L. (1998). Mechanistic-based model for predicting reflective cracking in asphalt concrete-overlaid pavements. *Transportation Research Record*, No.1629, pp 234–241.

Park, S.W., Kim, Y.R. and Schapery, R.A. (1996). A viscoelastic continuum damage model and its application to uniaxial behavior of asphalt concrete. *Journal of Mechanics of Materials*, Vol.24, No.4, pp.241–255.

Perl, M., Uzan, J. and Sides, A. (1983). Visco-elasto-plastic consititutive law for a bituminous mixture under repeated loading. *Transportation Research Record*, No.911, pp.20–27.

Perzyna, P. (1966). Fundamental problems in viscoplasticity. *Advances in Applied Mechanics*, Vol.9, pp.243–377. Academic Press, New York.

Schapery, R.A. (1981). On viscoelastic deformation and failure behavior of composite materials with distributed flaws. *Advances in Aerospace Structures and Materials*, AD-01, ASME, New York, pp.5–20.

Schapery, R. A. (1984). Correspondence principles and a generalized J-integral for large deformation and fracture analysis. International Journal of Fracture, Vol.25, pp.195–223.

Schapery, R.A. (1987). "Deformation and fracture characterization of inelastic composite materials using potentials. *Polymer Engineering and Science*, Vol.27, No.1, pp. 63–76.

Schapery, R.A. (1990). A theory of mechanical behavior of elastic media with growing damage and other changes in structure. *Journal of the Mechanics and Physics of Solids*, Vol.38, pp.215–253.

Schapery, R.A. (1999). Nonlinear viscoelastic and viscoplastic constitutive equations with growing damage. *International Journal of Fracture*, Vol.97, pp.33–66.

Smith, B.J. and Hesp, S.A.M. (2000). Crack pinning in asphalt mastic and concrete regular fatigue studies. *Transportation Research Record*, No.1728, pp.75–81.

Song, S.H., Paulino, G.H., and Buttlar, W.G. (2005). Cohesive zone simulation of mode i and mixed-mode crack propagation in asphalt concrete. *Geotechnical Special Publication*, No.130–142, pp.189–198, Geo-Frontiers.

Song, S.H., Paulino, G.H., and Buttlar, W.G. (2006). Simulation of crack propagation in asphalt concrete using an intrinsic cohesive zone model. *Journal of Engineering Mechanics*, Vol.132, No.11, pp.1215–1223.

Song, S.H., Wagoner, M.P., Paulino, G.H., and Buttlar, W.G. (2008). Crack opening displacement parameter in cohesive zone models: Experiments and simulations in asphalt concrete. *Fatigue and Fracture of Engineering Materials and Structures*, Vol.31, No.10, pp.850–856.

Sousa, J.B., Pais, J.C., Prates, M., Barros,R., Langlois, P. and Leclerc, A.M. (1998). Effect of aggregate gradation on fatigue life of asphalt concrete mixes. *Transportation Research Record*, No.1630, pp 62–68.

Tashman, L., Masad, E., Zbib, H., Little, D. and Kaloush, K. (2005). Microstructural viscoplastic continuum model for permanent deformation in asphalt pavements. *Journal of Engineering Mechanics*, Vol. 131, No.1, pp. 48–57.

Tayebali, A.A., Deacon, J.A. and Monismith,C. L. (1996) Development and evaluation of surrogate fatigue models for SHRP A-300A abridged mix design procedure. *Proceedings of the Association of Asphalt Paving Technologists*, Vol. 64, pp.340–366.

Tobita, Y. (1989). Fabric tensors in constitutive equations for granular materials, *Soils and Foundations*. Vol. 29, No. 4, pp.99–104.

Ullidtz, P. (2005). Simple model for pavement damage. *Transportation Research Record*, No.1905, pp.128–137.

Uzan, J. (1996). Asphalt concrete characterization for pavement performance prediction. *Journal of the Asscoiation of Asphalt Paving Technologiests*, Vol. 65, pp.573–607.

Uzan, J. (2007). Evaluation of fatigue cracking, *Transportation Research Record 1570*, pp.89–95.

Van Dijk, W. (1975). Practical fatigue characterization of bituminous mixes. Proceedings of the Association of Asphalt Paving Technologists, Vol. 44, 37–74.

CHAPTER 8

Finite Element Method and Boundary Element Method

8.1 Introduction

This chapter will discuss the application of Finite Element Method (FEM) and Boundary Element Method (BEM). The general philosophy is to present: 1) a brief review on the very fundamentals of the two methods; 2) implementation of constitutive models using implicit and semi-implicit methods for FEM; 3) implementation of interface models on FEM; and 4) some literature in modeling asphalt concrete (AC) using BEM. There are many references on using FEM for simulating AC behavior and the models discussed in Chapter 7 are mainly implemented on FEM. Therefore, the literature review will not cover those using FEM. Several general works are suggested for more details, such as Zienkiwicz (1977); Huebner et al. (1995); Scarpas (2004); Brebbia et al. (1984); and Gaul (2003). Many commercial programs such as ABAQUS, ANSYS, and ADINA are conveniently available and so no detailed discussions on general FEM are considered necessary. However, the application techniques such as interface element, infinite element, rigid element, which may find direct applications, are discussed instead. A special FEM, CAPA-3D, dedicated to modeling and simulation of pavements and paving materials developed by Scarpas and his colleagues (2004), is a unique reference for most readers.

8.2 Numerical Solution Approaches to Elasticity Problems, FEM

8.2.1 Theory

In elasticity, two commonly used approaches are the force method and the displacement method. FEM is typically a displacement-driven approach. This approach involves the following procedure: 1) obtaining strains from displacements; 2) obtaining stresses from strains; and 3) solving the equilibrium equations expressed in displacements and their partial derivatives with boundary conditions. A brief review is given as follows:

Strain-displacement relationship: $\varepsilon_{ij} = (\dfrac{\partial u_i}{\partial x_j} + \dfrac{\partial u_j}{\partial x_i})/2$ (8-1)

Stress-strain relationship (Hooke's Law): $\sigma_{ij} = D_{ijkl}\varepsilon_{kl}$ (8-2)

Equilibrium equations (static problem): $\dfrac{\partial \sigma_{ij}}{\partial x_i} + b_j = 0$ (8-3)

For isotropic elasticity: $D_{ijkl} = \lambda \delta_{ij}\delta_{kl} + 2\mu(\delta_{ik}\delta_{jl} + \delta_{il}\delta_{jk})$ (8-4)

λ and μ are *Lamé* constants.

The equilibrium equations represented by displacements will take the following format:

$$(\lambda + \mu)u_{kk,j} + \mu u_{j,ii} + b_j = 0$$ (8-5)

In a more general format, it can be expressed as the following operator format:

$$Lu_j + b_j = 0$$ (8-6)

Where L is a linear operator.

This set of equations can be also written in the vector format.

$$Lu + b = 0$$

The following is to obtain the specific format of the L operator.

Equations (8-3) and (8-1) can be written in the following format:

$$D_{ijkl}\frac{\partial \varepsilon_{kl}}{\partial x_i} + b_j = 0$$ (8-7)

$$\varepsilon_{kl} = (\frac{\partial u_k}{\partial x_l} + \frac{\partial u_l}{\partial x_k})/2$$ (8-8)

Therefore:

$$D_{ijkl}(\frac{\partial^2 u_k}{\partial x_i \partial x_l} + \frac{\partial^2 u_l}{\partial x_i \partial x_k})/2 + b_j = 0$$ (8-9)

Since $D_{ijkl} = D_{ijlk}$, one can prove that $D_{ijkl}\dfrac{\partial^2 u_k}{\partial x_i \partial x_l} = D_{ijkl}\dfrac{\partial^2 u_l}{\partial x_i \partial x_k}$ and $D_{ijkl}\dfrac{\partial u_k}{\partial x_l} = D_{ijkl}\dfrac{\partial u_l}{\partial x_k}$

So $D_{ijkl}\dfrac{\partial^2}{\partial x_i \partial x_l}u_k + b_j = 0$ (8-10)

Denote $L_{jk} = D_{ijkl}\dfrac{\partial^2}{\partial x_i \partial x_l}$, one has

$$L_{jk}u_k + b_j = 0$$ (8-11)

Here, L_{jk} is a linear operator.

For solving the equilibrium equations using the analytical approach, it is equivalent to finding displacement functions that satisfy the above equilibrium equations at any point and the corresponding boundary conditions. For relatively simple problems, analytical solutions can be obtained. For most of the complicated geometric and loading conditions, it is almost impossible to find analytical solutions. Numerical methods are often used to obtain approximate solutions. Approximate solutions may result in some residual forces:

$$L_{jk}u_k + b_j = R_j$$ (8-12)

Where R_j is the residual force.

While at each point it will result in some residual forces, one may release the strong satisfaction of Equation (8-11) with a weak satisfaction in the following integral format:

$$\int_{\Omega} R_j w_j d\Omega = 0 \tag{8-13}$$

$$\int_{\Omega} (Lu_j + b_j) w_j d\Omega = 0 \tag{8-14}$$

Where w is a weighting function or test function. It can be proved that if the above weak equations are satisfied for any chosen test function w_j, one will have the strong satisfaction. Equation (8-14) is the weak solution statement. It can be further represented in the following equation with domain decomposition. The different domains are finite in size and sum up to occupy the entire object.

$$\sum_{\Omega_1}^{\Omega_n} \int_{\Omega} (Lu_j + b_j) w_j d\Omega = 0 \tag{8-15}$$

These domains can be considered as finite meshes.

The following considers the integral of the first term in Equation (8-14):

$$L_{jk} = D_{ijkl} \frac{\partial^2}{\partial x_i \partial x_l}$$

$$\int_{\Omega} D_{ijkl} u_{k,il} w_j d\Omega = 0$$

Since

$$u_{k,il} w_j = (u_{k,i} w_j)_{,l} - u_{k,i} w_{j,l} \tag{8-16}$$

$$\int_{\Omega} D_{ijkl} u_{k,il} w_j d\Omega = \int_{\Omega} D_{ijkl} [(u_{k,i} w_j)_{,l} - u_{k,i} w_{j,l}] d\Omega \tag{8-17}$$

By using the divergence theorem, one has

$$\int_{\Omega} D_{ijkl} (u_{k,i} w_j)_{,l} d\Omega = \int_{\Gamma} D_{ijkl} u_{k,i} w_j n_l d\Gamma$$

$$\int_{\Omega} D_{ijkl} u_{k,il} w_j d\Omega = \int_{\Gamma} D_{ijkl} u_{k,i} w_j n_l d\Gamma - \int_{\Omega} D_{ijkl} u_{k,i} w_{j,l} d\Omega \tag{8-18}$$

Since

$$\sigma_{jl} = D_{ijkl} u_{k,i} \qquad \tau_j = \sigma_{jl} n_l \tag{8-19}$$

The surface integral of the first term on the right then becomes:

$$\int_{\Gamma} D_{ijkl} u_{k,i} w_j n_l d\Gamma = \int_{\Gamma} \tau_j w_j d\Gamma \tag{8-20}$$

$$\int_{\Omega} D_{ijkl} u_{k,il} w_j d\Omega = \int_{\Gamma} \tau_j w_j d\Gamma - \int_{\Omega} D_{ijkl} u_{k,i} w_{j,l} d\Omega \tag{8-21}$$

By using Equation (8-6) or (8-11), one will arrive at:

$$\int_{\Omega} D_{ijkl} u_{k,i} w_{j,l} d\Omega = \int_{\Gamma} \tau_j w_j d\Gamma + \int_{\Omega} b_j w_j d\Omega \tag{8-22}$$

Considering $D_{ijkl}u_{k,i} = \sigma_{jl}$, the above equation becomes an energy conservation equation: work by the stress over the strain is equal to work by surface forces and body forces over surface and body displacements. In general, it is actually the virtual work mechanism.

The strain energy of the entire system will be:

$$\Pi = \frac{1}{2}\int_{\Omega} D_{ijkl}u_{k,i}u_{j,l}\,d\Omega - \int_{\Gamma}\tau_j u_j\,d\Gamma - \int_{\Omega}b_j u_j\,d\Omega \tag{8-23}$$

It is understood that the minimization of the strain energy will result in equilibrium conditions.

If one subdivides Ω into a set of finite domains (finite elements) $\Omega_m(m = 1,N)$, each subdomain has r_m nodes, one can have the following domain sum:

$$\int_{\Omega_1}\cdots\,d\Omega + \cdots\int_{\Omega_m}\cdots\,d\Omega + \cdots\int_{\Omega_N}\cdots\,d\Omega\int_{\Omega}\cdots\,d\Omega \tag{8-24}$$

For the surface integral, the inter-element surface integral will cancel out, so only the external surface integral will play a role in the above equations. The above equation is the foundation for the FEM approach. The major techniques and procedures are presented as follows.

Step 1: Discretization of the Continuum

Considering the geometry of the object, the properties of the materials, the defects (cracking) and interface conditions, and the loading conditions, the object can be subdivided into subdomains. It is also directly related to Step 3: selection of element type. Usually, a fine element mesh is used for locations of concentrated loading, cracking tips, and interfaces.

Step 2: Selection of Interpolation Functions

Depending on the shapes of the discretized domains (elements), accuracy requirements and computational efficiency, there are a variety of interpolation functions (shape functions) to select. Typically, polynomials are selected for the convenience of integration and differentiation. Shape functions must satisfy requirements such as continuity on element or domain boundaries and permission for rigid movement of the element.

This step is actually to obtain the displacement field in any of these finite domains through interpolating by using the displacements of nodes:

$$u_1 = \sum_{i=1}^{r} N_i(x_1,x_2,x_3)u_{1i}^q,\ u_2 = \sum_{i=1}^{r} N_i(x_1,x_2,x_3)u_{2i}^q,\ u_3 = \sum_{i=1}^{r} N_i(x_1,x_2,x_3)u_{3i}^q \tag{8-25}$$

Where r is the number of nodes; u_1, u_2, and u_3 are the three displacements in the domain including edges and surfaces. Continuing, q represents the corresponding displacements of the nodes and N_i are shape functions. Denoting u and u^q as displacement vectors and N as a matrix, Equation (8-26) can be obtained.

$$u = \begin{Bmatrix} u_1 \\ u_2 \\ u_3 \end{Bmatrix} = \begin{Bmatrix} u \\ v \\ w \end{Bmatrix}$$

$$u^q = (u_1^q,v_1^q,w_1^q,...,u_r^q,v_r^q,w_r^q)^T \quad \text{or} \quad u^q = (u_{11}^q,u_{21}^q,u_{31}^q,...,u_{1r}^q,u_{2r}^q,u_{3r}^q)^T$$

$$N = \begin{bmatrix} N_1,0,0,N_2,0,0...N_r,0,0 \\ 0,N_1,0,0,N_2,0...0,N_r,0 \\ 0,0,N_1,0,0,N_2...0,0,N_r \end{bmatrix}$$

$$u = Nu^q \tag{8-26}$$

Step 3: Formulation of Element Properties

The objective of this step is to conduct element-level integration to associate node displacements with equivalent node forces. There are three typical approaches: the direct approach, the variational approach, and the weighted residual approach. The major technique is to represent the volume integration (energy, forces) of the element in terms of node variables.

Equation (8-1) can be represented in the following format:

$$\varepsilon = Lu \tag{8-27}$$

$$\varepsilon = (\varepsilon_{11},\varepsilon_{22},\varepsilon_{33},2\varepsilon_{12},2\varepsilon_{23},2\varepsilon_{13})^T$$

$$L = \begin{bmatrix} \dfrac{\partial}{x_1},0,0 \\[2ex] 0,\dfrac{\partial}{\partial x_2},0 \\[2ex] 0,0,\dfrac{\partial}{\partial x_3} \\[2ex] \dfrac{\partial}{\partial x_2},\dfrac{\partial}{\partial x_1},0 \\[2ex] 0,\dfrac{\partial}{\partial x_3},\dfrac{\partial}{x_2} \\[2ex] \dfrac{\partial}{\partial x_3},0,\dfrac{\partial}{\partial x_1} \end{bmatrix}$$

Please note the difference between Cauchy strain and the engineering strain for the shear components. The corresponding coefficients for Hooke's law should be adjusted by a factor of one-half.

Since $u = Nu^q$, $\varepsilon = Lu$.

Denote $B = LN$, one has:

$$\varepsilon = Bu^q \tag{8-28}$$

Denote the matrix format of the elasticity tensor as C (6×6 matrix); one has the following stress-strain relation (in vector format, $\sigma = (\sigma_{11},\sigma_{22},\sigma_{33},\sigma_{12},\sigma_{23},\sigma_{13})^T$):

$$\sigma = C\varepsilon \tag{8-29}$$

The total strain energy in Equation (8-23) can be considered as the sum of the strain energies of the elements.

$$\Pi = \sum_{m=1}^{M} \Pi_m = \frac{1}{2}\int_{\Omega^e} D_{ijkl} u_{k,i} u_{j,l}\, d\Omega - \int_{\Gamma^e} \tau_j u_j\, d\Gamma - \int_{\Omega^e} b_j u_j\, d\Omega \tag{8-30}$$

The minimization of the strain energy requires the following conditions to be satisfied:

$$\delta\Pi = \sum_{m=1}^{M} \delta\Pi_m = 0 \tag{8-31}$$

Which means that the displacement function should be properly selected so that the overall strain energy is minimized. A consistent requirement at the element level will need $\delta\Pi_m = 0$.

The following will deal with the element-level minimization through the discretization method:

$$\Pi_e = \frac{1}{2}\int_{\Omega^e} (u_e^q)^T B^T C B u_e^q\, d\Omega - \int_{\Gamma^e} N^T \tau u_e^q\, d\Gamma - \int_{\Omega^e} N^T b u_e^q\, d\Omega \tag{8-32}$$

The minimization conditions will require that $\dfrac{\partial \Pi_e}{\partial u_i^e} = 0$.
This will result in:

$$\int_{\Omega^e} B^T C B u_e^q\, d\Omega - \int_{\Gamma^e} N^T \tau\, d\Gamma - \int_{\Omega^e} N^T b\, d\Omega \tag{8-33}$$

or

$$K^e u_e^q = F^e \tag{8-34}$$

Where $F^e = F_B^e + F_S^e$: equivalent node forces.

$K^e = \int_{\Omega^e} B^T C B\, d\Omega$: element stiffness matrix.

$F_S^e = \int_{\Gamma^e} N^T \tau\, d\Gamma$: equivalent forces due to the surface tractions.

$F_B^e = \int_{\Omega^e} N^T b\, d\Omega$: equivalent forces due to distributed body forces.

The superscript e denotes element-level variables.

Step 4: Formulate the System Equation

The domain decomposition equation also presents a way to assemble the element properties into the overall property matrix, which relates the node displacements to the forces applied on the nodes. Since elements share nodes, a specific order is necessary. The general philosophy is to sum all the displacement contributed by the elements sharing that node. Different orders will result in different required computational time.

$$Ku^q = F \tag{8-35}$$

Where K is the overall stiffness matrix; u^q is the node displacement vector; and F is the equivalent node force element. This is the system-level equilibrium equation.

There are many methods to assemble the system matrix from the element matrices. They can be found in some of the reference books listed in Section 8.1.

Step 5: Application of the Boundary Conditions

There are basically two types of boundary conditions— displacement conditions and force conditions (or surface tractions). Mathematically, it is equivalent to the modification of the node displacement vector u^q and the node force vector F.

Step 6: Solution of the System Equation

The system equation may be linear or non-linear. There are various solution techniques to solve the equation efficiently. The books on FEM listed in Section 8.1 have excellent discussions on this topic.

Step 7: Check of Convergence Criteria

There are usually two types of convergence criterion — the displacement and the force convergence criteria. Most of the commercial FEM codes provide guides on selection of these criteria.

8.2.2   Example

The following will use an example to consistently illustrate the seven steps mapped into the formulations. The example is a 2D case, a beam of cantilever of $w \times h \times l = 0.1 \times 0.2 \times 1$ meter subjected a concentrated load of 1N (Figure 8.1). The material is isotropic and linear elastic with $E = 10^8$ Pa and $v = 0.25$. 2D elements are used to model it. The following only illustrates the problem setup process. It is suggested that readers accomplish the solution with the help of Matlab and compare it with the solution using ABAQUS or other commercial software.

Step 1: The discretization is illustrated in Figure 8.1. It consists of 30 triangle elements with a total of 24 nodes. There are a total of 48 unknown nodal displacements (u_i, v_i, $i = 1, 24$) Please note that the discretization manner is not unique.

Step 2: There are various interpolation functions from which to choose. Most of the books on FEM have detailed descriptions on the typical types of elements such as the criteria that the interpolation functions must satisfy and the corresponding characteristics. In this case, the simplest interpolation functions, the linear interpolation functions, were selected.

$$u = a_1 + b_1 x + c_1 y, \quad v = a_2 + b_2 x + c_2 y \tag{8-36}$$

For the three nodes, i, j, and k (Figure 8.2), one can have three sets of equations. From the three sets of equations, one can solve for the six coefficients.

$$u_i = a_1 + b_1 x_i + c_1 y_i, \quad u_j = a_1 + b_1 x_j + c_1 y_j, \quad u_k = a_1 + b_1 x_k + c_1 y_k$$

$$v_i = a_2 + b_2 x_i + c_2 y_i, \quad v_j = a_2 + b_2 x_j + c_2 y_j, \quad v_k = a_2 + b_2 x_k + c_2 y_k$$

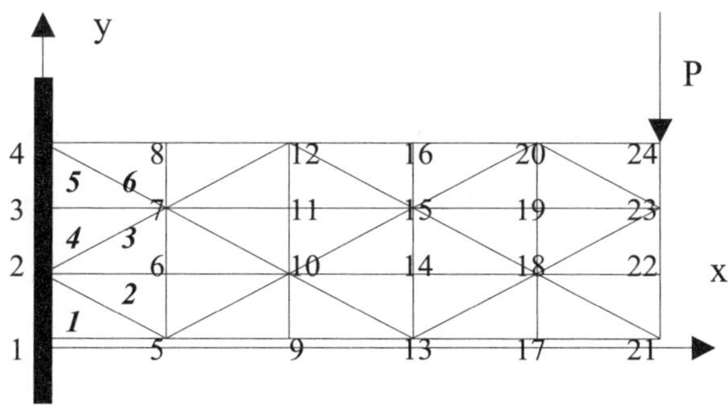

FIGURE 8.1   Illustration of the FEM mesh for the example problem. (The origin is at Node 1; the Z direction is perpendicular to the plane and point out the paper).

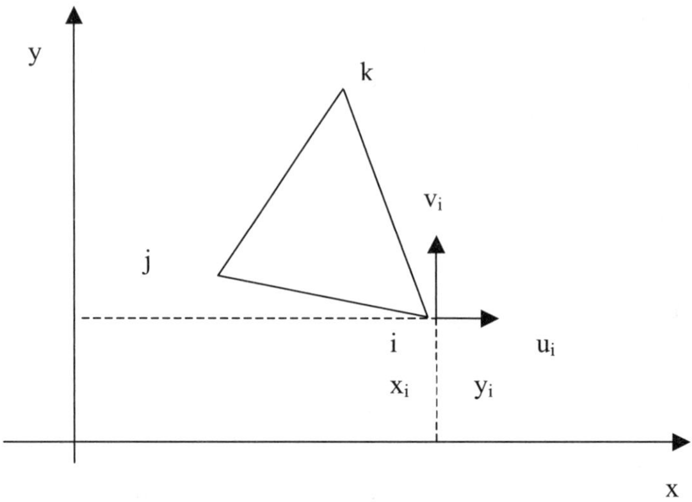

Figure 8.2 Illustration of the symbol scheme for a triangle element.

Conveniently, one can solve for the coefficients (a_1, b_1, c_1, a_2, b_2, c_2) and represent the interpolation function in the format by Zienkiewicz (1977).

$$u = \frac{1}{2A}[(\alpha_i + \beta_i x + \gamma_i y)u_i + (\alpha_j + \beta_j x + \gamma_j y)u_j + (\alpha_k + \beta_k x + \gamma_k y)u_k]$$

$$v = \frac{1}{2A}[(\alpha_i + \beta_i x + \gamma_i y)v_i + (\alpha_j + \beta_j x + \gamma_j y)v_j + (\alpha_k + \beta_k x + \gamma_k y)v_k] \tag{8-37}$$

$$\alpha_i = x_j y_k - x_k y_j$$
$$\beta_i = y_j - y_k$$
$$\gamma_i = x_k - x_j$$

$$N_i = (\alpha_i + \beta_i x + \gamma_i y)/2A$$
$$N_j = (\alpha_j + \beta_j x + \gamma_j y)/2A$$
$$N_k = (\alpha_k + \beta_k x + \gamma_k y)/2A \tag{8-38}$$

Where A is the area of the triangle.

The matrix format of the interpolation functions can be represented as:

$$\begin{Bmatrix} u \\ v \end{Bmatrix} = \begin{bmatrix} N_i,0,N_j,0,N_k,0 \\ 0,N_i,0,N_j,0,N_k \end{bmatrix} \begin{Bmatrix} u_i \\ v_i \\ u_j \\ v_j \\ u_k \\ v_k \end{Bmatrix}$$

The strains can be represented as:

$$L = \left\{ \begin{array}{l} \dfrac{\partial}{\partial x}, 0 \\[2mm] 0, \dfrac{\partial}{\partial y} \\[2mm] \dfrac{\partial}{\partial y}, \dfrac{\partial}{\partial x} \end{array} \right\}$$

$$B = \left\{ \begin{array}{l} \beta_i, 0, \beta_j, 0, \beta_k, 0 \\[1mm] 0, \gamma_i, 0, \gamma_j, 0, \gamma_k \\[1mm] \gamma_i, \beta_i, \gamma_j, \beta_j, \gamma_k, \beta_k \end{array} \right\}$$

Obviously, for a fixed triangle, B is a constant matrix (only related to the coordinates of the three nodes). In other words, the strains are constant in the element if the linear interpolation functions are used ($\varepsilon = B\ (u_i, v_i, u_j, v_j, u_k, v_k)^T$). Typically, for higher order polynomial interpolation functions, the strains in the element are not constant.

Step 3: This is actually the process to obtain the element stiffness matrix and the equivalent node forces due to the boundary forces and the body forces.

$$K^e = \int_{\Omega^e} B^T CB d\Omega$$

$$C = \frac{E}{1-v^2} \left\{ \begin{array}{l} 1, v, 0 \\[1mm] v, 1, 0 \\[1mm] 0, 0, \dfrac{1-v}{2} \end{array} \right\}$$

Apparently $B^T CB$ is a constant matrix and K^e is a 6×6 symmetric matrix.

$F_S^e = \int_{\Gamma^e} N^T \tau d\Gamma$: equivalent forces due to the surface tractions

$F_B^e = \int_{\Omega^e} N^T b d\Omega$: equivalent forces due to distributed body forces

$$\int_{\Omega^e} N^T b d\Omega$$

$$\begin{bmatrix} N_i, 0, N_j, 0, N_k, 0 \\ 0, N_i, 0, N_j, 0, N_k \end{bmatrix}$$

$$b = \left\{ \begin{array}{c} 0 \\ -\rho g \end{array} \right\}$$

$$Nb = (0, -\rho g N_i, 0, -\rho g N_j, 0, -\rho g N_k)^T$$

$$F_B^e = (0, -\rho g A h / 3, 0, -\rho g A h / 3, 0, -\rho g A h / 3)^T$$

$$F_S^e = \int_{\Gamma^e} N^T \tau d\Gamma$$

In this special case, the only non-zero component is $-P$ at Node 24.

Step 4: By summing up all the element equations, one obtains a set of equations with all the node displacements as independent variables. In this example, there are 48 unknown node displacements. It should be noted that the only differences for the element stiffness are differences in node coordinates for different elements. Therefore, one can develop a subroute to calculate the stiffness and simply replace i, j, and k coordinates through coordinate transfer in this case.

$$u = (u_1^1, u_2^2, u_1^2, u_2^2, ...u_{24}^1, u_{24}^2)$$

$$K = \left[48 \times 48 \right]$$

$$F = (F_1^1, F_2^2, F_1^2, F_2^2, ...F_{24}^1, F_{24}^2)$$

Step 5: In addition, the known displacements (for example, zero displacements) for all the nodes should be placed in the system equation. In this example, (u_1, v_1), (u_2, v_2), (u_3, v_3), and (u_4, v_4) are all equal to zero.

By surface integral, one can obtain: $F_S^e = \int_{\Gamma^e} N^T \tau d\Gamma$

$$F = (0, -2\rho g A h / 3, ..0, -P, ..., 0, -\rho g A h / 3)$$

Step 6: With the above preparations, everything is ready to solve Equation (8-35). This is actually a linear algebra equation. One can use MathCad or Matlab to solve it. As a matter of fact, there are quite a few books focusing on using Matlab to solve FEM problems (Kwon and Bang, 2000; Kattan, 2007).

$$Ku^q = F$$

Solving for the above system equation, one can obtain the displacements for all the nodes. Then one can calculate the strains and stresses.

Step 7: Check the Convergence Criteria
The author suggests that readers manually solve this problem so an in-depth understanding of the FEM procedure can be obtained. There are many books on FEM, so detailed descriptions on FEM will not be presented. Nevertheless, it is anticipated that concise presentations will allow readers to understand the fundamentals in FEM.

There are also matured methods for accounting non-linear problems, including both geometric and material non-linearity. Interested readers may refer to books by Bonet and Wood (2008) and Crisfield (1997). The following sections will focus on several topics of interest to researchers in the AC area, including: 1) implementation of interface models; 2) use of rigid element; 3) use of infinite element; and 4) implementing constitutive models (it involves a local integration process for material models only). More recent developments in areas such as XFEM (Moës et al., 1999; Mohammadi, 2008), Mesh-Free FEM (Belytschko and Chen, 2007), micromechanics FEM (Zohdi and Wriggers, 2005), and Multiscale FEM (Efendiev and Hou, 2008) will not be covered.

8.3   Interface Element

Interfaces often demonstrate different properties (including geometric characteristics) from those of the two bulk materials forming the interfaces. The deformation format for the interfaces is often different from those of the bulk materials. Sometimes it is necessary to develop interface elements to more efficiently and accurately model the interface properties. A brief but comprehensive introduction on interface elements and their application to pavement analysis and paving materials was given by Scarpas (2004). The work presented here is based on Wang et al. (2006). The various finite element techniques for analyzing an interface or joint proposed in recent years can be basically divided into two categories. The first category contains nodal or point interface elements that connect normal solid elements with each other using discrete springs (Beer, 1985). A special joint finite element with zero thickness, which was developed by Goodman et al. (1968), is a good example. Kaliakin and Li (1995) improved the zero-thickness interface element to eliminate deficiency of spurious stress oscillations, which might be caused by ill-conditioning of the interface element stiffness matrix due to large off-diagonal terms (Wilson, 1977). Research by Day and Potts (1994) showed that the ill-conditioning could be reduced by carefully selecting the size of the solid elements adjacent to the interfaces for 2D analyses. The idea of using continuous interface elements with a finite thickness based on continuum mechanics appears to be another choice (Desai et al., 1984), which falls in the second category to model joints/interfaces, such as the continuous interface elements with a tiny thickness (Zienkiewicz et al., 1970) and the thin elements (Desai et al., 1984) etc. Pande and Sharma (1979) found that little ill-conditioning is experienced with the application of a very thin 2D element. It is suggested that a thin element is numerically more reliable than an element with a zero thickness.

Frequently, shear strains are concentrated through those interfaces within AC, which produce large shear deformation along interfaces, but very small deformation across interfaces. As such, a very small rigid body rotation will be generated if interfaces are nearly straight, which implies the possibility of the use of infinitesimal strain theories. The localized large shear strains along interfaces will inevitably lead to dramatic distortion of the original interface elements, and the distortion will cause poor element configuration if the mesh is kept unchanged. The distorted mesh is very likely to result in incorrect displacements and stresses in these interface elements, and even analysis failures due to the distortion-induced ill-conditioning of the stiffness matrix.

This section presents the development of interface elements using the work by Wang and Wang (2006). In this special development, motions of an interface with assumed thickness are treated as large deformation problems to which the continuum finite element analysis might be applied. All the node coordinates are updated at each load step, and continuous interface elements are reconstructed based on the concept of the contact band element approach developed by Wang et al. (1995, 2002) after converged solutions are achieved at each load step. Infinitesimal strain theories are still employed within each load step. As an example, strain descriptions in three-node and four-node interface elements are formulated to harmonize the normal and shear stress-strain relationships.

This continuous interface element is different from the thin-layer element (Desai et al., 1984) in the sense that the continuous interface elements are continuously renewed based on their current interface configuration subject to large shear deformation. Unlike those rectangular thin-layer elements, they could be quadrilateral and triangular elements. The continuous interface elements are equipped with the proposed strain formulation and specific anisotropic Mohr-Coulomb yield criterion.

8.3.1   Special Strain Descriptions for Continuous Interface Elements

Occurrence of shear strain localization along an interface indicates the concentration of shear deformation with high gradients within the narrow interface zone. High gradients occur only in the direction normal to the interface. In the direction parallel to the interface, shear strains do not vary significantly. In conventional finite element analysis with infinitesimal strain assumption, the strain-displacement relations for a 2D problem, also used for a continuous interface element, are written as:

$$\begin{Bmatrix} \gamma_{nt} \\ \varepsilon_{nn} \\ \varepsilon_{tt} \end{Bmatrix} = \begin{Bmatrix} \dfrac{\partial u}{\partial n} + \dfrac{\partial v}{\partial t} \\ \dfrac{\partial v}{\partial n} \\ \dfrac{\partial u}{\partial t} \end{Bmatrix} \tag{8-39}$$

Here, n denotes the direction normal to the interface, and t the direction parallel to the interface, as shown in Figure 8.3(a). Displacements u and v are parallel and normal to the interface, respectively. Displacement u has much higher gradient in n direction than v has in t direction, which is mathematically expressed as:

$$\left| \frac{\partial u}{\partial n} \right| \gg \left| \frac{\partial v}{\partial t} \right| \tag{8-40}$$

So, it is accurate enough to keep the first term for the shear strain on the right-hand side of Equation (8-39). Due to the very thin interface, γ_{nt} and ε_{nn} can be written as:

$$\begin{Bmatrix} \gamma_{nt} \\ \varepsilon_{nn} \end{Bmatrix} \approx \begin{Bmatrix} \dfrac{\Delta u}{d} \\ \dfrac{\Delta v}{d} \end{Bmatrix} \tag{8-41}$$

Where Δu and Δv are tangential and normal relative displacements across the interface, respectively, and d is thickness of the interface. The normal strain in the direction parallel to the interface, ε_t, in any continuous interface element, takes the formulation as in regular solid elements. Therefore, the mathematical expression for the strains in continuous interface element is given by:

$$\begin{Bmatrix} \gamma_{nt} \\ \varepsilon_{nn} \\ \varepsilon_{tt} \end{Bmatrix} = \begin{Bmatrix} \dfrac{\Delta u}{d} \\ \dfrac{\Delta v}{d} \\ \dfrac{\partial u}{\partial t} \end{Bmatrix} \tag{8-42}$$

There are two advantages to the special strain description in Equation (8-42). First, the experimental relationship between stresses and relative displacements across an interface can be used directly for interface elements. Second, it provides a fundamental approach to describe strains in three-node triangular and four-node quadrilateral

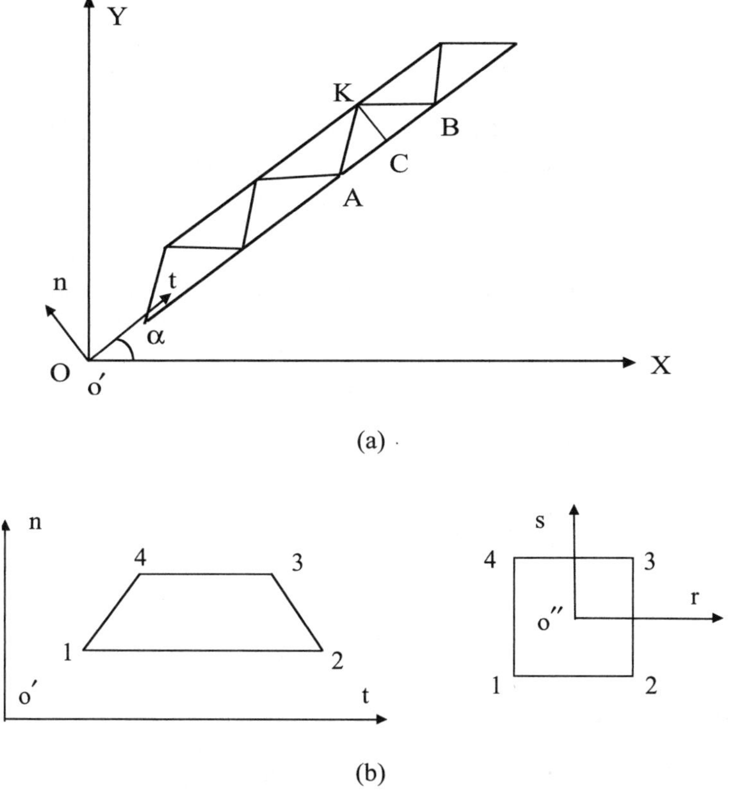

Figure 8.3   Continuous interface elements, (a) triangular elements under different coordinate systems, (b) rectangular elements under different coordinate system.

elements using relative displacements (Wang, 1998). Those three-node triangular and four-node quadrilateral elements with special strain descriptions are necessary in the analyses of interfaces subject to large shear deformation. It will be seen that the high displacement gradients in continuous interface element are simulated with high accuracy if the special strain description is used, even though the continuous interface element keeps a normal size in the direction parallel to the interface, which is much longer than its thickness (i.e., the continuous interface element does not need to be well refined in the direction parallel to the interface).

With the continuous interface elements updated, new interface elements may appear after the removal of old ones. It is not possible to directly inherit stresses and strains at the same traditional Gaussian points of each element for a new load step. However, the number of nodal points will never change. Therefore, it is necessary to calculate and build up stresses and strains and some other parameters if present at nodal points. At the beginning of every load step, stresses and strains at Gaussian points of new continuous interface elements are interpolated from nodal values (Wang et al., 2002). The opening of continuous interface elements can be evaluated based on the normal forces acting on element boundaries. Stiffness of the elements, stresses, and strains at corresponding nodal points will be released if they are in tension. The continuous interface elements or interface elements denote those with a finite thickness and the special strain descriptions. It is noteworthy that those continuous interface elements are renewable based on the current configurations of interfaces.

8.3.2 Formulation of Strains in Continuous Interface Elements

As noted previously, with the emergence of large shear deformation along an interface, distortions of original continuous interface elements are very likely to occur if they remain unchanged. Therefore, interface elements need to be renewed based on the updated coordinates. The renewal is different from the technique for the simulation of shear band inception and propagation, in which new nodal points lead to the change in global degrees of freedom (Wang et al., 2003). In this method, for the sake of simplicity, triangular and quadrilateral interface elements are automatically generated to circumvent the element distortions. The contact band element technique, which is used to simulate shear band inception and propagation by Wang et al. (1995, 2002) is employed to update interface elements. Contact band elements are generated through regular solid elements by breaking those solid elements, and are not prescribed. They also differ from continuous interface elements in constitutive descriptions. A brief description of the formation of new continuous interface elements is included.

A portion of an interface is shown in Figure 8.4. With an increase in the applied shearing load, configuration of the original interface elements gets greatly distorted as shown in Figure 8.4(a). As a result, numerical solutions will be drastically degraded due to the large distortion. However, if coordinates of the nodal points are updated, and they are reorganized to produce new interface elements based on the updated coordinates, the renewed interface elements have much better configurations, as illustrated in Figure 8.4(b). Consequently, they give a more accurate description of displacements and strains in the interface elements with three or four nodes. The following example is given to illustrate the formation of new interface elements.

As shown in Figure 8.4 (a) or (b), the interface is originally discretized with several initial continuous interface elements. The discretization gives m nodes on the upper boundary, S_{IJ}, and n nodes on its lower boundary, S_{JI}, which are denoted by:

$$ND := \left\{ N_u^1, N_u^2, ..., N_u^m ; \ N_l^1, N_l^2, ..., N_l^n \right\} \tag{8-43}$$

Where N_u^i denotes point i on the upper boundary, and N_l^j represents point j on the lower boundary. Considering a nodal point K on S_{IJ} at time t, the distance from node K to any node L on S_{JI} is defined by:

$$ {}^t d_{KL} = \left\| {}^t X_K^{S_{IJ}} - {}^t X_L^{S_{JI}} \right\|_2 \tag{8-44}$$

Node K forms two angles α and β with line segment MN on S_{JI}:

$$ \alpha = \cos^{-1} \left[\frac{\left({}^t X_K^{S_{IJ}} - {}^t X_M^{S_{JI}} \right)^T \left({}^t X_N^{S_{JI}} - {}^t X_M^{S_{JI}} \right)}{\left\| {}^t X_K^{S_{IJ}} - {}^t X_M^{S_{JI}} \right\|_2 \left\| {}^t X_N^{S_{JI}} - {}^t X_M^{S_{JI}} \right\|_2} \right] \tag{8-45}$$

$$ \beta = \cos^{-1} \left[\frac{\left({}^t X_K^{S_{IJ}} - {}^t X_N^{S_{JI}} \right)^T \left({}^t X_M^{S_{JI}} - {}^t X_N^{S_{JI}} \right)}{\left\| {}^t X_K^{S_{IJ}} - {}^t X_N^{S_{JI}} \right\|_2 \left\| {}^t X_M^{S_{JI}} - {}^t X_N^{S_{JI}} \right\|_2} \right] \tag{8-46}$$

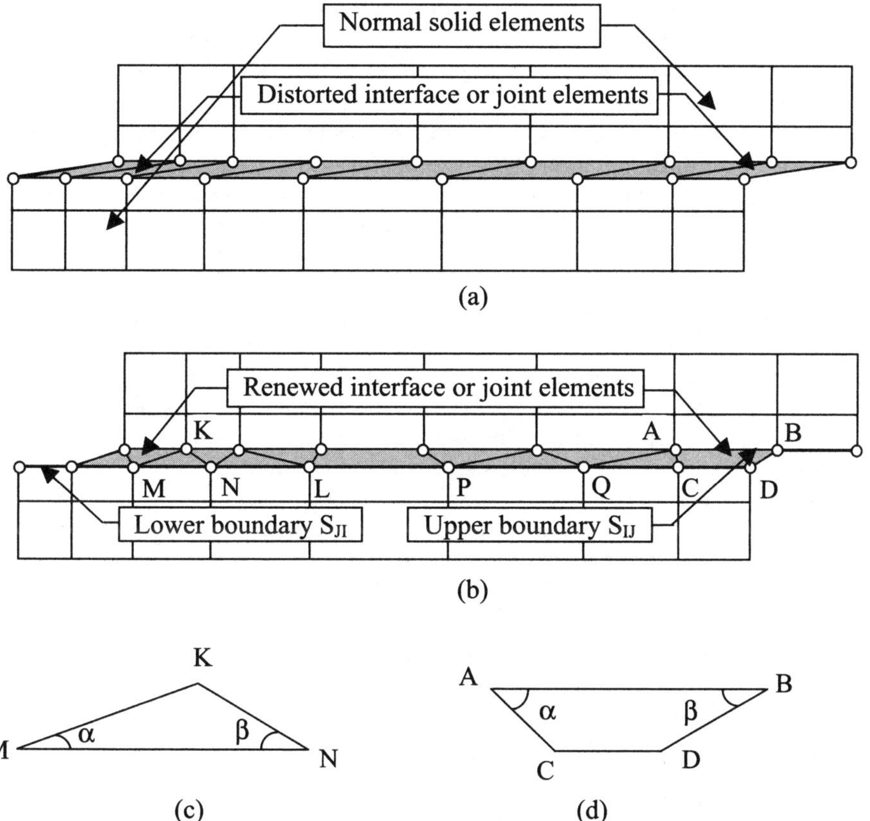

Figure 8.4 The renewal of the interface or joint elements, (a) distortion of the conventional interface elements, (b) the reconstruction of the interface elements, (c) a triangular interface element, (d) a rectangular interface element.

Where $^{t}X^{S_{IJ}(or S_{JI})}$ denotes the position vector of any point on S_{IJ} or S_{JI} under the global coordinate system at time t; α and β are shown in Figure 8.4(c) and Figure 8.4(d), and $\|\ \|_2$ indicates the length of a line segment. If α and β are either equal to or less than 90° and the following inequality holds for all the nodal points on S_{JI}:

$$\left\|^{t}X_K^{S_{IJ}} - {}^{t}X_M^{S_{JI}}\right\|_2 \leq \left\|^{t}X_K^{S_{IJ}} - {}^{t}X_N^{S_{JI}}\right\|_2 \leq \dots \leq \left\|^{t}X_K^{S_{IJ}} - {}^{t}X_P^{S_{JII}}\right\|_2 \leq \left\|^{t}X_K^{S_{IJ}} - {}^{t}X_Q^{S_{JI}}\right\|_2 \leq \dots \tag{8-47}$$

Then, a new interface element, e_{KMN}, is created, as depicted in Figure 8.4b. The same strategy is conducted for every nodal point $N_u^i \in ND$ ($i = 1, 2, \dots m$) on the upper boundary, S_{IJ}, and every nodal point $N_l^j \in ND$ ($j = 1, 2, \dots n$) on the lower boundary, S_{JI}, to get all the new interface elements. If two nodes, C and D, share the same segment, AB, as shown in Figure 8.4b or 8.4d, a new four-node interface element, e_{ABCD}, will be found. Any nodal points on the upper and lower surfaces that satisfy condition (8-47), and where corresponding angles α and β are both less than or equal to 90°, will form a new interface element. It is noted that a few continuous interface elements at both ends of an interface will disappear at different time steps if the interface is subject to larger and larger deformation, as shown in Figure 8.4b. The number of nodal points for an interface element varies and is dependent on the locations of the nodal points. The continuous interface elements are re-formed after converged results are obtained at each load step with node coordinates updated in the wake of each global equilibrium iteration. Extra equilibrium iterations are performed before a new load increment is applied for the new interface elements. Due to the continuous updates of the interface elements, stresses, strains, and other state variables at each nodal point should be tracked down. Initial stresses, strains, etc. at Gaussian points within a new interface element will be obtained by interpolating from those nodal values at the beginning of each load step (Wang et al., 2002).

After the formation of the continuous interface elements, the strains need to be formulated for triangular and rectangular interface elements, respectively. For large deformation analyses of interfaces or joints, triangular and quadrilateral interface elements are necessary, but not well documented. No considerable attention has been paid to characterize their strain distributions. In the next subsection, strains in the triangular and quadrilateral interface elements are defined and derived. The discussions in the previous section are fundamentals to the new strain descriptions. Shear and normal strains across interfaces are related to relative displacements between the upper and lower surfaces. The strains stated above are assumed to be uniform in the directions normal to an interface.

8.3.3 Triangular Interface Element

As shown in Figure 8.3a, at certain load steps, points K, A, and B form a new triangular element. The relative displacement rates of point K to segment AB at solution time $t + \Delta t$ are defined as:

$$
{}^{t+\Delta t}\left\{\begin{array}{c} \Delta \dot{W}_t \\ \Delta \dot{W}_n \end{array}\right\} = {}^{t+\Delta t}\left\{\begin{array}{c} \dot{U}_K \\ \dot{V}_K \end{array}\right\} - \begin{bmatrix} {}^{t+\Delta t}N_A^C & 0 & {}^{t+\Delta t}N_B^C & 0 \\ 0 & {}^{t+\Delta t}N_A^C & 0 & {}^{t+\Delta t}N_B^C \end{bmatrix} {}^{t+\Delta t}\left\{\begin{array}{c} \dot{U}_A \\ \dot{V}_A \\ \dot{U}_B \\ \dot{V}_B \end{array}\right\}
$$

$$
= \begin{bmatrix} -{}^{t+\Delta t}N_A^C & 0 & -{}^{t+\Delta t}N_B^C & 0 & 1 & 0 \\ 0 & -{}^{t+\Delta t}N_A^C & 0 & -{}^{t+\Delta t}N_B^C & 0 & 1 \end{bmatrix} {}^{t+\Delta t}\begin{bmatrix} \dot{U}_A \\ \dot{V}_A \\ \dot{U}_B \\ \dot{V}_B \\ \dot{U}_K \\ \dot{V}_K \end{bmatrix} \qquad (8\text{-}48)
$$

$$
= {}^{t+\Delta t}\left[N_{int} \right]^{t+\Delta t}\left\{ \dot{u} \right\}
$$

Where Δ denotes the difference between the upper and lower surface quantities, $\cdot$ denotes the rate of quantities, $\dot{U}$ and $\dot{V}$ are the tangential and normal displacement rates under coordinate system t-o'-n, $\Delta \dot{W}_t$ and $\Delta \dot{W}_n$ are the relative tangential and normal displacement rates, respectively. $^{t+\Delta t}N_A^C$ and $^{t+\Delta t}N_B^C$ are given by:

$$^{t+\Delta t}N_A^C = 1 - \,^{t+\Delta t}r_c \qquad ^{t+\Delta t}N_B^C = \,^{t+\Delta t}r_c \tag{8-49}$$

In which $^{t+\Delta t}r_c$ is simply obtained as the ratio of the length of segment AC to the length of segment AB. Here, point C is the image of point K on boundary AB. The location of point C is determined by assuming that KC and AB in Figure 8.3(a) are intersected at a right angle. As a result, the corresponding strain rates in the element as described in Section 2 take the form:

$$^{t+\Delta t}\left\{ \begin{array}{c} \dot{\varepsilon}_{nt} \\ \dot{\varepsilon}_{nn} \end{array} \right\} = \frac{1}{^{t+\Delta t}d} \,^{t+\Delta t}\left[N_{int} \right] \,^{t+\Delta t}\left\{ \dot{u} \right\} \tag{8-50}$$

Where thickness $^{t+\Delta t}d$ of the interface element, written as d hereafter, is calculated from current coordinates of points K and C, and will be updated continuously. As mentioned before, the normal strain parallel to the interface is taken as:

$$^{t+\Delta t}\dot{\varepsilon}_{tt} = \left[\frac{\partial N_A}{\partial t} \quad 0 \quad \frac{\partial N_B}{\partial t} \quad 0 \quad \frac{\partial N_K}{\partial t} \quad 0 \right]^{t+\Delta t} \left\{ \dot{u} \right\} \tag{8-51}$$

N_A, N_B, and N_K, different from $^{t+\Delta t}N_A^C$ or $^{t+\Delta t}N_B^C$, are regular shape functions for a triangular element. Therefore, for a triangular interface element, the three strain rates are given by:

$$^{t+\Delta t}\left\{ \begin{array}{c} \dot{\varepsilon}_{nt} \\ \dot{\varepsilon}_{nn} \\ \dot{\varepsilon}_{tt} \end{array} \right\} = \frac{1}{d} \left[\begin{array}{cccccc} -\,^{t+\Delta t}N_A^C & 0 & -\,^{t+\Delta t}N_B^C & 0 & 1 & 0 \\ 0 & -\,^{t+\Delta t}N_A^C & 0 & -\,^{t+\Delta t}N_B^C & 0 & 1 \\ d\left(\frac{\partial N_A}{\partial t}\right) & 0 & d\left(\frac{\partial N_B}{\partial t}\right) & 0 & d\left(\frac{\partial N_K}{\partial t}\right) & 0 \end{array} \right] \,^{t+\Delta t}\left\{ \begin{array}{c} \dot{U}_A \\ \dot{V}_A \\ \dot{U}_B \\ \dot{V}_B \\ \dot{U}_K \\ \dot{V}_K \end{array} \right\}$$

$$= \frac{1}{d} \,^{t+\Delta t}\left[N' \right] \,^{t+\Delta t}\left\{ \dot{u} \right\} \tag{8-52}$$

8.3.4 Quadrilateral Interface Element

Sometimes it would be more suitable to use quadrilateral elements to accommodate deformed configurations of interfaces. As shown in Figure 8.3b, relative displacement rates at time $(t+\Delta t)$ between the upper and lower surfaces at any point in the element are only the functions of local coordinate r, and are irrelevant to the other local coordinate, s. They can be written as:

$$
{}^{t+\Delta t}\left\{\begin{array}{c} \Delta \dot{W}_{nt} \\ \Delta \dot{W}_{nn} \end{array}\right\} = \begin{bmatrix} N_3 & 0 & N_4 & 0 & -N_1 & 0 & -N_2 & 0 \\ 0 & N_3 & 0 & N_4 & 0 & -N_1 & 0 & -N_2 \end{bmatrix} {}^{t+\Delta t}\left\{\begin{array}{c} \dot{U}_3 \\ \dot{V}_3 \\ \dot{U}_4 \\ \dot{V}_4 \\ \dot{U}_1 \\ \dot{V}_1 \\ \dot{U}_2 \\ \dot{V}_2 \end{array}\right\}
\tag{8-53}
$$

Let ${}^{t+\Delta t}\left\{\dot{U}\right\}$ denote matrix ${}^{t+\Delta t}\left[\dot{U}_3 \ \ \dot{V}_3 \ \ \dot{U}_4 \ \ \dot{V}_4 \ \ \dot{U}_1 \ \ \dot{V}_1 \ \ \dot{U}_2 \ \ \dot{V}_2\right]$, the corresponding strain rates would read:

$$
{}^{t+\Delta t}\left\{\begin{array}{c} \dot{\varepsilon}_{nt} \\ \dot{\varepsilon}_{nn} \end{array}\right\} = \frac{1}{d} {}^{t+\Delta t}\begin{bmatrix} N_3 & 0 & N_4 & 0 & -N_1 & 0 & -N_2 & 0 \\ 0 & N_3 & 0 & N_4 & 0 & -N_1 & 0 & -N_2 \end{bmatrix}^{t+\Delta t}\left\{\dot{U}\right\}^T
\tag{8-54}
$$

On the upper and lower boundaries of the continuous interface element, local coordinate s is equal to 1 and -1, respectively. Hence,

$$
N_1 = \frac{1-r}{2} = N_4; \ N_2 = \frac{1+r}{2} = N_3
\tag{8-55}
$$

The normal strain in the direction of the interface length is still taken as in a regular solid element. Therefore, total strain rates in continuous quadrilateral interface elements are described as:

$$
{}^{t+\Delta t}\left\{\begin{array}{c} \dot{\varepsilon}_{nt} \\ \dot{\varepsilon}_{nn} \\ \dot{\varepsilon}_{tt} \end{array}\right\} = \frac{1}{d} {}^{t+\Delta t}\begin{bmatrix} N_3 & 0 & N_4 & 0 & -N_1 & 0 & -N_2 & 0 \\ 0 & N_3 & 0 & N_4 & 0 & -N_1 & 0 & -N_2 \\ d\dfrac{\partial N_3^s}{\partial t} & 0 & d\dfrac{\partial N_4^s}{\partial t} & 0 & d\dfrac{\partial N_1^s}{\partial t} & 0 & d\dfrac{\partial N_2^s}{\partial t} & 0 \end{bmatrix} {}^{t+\Delta t}\left\{\dot{U}\right\}
$$

$$
= \frac{1}{d} {}^{t+\Delta t}\left[N^{'}\right] {}^{t+\Delta t}\left\{\dot{U}\right\}
\tag{8-56}
$$

Where N_1^s, N_2^s, N_3^s and N_4^s are regular shape functions for a quadrilateral solid element.

8.3.5 Elasto-Plastic Constitutive Model for the Continuous Interface Elements

In a continuous interface element, stress rates $^{t+\Delta t}\left\{\dot{\sigma}\right\}$ corresponding to the strain rates described above are written as:

$$^{t+\Delta t}\left\{\dot{\sigma}\right\} = {}^{t+\Delta t}\left\{\begin{array}{c}\dot{\tau}_{nt}\\ \dot{\sigma}_{nn}\\ \dot{\sigma}_{tt}\end{array}\right\} = {}^{t+\Delta t}\left[D_{ep}\right]\,{}^{t+\Delta t}\left\{\dot{\varepsilon}\right\} = \frac{1}{d}\,{}^{t+\Delta t}\left[D_{ep}\right]\,{}^{t+\Delta t}\left[N'\right]\,{}^{t+\Delta t}\left\{\dot{u}\right\} \tag{8-57}$$

Where $[D_{ep}]$ is the elastoplastic matrix of interface materials. Stress state of an interface is always expressed as (σ_n, τ) in laboratory tests, where σ_n and τ are normal and shear stresses, respectively. They are obtained by dividing applied forces by the interface area subject to the forces. For the same reason as stated in Section 2, one must have $\dot{\tau}_{nt}$ equal to $\dot{\tau}$ and $\dot{\sigma}_{nn}$ equal to $\dot{\sigma}_n$.

Physical explanation of $[D_{ep}]$ is different from the traditional elastoplastic matrix in a regular solid element. Its elastic part should be set to coincide with the stress-strain response of the interface:

$$^{t+\Delta t}\left\{\begin{array}{c}\dot{\tau}\\ \dot{\sigma}_n\end{array}\right\} = \begin{bmatrix}k_t & 0\\ 0 & k_n\end{bmatrix}\,{}^{t+\Delta t}\left\{\begin{array}{c}\Delta\dot{u}\\ \Delta\dot{v}\end{array}\right\} \tag{8-58}$$

Where $\Delta\dot{u}$ and $\Delta\dot{v}$ are relative tangential and normal displacement rates, respectively. Variables k_t and k_n are the shear and normal stiffness of the interface, respectively. Also, in the traditional interface analyses, normal stress-strain response in the interface length direction is always neglected regardless of the response across the interface. Considering all these facts in experimental analyses of interfaces, an anisotropic model must be applied for the continuous interface element. In this section, a specific cross-anisotropic elastic model is employed.

8.3.6 The Cross-Anisotropic Elastic Model

The cross-anisotropic elastic model is expressed as:

$$\left[D_e\right] = \begin{bmatrix}D_{11} & 0 & 0\\ 0 & D_{22} & D_{23}\\ 0 & D_{32} & D_{33}\end{bmatrix} \tag{8-59}$$

Components of $[D_e]$ matrix are (Desai and Christian, 1977):

$$D_{11} = 2G_{nt}$$
$$D_{22} = A\left(1-\mu_{nt}^2\right)$$
$$D_{23} = D_{32} = A\mu_{nt}\left(1+\mu_{tt}\right)$$
$$D_{33} = A\left(\frac{E_{tt}}{E_{nn}}-\mu_{nt}^2\right) \tag{8-60}$$

Where $A = E_{tt}/\left[\left(1+\mu_{tt}\right)\left(E_{tt}/E_{nn}\left(1-\mu_{tt}\right)-2\mu_{nt}\right)\right]$. E_{tt} and E_{nn} are interpreted as the Young's moduli for loading in the tangential plane (t-o'-n), E_{tt} being along the axis t and E_{nn} along the normal axis n, respectively. The variable μ_{tt} is the Poisson's ratio relating loading along one axis in the tangential plane to the strain along the other axis in the plane. The relationship between extensional strains in the tangential plane and normal loading, or between normal extensional strains and tangential loading, is controlled by the other Poisson's ratio μ_{tn} or μ_{nt}. The variable G_{nt} relates shear stresses out of the tangential plane to shear strains out of the tangential plane, E_{tt} and μ_{tt} can be obtained from conventional laboratory tests. E_{nn} and G_{nt} are calculated approximately from the relations (Sharma and Desai, 1992):

$$E_{nt} = k_n d, \text{ and } G_{nt} = k_t d \tag{8-61}$$

The contribution to the in-plane stress σ_{tt} from normal strain ε_{nn} is often neglected (Sharma and Desai, 1992). Therefore, μ_{nt} is set to be zero. The special Mohr-Coulomb yielding criterion with the associated flow rule is employed, which is an elasto-perfectly plastic model, and is literally described by Chan et al. (1997).

As described at the beginning, tangential shearing along an interface is the preferred direction of yielding due to the anisotropy of interface materials (Chan et al., 1997). This implies that a general yielding criterion in which yielding may take place in any direction is not applicable to the continuous interface modeling. Yielding along the tangential direction of the interface should be prescribed in the yield criterion for the continuous interface elements.

8.4   Infinite Element and Rigid Element

8.4.1   Infinite Element

In solving the boundary value problems of an infinite domain or problems whose far-field solutions may not so significantly affect the region of interest (ROI), those far-field areas may be represented by infinite elements. The ABAQUS manual and the manual for 3D computer aided pavement analysis system (CAPA-3D) by Scarpas present excellent references.

8.4.2   Rigid Element

The principal advantage to represent portions of a model with rigid bodies rather than deformable finite elements is computational efficiency. Element-level calculations are not performed for elements that are part of a rigid body. Although some computational effort is required to update the motion of the nodes of the rigid body and to assemble concentrated and distributed loads, the motion of the rigid body is determined completely by a maximum of six degrees of freedom at the reference node.

8.5   Constitutive Model Implementation

8.5.1   Mathematics Representation of the FEM Solution Process

In the FE solution for a stress-strain problem, equilibrium and boundary conditions are properly combined into the weak-form boundary value problem through the principle of virtual work. The spatial discretization of the weak form equations uses finite element and some spatial interpolation scheme for the displacement u, i.e.:

$$u = Nu^q \qquad (8\text{-}62)$$

Where u^q is the displacement at the nodes, N is the interpolating function. Through the strain-displacement relation and constitutive model, the equilibrium equation could be represented as a set of equations with node displacements as independent variables, or:

$$g(u^q) = 0 \qquad (8\text{-}63)$$

For a quasi-static initial boundary value problem in a specific time interval, the solution of Equation (8-63) is obtained incrementally and successively by dividing the time interval into an appropriate number of time steps. For each step, the solution procedure for a displacement driving FE system could be summarized as the following three steps:

1. Global computation (equilibrium equations). Under the defined material behavior, loading condition, kinematic conditions, Equation (8-63) could be solved iteratively. For example, for the FE code ABAQUS, Newton's method is used. Using Δu, the strain increment $\Delta \varepsilon$ can be obtained.

2. Local computation (constitutive equations). At each material calculation point, stress and the set of internal state variables are integrated and updated under specified initial conditions and given strain increment $\Delta \varepsilon$. For ABAQUS and user's materials, this is accomplished through UMAT. The following sections deal with this computation.

3. Residual force computation and convergence check. Determine the residual stress of the FE equilibrium equations and check convergence condition. If not reached, repeat 1 and 2.

Two basic strategies exist for the step 1 and 2 computations: (1) simultaneous solution of the equilibrium and constitutive equations, or (2) separate solution of the two sets of equations requiring iteration procedures at the global level. Among the techniques of the first type, the method by Kanchi et al. (1978) and Marques et al. (1983) and the method by Hughes et al. (1985) are two different possibilities. The technique by Kanchi and Marques evaluates state variables at time t_{k+1} by Taylor series expansion at time t_k without further iteration, while the technique by Hughes uses Newton-Raphson iterations. Among the techniques of the second type, there are two techniques also: (a) use the same time step for both equilibrium and constitutive equations (Snyder et al., 1981; Heaps et al., 1986) and (b) use different time steps for equilibrium and constitutive equations with smaller time steps for constitutive equations (Ottosen and Gunneskov 1985). Most of these techniques could be applied to local computations.

In local computations, FE solution could be treated as the following mathematical equivalency:

Given at Step k: state variables, σ_k and $\Delta \varepsilon$, Δt, t_k

Find at Step $k+1$: state variables and $\Delta \sigma$, $\sigma_{k+1} = \sigma_k + \Delta \sigma$, $t_{k+1} = t_k + \Delta t$

So that all the state variables satisfy constitutive equations. Both this update process and the global update process are called time integration as the solution is concerned with the integration of differential equations driven by fictitious or real time.

8.5.2   Time Integration Algorithms

A general method concerning the solution of a partial differential Equation (8-64) is the generalized Euler method represented by Equation (8-65) (Lubliner, 1990).

$$\dot{\xi} = \phi(\xi, t) \tag{8-64}$$

$$\Delta\xi = \Delta t[(1-\theta)\phi(\xi, t) + \theta\phi(\xi + \Delta\xi, t + \Delta t)] \tag{8-65}$$

where ξ – could be a scalar, a vector or a tensor, or combined quantities
      t – could be time or fictitious time
      θ – weight parameter

$\theta = 0$, Explicit method; $\theta \neq 0$ Implicit method. In implicit methods, several well known methods include: Backward Euler method ($\theta = 1$), Trapzoidal method ($\theta = 0.5$), and Galerkin method ($\theta = 2/3$).

An iteration method producing faster convergence is the Newton-Raphson method, by which the equation will be solved with the following iteration:

Define
$$\psi(\Delta\xi, \xi, t, \Delta t) \equiv \dot{\xi} - \phi(\xi, t) \tag{8-66}$$

$$J = \left.\left(\frac{\partial\psi}{\partial\Delta\xi}\right)\right|_{\Delta\xi^0} = I - \theta\Delta t\left.\left(\frac{\partial\phi}{\partial\xi}\right)\right|_{\xi+\Delta\xi^0} \tag{8-67}$$

$$\Delta\xi^{(1)} = \Delta\xi^{(0)} + J^{-1}\psi(\Delta\xi^{(0)}) \tag{8-68}$$

Where I is a unit matrix. The Newton-Raphson method is the most commonly used method in implicit FE algorithm.

Generally, the numerical implementation of the solution to a set of constitutive equations requires the consideration of accuracy, stability, and efficiency. A compromise should be reached between these factors for a large scale or a long sequence FE problem. Ortiz et al. (1985) summarized several important findings about the generalized Euler method:

    a. ($\theta > 0.5$), unconditionally stable; $\theta < 0.5$, conditionally stable

    b. $\theta = 0.5$, better accuracy

    c. The equivalency of small-scale stability in energy normal is equivalent to large-scale stability in the associated geodesic distance. Stability analysis may be confined to the assessment of small-scale stability.

Although viscoplasticity constitutive equations are usually highly nonlinear and numerically very stiff, a small variation of the independent variables will produce a great change of the dependent variables (Krieg, 1977; Shin et al., 1977). Both explicit and implicit methods could be used for the time integration. Successful examples of applying explicit method were presented by Zienkiewics et al. (1974), however, the stability condition for this method was found by Cormeau (1975) to be very critical and requires very small time steps. Another scheme using explicit method with subincrementation to enhance the accuracy was successfully applied by Kumar et al. (1980) and Banthia et al. (1985). This scheme usually has to combine with error checking and automatic time stepping. As both time stepping and error checking are model and loading process dependent, the criteria for time stepping and error checking should be established for different problems.

The fully implicit method is believed to be the most accurate method. The consistent tangent moduli method by Simo et al. (1985) for rate-independent elastoplasticity and the method by Lush et al. (1989) are quite popular. However, for this method, the backward operator requires the solution of a set of nonlinear equations with a Newton-type iteration procedure and can be computationally intensive. For large-scale FE simulation or long-sequence FE simulation, computation expenses could be critical. Also, according to Busso et al. (1994), the complexity of the iteration expressions can limit its implementation into finite element codes for some elaborate constitutive models with tensorial internal variables.

To avoid the problems encountered in both explicit method and implicit method, the implicit method combined with Taylor series approximation (Kanchi et al., 1978; Marques et al., 1983) can be used. By this approach, the finite incremental constitutive equations were derived from first order Taylor series expansion. The simultaneous solution of the evolution equations placed this procedure on a robust foundation. Examples of this procedure were given by Dombrovsky (1992), where he applied this procedure to derive the finite incremental stress-strain relation for two popular viscoplasticity models: the Miller model (Miller, 1976) and the Bodner-Partom model (Bodner et al., 1975). This method was proved to be stable, accurate, and efficient. It allows a much larger time step than the explicit method without encountering stability problems and is computationally less intensive than the fully implicit method.

8.6   Semi-Implicit Implementation of the SHRP Viscoplasticity Deformation Model

8.6.1   Introduction to the SHRP Model

AC has long been recognized as a viscoplasticity material (Lai et al., 1973; Huschek, 1977; Brown and Bell 1979). However, in the conventional design and analysis, it is still treated either as a nonlinear elasticity material or as a viscoelasticity material. The lack of a proper constitutive model and its corresponding numerical implementation are the principal barriers to advance in this field. In the last decade, a research program under SHRP (SHRP-A-415, 1994) investigated the permanent deformation mechanism of AC and identified the following significant permanent deformation characteristics:

- Dilation
- Shearing stiffening under hydrostatic pressure
- Temperature dependence
- Residual permanent deformation
- Negligible volumetric creep deformation

A unified viscoplasticity constitutive model with combined isotropic and kinematic hardening was developed to include these properties. The constitutive model is a very complicated model that had some realistic performance prediction for AC. With this constitutive model the analysis of the scaling effect of laboratory simulation tests becomes feasible and realistic. This section presents the implementation of the SHRP constitutive model on the FE code ABAQUS, the time integration algorithm, the incremental constitutive equations, and some validations.

8.6.2 The Constitutive Equations

In the SHRP model, dilatancy and hardening were considered to be related to the aggregate skeleton and were modeled by a nonlinear spring of third order Green hyperelastcity (Desai and Siriwardane, 1984) with elastic strain energy up to the fourth order. Temperature and rate dependence were considered to be associated with the binder and were modeled by a non-linear dashpot. Residual permanent deformation was modeled by plasticity of Mises type with associated flow, and combined isotropic and kinematic hardening. A graphical representation of the model is illustrated in Figure 8.5. It is understood that the viscoelasticity component and the elastoplasticity component share the same strain, and the stress is the sum of the stresses of the two components. Figures 8.5a and 8.5b respectively represent the model in total and incremental stress-strain format.

The Elastic Component

The elastic component for both viscoelasticity and elasoplasticty components are the same. The elastic strain energy is a function of the invariants of the elastic strain tensor.

$$W(\varepsilon^e) = \frac{1}{2}C_1 I_1^2 + C_2 I_2 + \frac{1}{3}C_3 I_1^3 + C_4 \bar{I}_1 I_2 + C_5 I_3 + \frac{1}{4}C_6 I_1^4 + C_7 I_1^2 I_2 + C_8 I_1 I_3 + \frac{1}{2}C_9 I_2^2 \quad (8\text{-}69)$$

Where ε^e is the elastic strain tensor and I_1, I_2, I_3 are the invariants of the elastic strain tensor. $W(\varepsilon^e)$ is the elastic strain energy.

$$I_1 = trace(\varepsilon^e), \ I_2 = \frac{1}{2}(I_1^2 - \varepsilon^e : \varepsilon^e), \ I_3 = det(\varepsilon^e) \quad (8\text{-}70)$$

And

$$\varepsilon^e = \varepsilon - \varepsilon^i \quad (8\text{-}71)$$

Where ε^i = the inelastic strain tensor and ε is the total strain tensor. C_i, $i = 1,9$ are material constants. ε^i = creeping strain in the viscoelasticity component and is plastic strain in the elastoplasticity component.

The stress tensor is:

$$\sigma_{ij} = \frac{\partial W}{\partial \varepsilon_{ij}^e} \quad (8\text{-}72)$$

The elasticity tensor is:

$$C_{ijkl} = \frac{\partial^2 W}{\partial \varepsilon_{ij}^e \partial \varepsilon_{kl}^e} \quad (8\text{-}73)$$

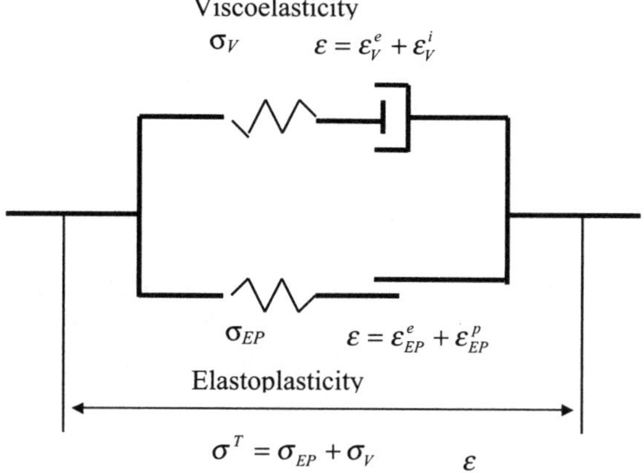

a. Constitutive Model in Total Stress and Strain

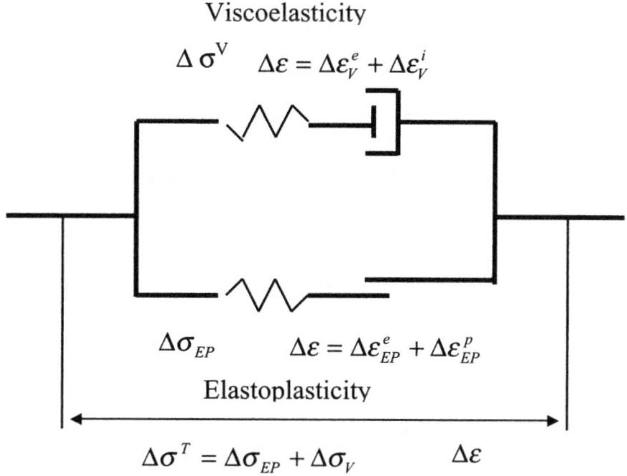

b. Constitutive Model in Incremental Stress and Strain
(where V denotes viscoelasticity component, EP denotes elastoplasticity component)

Figure 8.5   The SHRP viscoplasticity model.

Thus,

$$C_{ijkl} = C_{ijlk} = C_{jikl} = C_{klij} \tag{8-74}$$

The detailed derivation of the elasticity tensor was not presented.

The Viscous Component

$$\dot{\varepsilon}^i = (1 - \frac{b^n}{\rho^n})^{-1}\zeta^{-1}\partial_{\varepsilon^e}W(\varepsilon^e) \tag{8-75}$$

Equation (8-75) basically states that the current inelastic strain rate is proportional to the current stress.

Where
$$b = \max(b_\tau),\ 0 \le \tau \le t \tag{8-76}$$

$$b_\tau = (\varepsilon^i : \zeta : \varepsilon^i)^{\frac{1}{2}} \tag{8-77}$$

And
$$\zeta_{ijkl} = \lambda^v \delta_{ij}\delta_{kl} + \mu^v (\delta_{ik}\delta_{jl} + \delta_{il}\delta_{jk}) \tag{8-78}$$

ζ_{ijkl} is a Rank 4 isotropic tensor, δ_{ij} is the Kroneker delta. τ is real time. $\lambda^v, \mu^v, \rho, n$ are temperature-dependent material constants. Under a "thermorheologically simple" assumption, the viscosity at different temperatures are related by the following relation:

$$\zeta = \zeta_0 e^{\frac{C_T(T-T_0)}{T_0 T}} \tag{8-79}$$

T_0 = reference temperature at which material constants are characterized; T-current temperature; C_T = material constant.

Plasticity Component

The classical rate independent Mises plasticity model with associated flow rule and linear isotropic and kinematic hardening was adopted for the plasticity component.

The yield function:

$$f(\sigma, q, \alpha) = \|\eta\| - \sqrt{\frac{2}{3}} K(\alpha) \tag{8-80}$$

The flow law:
$$\dot{\varepsilon}^p = \gamma \frac{\eta}{\|\eta\|} \tag{8-81}$$

The evolution of the equivalent plastic strain:

$$\dot{\alpha} = \sqrt{\frac{2}{3} \dot{\varepsilon}^p : \dot{\varepsilon}^p} \tag{8-82}$$

The evolution of back stresses (kinematic hardening):

$$\dot{q} = \frac{2}{3} H'(\alpha) \frac{\eta}{\|\eta\|} \tag{8-83}$$

The other equations were also presented as follows:

$$\alpha = \int_0^t \sqrt{\frac{2}{3}} \left\| \dot{\varepsilon}^p(\tau) \right\| d\tau \tag{8-84}$$

$$\eta = dev[\sigma] - q \tag{8-85}$$

$$tr[q] = 0 \tag{8-86}$$

$$H'(\alpha) = (1-\beta)\bar{H} \tag{8-87}$$

$$K(\alpha) = \sigma_y + \beta \bar{H} \alpha \tag{8-88}$$

where q = the back stress or kinematic hardening center
$\qquad \varepsilon^p$ = plastic strain
$\qquad \alpha$ = the equivalent plastic strain for isotropic hardening
$\qquad \sigma$ = Cauchy stress tensor
$\qquad \dot{q}\ \dot{\varepsilon}^p\ \dot{\alpha}$ = the corresponding rate
$\qquad \sigma_y, \bar{H}, \beta$ = material constants

The constitutive model is a unified viscoplasticity model (Krempl, 1987). It might not be applicable to pressure-sensitive materials such as granular material and bonded granular materials. The Drucker-Prager (Drucker et al., 1952) yield function might be better suited for this material. However, the discussion of the suitability and improvement on the constitutive model are beyond the purpose of this section whose main focus is to illustrate the semi-implicit method. Implicit implementation of the Drucker-Prager type of model using implicit method can be found in Wang et al. (2004). As this constitutive model is one of the most advanced available constitutive models with detailed material characterization, its implementation constitutes a good example. The SHRP constitutive model is a very complicated model, mathematically, and the set of equations has the following features:

- High nonlinearity
- Rate dependency
- Numerical stiffness

The implementation of this constitutive model is quite representative of the implementation of other viscoplasticity and elastoplasticity models. An analysis of the physical structure of the model is helpful. For example, the elastoplasticity model with linear elasticity could be deduced from the general model by setting $C_i = 0, i = 3-9$ [see Equation (8-69) above] and canceling out the viscous component.

8.6.3 Finite Incremental Constitutive Equations

Elastoplasticity Component
Equation 8-64 is a generalized differential equation. The actual equations of evolution in this case were Equations (8-81), (8-82), and (8-83). The several variables requiring updates are stresses, kinematic and isotropic hardening variables, plastic strains, etc. The three sets of equations could be generalized in the following format:

$$\dot{\varepsilon}^p = f_1(\sigma, q, \alpha) \tag{8-89}$$

$$\dot{q} = f_2(\sigma, q, \alpha) \tag{8-90}$$

$$\dot{\alpha} = f_3(\sigma, q, \alpha) \tag{8-91}$$

Where $\sigma, q, \alpha, \varepsilon^p$ are the stress tensor, kinematic center, equivalent plastic strain, and plastic strain tensor. $\dot{\varepsilon}^p$, $\dot{q}$, $\dot{\alpha}$ are correspondingly the rate of the variables.

The unified implicit solution scheme could be represented in Equations (8-92), (8-93), and (8-94).

$$\Delta \varepsilon^p = \Delta t[(1-\theta)\dot{\varepsilon}_k^p + \theta \dot{\varepsilon}_{k+1}^p] \tag{8-92}$$

$$\Delta q = \Delta t[(1-\theta)\dot{q}_k + \theta \dot{q}_{k+1}] \tag{8-93}$$

$$\Delta \alpha = \Delta t[(1-\theta)\dot{\alpha}_k + \theta \dot{\alpha}_{k+1}] \tag{8-94}$$

If $\dot{\varepsilon}_{k+1}^p$, $\dot{q}_{k+1}$ and $\dot{\alpha}_{k+1}$ are expressed as a Taylor series to the first order, the following equations could be obtained:

$$\dot{\varepsilon}_{k+1}^p \approx \dot{\varepsilon}_k^p + \frac{\partial \dot{\varepsilon}^p}{\partial \sigma} : \Delta\sigma + \frac{\partial \dot{\varepsilon}^p}{\partial q} : \Delta q + \frac{\partial \dot{\varepsilon}^p}{\partial \alpha} \bullet \Delta\alpha \tag{8-95}$$

$$\dot{q}_{k+1} \approx \dot{q}_k + \frac{\partial \dot{q}}{\partial \sigma} : \Delta\sigma + \frac{\partial \dot{q}}{\partial q} : \Delta q + \frac{\partial \dot{q}}{\partial \alpha} \bullet \Delta\alpha \tag{8-96}$$

$$\dot{\alpha}_{k+1} \approx \dot{\alpha}_k + \frac{\partial \dot{\alpha}}{\partial \sigma} : \Delta\sigma + \frac{\partial \dot{\alpha}}{\partial q} : \Delta q + \frac{\partial \dot{\alpha}}{\partial \alpha} \bullet \Delta\alpha \tag{8-97}$$

Assume the stress increment could be approximated as:

$$\Delta\sigma = C : (\Delta\varepsilon - \Delta\varepsilon^p) \tag{8-98}$$

Then one would have the following equations:

$$[C^{-1} + \Delta t\theta \frac{\partial \dot{\varepsilon}^p}{\partial \sigma}] : \Delta\sigma = \Delta\varepsilon - \Delta t\,\dot{\varepsilon}_k^p - \Delta t\theta \frac{\partial \dot{\varepsilon}^p}{\partial q} : \Delta q - \Delta t\theta \frac{\partial \dot{\varepsilon}^p}{\partial \alpha} \bullet \Delta\alpha \tag{8-99}$$

$$[I - \Delta t\theta \frac{\partial \dot{q}}{\partial q}] : \Delta q = \Delta t\,\dot{q}_k + \Delta t\theta \frac{\partial \dot{q}}{\partial \sigma} : \Delta\sigma + \Delta t\theta \frac{\partial \dot{q}}{\partial \alpha} \bullet \Delta\alpha \tag{8-100}$$

$$[1 - \Delta t\theta \frac{\partial \dot{\alpha}}{\partial \alpha}] \bullet \Delta\alpha = \Delta t\,\dot{\alpha}_k + \Delta t\theta \frac{\partial \dot{\alpha}}{\partial \sigma} : \Delta\sigma + \Delta t\theta \frac{\partial \dot{\alpha}}{\partial q} : \Delta q \tag{8-101}$$

In the SHRP model, $\dfrac{\partial \dot{\varepsilon}^p}{\partial \alpha} = 0$, $\dfrac{\partial \dot{q}}{\partial \alpha} = 0$, $\dfrac{\partial \dot{\alpha}}{\partial \alpha} = 0$ and thus one has:

$$[C^{-1} + \Delta t\theta \frac{\partial \dot{\varepsilon}^p}{\partial \sigma}] : \Delta\sigma = \Delta\varepsilon - \Delta t\,\dot{\varepsilon}_k^p - \Delta t\theta \frac{\partial \dot{\varepsilon}^p}{\partial q} : \Delta q \tag{8-102}$$

$$[I - \Delta t\theta \frac{\partial \dot{q}}{\partial q}] : \Delta q = \Delta t\,\dot{q}_k + \Delta t\theta \frac{\partial \dot{q}}{\partial \sigma} : \Delta\sigma \tag{8-103}$$

$$\Delta\alpha = \Delta t\,\dot{\alpha}_k + \Delta t\theta \frac{\partial \dot{\alpha}}{\partial \sigma} : \Delta\sigma + \Delta t\theta \frac{\partial \dot{\alpha}}{\partial q} : \Delta q \tag{8-104}$$

Note: All the derivatives are evaluated at status k and $\Delta\varepsilon = \dot{\varepsilon}\Delta t$.

From the above equations, one can see that $\Delta\alpha$ is decoupled (linear isotropic hardening) from the first two equations and the first two equations could be solved relatively easily. Rearranging the first two equations, one obtains the following equations:

$$[C^{-1} + \Delta t\theta\frac{\partial\dot{\varepsilon}^p}{\partial\sigma}] : \Delta\sigma + \Delta t\theta\frac{\partial\dot{\varepsilon}^p}{\partial q} : \Delta q = \Delta\varepsilon - \Delta t\,\dot{\varepsilon}_k^p \tag{8-105}$$

$$\Delta t\theta\frac{\partial\dot{q}}{\partial\sigma} : \Delta\sigma - [I - \Delta t\theta\frac{\partial\dot{q}}{\partial q}] : \Delta q = \Delta t\,\dot{q}_k \tag{8-106}$$

In computer implementation, tensors such as σ, ε, q, ε^p, $\Delta\sigma$, $\Delta\varepsilon$, Δq, $\Delta\varepsilon^p$ are transformed into six dimensional vectors, while the elasticity tensor C and viscosity tensor η are transformed into 6×6 matrices. By this convention, close form solution to the above equations can be obtained quite conveniently:

Denote:

$$C^{-1} + \Delta t\theta\frac{\partial\dot{\varepsilon}^p}{\partial\sigma} = X_1, \qquad \Delta t\theta\frac{\partial\dot{\varepsilon}^p}{\partial q} = Y_1, \qquad \Delta\varepsilon - \Delta t\,\dot{\varepsilon}_k^p = Z_1$$

$$\Delta t\theta\frac{\partial\dot{q}}{\partial\sigma} = X_2, \qquad I - \Delta t\theta\frac{\partial\dot{q}}{\partial q} = Y_2, \qquad -\Delta t\,\dot{q}_k = Z_2$$

Where X_1, X_2, Y_1, Y_2 are 6×6 matrices, while Z_1 and Z_2 are six-dimensional vectors (this implementation is not valid for plane stress problem). Equations (8-105) and (8-106) are then simplified as:

$$X_1 : \Delta\sigma + Y_1 : \Delta q = Z_1 \tag{8-107}$$

$$X_2 : \Delta\sigma - Y_2 : \Delta q = Z_2 \tag{8-108}$$

$$\Delta\sigma = (Y_1^{-1}Z_1 + Y_2^{-1}Z_2) / (Y_2^{-1}X_1 + Y_2^{-1}X_2) \tag{8-109}$$

$$\Delta q = (X_1^{-1}Z_1 - X_2^{-1}Z_2) / (X_1^{-1}Y_1 + X_2^{-1}Y_2) \tag{8-110}$$

$$\Delta\alpha = \Delta t\,\dot{\alpha}_k + \Delta t\theta\frac{\partial\dot{\alpha}}{\partial\sigma} : \Delta\sigma + \Delta t\theta\frac{\partial\dot{\alpha}}{\partial q} : \Delta q \tag{8-111}$$

Note: Equation (8-111) is a duplication of Equation (8-104).

So far, the finite incremental constitutive equations have been obtained and the computer implementation is quite straightforward from now on. It should be noted that the above implementation refers to the elastoplastic branch only.

The Viscous Component

For simple illustration, the nonlinear dashpot could be replaced with a linear dashpot and thus $(1 - \frac{b^n}{\rho^n})^{-1} = 1$.

$$\Delta\varepsilon^i = \int_{t_n}^{t_{n+1}} \xi^{-1} : \sigma\,dt \tag{8-112}$$

Using the Euler method, this integral could be estimated as follows:

$$\Delta\varepsilon^i = M : [\sigma_n(1-\theta) + \sigma_{n+1}\theta]\Delta t \tag{8-113}$$

Where $M = \eta^{-1}$ and η is the viscosity tensor.

One can expand σ_{n+1} around σ_n as a Taylor series and truncate the second and above terms, and then have the following expression:

$$\sigma_{n+1} = \sigma_n + C : (\Delta\varepsilon - \Delta\varepsilon^i) \tag{8-114}$$

Thus, one has:

$$\Delta\varepsilon^i = M : [\sigma_n + \theta C : (\Delta\varepsilon - \Delta\varepsilon^i)]\Delta t \tag{8-115}$$

$$\Delta\varepsilon^i = \frac{M : \sigma_n\Delta t + \theta M : C : \Delta\varepsilon\Delta t}{I + \theta M : C\Delta t} \tag{8-116}$$

Where I is a rank 4 unit tensor.

For the strain rate equation, the principle of objectivity requires use of the Jaumann derivative of Cauchy stress (see Lush et al., 1989), however, for small strain problems with small rotation, the Cauchy stress can be used.

8.6.4   Implementation of Constitutive Model on ABAQUS

In selecting an FE code, robust algorithm, wide validation, and low cost are critical. ABAQUS (1984) is a good choice in terms of these criteria. Two of the many functions included with ABAQUS are User's Material and User's Element interfaces which are designed for the implementation of self-developed constitutive models, and interface models. Once the constitutive models or interface models are implemented, they become an integrated part of ABAQUS and allow for different material combinations.

ABAQUS is a displacement-driving FE code. In solving the global equilibrium equations it uses an implicit method. Implementation of constitutive models on ABAQUS through its user-defined material (UMAT) is not a new exercise Lush et al., 1989), however, due to the implicit method adopted for global computations, the step used for time integration might be larger than suitable for local integration (constitutive integration). Subincrementation was widely used in combination with either the explicit or implicit method. A robust scheme for the local integration is crucial for accuracy and efficiency.

One technical tip involving the update of the Jacobian matrix is proper to mention. The stress Jacobian Jac or the consistent tangent stiffness is defined as:

$$Jac = \frac{\Delta\sigma}{\Delta\varepsilon} \tag{8-117}$$

A feature concerning the stress Jacobian as emphasized by Hughes (1985) and Lush et al. (1989) is that it enhances the rate of convergence in the search for the incremental deformation that leads to the satisfaction of the momentum balance equation, but in the end has no effect on the accuracy of the solution. The implication of this feature is the acceptability of a proper approximation of the Jacobian in case of difficulties in obtaining an accurate expression. In this case (Wang and Li, 2008) the stress Jacobian is additive for the viscous component and the elastoplasticity component. For the viscous component, the average of the elasticity modulus at the start and at the end of an interval was used, while for the elastoplasticity component the average elastoplasticity modulus at the start and at the end of an interval was used. Wang and Li (2008) demonstrated the effectiveness of the semi-implicit method.

8.7 Full Implicit Implementation of the Druker-Plager Model

In this section, the implementation mechanisms for fully implicit methods are demonstrated based on Wang et al. (2004). The additive decomposition of strain rate is adopted for the infinitesimal strain assumption. For the closest-point projection approximation of a rate-independent problem, the behavior of an elastoplastic material can be generally characterized by the following equations:

$$d\varepsilon = d\varepsilon^e + d\varepsilon^p \tag{8-118}$$

$$\sigma = C^e : \varepsilon^e \tag{8-119}$$

$$\phi(\sigma, H_\alpha) = \phi(\sigma_m, S, H_\alpha) = \phi(I_1, J_2, J_3, H_\alpha) = 0 \tag{8-120}$$

$$d\varepsilon^p = \Delta\lambda \frac{\partial g}{\partial \sigma} \tag{8-121}$$

$$dH_\alpha = h(d\varepsilon^p, \sigma, {}^t H_\beta) = h(d\varepsilon^p, I_1, J_2, J_3, {}^t H_\beta) \tag{8-122}$$

Where ε^e and ε^p are elastic and plastic strains, respectively, while ϕ and g denote yield and plastic potential functions, respectively. H_α is a set of scalar state variables governing the hardening yield surface, and ${}^t H_\beta$ denotes the state variables at time t. They are both assumed twice differentiable functions in order to get second derivatives, which are necessary in the stress integration process. $\Delta\lambda$, a positive multiplier, is called a plastic consistency parameter satisfying the Kuhn-Tucker complementary conditions (Simo and Hughes, 1998):

$$\Delta\lambda \geq 0, \quad \phi(\acute{o}, H_\alpha) \leq 0, \quad \Delta\lambda\phi(\acute{o}, H_\alpha) = 0 \tag{8-123}$$

C^e represents the fourth order elasticity tensor. It is given by:

$$C^e_{ijkl} = \lambda\delta_{ij}\delta_{kl} + 2G\delta_{ik}\delta_{jl} \tag{8-124}$$

Where $\lambda = K - \frac{2}{3}G$, K and G are elastic bulk and shear modulus, respectively, and δ_{ij} is the Kronecker delta.

In the closest point return mapping algorithm or the Euler backward algorithm, for any given strain increments $\Delta\varepsilon$, the corresponding stress increments must be computed iteratively. The stresses at time $t+\Delta t$ can be written in the following form:

$${}^{t+\Delta t}\sigma = {}^t\sigma + \Delta\sigma = {}^{t+\Delta t}\sigma^{tr} - C^e : \Delta\varepsilon^p \tag{8-125}$$

Where $\Delta \varepsilon^p$ is plastic strain increment, from time t to time $t + \Delta t$. $^{t+\Delta t}\sigma^{tr}$, the stress predictor, takes the form:

$$^{t+\Delta t}\sigma^{tr} = {}^t\sigma + C^e : \Delta\varepsilon \tag{8-126}$$

If the full Newton scheme is employed to solve the nonlinear equilibrium equations, one needs to calculate the so-called tangential linearization moduli right after the stress calculation, which is consistent with the stress integration algorithm by applying total stresses at time $t+\Delta t$. It is simply written as:

$$^{t+\Delta t}C = {}^{t+\Delta t}\left(\frac{\partial\sigma}{\partial\varepsilon}\right) \tag{8-127}$$

One can decompose the stress tensor σ into hydrostatic and deviatoric components:

$$\sigma = \sigma_m I + S = -pI + S \tag{8-128}$$

Where I is the identity tensor and σ_m denotes the mean stress, i.e., the average of the three normal stresses. Pressure p, holding an opposite sign with the mean stress, and deviatoric stress components S are expressed as (Wang, et al., 2004):

$$^{t+\Delta t}p = {}^{t+\Delta t}p^{tr} + \frac{1}{a_M}\Delta\varepsilon_m^p \tag{8-129}$$

$$^{t+\Delta t}S_{ij} = {}^{t+\Delta t}S_{ij}^{tr} - \frac{1}{a_E}\Delta e_{ij}^p \quad (i, j = 1, 2, 3) \tag{8-130}$$

Stress predictors $^{t+\Delta t}p^{tr}$ and $^{t+\Delta t}S_{ij}^{tr}$ are computed as:

$$^{t+\Delta t}p^{tr} = \frac{1}{a_M}{}^{t+\Delta t}\varepsilon_m^{''} \tag{8-131}$$

$$^{t+\Delta t}S_{ij}^{tr} = \frac{1}{a_E}{}^{t+\Delta t}e_{ij}^{''} \tag{8-132}$$

Where $^{t+\Delta t}\varepsilon_m^{''} = {}^{t+\Delta t}\varepsilon_m - {}^t\varepsilon_m^p$, $^{t+\Delta t}e_{ij}^{''} = {}^{t+\Delta t}e_{ij} - {}^te_{ij}^p$ are known values by which the elastic trial stresses are defined. $^{t+\Delta t}\varepsilon_m$ and $^{t+\Delta t}e_{ij}$ are total mean strain and deviatoric strains at time $t + \Delta t$, respectively. Both a_M and a_E are elastic constants, which are written as:
$a_M = (1-2v)/E$, $a_E = (1 + v)/E$, respectively. $^te_m^p$ and $^te_{ij}^p$ are plastic mean strain and plastic deviatoric strain components at time t, respectively. Their incremental forms are denoted by $\Delta\varepsilon_m^p$ and Δe_{ij}^p, respectively. The use of the non-associated flow rule enables one to express the plastic strain increment $\Delta\varepsilon^p$ as:

$$\Delta\varepsilon^p = \Delta\lambda\left(\frac{\partial g}{\partial\sigma}\right)$$

For more details on the implicit implementation, one can see Wang et al. (2004) on the example of the $J3$ dependent constitutive model.

8.8   Boundary Element Method

8.8.1   Theoretical Basis

The basic FEM Equation (8-18) is copied as follows:

$$\int_{\Omega} D_{ijkl} u_{k,il} w_j \, d\Omega = \int_{\Gamma} D_{ijkl} u_{k,i} w_j n_l \, d\Gamma - \int_{\Omega} D_{ijkl} u_{k,i} w_{j,l} \, d\Omega \tag{8-18}$$

Considering the identity $u_{k,i} w_{j,l} = (u_k w_{j,l})_{,i} - u_k w_{j,li}$ and applying the divergence theorem, the following equation can be derived:

$$\int_{\Omega} D_{ijkl} u_{k,i} w_{j,l} \, d\Omega = \int_{\Gamma} D_{ijkl} u_k w_{j,l} n_i \, d\Gamma - \int_{\Omega} D_{ijkl} u_k w_{j,li} \, d\Omega \tag{8-133}$$

Replacing the second term on the right side of Equation (8-18) with the right side of 8-133, the following equation can be derived:

$$\int_{\Omega} D_{ijkl} (u_{k,il} w_j - u_k w_{j,li}) \, d\Omega = \int_{\Gamma} D_{ijkl} (u_{k,i} w_j n_l - u_k w_{j,l} n_i) \, d\Gamma \tag{8-134}$$

If one selects the fundamental solution (corresponding to the displacement field caused by a concentrated force) as the weighting function $w = u^*$ and denotes the corresponding field variables as σ^*, τ^*, ε^*, and so on, and considering:

$D_{ijkl} u_{k,il} = \sigma_{ki,l}$ $D_{ijkl} w_{j,li} = \sigma^*_{ji,l}$ and the stress-surface traction relationship, one has:

$$\int_{\Omega} D_{ijkl} (u_{k,il} w_j - u_k w_{j,li}) \, d\Omega = \int_{\Omega} (\sigma_{ki,l} u^*_j - \sigma^*_{ji,l} u_k) \, d\Omega$$

And

$$\int_{\Omega} (\sigma_{ki,l} u^*_j - \sigma^*_{ji,l} u_k) \, d\Omega = \int_{\Gamma} (\tau_j u^*_j - u_k \tau^*_k) \, d\Gamma \tag{8-135}$$

It can be conveniently proved that this is actually Betti's reciprocal work principle, which is concerned with the two states of stress and strain within a domain (Ω, Γ) contained in a large domain (Ω^*, Γ^*). Betti's principle states that the virtual work (virtual strain energy) done by σ_{ki} over ε^*_{ki} is equal to the virtual work done by σ^*_{ki} over ε_{ki}, or:

$$\sigma_{ki} \varepsilon^*_{ki} = \varepsilon_{ki} \sigma^*_{ki} \tag{8-136}$$

This can be conveniently proved as:

$$\sigma_{ki} = D_{ijkl} \varepsilon_{jl} \text{ and } \sigma^*_{ki} = D_{ijkl} \varepsilon^*_{jl}$$

And thus one has:

$$\int_{\Omega} \sigma_{ki} \varepsilon^*_{ki} \, d\Omega = \int_{\Omega} \sigma^*_{ki} \varepsilon_{ki} \, d\Omega$$

or

$$\int_{\Omega} \sigma_{ki} u^*_{k,i} \, d\Omega = \int_{\Omega} \sigma^*_{ki} u_{k,i} \, d\Omega \tag{8-137}$$

By identities $\sigma_{ki} u^*_{k,i} = (\sigma_{ki} u^*_k)_{,i} - \sigma_{ki,i} u^*_k$ and $\sigma^*_{ki} u_{k,i} = (\sigma^*_{ki} u_k)_{,i} - \sigma^*_{ki,i} u_k$

And using the Gauss theorem, one can convert Equation (8-137) into the following equation:

$$\int_\Omega \sigma_{ki} \dot{u}_{k,i} d\Omega = \int_\Gamma \sigma_{ki} \dot{u}_k n_i d\Gamma - \int_\Omega \sigma_{ki,i} \dot{u}_k d\Omega$$

$$\int_\Omega \dot{\sigma}_{ki} u_{k,i} d\Omega = \int_\Gamma \dot{\sigma}_{ki} u_k n_i d\Gamma - \int_\Omega \dot{\sigma}_{ki,i} u_k d\Omega \tag{8-138}$$

Therefore, the following equations can be obtained:

$$\int_\Gamma \sigma_{ki} \dot{u}_k n_i d\Gamma - \int_\Omega \sigma_{ki,i} \dot{u}_k d\Omega = \int_\Gamma \dot{\sigma}_{ki} u_k n_i d\Gamma - \int_\Omega \dot{\sigma}_{ki,i} u_k d\Omega \tag{8-139}$$

Or

$$\int_\Omega (\sigma_{ki,l} \dot{u}_j - \dot{\sigma}_{ji,l} u_k) d\Omega = \int_\Gamma (\tau_j \dot{u}_j - u_k \dot{\tau}_k) d\Gamma \tag{8-140}$$

It is the same as Equation (8-135). This is actually the basis equation for BEM.

8.8.2 Fundamental Solution

The fundamental solution is the solution corresponding to an infinite domain subjected to a unit force at point ξ or $\delta(x - \xi)n_j$ (the unit force in j direction). Since there are three displacement components and three unit force directions, the displacement responses can be represented as a tensor $\dot{u}_{ij}$ (displacement in the ith direction due to the forces applied in the jth direction) or:

$$b_j^* = \delta(x, \xi)n_j \ (\Omega^* \to \infty \ \Gamma^* \to \infty)$$

Considering the two equilibrium equations (static situations for less complexity in understanding the fundamental mechanisms):

$$D_{ijkl} u_{k,li} = -b_j, \ D_{ijkl} \dot{u}_{k,li} = -b_j^* \tag{8-141}$$

Then

$$\int_\Omega \sigma_{ki,i} \dot{u}_k d\Omega = \int_\Omega b_k \dot{u}_k d\Omega, \ \int_\Omega \dot{\sigma}_{ki,i} u_k d\Omega = -\int_\Omega b_k^* u_k d\Omega \tag{8-142}$$

Equation (8-140) $\int_\Omega (\sigma_{ki,l} \dot{u}_j - \dot{\sigma}_{ji,l} u_k) d\Omega = \int_\Gamma (\tau_j \dot{u}_j - u_k \dot{\tau}_k) d\Gamma$ becomes:

$$\int_\Gamma (\tau_k \dot{u}_k - \dot{\tau}_k u_k) d\Gamma = \int_\Omega (b_j^* u_j - b_j u_j^*) d\Omega$$

Since $b_j^* = \delta(x, \xi)n_j$, and considering the identity $\int_\Omega b_j^* u_j d\Omega = u_j(\xi)n_j$, one has:

$$u_j(\xi)n_j = \int_\Gamma (\tau_k \dot{u}_k - \dot{\tau}_k u_k) d\Gamma + \int_\Omega b_j u_j^* d\Omega \tag{8-143}$$

Equation (8-143) is the well-known Somigliana's identity. It is basically the jth displacement component at point ξ caused by a unit force in the jth (1, 2, 3) direction. Considering that there are three components of the displacements and forces, Equation (8-143) actually has nine equations and displacements can be represented as $u_{ij}(\xi)$, displacements in i direction ($i = 1, 2, 3$) due to the forces in the j direction ($j = 1, 2, 3$). The fundamental solution $u_j^*(x,\xi)$ should also have nine components as $u_{ij}^*(x,\xi)$, which is the solution to the equation $L_{ik}u_{ij}^*(x,\xi) = -\delta(x,\xi)\delta_{ij}$.

$$u_i(\xi) = \int_\Gamma (\tau_k u_{ik}^* - \tau_{ik}^* u_k)d\Gamma + \int_\Omega b_j u_{ij}^* d\Omega \tag{8-144}$$

For simplification, this discussion is stopped here. Interested readers can read more details in many excellent books referred in Section 8.1.

Equation (8-144) is the basis for BEM approach. However, it still has a domain integral (it can be resolved conveniently as b_j and u_{ij}^* are known (the fundamental solution is the Kevin solution).

There are the multiple reciprocity method, the dual reciprocity method, and the direct transformation method (using higher order solutions) to convert the domain integral to the boundary integral. For simplification, one may neglect the body force term if it is not so critical (for example, the traffic loading is much larger than the weight of the materials for the surface course). For simplification, the body force in Equation (8-144) will be dropped out.

Equation (8-144) calculates the displacement for a point within the domain. It can be extended to the calculation of the displacement of points on the boundary. It can be estimated in such a way as to extend a small boundary of a circle of a sphere that is centered on that point (Figure 8.6). If the additional boundary integral is assessed and the circle or sphere radius approaches zero, the solution for the displacement of the point on the boundary is obtained.

For the fundamental solution (the Kelvin solution, see Chapter 1), it can be shown that $u_{ij}^* \sim \dfrac{1}{r}$ (weakly singular), $t_{ij}^* \sim \dfrac{1}{r^2}$ (strongly singular) for $r \to 0$. By evaluating the integrals around the point on the surface, one can obtain the following boundary integral equation (BIE):

$$c_{ij}(\xi)u_j(\xi) + \int_\Gamma t_{ij}^*(x,\xi)u_j(x)d\Gamma = \int_\Gamma u_{ij}^*(x,\xi)t_j(x)d\Gamma \tag{8-145}$$

$\int_\Gamma t_{ij}^*(x,\xi)u_j(x)d\Gamma$ is a Cauchy principal value integral.

$$c_{ij}(\xi) = \delta_{ij} + \lim_{r \to 0}\int_\Gamma t_{ij}^*(x,\xi)d\Gamma \tag{8-146}$$

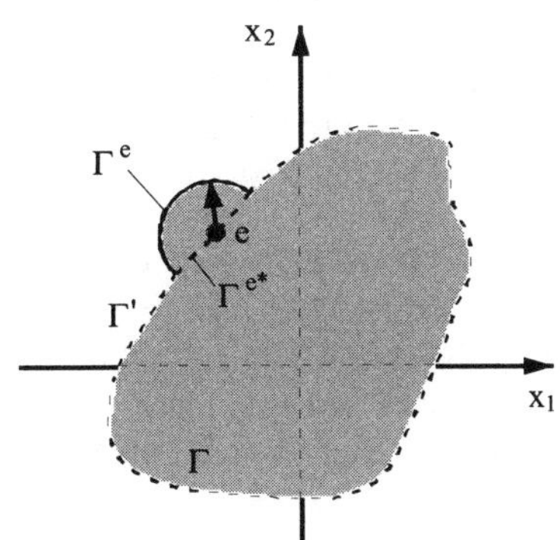

FIGURE 8.6   Illustration of the scheme to obtain the displacements on the boundary.

Therefore, the displacements in the domain can be calculated using Equation (8-144) and those on the boundary can be calculated using Equation (8-145). Once the displacement is obtained, the strains and stresses can be calculated. The major efforts for BEM are therefore the evaluation of the surface integrals $\int_\Gamma (\tau_k u_{ik}^* - \tau_{ik}^* u_k) d\Gamma$, $\int_\Gamma t_{ij}^*(x,\xi) u_j(x) d\Gamma$ and $\int_\Gamma u_{ij}^*(x,\xi) t_j(x) d\Gamma$. There are many numerical methods to develop such evaluations. Gaul et al. (2003), Becker (1992), Paris and Canas (1997), and Brebbia et al. (1984) have excellent descriptions on these techniques. By far the fundamental mechanisms of BEM have been presented.

8.8.3   BEM Applications to Modeling AC

BEM application to AC is very limited. The major reason may be due to the lack of commercially available powerful software to allow convenient applications. Another barrier might be the difficulties in finding the fundamental solutions. Nevertheless, it may have significant advantages in modeling AC behavior with multiple aggregate inclusions.

The displacement discontinuity boundary element method (DD-BEM) has the potential to simulate cracking in granular materials such as AC. In comparison to the continuum-based fracture-mechanics-based approaches, this method explicitly models the crack initiation and propagation in AC.

The DD-BEM method simulates cracking by creating a numerical model with two types of elements: exterior boundary elements and potential crack elements (Sangpetngam, 2003). The exterior boundary elements along the boundary of a problem simulate the edge of specimen and random potential crack elements are placed inside a specimen to simulate predefined crack paths that are assumed to occur along a grain boundary or pass through a grain.

When applied, DD-BEM is usually coupled with various tessellation techniques that have been widely used to represent granular structure in simulation of fracture process. Two basic tessellation schemes, Delaunay and Voronoi, have been used to simulate granular structure of brittle rocks by Peirce and Napier (1995) and Steen et al. (2001), and produced realistic failure patterns. Peirce and Napier (1995) developed a new boundary element solution technique (the multi-pole method) using the two tessellation schemes (Delaunay and Voronoi) for three levels of grain densities. The Voronoi assemblies outperform the Delauney triangulations in that the former are less prone to shed load.

Birgisson et al. (2002a) applied the 2D DD-BEM to simulate crack growth in Superpave IDT specimens, using exterior boundary elements with randomly positioned potential crack elements (Voronoi tessellation) inside the specimen. The numerical model can realistically capture stress-strain responses and crack pattern with an appropriate set of material parameters for local failure at potential crack elements. Birgisson et al. (2002b) further found that Voronoi tessellation with internal fracture path provided realistic simulation of crack growth in asphalt mixture as aggregate particles are allowed to break down at high load level. The two crack growth rules evaluated, i.e., the sequential and parallel, were found to give similar results. Using the same numerical schemes as Birgisson et al. (2002b), it was found that the method was capable of predicting results that matched vertical compressive stress-strain curve, horizontal tensile stress-strain curve, crack pattern, tensile strength, and fracture energy of the mixes during an IDT test.

Sangpetngam (2003) adopted a crack growth model developed by Zhang et al. (2001) into a DD-BEM-based pavement fracture simulator. He used crack tip elements ahead of crack tips to improve accuracy of stress distribution and replaced the process zone with yield elements. The upgrade model requires the determination of four fundamental mixture parameters that can be obtained from less than one hour of testing using the SuperPave IDT. These parameters can account for micro-damage, crack propagation, and healing for stated loading conditions, temperatures, and rest periods. The implementation of crack growth model in the numerical scheme increases the application of fracture mechanics for predicting crack growth in asphalt mixture for various geometries, loading, and boundary conditions.

Using the validated DD-BEM method with a proper tessellation scheme, Birgisson et al. (2003) observed that cracks in AC could start as micro-cracks and later propagate and coalesce to form macro-cracks as the mixture is subjected to sustained tensile stresses, shear stresses, or a combination of both. Using the same model, Birgisson et al. (2004) simulated the IDT tests and showed agreement with laboratory observations. The improved understanding of the mechanism of cracking can be expected to lead to more reliable prediction of the field performance of asphalt mixtures. Recognizing that both crack initiation and propagation processes are directly related to stress-strain fields in asphalt layers, researchers recently have focused on developing more realistic constitutive law that can describe the cracking response of asphalt mixtures under realistic traffic conditions composed of multiple load levels and random rest periods. Sangpetngam et al. (2004) extended the DD-BEM to simulating layered asphalt pavements. Birgisson et al. (2006, 2007, 2008) compared the DD-BEM predicted crack patterns in hot mix asphalt (HMA) mixtures to those measured using a digital image correlation (DIC) system that could accurately obtain displacement and strain fields and detect crack patterns. The comparisons indicated the promising features of the DD-BEM simulations.

In summary, the DD-BEM approach shows potential applicability in identifying critical factors that determine the cracking mechanism of asphalt mixture, and therefore can be used for optimizing or improving the crack resistance of asphalt mixtures. Nevertheless, only very few researchers are engaging in BEM applications research. A combined FEM and BEM approach may demonstrate significant advantages.

Suggested Readings

The theory part of Section 8.2 and Section 8.8 are based on consolidated teaching notes of the author. These notes are prepared following the presentation from an excellent book by Gaul et al. (2003). In this chapter, the presentation for general dynamics problems in Gaul et al. (2003) is reduced to the statics problems for simplification. If readers need more backgrounds, please read this book. For convenience and connections, the symbols adopted in these two sections are consistent with those used in Gaul et al. (2003).

References

ABAQUS (1984). Abaqus User's Manual, *Hibbett, Karlsson and Sorenson, Inc.*

Banthia, V. and Mukherjee, S. (1985). On an improved time integration scheme for stiff constitutive models of inelastic deformation. *Journal of Engineering Materials*, Vol.107, pp.282–285.

Becker, A.A. (1992). The Boundary Element Method in Engineering: A Complete Course, *McGraw-Hill*, London.

Beer, G. (1985). An isoparametric joint/interface element for finite element analysis. *International Journal for Numerical Method in Engineering*, Vol.21, pp.585–600.

Belytschko, T. and Chen, J.S. (2007). Meshfree and Particle Methods. *John Wiley and Sons Ltd.*

Birgisson, B., Soranakom, C., Napier, J.A.L. and Roque, R. (2002a). Simulation of the cracking behavior of asphalt mixtures using random assemblies of displacement discontinuity boundary elements. *The 15th ASCE Engineering Mechanics Conference*, Columbia University, New York.

Birgisson, B., Sangpetngam, B. and Roque, R. (2002b). Prediction of the viscoelastic response and crack growth in asphalt mixtures using the boundary element method. *Transportation Research Record*, No.1789, pp.129–35.

Birgisson, B., Soranakom, C., Napier, J.A.L. and Roque, R. (2003). Simulation of fracture initiation in hot-mix asphalt mixtures. *Transportation Research Record*, No.1849, pp.183–190.

Birgisson, B., Soranakom, C., Napier, J.A.L. and Roque, R. (2004). Microstructure and fracture in asphalt mixtures using a boundary element approach. *Journal of Materials in Civil Engineering*, Vol.16, pp.116–121.

Birgisson, B., Montepara, A., Napier, J.A.L., Romeo, E., Roncella, R. and Tebaldi, G. (2006). Micromechanical analyses for measurement and prediction of HMA fracture energy. *Transportation Research Record*, No.1970, pp.186–195.

Birgisson, B., Montepara, A., Romeo, E., Roque, R., Roncella, R. and Tebaldi G. (2007). Determination of fundamental tensile failure limits of mixtures. *Journal of the Association of Asphalt Pavement Technologists*, Vol.76, pp.303–344.

Birgisson, B., Montepara, R.A.E., Roncella, R., Napier, J.A.L. and Tebaldi, G. (2008). Determination and prediction of crack patterns in hot mix asphalt (HMA) mixtures. *Engineering Fracture Mechanics*, Vol.75, pp.664–673.

Bodner, S.R. and Partom, Y. (1975). Constitutive equations for elastic-viscoplastic strain-hardening materials. *Journal of Applied Mechanics*, Vol.42, pp.385–389.

Bonet, J. and Wood, R.D. (2008). Nonlinear continuum mechanics for finite element analysis. 2nd Edition, *Cambridge University Press*, Cambridge.

Brebbia, C.A., Telles, J.C.F. and Wrobel, L.C. (1984). Boundary Element Techniques Theory and Application in Engineering. *Springer-Verlag*, Berlin.

Brown, S.F. and Bell, C.A. (1979). The prediction of permanent deformation in asphalt pavements. Proceedings, *Journal of the Association of Asphalt Paving Technologists*, Vol.48, pp.438–476.

Busso, E.P., Kitano, M. and Kumazawa, T. (1994). A forward gradient time integration procedure for an internal variable constitutive model of Sn-Pb solder. *International Journal for Numerical Methods in Engineering*, Vol.37, pp.539–558.

Chan, D., Wang, X. and Morgenstern, N. (1997). Kinematic modeling of shear zone deformation. *Deformation and Progressive Failure in Geomechanics, IS-Nagoya'97*, pp.389–394.

Cormeau, I.C. (1975). Numerical stability in quasi-static elasto/visco-plasticity. *International Journal for Numerical Methods in Engineering*, Vol.9, pp.109–127.

Crisfield, M.A. (1997). Non-Linear Finite Element Analysis of Solids and Structures, Vol.2, Advanced Topics. *John Wiley & Sons, Inc*, New York.

Day, R.A. and Potts, D.M. (1994). Zero thickness interface element-numerical stability and application. *International Journal for Numerical and Analytical Methods in Geomechanics*, Vol.18, pp.689–708.

Desai, C.S., and Christian, J.T. (1977). (Editors and Contributors of Six Chapters), Numerical Methods in Geotechnical Engineering, *McGraw Hill Book Co.*, New York.

Desai, C.S. and Siriwardane, H.J. (1984). Constitutive Laws for Engineering Materials with Emphasis on Geologic Materials. *Prentice-Hall, Inc.*, Englewood Cliffs, New Jersey.

Desai, C.S., Zaman, M.M., Lightner, J.G. and Sirzwardane, H.J. (1984). Thin-layer element for interfaces and joints. *International Journal for Numerical and Analytical Methods in Geomechanics*, Vol.8, pp.19–43.

Dombrovsky, L.A. (1992). Incremental constitutive equations for Miller and Bodner-Partom viscoplastic models. *Computers and Structures*, Vol.44, No.5, pp.1065–1072.

Drucker, D.C., and Prager, W. (1952). Soil mechanics and plastic analysis or limit design. *Quart Applied Mathematics*, Vol.10, No.2, pp.157–165.

Efendiev, Y. and Hou, T.Y. (2008). Multiscale Finite Element Methods: Theory and Applications. *Springer-Verlag*, Berlin.

Gaul, L., Kogl, M. and Wagner, M. (2003). Boundary Element Methods for Engineers and Scientists : An Introductory Course with Advanced Topics. *Springer*, Berlin.

Goodman, R.E., Taylor, R.L. and Brekke, T.L. (1968). A model for the mechanics of jointed rock. *Journal of Soil Mechanics and Foundations Division*, Vol.94, pp.637–659.

Heaps, C.W. and Mansfield, L. (1986). An improved solution procedure for creep problems. *International Journal of Numerical Methods in Engineering*, Vol.23, pp.525–532.

Huebner, K.H., Thornton, E.A. and Byrom, T.G. (1995). The Finite Element Method for Engineers. *John Wiley*, New York.

Hughes, T.J.R. (1985). Numerical implementation of constitutive models: Rate-independent deviatoric plasticity. S. Nemat-Nasser, R.J Asaro and G.A. Hegemier (Eds). *Theoretical Foundation for Large Scale Computations of Non-linear Material Behavior*, Martinus Nijhoff Publishers, Boston, pp.29–57.

Huschek, S. (1977). Evaluation of rutting due to viscous flow in asphalt pavements. *Proceedings, Fourth International Conference on the Structural Design of Asphalt Pavements*, Vol.I, Ann Arbor, pp.497–508.

Kaliakin, V.N. and Li, J. (1995). Insight into deficiencies associated with commonly used zero-thickness interfaces elements. *Computers and Geotechnics*, Vol.17, pp.225–252.

Kanchi, M.B., Zienkiewicz, O.C. and Owen, D.R.J. (1978). The visco-plastic approach to problems of plasticity and creep involving geometric non-linear effects. *International Journal of Numerical Methods in Engineering*, Vol.12, pp.169–181.

Kattan, P.I. (2007). MATLAB Guide to Finite Elements: An Interactive Approach. *Springer-Verlag*, Berlin.

Krempl, E. (1987). Models of viscoplasticity, some comments on equilibrium (back) stress and drag stress. *Acta Mechanica*, Vol.69, pp.25–42.

Krieg, R.D. (1977). Numerical integration of new unified plasticity-creep formulations. *Proceedings of the 4th International Conference on Structural Mechanics in Reactor Technology*, San Francisco, Paper M6/4.

Kumar, V., Morjaria, M. and Mukherjee, S. (1980). Numerical integration of some stiff constitutive models of inelastic deformation. *Journal of Engineering Materials*, Vol.102, pp.92–96.

Kwon, Y.W. and Bang, H. (2000). The Finite Element Method using MATLAB. *CRC Press*, Boca Raton.

Lai, J.S. and Anderson, D. (1973). Irrecoverable and recoverable nonlinear viscoelastic properties of asphalt concrete. *Transportation Research Record*, No.468, pp.73–88.

Lubliner, J. (1990). Plasticity Theory. *Macmillan Publishing Company*, New York.

Lush, A.M., Weber, G. and Anand, L. (1989) An implicit time-integration procedure for a set of internal variable constitutive equations for isotropic elasto-viscoplasticity. *International Journal of Plasticity*, Vol.5, pp.521–549.

Marques, J.M.M.C. and Owen, D.R. (1983). Strain hardening representation for implicit quasi-static elasto-viscoplastic algorithms. *Computers and Structures*, Vol.17, pp.301–304.

Miller, A.K. (1976). An inelastic constitutive model for monotonic, cyclic and creep deformation. *Journal of Engineering Materials*, Vol.98, pp.97–113.

Moës, N., Dolbow, J. and Belytschko, T. (1999). A finite element method for crack growth without remeshing. *International Journal for Numerical Methods in Engineering*, Vol.46, No.1, pp.131–150.

Mohammadi, S. (2008). Extended Finite Element Method: for Fracture Analysis of Structures. *Blackwell Publishing*, Oxford.

Ortiz, M. and Popov, E.P. (1985). Accuracy and stability of integration algorithms for elastoplastic constitutive relations. *International Journal for Numerical Methods in Engineering*, Vol.21, pp.1561–1567.

Ottosen, N.S. and Gunneskov, O. (1985). Nonlinear subincremental method for determination of elastic-plastic creep behavior. *International Journal for Numerical Methods in Engineering*, Vol.21, pp.2237–2256.

Pande, G.N. and Shama, K.G. (1979). On joint/interface elements and associated problems of numerical ill-conditioning. *International Journal for Numerical and Analytical Methods in Geomechanics*, Vol.3, pp.293–300.

Paris, F. and Canas, J. (1997). Boundary Element Method. *Oxford University Press*, Oxford.

Peirce, A.P. and Napier, J.A.L. (1995). A spectral multipole method for efficient solution of large-scale boundary element models in elastostatics. *International Journal for Numerical Methods in Engineering*, Vol.38, No.23, pp.4009–4034.

Sangpetngam, B. (2003). Development and Evaluation of a Viscoelastic Boundary Element Method to Predict Asphalt Pavement Cracking. *Ph.D. Thesis*, University of Florida.

Sangpetngam, B., Birgisson, B. and Roque, R. (2003).Development of an efficient crack growth simulator based on hot mix asphalt fracture mechanics. *Transportation Research Record*, No.1832, pp.105–112.

Sangpetngam, B., Birgisson, B. and Roque, R. (2004). Multilayer boundary-element method for evaluating top-down cracking in hot-mix asphalt pavements. *Transportation Research Record*, No.1896, pp.129–137.

Scarpas, T. (2004). A Mechanics based Computational Platform for Pavement Engineering. *Delft University of Technology*, The Netherland.

Sharma, K.G. and Desai, C.S. (1992). Analysis and implementation of thin-layer element for interfaces and joints. *Journal of Engineering Mechanics,* Vol.118, pp.2442–2462.

Shin, C.F., Delorenzi, H.G. and Miller, A.K. (1977). A stable computational scheme for stiff time dependent constitutive equations. *Proceedings of the 4th International Conference on Structural Mechanics in Reactor Technology*, paper L2/2.

SHRP-A-415 (1994). Permanent deformation response of asphalt aggregate mixes. *Strategic Highway Research Program*, National Research Council, Washington, D.C.

Simo, J.C. and Taylor, R.L. (1985). Consistent tangent operators for rate-independent elastoplasticity. *Computer Methods in Applied Mechanics and Engineering*, Vol.48, pp.101–118.

Simo, J.C. and Hughes, T.J.R. (1998). *Computational Inelasticity* (Interdisciplinary Applied Mathematics), Vol.7, *Springer*, New York.

Snyder, M.D. and Bathe, K.J. (1981). A solution procedure for thermo-elastic and creep problems. *Nuclear Engineering Design*, Vol.64, pp.49–80.

Steen, B.V.D., Vervoort, A. and Napier, J.A.L. (2001). Numerical modeling of fracture initiation and propagation in biaxial tests on rock samples. *International Journal of Fracture*, Vol.108, pp.165–191.

Wang, X. (1998). Numerical simulation of the onset and propagation of shear band localization. *Ph.D. Dissertation, Department of Civil and Environmental Engineering, University of Alberta*, Edmonton, Canada.

Wang, X. Chan, D.H. and Morgenstern, N.R. (1995). A numerical scheme for modeling kinematic propagation of shear bands. *Numerical Models in Geomechanics, NUMOG V*, G.N. Pande & S. Pietruszczak (Eds.), Balkema, Davos, Switzerland, pp.215–222.

Wang, X., Chan, D. and Morgenstern, N. (2002). Numerical modeling of shear bands by element bands. *International Journal for Numerical Methods in Engineering*, Vol.54, pp.1131–1159.

Wang, X., Chan, D. and Morgenstern, N. (2003). Kinematic modelling of shear band localization using discrete finite elements. *International Journal for Numerical and Analytical Methods in Geomechanics*, Vol.27, No.4, pp.289–324.

Wang, X. and Wang, L. (2006). Continuous interface elements subject to large shear deformations. *International Journal of Geomechanics*, Vol.6, No.2, pp.97–107.

Wang, X., Wang, L. and Xu, L.M. (2004). Formulation of the return mapping algorithm for the elasto-plastic models for soils. *Computers and Geotechnics*, Vol.31, No.4, pp.315–338.

Wang, L. and Li, Q.B. (2008). Finite incremental constitutive equations for the SHRP viscoplasticity model. *Geotechnical Special Publication*, No. 184, pp.108–121.

Wilson, E.L. (1977). Finite elements for foundations, joints and fluids. *Finite Elements in Geomechanics Gudefius, A. G. (Ed.)*, Wiley, New York, pp.319–350.

Zhang, Z., Roque, R., Birgisson, B. and Sangpetngam, B. (2001). Identification and verification of a suitable crack growth law. *Journal of the Association of Asphalt Paving Technologists*, Vol.70, pp.206–241.

Zienkiewicz, O.C., Best, B., Dullage, C., and Stagg, K.G. (1970). Analysis of nonlinear problem with particular reference to jointed rock systems. *Proceedings of the 2nd Congress of the International Society for Rock Mechanics*, pp.8–14.

Zienkiewicz, O.C. and Cormeau, I.C. (1974). Visco-plasticity—plasticity and creep in elastic solids-a unified numerical solution approach. *International Journal of Numerical Methods in Engineering*, Vol.8, pp.821–845.

Zienkiewicz, O.C. (1977). The Finite Element Method. *McGraw-Hill Book Company (UK) Limited*.

Zohdi, T.I. and Wriggers, P. (2005). Introduction to computational micromechanics. *Lecture Notes in Applied and Computational Mechanics*, Vol.20, *Springer-Verlag*, Berlin.

CHAPTER 9

Applications of Discrete Element Method

9.1 Introduction

The distinct element method, or discrete element method (DEM), was first introduced by Cundall in the early 1970s. It was originally applied to rocks and then extended to granular materials (Cundall and Strack, 1979). The basic idea of DEM is straightforward: application of Newton's second law to characterize movements of multiple interacting rigid bodies. The general philosophy follows the following steps: 1) assign the initial locations and velocities to individual particles; 2) detect whether adjacent particles are in contact or not; 3) calculate the forces and moments due to contacts and other sources; 4) compute acceleration via Newton's second law; 5) integrate the acceleration to update the velocity; 6) further integrate the velocity to update the positions; and 7) iterate to obtain the dynamic equilibrium conditions for all the particles. DEM has been widely used in rock and soil, geotechnical engineering, granular materials, fluids, and concrete. It has found more applications in asphalt concrete (AC) in recent years. Most applications in asphalt have made use of Particle Flow Code PFC 3D or PFC 2D. The following provides brief descriptions of some basic elements of how PFC 3D solves a problem. There are other DEM platforms that can address irregular shapes (such as ellipsoids and polygons) and more complicated contact models. These platforms are not described as few researchers in AC are using them.

9.2 Fundamentals of DEM

The initial purpose of developing the DEM approach was for the verification of constitutive models for granular soils. The model developed by Cundall assumes the following:

1. The particles are circular/spherical and treated as rigid bodies.

2. The contact points between particles occur over an infinitesimally small area.

3. The particles are allowed to overlap slightly at contact points.

4. The magnitude of the overlap is linearly proportional to the contact force, unless slip or separation occurs.

5. The slip condition is provided by Mohr-Coulomb law.

Obviously, there are quite a few problems with these assumptions in relation to realistic granular materials and their behavior.

1. Particles are not circular or spherical. They have complicated shapes, angularity, and textures. As shape, angularity, and texture affect the contact properties significantly, the original DEM has some limitations in quantitative predicting the behavior of granular materials.

2. A lack of rules guides the relationship between slip and residual forces. In other words, neither hardening nor softening is considered.

3. The linear contact model is not valid when stress level is high.

Since the pioneering development of DEM by Cundall, numerous publications and improvements have been developed in the areas of soil mechanics or granular materials. Improvements include the use of polygons and ellipsoids to represent more complicated shapes, particle fluids interactions, and more complicated contacts such as Hertz contact and viscoelastic contact. Two major issues involving the use of DEM for modeling AC are 1) quantifying the effects due to surrounding materials such as mastics or binder; and 2) microscopic parameter characterization. Most current practice uses back-calculation to obtain the microscopic parameters. Future research to quantify the effects due to the surrounding medium is necessary.

9.2.1 Force and Displacement Analysis

For a set of particles, the core of DEM is represented by Equation 9-1:

$$M\Delta \ddot{X} + C\Delta \dot{X} + S\Delta X = \Delta R \tag{9-1}$$

C = damping matrix
M = mass stiffness matrix
S = stiffness matrix
ΔR = incremental force referring to a dynamically equilibrium status
ΔX = incremental displacement

The mass should be generically understood as mass and mass momentum. The displacement could be translational and rotational. The forces may include body forces, surface tractions, distributed momentums (i.e., due to electromagnetic forces), and other force-induced momentums. Figure 9.1 illustrates a 2D case and Equations 9-2a, b, and c represent the generic force vector, the mass matrix, and the displacement vector in 2D cases.

$$\Delta R^A = (\Delta F_1^A, \Delta F_2^A, \Delta M^A / r^A)^T \tag{9-2a}$$

$$M = \begin{pmatrix} M & 0 & 0 \\ 0 & M & 0 \\ 0 & 0 & I/r \end{pmatrix} \tag{9-2b}$$

$$\Delta X = (\Delta X_1, \Delta X_2, \omega)^T \tag{9-2c}$$

The major procedures for DEM focus on defining and updating the individual terms such as ΔR and ΔX. Determination of ΔR is mainly addressed by the determination of

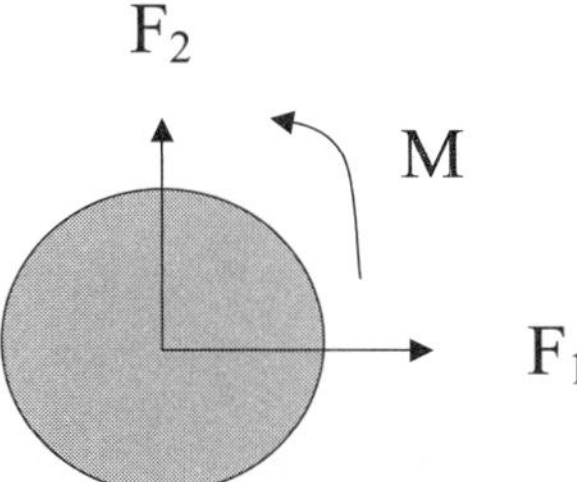

FIGURE 9.1 Forces applied to a particle.

contact forces. Determination of ΔX is mainly addressed by the solution of Equation 9-1. Figure 9.2 illustrates the determinations of the relative displacements between two particles, and the forces resulted from contacts on particle A in relation to Particle B.

The overall forces on Particle A due to contacts are:

$$\Delta R^A = K^A \Delta U^A - \sum_B K'^{AB} \Delta U^B \tag{9-3}$$

Where:

$$K^A = \sum_B K^{AB}$$

$$K^{AB} = \begin{bmatrix} k_n(n_1^{AB})^2 + k_n(n_2^{AB})^2 & (k_n - k_t)n_1^{AB}n_2^{AB} & -k_t n_2^{AB} \\ (k_n - k_t)n_1^{AB}n_2^{AB} & k_n(n_2^{AB})^2 + k_n(n_1^{AB})^2 & k_t n_1^{AB} \\ -k_t n_2^{AB} & k_t n_1^{AB} & k_t \end{bmatrix}$$

$$K'^{AB} = \begin{bmatrix} k_n(n_1^{AB})^2 + k_n(n_2^{AB})^2 & (k_n - k_t)n_1^{AB}n_2^{AB} & k_t n_2^{AB} \\ (k_n - k_t)n_1^{AB}n_2^{AB} & k_n(n_2^{AB})^2 + k_n(n_1^{AB})^2 & -k_t n_1^{AB} \\ k_t n_2^{AB} & -k_t n_1^{AB} & -k_t \end{bmatrix}$$

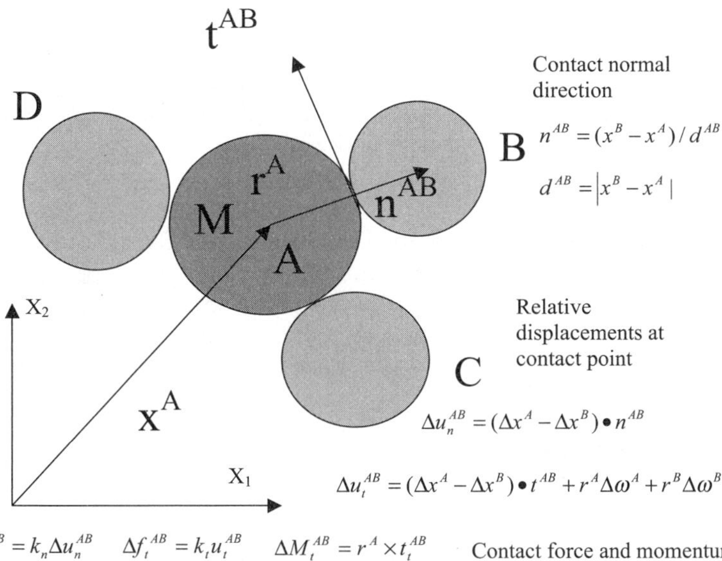

FIGURE 9.2 Illustration of the determination of forces and displacements.

$$\Delta U^A = (\Delta x_1^A, \Delta x_2^A)$$

$$\Delta f_n^{AB} = (k_n \Delta U^A \bullet n)n$$

Its component in x_1 and x_2 directions:

$$\Delta f_n^{AB}(x_1) = (k_n \Delta U^A \bullet n)n_1 = k_n (n_1^{AB})^2 \Delta U^A$$

$$\Delta f_n^{AB}(x_2) = (k_n \Delta U^A \bullet n)n_2 = k_n (n_2^{AB})^2 \Delta U^A$$

For the cosines in the t direction:

$$n_1(t) = -n_2(n)$$

$$n_2(t) = n_1(n)$$

$$\Delta f_t^{AB} = (k_t \Delta U^A \bullet n(t))n(t)$$

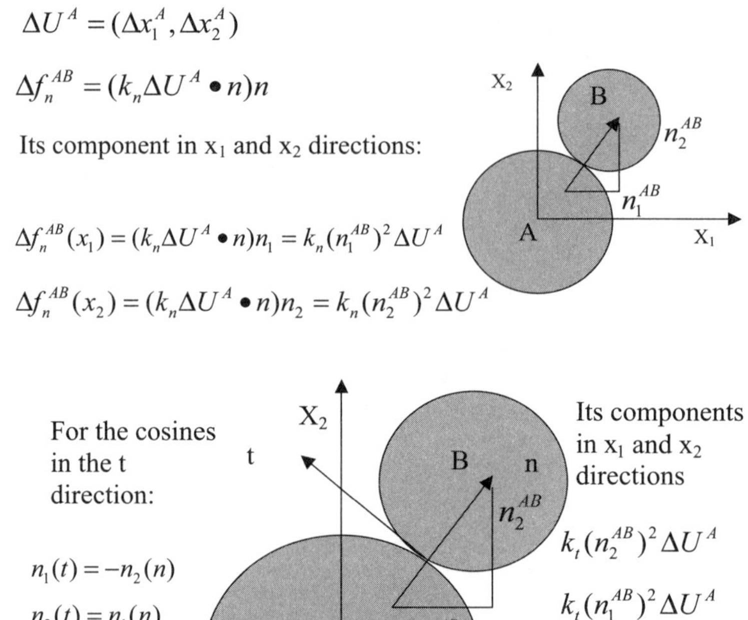

FIGURE 9.3   Force components for pair of particles in contact.

The derivation for the components (first term as an example) in the matrix can be illustrated in Figure 9.3. This approach is named "granular element method" (Oda and Iwashita, 1999). Other terms can be derived in a similar approach.

9.2.2   Contact Models

There are quite a few contact models that are widely used in DEM simulations. They include the bilinear model, Hertz-Mindlin contact model, the parallel bond contact, and the viscoelastic contact model. There are more complicated models that incorporate rough contacts, however, these models are barely used in the DEM simulation for AC. Nevertheless, models that can consider the rough contact will present better approximations to accurate solutions.

9.2.2.1   Bilinear Contact Model

For two particles in contact, a series contact model will have the following relationships.

$$k_n^{[A]} \Delta U_n^A = k_n^{[B]} \Delta U_n^B = k^n (\Delta U_n^A + \Delta U_n^B)$$

This relationship is also applicable to the force-displacement relationship in the tangent direction. From this relationship the following equations can be obtained:

$$k^n = \frac{k_n^{[A]}k_n^{[B]}}{k_n^{[A]} + k_n^{[B]}} \qquad k^s = \frac{k_s^{[A]}k_s^{[B]}}{k_s^{[A]} + k_s^{[B]}} \tag{9-4a, b}$$

9.2.2.2 Hertz Contact Model

Contact normal secant stiffness:

$$K^n = \left(\frac{2\langle G\rangle\sqrt{2\tilde{R}}}{3(1-\langle v\rangle)}\right)\sqrt{U^n} \tag{9-5}$$

Contact shear tangent stiffness:

$$k^s = \left(\frac{2\left(\langle G\rangle^2\, 3(1-\langle v\rangle)\tilde{R}\right)^{1/3}}{2-\langle v\rangle}\right)\left|F_i^n\right|^{1/3} \tag{9-6}$$

More details about the formulations can be found from the PFC3D manuals.

9.2.2.3 Parallel Bond Model

Bonding model is another key model in PFC3D. It is only formed between proximate particles (not between balls and walls), and continues to exist until broken, which happens when its strengths are exceeded.

PFC3D 3.0 supports two bonding models: contact bond model and parallel bond model. The parallel bond model can transfer momentums.

The parallel bond acts over a circular cross-section as shown (Figure 9.4), which enables it to transmit both a force and a momentum. Another important feature of parallel

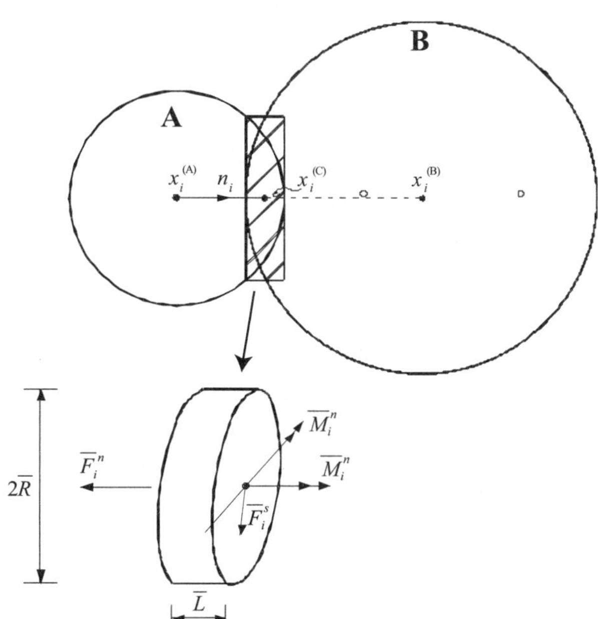

Figure 9.4 Schematic of parallel bond (Itasca Consulting Group, 2005).

bond is that it can act in parallel with a slip model. Below are the equations governing the normal and shear components for both the force and moment:

$$\overline{F}_i = \overline{F}_i^{\,n} + \overline{F}_i^{\,s} \tag{9-7}$$

$$\overline{M}_i = \overline{M}_i^{\,n} + \overline{M}_i^{\,s} \tag{9-8}$$

The elastic properties and the limit of the bond can be set up by designating the following factors: normal stiffness $\overline{k}^n$; shear stiffness $\overline{k}^s$; normal strength $\overline{\sigma}_c$; shear strength $\overline{\tau}_c$; and bond disk radius $\overline{R}$. The equations for defining the above variables are listed in Table 9.1 (Itasca Consulting Group, 2005).

The maximum tensile and shear stresses of the bond periphery are calculated below to compare with the strengths; if exceeded, the bond would break.

$$\sigma_{\max} = \frac{-\overline{F}^{\,n}}{A} + \frac{\left|\overline{M}_i^{\,s}\right|}{I}\overline{R} \tag{9-9}$$

$$\tau_{\max} = \frac{\left|\overline{F}_i^{\,s}\right|}{A} + \frac{\left|\overline{M}^{\,n}\right|}{I}\overline{R} \tag{9-10}$$

Table 9.2 summarizes the key factors and corresponding variables for some constitutive models that are used in granular materials and AC. Figure 9.5 describes conditions when those models are no longer valid. Note that the contact-stiffness model always exists (may not be the bi-linear model); the slip model would change from a normal-force-related status to the maximum force status as the shear force increases; and the parallel bond model would break when either the normal or shear force reaches its strength.

Force	Normal	$\Delta\overline{F}_i^{\,n} = (-\overline{k}^n A\Delta U^n)n_i$	$\Delta U_i = V_i\Delta t$
	Shear	$\Delta\overline{F}_i^{\,s} = -\overline{k}^s A\Delta U_i^{\,s}$	
Moment	Normal	$\Delta\overline{M}_i^{\,n} = (-\overline{k}^s J\Delta\theta^n)n_i$	$\Delta\theta_i = (w_i^{[B]} - w_i^{[A]})\Delta t$
	Shear	$\Delta\overline{M}_i^{\,s} = -\overline{k}^n I\Delta\theta_i^{\,s}$	
Bond disk area		$A = \pi\overline{R}^2$	
Polar moment of inertia of the disk cross-section		$J = \dfrac{1}{2}\pi\overline{R}^4$	
Moment of inertia of the disk cross-section		$I = \dfrac{1}{4}\pi\overline{R}^4$	

TABLE 9.1 Force and moment for parallel bond and the corresponding micro parameters.

Model	Key Factors	Controlling Variables in PFC3D
Contact-stiffness model (Linear)	Normal stiffness	md_kn
	Shear stiffness	md_ks
Slip model	Friction factor	md_fric
Bonding model (parallel)	Normal stiffness	pb_kn
	Shear stiffness	pb_ks
	Normal strength	pb_sn
	Shear strength	pb_ss
	Radius	pb_radmult

TABLE 9.2   Key micro-variables in different contact constitutive models in PFC3D.

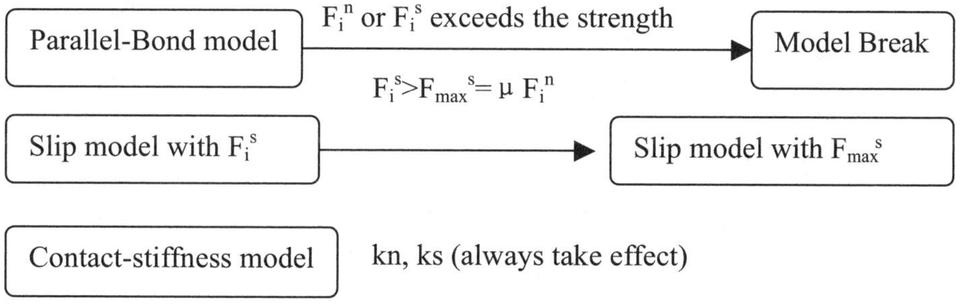

FIGURE 9.5   Existing condition of different contact models in DEM of PFC3D.

9.2.2.4   The Burger's Model

The Burger's model has been implemented in PFC3D. The microscopic components of the model are illustrated in Figure 9.6 (Itasca Consulting Group, 2005). The governing equations for the Maxwell element and the Kevin element are presented in Equations 9-11 and 9-12.

$$\dot{\varepsilon}_M = \frac{\sigma}{\eta_M} + \frac{\dot{\sigma}}{R_M} \tag{9-11}$$

$$\sigma = R_K \varepsilon_K + \eta_K \dot{\varepsilon}_K \tag{9-12}$$

σ = stress carried by either mechanistic model

$\dot{\sigma}$ = first derivative of stress carried by the mechanistic model

ε_M = strain within the Maxwell model

$\dot{\varepsilon}_M$ = first derivative of strain within the Maxwell model

ε_K = strain within the Kelvin model

$\dot{\varepsilon}_K$ = first derivative of strain within the Kelvin model

R_M and R_K = Maxwell and Kelvin springs stiffnesses, respectively

η_M and η_K = Maxwell and Kelvin dashpot damping coefficients, respectively

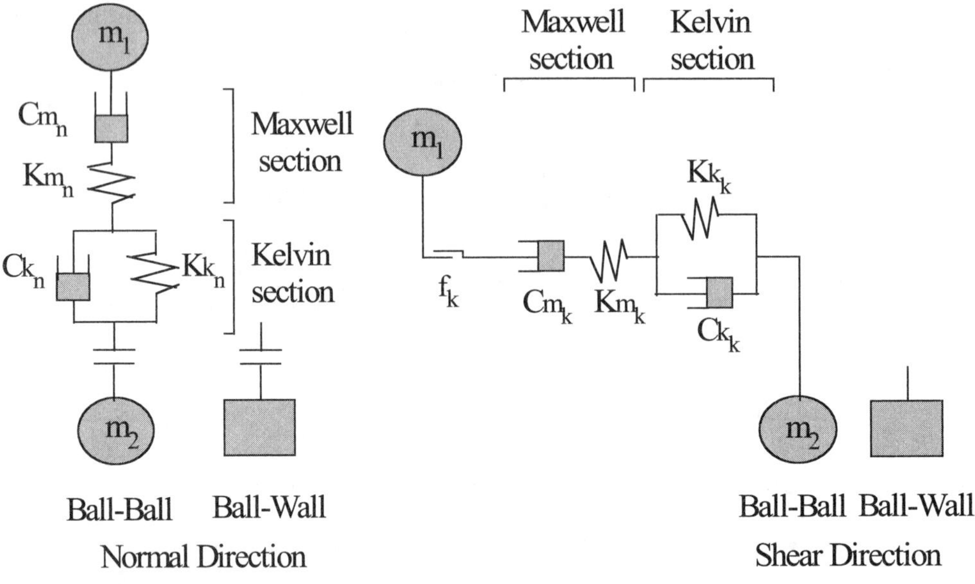

FIGURE 9.6   The Burger model (normal and shear stiffness) used in DEM modeling (Itasca, 2005).

The overall controlling equation between the contact force f and displacement u of Burger's model (a second-order differential equation) is presented in Equation 9-13. Using the Laplace transform method from Chapter 6, the equation can be conveniently solved.

$$f + \left[\frac{C_k}{K_k} + C_m \left(\frac{1}{K_k} + \frac{1}{K_m} \right) \right] \dot{f} + \frac{C_k C_m}{K_k K_m} \ddot{f} = \pm C_m \dot{u} \pm \frac{C_k C_m}{K_k} \ddot{u} \tag{9-13}$$

9.2.3   Solution Scheme

With the above preparations, the two sets of equations, including the translational motion (Equation 9-14) and rotational motion (Equation 9-15), can be solved.

$$F_i = m(\ddot{x}_i - g_i) \tag{9-14}$$

$$M_1 = I_1 \dot{\omega}_1 + (I_3 - I_2)\omega_3 \omega_2$$

$$M_2 = I_2 \dot{\omega}_2 + (I_1 - I_3)\omega_1 \omega_3$$

$$M_3 = I_3 \dot{\omega}_3 + (I_2 - I_1)\omega_2 \omega_1 \tag{9-15}$$

$$M_i = I \dot{\omega}_i = (\tfrac{2}{5} mR^2)\omega_i \quad \text{(rotational motion)} \tag{9-16}$$

A finite difference method is usually applied to solve the above equations.

$$\ddot{x}_i^{(t)} = \frac{1}{\Delta t}\left(\dot{x}_i^{(t+\Delta t/2)} - \dot{x}_i^{(t-\Delta t/2)}\right)$$

(9-17)

$$\dot{\omega}_i^{(t)} = \frac{1}{\Delta t}\left(\dot{\omega}_i^{(t+\Delta t/2)} - \dot{\omega}_i^{(t-\Delta t/2)}\right)$$

(9-18)

From the known accelerations and initial velocities, it is possible to solve for the velocities.

$$\dot{x}_i^{(t+\Delta t/2)} = \dot{x}_i^{(t-\Delta t/2)} + \left(\frac{F_i^{(t)}}{m} + g_i\right)\Delta t$$

(9-19)

$$\dot{\omega}_i^{(t+\Delta t/2)} = \dot{\omega}_i^{(t-\Delta t/2)} + \left(\frac{M_i^{(t)}}{I}\right)\Delta t$$

(9-20)

From the velocities, the position can be obtained.

$$x_i^{(t+\Delta t)} = x_i^{(t)} + \dot{x}_i^{(t+\Delta t/2)}\Delta t$$

(9-21)

9.2.4   Contact Detection

Contact detection is a complicated procedure. As PFC3D uses only spheres, the contact detecting becomes much simpler: as long as the distance between the centers of two balls is smaller or equal to the sum of the radii of the two balls, the two balls are considered in contact (Figure 9.7).

Between balls:

$$D^{A-B} \leq r_A + r_B$$

(9-22)

Between ball and wall:

$$D^{A-W} \leq r_A$$

(9-23)

There are more complicated contact-detecting mechanisms for complicated shapes (Majiza, 2002). Research in this area is fundamental in nature and there is no commercially available software; progress in this area is slow. The Clump mechanism (bonding small balls together, Itasca Consulting Group, 2005) provided by PFC allows the use of bonding small balls to represent more complicated shapes.

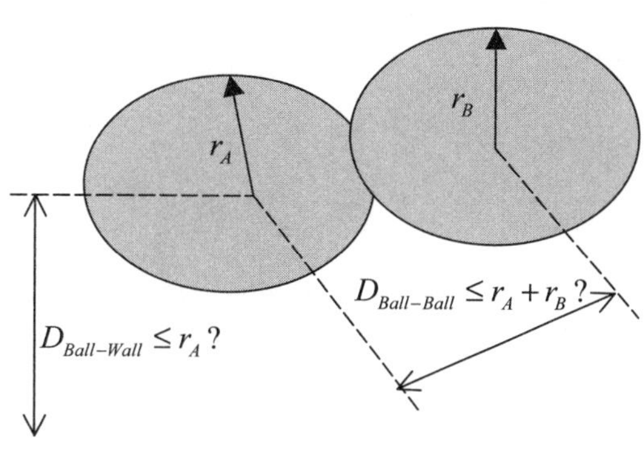

FIGURE 9.7 Contact detecting algorithm.

9.3   Shape Representation and Computational Efficiency Enhancement

Due to the complexity to model the real particles in 3D, spheres, ellipsoids, or discs are usually used in DEM simulations instead of complex particles. However, particle shape is an important factor that affects the interlock among the particles and other properties of the materials. It is necessary to build more sophisticated models that consider the particle shape effects. In recent years, some research has been carried out to study the effect of particle shape. Attempts were made to simulate elliptical systems both in 2D (Rothernburg, 1992) and 3D (Lin and Ng, 1997). More complicated particle shapes may be modeled using superquadrics (Williams and Pentland, 1991) or bonding a number of spheres together (Walton and Braun, 1993). Jensen et al. (2001a, b) presented a detailed computational investigation of the effect of particle shape on the interface shear behavior of granular materials. Clusters made of several discs were applied to model the rough particles using DEM simulation. Seven cluster shapes of varying degrees of roughness were presented in his paper. However, simulating real materials made up of a very large number of particles with complex shapes is currently still a challenge in DEM modeling. The method presented in this section provides a practical way to represent the particles with complex shapes, which makes it possible to build more sophisticated models that can consider the particle shape effects.

9.3.1   Basic Idea of Clustering

A particle with a complex shape can be modeled by combining smaller discs in 2D or spheres in 3D into a cluster with a similar shape, shown in Figure 9.8.

In PFC^{3D}, spheres within a cluster are allowed to overlap to any extent, but each cluster still behaves as a rigid body with deformable boundaries. Contacts internal to the cluster are skipped and not affected during the calculation cycle. Contact forces are not generated between those spheres within the cluster and the contact forces that exist when a cluster is created or when a sphere is added to the cluster remain unchanged during computational processes. For more details of the PFC^{3D} program, please see the Itasca Manual 5.0 (Itasca, 2005).

9.3.2   Idea of Representing Particles Using X-ray Tomography Imaging

As illustrated in Chapter 3, X-ray tomography imaging is a powerful tool to characterize the inside structure of materials. It has been widely applied to study the microstructure of geomaterials in recent years. The specimen is scanned using X-ray layer by layer (or cone beam) and the location of each particle in each layer is recorded pixel by pixel.

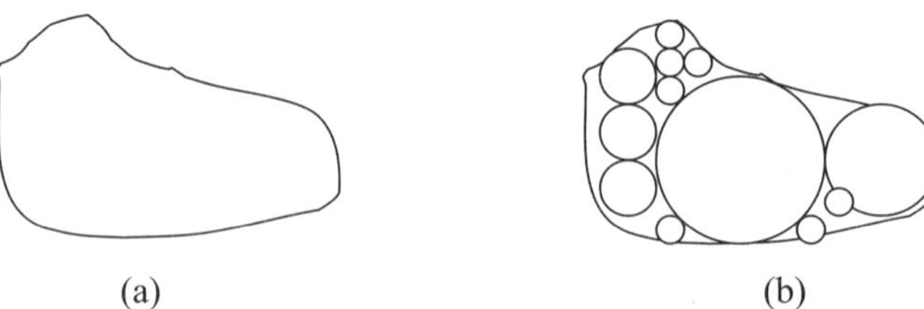
(a)   (b)

(a) Outline of a Real Particle with Complex Shape (b) Cluster Made up of Smaller Balls

FIGURE 9.8   Illustration of clustering.

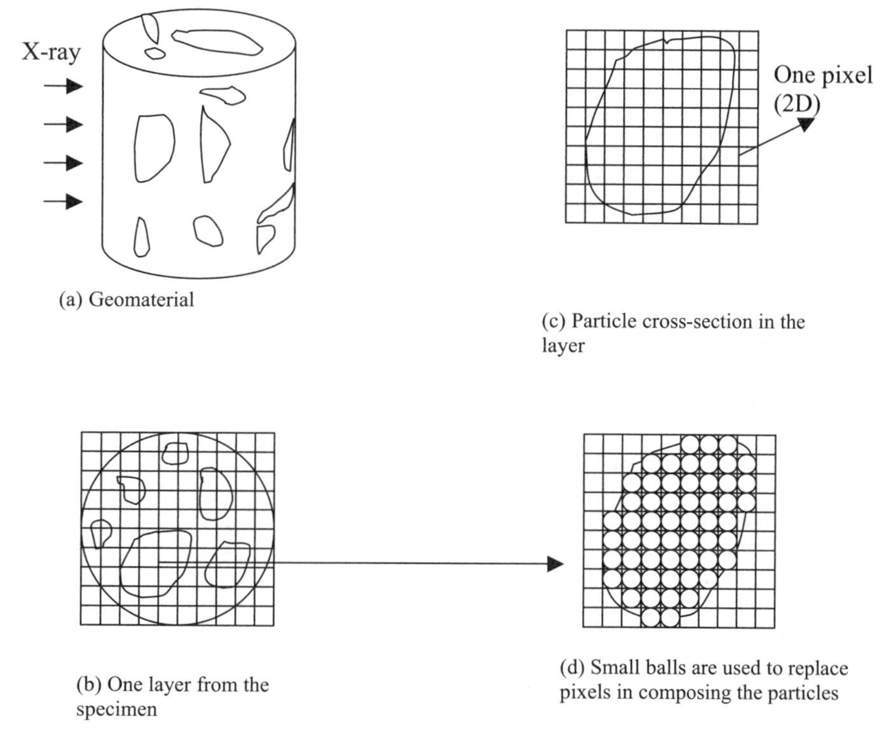

Figure 9.9   Process of rebuilding the particles using XCT imaging.

For each layer, a small ball is used to represent each pixel that is composed of the particle. Then the particle in 2D can be represented by those small balls. This process is shown in Figure 9.9.

Combining all of the small balls in different layers together, the entire aggregate skeleton of the specimen can be rebuilt in 3D.

9.3.3   Burn Algorithm 1 to Reduce the Number of Balls

A very large number of small balls are needed to represent the real particles using the above method. As a result, huge memory and time are required in the calculation process. To avoid this, it is necessary to apply certain algorithms to reduce the number of required balls, as shown in Figure 9.10. To facilitate a demonstration, the 2D process for one particle is shown here. Similar algorithms can be applied to a 3D case.

The procedure consists of the following steps.

Step 1. For a given cluster of circles used to represent a real particle, determine the circles on the boundary of the cluster and record their center coordinates one by one.

Step 2. Scan all of the existing circles and record the coordinates of the centers of those circles. For each center, try different radii; find the largest circle within the boundary of the cluster. Compare those circles and find the largest one, namely C1.

Step 3. Remove the small circles whose centers lie in or on the boundary of C1.

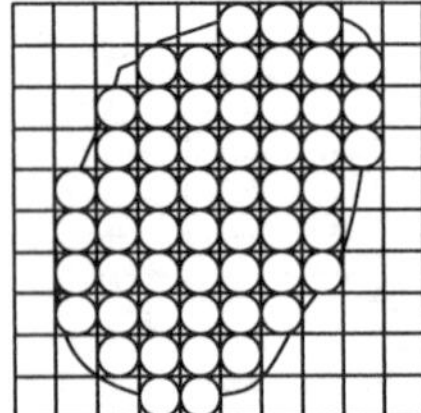

(a) Cluster of small circles with equal radius is used to represent the particle

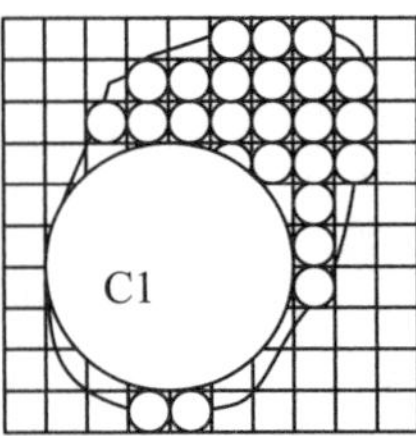

(b) Remove small circles contained in the first big circle C1

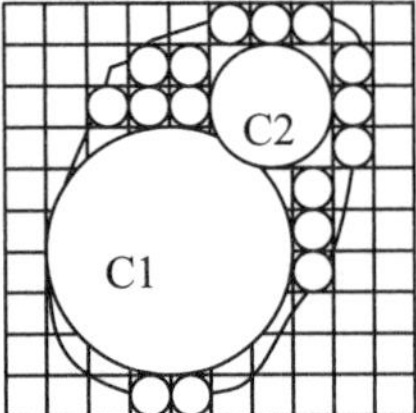

(c) Continue removing small circles in the second big circle C2

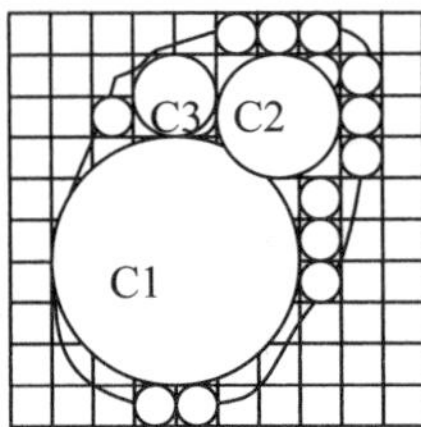

(d) Generate circles C1, C2, C3…

FIGURE 9.10   Process of replacing small circles with larger ones (Burn algorithm 1).

Step 4. For the remaining small circles, repeat Step 1 while determining the boundary of the reamaining cluster circles, then repeat the second step and the third step, finding Circle C_2 and removing small circles whose center lies in or on the boundary of C_2.

Step 5. Repeat the fourth step and find C_n until the radius of C_n is not larger than the radius of small circles.

Step 6. Generate new circles using the center and radius of $C_1, C_2, …C_n$, respectively. Most of the small circles are then replaced by those larger circles.

Run a similar process for each particle, so the entire aggregate skeleton can be rebuilt in this way.

For a 3D case, assume the number of small balls is N_s, the number of balls after applying the Burn Algorithm is N_l, then the reduced number of balls N_r can be calculated as: $N_r = N_s - N_l$, which is an appropriate measurement of the efficiency of the algorithm.

9.3.4   Burn Algorithm 2 to Reduce the Number of Balls

Another algorithm can be effective in reducing the number of required circles or balls. It is named burn algorithm 2. Figure 9.11 presents the illustration in 2D. The method is applicable to 3D. The procedure consists of the following steps.

Step 1. Scan all the circles used to represent a certain particle (cross-section). Find four-circle groups and the centers of the groups. Use the center coordinates and double radius of the small circles to generate one bigger circle to replace the four small circles.

Step 2. Repeat Step 1 until no four-circle group can be found and replaced.

For the 3D case (Figure 9.12), the process is similar but every eight-ball group (four balls in the top and four balls in the bottom) needs to be found and replaced. If the

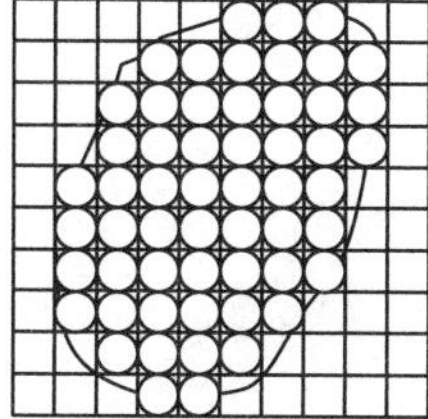

(a) Cluster of small circles with equal radius is used to represent the particle

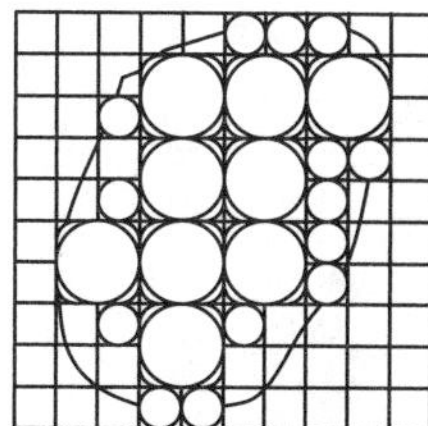

(b) Every four circles is replaced by a bigger circle

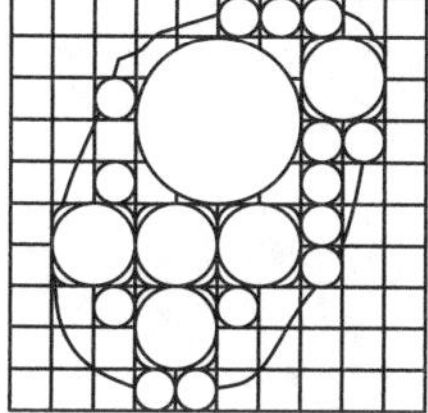

(c) Continue the above step until no four-circle-group with the same radius can be found

FIGURE 9.11   Process of replacing small circles with larger ones (Burn Algorithm 2).

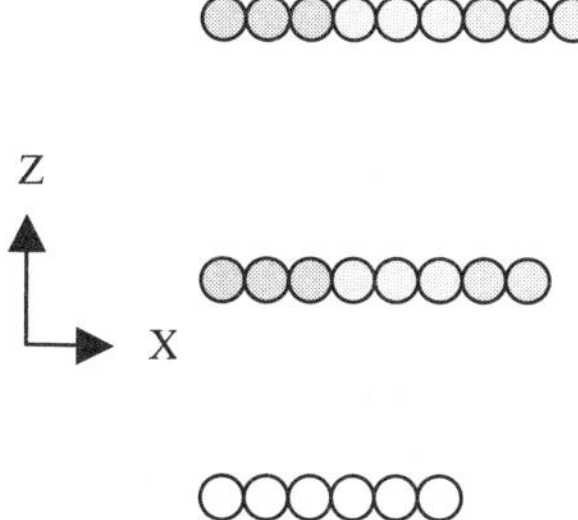

a) Small balls shown in z-x plane

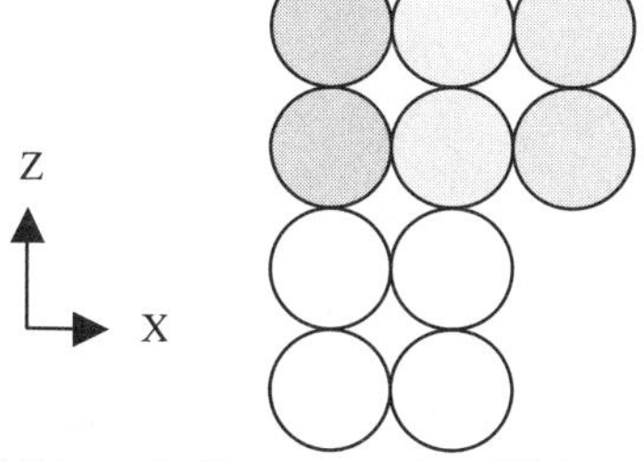

c) Bigger balls generated to fill the space between two layers (shown in z-x plane)

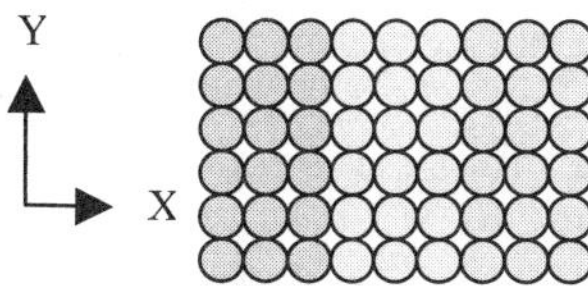

b) Small balls shown in x-y plane

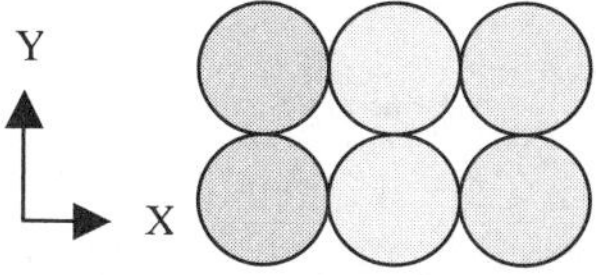

c) Bigger balls generated to fill the space between two layers (shown in y-x plane)

FIGURE 9.12   Bigger balls are generated to fill the space between two separated layers.

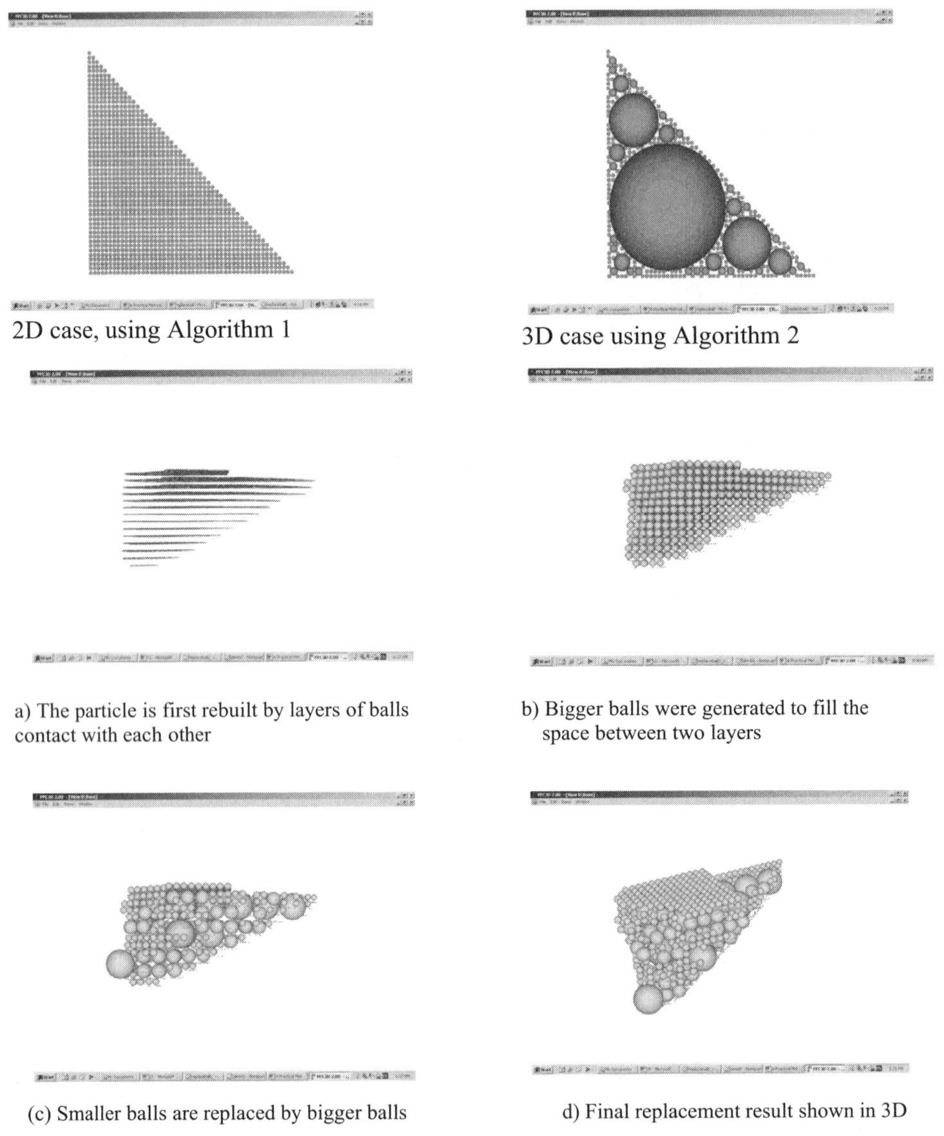

Figure 9.13 The application examples using the burn algorithms.

particle is rebuilt by layers of balls and those layers do not come in contact with each other (meaning part of the particle is lost), the lost part must be filled before applying the burn algorithm. For example, if the distance between those layers is 6 mm and the radius of small ball is 0.5 mm, two bigger balls can be generated between the space of two layers at distances 1.5 mm and 4.5 mm, respectively. The x, y center coordinates of the balls should be the center small ball of the nine-ball group and the radius should be three times that of the small ball. Compared to Algorithm 1, this method requires fewer loop cycles and much less time to complete. However, it is only suitable for more regularly packed clusters. More details can be found in Fu (2005).

Burn Algorithm	Case No.	Required Number of Balls		Reduced Number of Balls	Reduced Simulation Time (%)
		Before	After		
1	1, 2D	1308	262	1046	79%
	2, 2D	636	177	459	72%
	3, 3D	679	391	288	42%
2	4, 3D	56743	512	56231	99%

TABLE 9.3 PFC3D simulation result analysis applying burn algorithm.

9.3.5 Examples Using the Burn Algorithms

Figure 9.13 shows the results after applying the burn algorithms. Table 9.3 presents four cases on how the number of particles is reduced using the burn algorithms. The significant reduction in the number of balls to represent a particle indicates the saving of both memory and computational time. As for PFC3D, the approximate number of particles that can be created for different sizes of RAM is approximately a linear relation. The reduction of computational time is complicated but also very significant. Algorithm 2 is more efficient in reducing the number of balls and increasing computational efficiency for rectangular particles.

9.4 Validation of DEM Predictions at a Microscopic Level

While most DEM studies report macroscopic results consistent with experimental observation, few studies have been devoted to the microscopic level comparisons and the shape factor. Fu (2005), Fu et al. (2007), and Fu et al. (2010) demonstrated the DEM application of clustering approach in the simulation of a compression test and a direct shear test. The following is a summary of her studies. It also serves as the validation of DEM at a microscopic level.

9.4.1 Compression Test

9.4.1.1 Materials and Experimental Setup

A compression test with lateral sponge confinement on coarse aggregates was conducted to induce particle movements and structural deformations. Limestone aggregates passing a ½ in sieve but retained on a ⅜ in sieve were used. The aggregates were placed into a transparent cylindrical container (100 mm high, 103 mm diameter), which was specially designed for the convenience of X-ray scanning. A piece of sponge was placed along the inside wall of the container to allow some lateral displacements of individual aggregates to be discernable. Using X-ray tomography imaging, 2D sectional images were acquired. Then an axial load was applied on top of the aggregates. The physical and mechanical properties of the material, load magnitude, and specimen sizes are presented in Table 9.4. The specimen was scanned with XCT again and images of the deformed microstructure were acquired after the test. Based on the images acquired before and after testing, material microstructure, particle kinematics, and local strains can be obtained using the following methodology.

9.4.1.2 Particle Reconstruction

Particle reconstruction involves the recognition of particle cross-sections on adjacent slices and registration of the cross-sections to the same particle. After gray images are

Limestone Aggregates				Sponge			
Size (mm)	Specific Gravity	Aggregate Number	Young's Modulus (Gpa)	Thickness (mm)	Stiffness (MPa)	Container Size (mm)	Axial Load (g)
9.5~12.7	2.674	173	37	24	36.2	100H,103D	39.1 kN

TABLE 9.4 Properties of materials, specimen size, and load magnitude.

acquired by XCT, they can be converted into binary images and the digitized geometric characteristics of individual particle cross-sections, such as the mass center coordinates and area, can be obtained. Each particle cross-section (i) in each slice (z) was denoted as PC(i, z). The next step was to identify which particle cross-sections in sectional slices belong to a specific particle, namely a 3D reconstruction procedure. A similarity index (SI) method to identify individual particles from granular media was presented elsewhere (Wang et al., 2004, 2007). Further research has shown that some enhancement could be made to improve the efficiency of identification. Recognizing that the particle identification is very sensitive to mass center coordinates, a modified similarity index method was employed in this study. Two particle cross-sections in two adjacent slices were considered to belong to the same particle, if a minimum SI (SI_{min}) was obtained using the following equation:

$$SI(i,z)_{min} = \min(\left|x_{i,z} - x_{j,z+1}\right| + \left|y_{i,z} - y_{j,z+1}\right|)$$
$$i = 1, 2, \cdots n$$
$$j = 1, 2, \cdots m$$

(9-24)

Where n and m represent the number of particle cross-sections in adjacent slices z and $z+1$; x and y are mass center coordinates of the ith particle in z^{th} slice or j^{th} particle in $(z+1)^{th}$ slice. Using this equation, each particle cross-section in slice z could find its corresponding cross-section in slice $z+1$.

There are three possibilities that could confuse the computer judgment in individual particle cross-section identification:

1. A new particle emerges in the $(z+1)^{th}$ slice. In this case, no corresponding particle cross-section could be found in the zth slice (Figure 9.14a). Then a new code is assigned to this new particle.

2. Existing particle cross-sections disappear in the next slice ($z+1$). In this case, $n>m$. One particle cross-section in slice ($z+1$) would correspond to two or more particle cross-sections in slice z with SI_{min}. Then, the pair of cross-sections PC(i,z) and PC($j,z+1$) with the smallest SI_{min} are considered to belong to the same particle. Any other particle cross-sections in slice z corresponding to the particle cross-section PC($j,z+1$) are those that disappear in slice ($z+1$) (Figure 9.14b).

3. An existing particle disappears in the $(z+1)^{th}$ slice while a new particle emerges in that slice almost at the same location, having similar mass center coordinates (Figure 9.14c). Then the trend of the area change of the particle cross-section at slice $z-1, z, z+1, z+2$ is traced. In this case, the trend must satisfy the relationship $Area_{z-1,i} > Area_{z,i}$ & $Area_{z+1,i} < Area_{z+2,i}$, otherwise, those cross-sections are still considered to belong to the same particle.

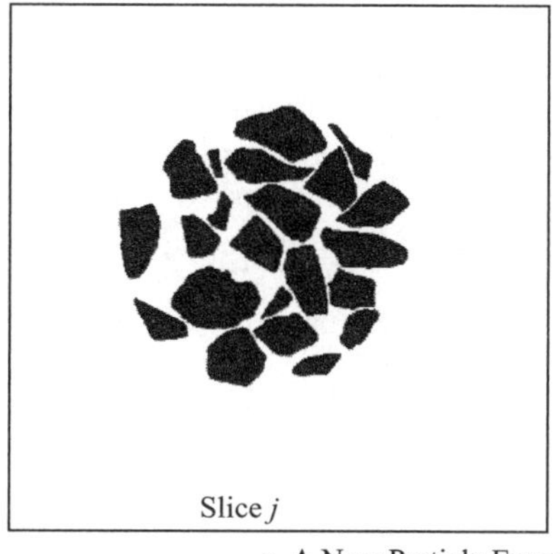
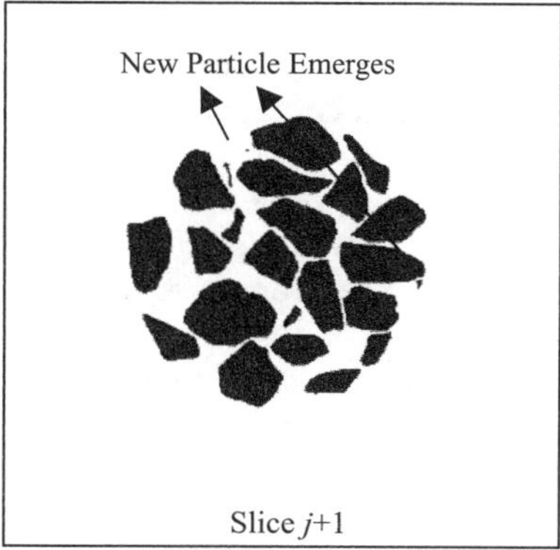

a. A New Particle Emerging in the Next Slice

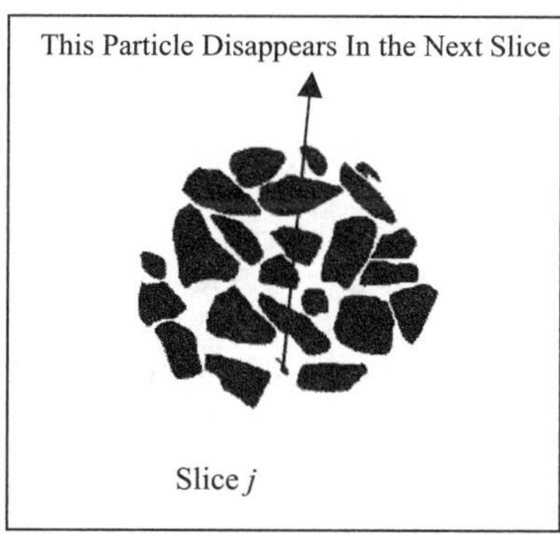
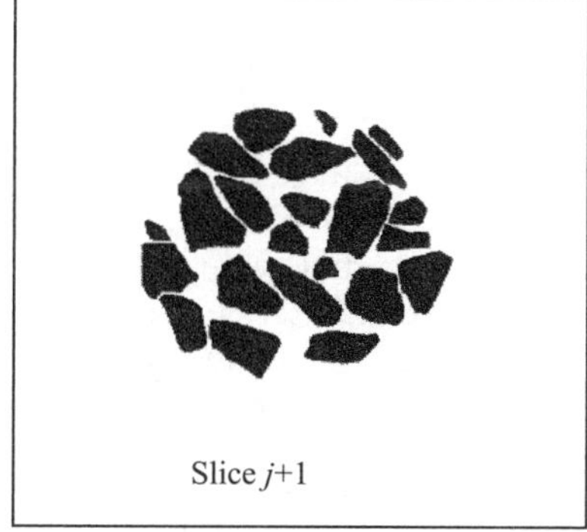

b. An Existing Particle Disappearing in the Next Slice

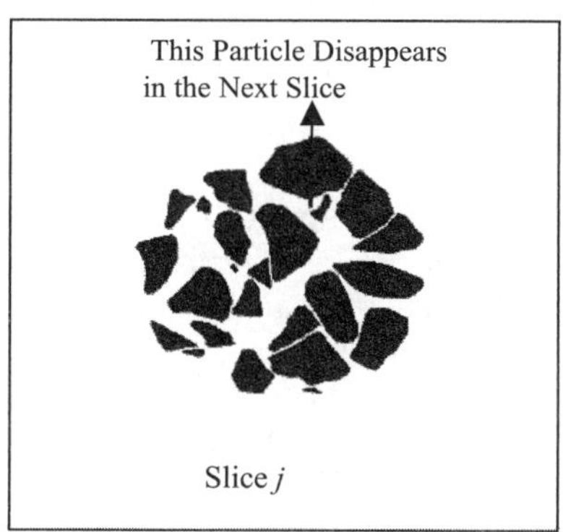
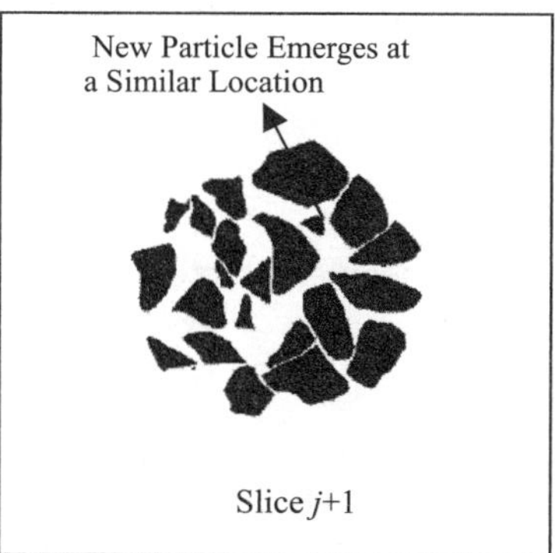

c. An Existing Particle Disappearing and New Particle
Emerging at a Similar location

FIGURE 9.14   Illustration of the complexity in particle reconstruction.

The details of the whole procedure were presented in Fu et al. (2007). Based on this procedure, 3D particles can be reconstructed. Once a particle is reconstructed, its volume, mass center coordinates, and particle orientation can be calculated following the procedure presented in Chapters 3 and 4.

9.4.1.3 Particle Identification

After being sustained under a compression force, the microstructure of the granular material would change accordingly. If the change of locations and orientations of the individual particles were recorded, the kinematics of the particles can then be calculated based on the mathematical relationships. For the confined compression test, the particle displacements were relatively small compared to the size of the aggregates, and the coordinates of the mass center of individual particles changed little in radial direction. Judgment can still be made based on the SI method for individual particles. The particle similarity index (SI_p) is defined as follows:

$$ SI_p = \left| x_b - x_a \right| + \left| y_b - y_a \right| + \left| z_b - (z_a + \frac{\Delta h}{h} * z_a) \right| \tag{9-25} $$

where x_b, y_b, z_b = mass center coordinates of individual particles before test

x_a, y_a, z_a = mass center coordinates of individual particles after test

Δh = vertical global deformation of the specimen

h = height of the specimen

The pair of particles giving the smallest SI_p is considered to be the same particle. If two or more particles are paired to the same original particle after testing, then their volumes are compared and the one whose volume is closest to that original particle is recognized. The same procedure of finding another original particle giving the smallest SI_p is followed for the remaining particles. This process is repeated until all the particles after testing are recognized.

9.4.1.4 Quantification of Particle Translational Kinematics in 3D

Once a particle is identified, the magnitude of particle translations (for the kth particle) can then be calculated as:

$$ DISP_k = \sqrt{u_k^2 + v_k^2 + w_k^2} \tag{9-26} $$

Where $u = x^a - x^b$; $v = y^a - y^b$; $w = z^a - z^b$, and the superscript a denotes after the testing and b denotes before testing. The rotation can be determined following the same procedure presented in Chapter 4. Due to the insignificant rotations in the compression test, only translation displacements are presented.

9.4.1.5 Computation of Micro Motions and Local Strains in 3D

With the displacement information, strains at the micro-level can be calculated. The displacements thus quantified can be used for comparison with numerical simulations. The procedures followed those presented in Chapter 4.

The global strains of the entire specimen are measured by:

$$ \varepsilon_1 = \frac{\Delta h}{h} \tag{9-27a} $$

$$ \varepsilon_2 = \varepsilon_3 = \frac{\Delta r}{r} \tag{9-27b} $$

$$ \varepsilon_{12} = \varepsilon_{13} = \frac{\varepsilon_1 - \varepsilon_2}{2} \tag{9-27c} $$

Where $\Delta h, h$ = vertical displacement and the height of the specimen; $\Delta r, r$ = displacement in the radial direction and the radius of the specimen.

The global volumetric strain is defined as:

$$\varepsilon_V = \varepsilon_1 + \varepsilon_2 + \varepsilon_3 = \varepsilon_1 + 2\varepsilon_3 \qquad (9\text{-}28)$$

The global volumetric strain can also be calculated from the sum of weighted local volumetric strain by the following equation:

$$\varepsilon_V' = \sum V_{fi}\varepsilon_{Vi}^l = \sum V_{fi}(\varepsilon_{1i}^l + \varepsilon_{2i}^l + \varepsilon_{3i}^l) \qquad (9\text{-}29)$$

where ε_V' = global volumetric strain calculated from local volume strain

$\quad V_{fi}$ = volumetric fraction of each local volume to the total volume

$\quad \varepsilon_i^l$ = local strain in tetrahedron i

By comparing the global volumetric strain calculated by direct measurements and that calculated by weighted local volumetric strains, the local strain quantification methodology can be evaluated.

9.4.1.6   Analysis of Experimental Results

The measured mass center coordinates and the displacements for 10 individual particles are presented in Table 9.5. The local macro strains in each tetrahedron, calculated by Equation 4-21, are presented in Table 9.6. It can be found from these values that complicated patterns of contraction and dilation exist in the material at microscopic level. This is consistent with logic that dilation and contraction are very much related to local structure. By using the strain in the z direction on a vertical line in every 10 mm, (i.e., $dz = 10$ mm), total deformation in z direction can be calculated as $\Delta z = \sum \varepsilon_{zi} dz = -9.2 mm$, which is very close to the measured vertical deformation, -9.8 mm. The global volumetric strain calculated by Equation 9-28 is 0.047 and that calculated by Equation 9-29 is 0.052, which are close to one another. Thus, the validations of the methods to quantify the local strains are acceptable.

Agg.No.	Coordinates of the Mass Center (mm)			Displacement in x, y, z Directions (mm)		
	x	y	z	x	y	z
1	46.15	29.55	4.36	0.68	0.45	0.03
2	56.19	34.22	5.33	0.65	1.27	1.18
3	39.01	33.28	7.27	0.38	0.28	–0.16
4	29.51	37.41	5.89	0.23	0.75	–0.02
5	65.51	37.97	5.04	–1.01	0.58	0.57
6	50.78	47.22	3.25	–0.04	0.35	–0.03
7	72.69	47.58	4.34	–0.81	0.22	0.13
8	30.07	48.18	5.62	–0.35	0.02	–0.03
9	41.34	50.11	3.91	–0.21	0.12	–0.03
10	61.13	52.16	2.22	–0.52	0.24	0.07

TABLE **9.5**   Mass center coordinates and the displacements for individual particles obtained from experimental testing (in 10 particles).

No.	Macro-Strain					
	ε_x	ε_y	ε_z	ε_{xy}	ε_{yz}	ε_{xz}
1	0.00	−0.05	−0.05	−0.04	0.02	−0.04
2	0.70	−1.10	−0.80	−0.43	−0.08	−0.97
3	0.07	−0.08	0.00	0.10	0.11	0.30
4	−0.14	0.03	3.35	0.15	0.33	−0.42
5	0.17	−0.26	−0.04	0.10	0.05	−0.09
6	0.02	0.12	−0.21	0.05	0.09	−0.05
7	0.01	0.03	−0.09	−0.01	0.05	−0.04
8	0.20	0.07	0.02	0.00	0.14	−0.03
9	−0.04	0.07	0.45	−0.04	0.03	−0.16
10	2.32	−1.03	−0.14	−2.11	−1.19	2.58

TABLE 9.6 Local macro-strains from experimental measurements (in 10 tetrahedrons).

9.4.1.7 PFC³ᴰ and DEM Simulation

The PFC³ᴰ was used for simulating the above test in order to investigate the particle shape effect on the micro-macro behavior of granular materials. It was used to conduct DEM simulation on irregular particles and on spherical particles, respectively. The actual microstructure used in the confined compression test was incorporated into the simulation. Irregular particles were represented by clusters of balls. Each cluster behaves as a rigid particle with deformable boundaries in PFC³ᴰ (Itasca, 2005). A burn algorithm (see Section 9.3) was applied to reduce the number of balls so that the calculation efficiency could be greatly improved. The virtual (digital) specimen composed of 173 particles in PFC³ᴰ is illustrated in Figure 9.15a. The spherical particles were generated using the particle mass centers and volumes the same as those of irregular particles. The virtual specimen using spherical particles is depicted in Figure 9.15b.

In the simulation, the particles were virtually stacked in a cylindrical container and a rigid load plate was placed on top of the aggregates. The load plate was made from hundreds of strongly bonded balls that would not break apart during the simu-

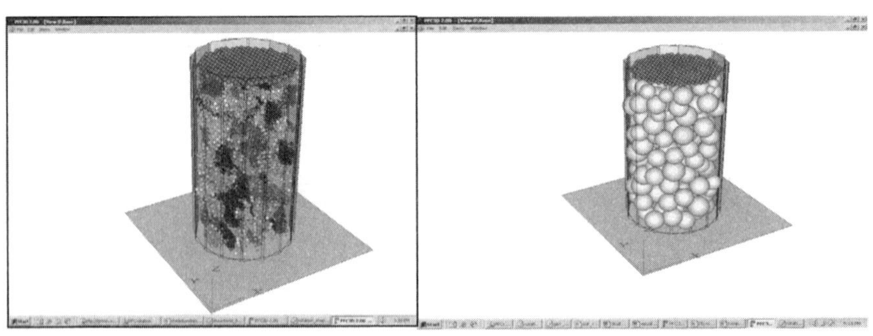

a. Irregular Particle　　　　　　　　b. Spherical Particle

FIGURE 9.15 Visualization of DEM simulation (irregular particles are represented by clusters of balls).

a. Experimental observation   b. Simulation using irregular particles   c. Simulation using spheres

FIGURE 9.16   Illustration of particle translational movements.

lation. An axial force was applied directly on the load plate. The physical properties of particles, boundary, and compact force were assigned similar to those in the confined compression test.

9.4.1.8   Simulation Results and Comparison to Experimental Measurements

The particle translational movements by simulations and by the actual experiment are illustrated in Figures 9.16a, b, and c (particle translational vectors). Most of the simulated irregular particle movements occurring in the vertical direction and at the lower part of the sample were very small, which is similar to the trend of experimental observations. However, simulated movements of spherical particles are more erratic. The simulated magnitudes of particle translations versus experimental measurements were plotted in Figures 9.17a and b. Again, the simulation using the irregular particles yielded results more consistent with experimental observations.

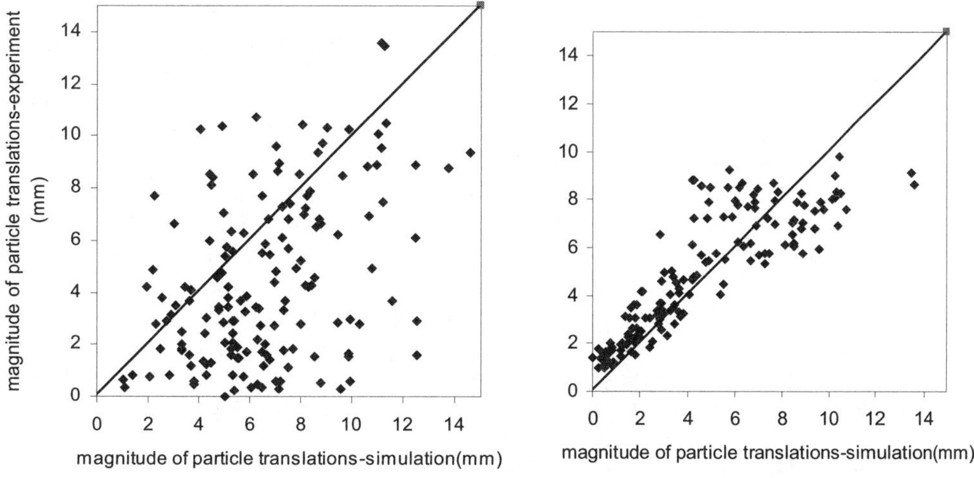

FIGURE 9.17   Simulated magnitudes of spherical particle translations versus experimental measurements.

Results	Experiment Irregular Particle	DEM Irregular Particle	% Relative Diff.	DEM Sphere	% Relative Diff.
Vertical contraction(mm)	9.800	9.230	5.816	12.300	25.510
Radial dilation(mm)	0.780	0.730	6.410	1.090	39.744
Global volume strain	−0.041	−0.039	4.368	−0.044	6.652
Change of porosity	0.024	0.023	5.859	0.027	13.858

TABLE 9.7 Macroscopic properties by DEM simulation and experimental measurements.

The comparison between the simulated macro-properties and the experimental results is presented in Table 9.7. The consistency between the simulation results based on irregular particles with those of experimental observations suggests that the DEM simulation incorporating particle shapes is a valid approach to predict the deformation of granular materials. Although DEM simulation using spheres may have acceptable overall macroscopic results, it cannot predict the kinematics of particles at the microscopic levels.

9.4.2 Simulation of Direct Shear Test and Shear Banding

Shear localization is a phenomenon encountered in granular materials when the deformation localizes suddenly into a narrow zone, and the shear stress reaches a peak value and then drops sharply to a residual state. In engineering practice, Coulomb's failure law is usually assumed to govern the evolution of the shear band; however, it may not truly describe the mechanism of shear banding. In fact, the basic micromechanism leading to the formation of a shear band is not well understood, even though research has been carried out in this area by both experimental studies and numerical methods for more than two decades (Cundall et al., 1982; Yoshida et al., 1994; Oda and Kazama, 1998; Iwashida and Oda, 1998).

DEM has been increasingly used in the study of the behavior of granular materials because of its ability to investigate granular materials at the microscopic level since the 1970s (Cundall and Strack, 1979; Ghaboussi, 1990; Thornton, 1992). Though it has been recognized as a promising tool, the difficulty in modeling the real microstructures has limited DEM's application. Most of the historical DEM simulations were performed using idealized particle systems such as spheres and ellipsoids (Cundall and Hart, 1992; Routhenberg, 1989; Lin and Ng, 1997). Though efforts have been made to simulate polygon-shaped particles or computer-generated irregular particles (Ni et al., 2000; Mirghasemi et al., 2002; Golchert, 2004), few DEM codes were developed based on the real microstructure of a material. Furthermore, the accuracy of those simulation results was unknown because almost no experimental measurements of the micro-quantities were available, and the validity of a numerical model is doubtful if there are no corresponding experimental data to support the simulation results. The 3D clustering DEM model presented in previous sections incorporated the real microstructure acquired by X-ray tomography imaging technology. The capability of the model was evaluated by the experimental observations at both micro- and macro-scales.

9.4.2.1 Materials and Experimental Setup

Crushed river gravel that passed the ½ in sieve and was retained on the ⅜ in sieve was used in this example. The rock's physical properties are presented in Table 9.8. A Direct

Material	Agg. Size(mm)	Agg. Specific Gravity	Young's Modulus (Gpa)	Poisson's Ratio	Weight(g)	Porosity (%)
Crushed river gravel	9.5~12.7	2.528	60	0.2	1878	44.4

TABLE 9.8   Physical properties of the material.

Shear Apparatus EL28-007 was utilized for the test. Normal forces applied were 20 N (Newton), 120 N, 320 N, and 520 N, respectively. The displacement rate was 0.12 mm/s.

A shear box with a size of 5 × 5 × 2.5 in (127 × 127 × 63.5 mm) was specially built for the test. It was made from plexiglass which can be penetrated easily by X-rays (Figure 9.18a). Aggregates were placed into the shear box and compacted by a 5-pound compact hammer until the desired unit weight was obtained. A normal force was then applied on top of the specimen. The top box was fixed, and the bottom box was pushed horizontally. The relative lateral displacement and shear force between the two boxes were monitored automatically during shearing by a data acquisition system: Wave book/512™ 12-bit 1MHZ. The maximum relative displacement between the top box and the bottom box is 9 mm.

The specimen (the shear box together with compacted aggregates) was scanned by XCT before and after shearing. Sectional images representing aggregate skeletons were obtained. The space between two adjacent images was 1.3 mm, about one-tenth of the maximum particle size. Based on the sectional images, the individual particle mass centers were quantified following the 3D digital analysis method developed by Fu et al. (2007), which was also presented in Chapter 4. The specimen porosity calculated by digital analysis was 44.8%, which was very close to the measured value 44.4%. Figure 9.18b presents some sectional gray images acquired by XCT. Based on the sectional images obtained before and after shearing, the particle translations, rotations, and local strains can be quantified.

a. The shear box and the material
used in the direct shear test

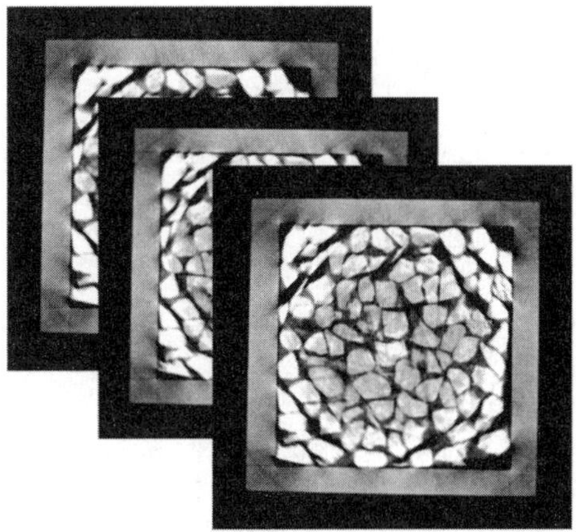

b. The gray images acquired by XCT

FIGURE 9.18   Illustration of the test setup and scanned sectional images.

9.4.2.2 Quantitative Evaluation of Strain Localization

Strain localization was evaluated by the ratio of the local strains to the global strains of the entire specimen. The global strains were calculated by the following equations:

$$\varepsilon_1 = \frac{\Delta h}{h}; \varepsilon_2 = \frac{\Delta l_l}{l_l}; \varepsilon_3 = \frac{\Delta l_w}{l_w} \tag{9-30a}$$

$$\varepsilon_{12} = \frac{\varepsilon_1 - \varepsilon_2}{2}; \varepsilon_{13} = \frac{\varepsilon_1 - \varepsilon_3}{2}; \varepsilon_{23} = \frac{\varepsilon_2 - \varepsilon_3}{2} \tag{9-30b}$$

where Δh = vertical deformation of the specimen

h = height of the specimen

$\Delta l_l, \Delta l_w, l_l, l_w$ = lateral deformation and length of the specimen ε_2 and ε_3 are nominal strains and are almost zero as Δl_l and Δl_w are almost zero.

The degree of strain localization is defined as the effective local strains normalized by the effective global strain (defined as $\varepsilon = \sqrt{\varepsilon_1^2 + \varepsilon_2^2 + \varepsilon_3^2 + \varepsilon_{12}^2 + \varepsilon_{23}^2 + \varepsilon_{13}^2}$). The quantified local strains and degree of strain localization are presented in Table 9.9. The strain was found very much localized in the specimen.

No.	Macro Local Strain						Effective Strain	Degree of Strain Localization
	ε_x	ε_y	ε_z	ε_{xy}	ε_{yz}	ε_{xz}	ε_{eff}	
1	0.01	0.00	0.01	0.00	0.01	−0.01	0.02	0.18
2	−0.01	0.00	−0.06	0.02	0.01	0.01	0.07	0.52
3	0.00	−0.02	0.04	0.00	−0.02	−0.02	0.06	0.46
4	−0.02	−0.02	−0.02	0.03	0.04	−0.04	0.10	0.74
5	0.01	0.01	0.01	0.01	−0.01	0.00	0.02	0.15
6	0.00	−0.04	0.22	0.02	−0.06	0.06	0.26	1.95
7	−0.01	−0.02	0.00	0.01	−0.01	0.00	0.03	0.24
8	0.00	−0.89	−0.02	−0.03	−0.02	−0.31	1.00	7.49
9	−0.01	−0.01	0.08	0.00	0.00	−0.01	0.08	0.61
10	0.01	0.00	0.04	0.01	0.01	−0.02	0.05	0.39
11	0.01	0.00	0.03	0.00	0.03	−0.02	0.06	0.43
12	0.01	0.00	0.02	0.00	0.01	0.01	0.03	0.23
13	−0.01	0.00	0.05	0.00	−0.01	−0.01	0.05	0.41
14	0.02	0.00	0.02	0.00	0.03	−0.01	0.05	0.39
15	−0.05	0.02	−0.12	0.01	−0.08	0.01	0.17	1.31
16	0.03	0.04	−0.07	0.04	0.01	−0.02	0.11	0.86
17	0.02	−0.01	0.01	0.01	−0.01	0.00	0.03	0.20
18	0.16	−0.67	0.53	−0.14	0.30	−0.56	1.26	9.48
19	−0.24	−0.01	−0.03	−0.06	0.11	0.02	0.30	2.24

TABLE 9.9 Local strains and strain localization by experimental measurements.

9.4.2.3 DEM Simulation Setup

PFC^{3D} was used to perform DEM simulations on irregular particles and on spherical particles respectively. The microstructure of the specimen used in the direct shear test was captured and incorporated into the simulations using digital image analysis methods. Irregular particles were represented by clusters of balls (Section 9.3) in the clustering DEM model. Each cluster behaved as a rigid particle with deformable boundaries, and contacts of balls within a cluster were ignored in the simulation. The Burn Algorithm 2 (Section 9.3) was applied to reduce the required number of balls. The rebuilt specimen was composed of 1280 particles in PFC^{3D} (Figure 9.19a). Approximately 200,000 small balls were used to represent those irregular particles. The spherical particles were generated using the particle mass centers and volumes calculated through image analysis for individual particles. The specimen represented by spherical particles was presented in Figure 9.19b. More details on the simulations can be found in Fu (2005) and Fu et al. (2010). The configurations after testing for the two digital specimens are presented in Figure 9.19c and d.

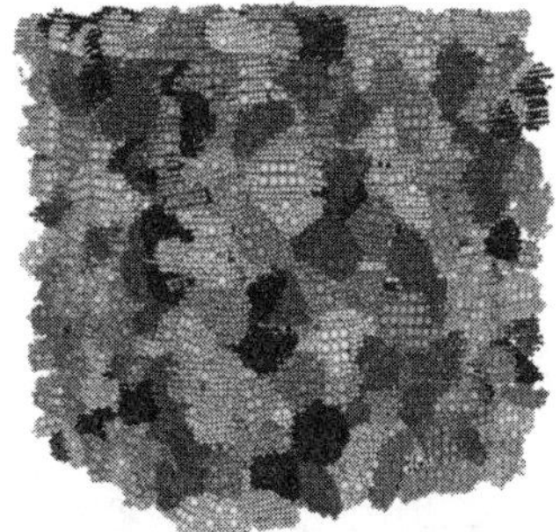

a. 3D visualization of microstructure of the material composed of irregular particles

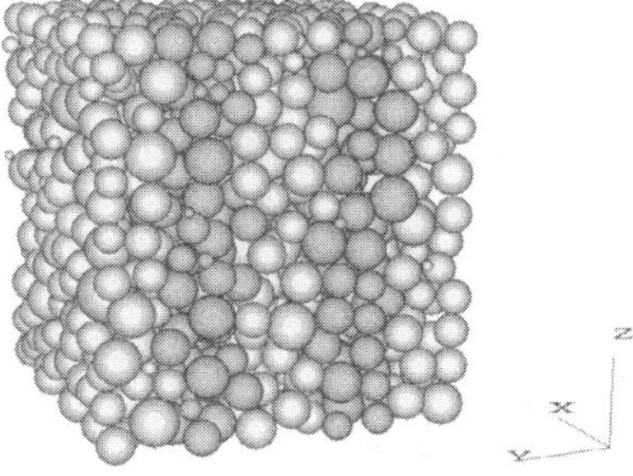

b. 3D visualization of microstructure of the material composed of spheres

c. Visualization of the sample after shearing, irregular particles

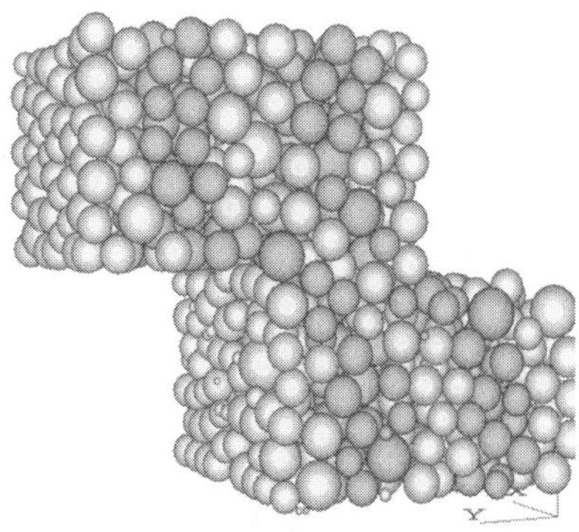

d. Visualization of the sample after shearing, spheres

FIGURE 9.19 Visualization of the particle configurations.

Results	Experiment	DEM-Irregular Particle	% Relative Difference	DEM-Sphere	% Relative Difference
Vertical dilation (mm)	5.00	4.50	10.00%	−3.70	174.00%
Change of porosity	0.02	0.02	10.44%	−0.02	175.99%
Peak shear force	63.47	67.63	6.55%	41.85	34.06%
Residue shear force	21.60	25.31	17.18%	32.77	51.71%
Peak friction angle	40.15	40.51	0.90%	39.96	0.48%
Residue friction angle	34.26	32.55	5.00%	30.84	9.99%

TABLE 9.10   Global macro-properties of experimental measurement and simulations.

9.4.2.4 Macro-Behavior: Simulation Results and Experimental Measurements

The simulations predicted both macro-properties and micro-properties of the assemblies. The global macro-properties discussed in this study were related to the features of the entire specimen such as the shear strength, global deformation, and bulk friction angle. The local macro-properties included the characteristics monitored in the local volumes, which were either tetrahedron formed by every four adjacent particles or the measuring layers in the specimen. The micro-properties included the kinematics and contacts of individual particles. The comparisons of global macro-properties between the experimental measurements and simulation results are presented in Table 9.10. The comparisons of other properties are referred to in Fu et al. (2010).

9.4.2.5 Micro Behavior: Simulation Results and Experimental Measurements

Particle Kinematics
Figures 9.20a, b, and c illustrate the experimentally measured, simulated irregular, and simulated spherical particle translational movements, respectively. In the figure showing the experimental results, particles in the lower box moved mostly along the horizontal direction (the direction along which two boxes had relative movement), except

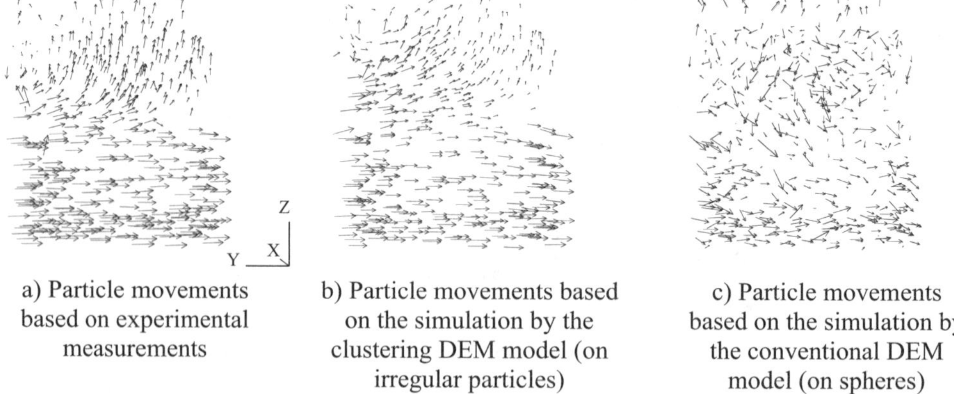

a) Particle movements based on experimental measurements

b) Particle movements based on the simulation by the clustering DEM model (on irregular particles)

c) Particle movements based on the simulation by the conventional DEM model (on spheres)

FIGURE 9.20   Particle translational movement.

those located very close to the shear plane. Particle movements in the upper box displayed significant vertical dilation. The simulations based on irregular particles showed the same trend as the experimental observations, while those based on spheres showed very erratic particle translational movements.

Particle Kinematics in the Shear Zone
To evaluate the predictions of the shear zone by the two DEM models, the micro-kinematics including magnitudes of particle displacements and magnitudes of particle rotations in measuring columns (refer to the location of the monitoring system; the colored sphere columns in Figure 9.19) were quantified. For simplification, only particle rotations are presented in Figure 9.21a (for spheres), Figure 9.21b (for irregular particles) and Figure 9.21c (for measured). In Figure 9.21a, the spherical particle rotations are erratic. Large particle rotations occurred not only near the shear plane but also at other areas. To summarize, no significant shear zone has been formed in the simulation based on spheres. In Figure 9.21b, the irregular particle rotation is small at the upper and

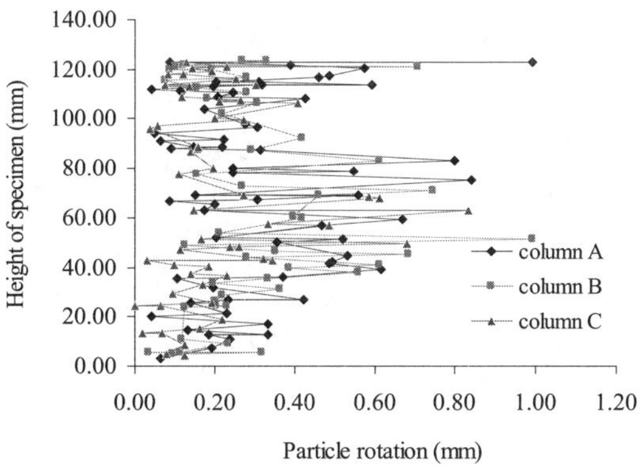

a) Particle rotation distribution along the height of the specimen, simulation using spherical particles

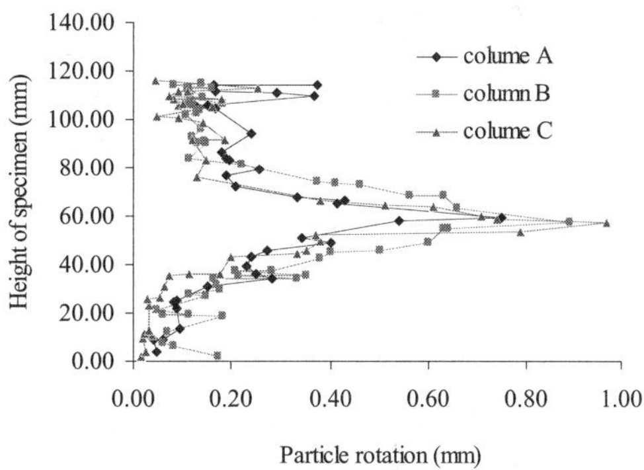

b) Particle rotation distribution along the height of the specimen, simulation using irregular particles

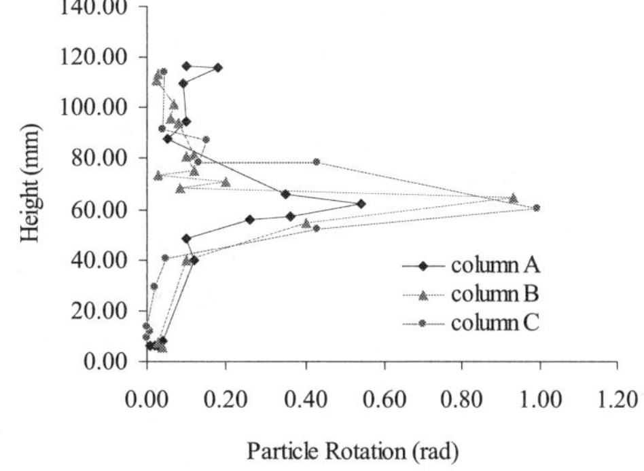

c) Measured particle rotation distribution along the height of the specimen

FIGURE 9.21   Particle rotational movement.

lower part of the specimen, but reaches the peak value near the shear plane, which agrees with the experimental observations (Figure 9.21c).

The comparison between simulations and experimental measurements of both displacement and rotation indicate that particle shape is a significant factor for DEM simulations. This is consistent with the DEM rotational formulations.

9.5   DEM Applications for AC

9.5.1   Historic Use of DEM

Since 1971, when Cundall (1971) originally introduced DEM and used it in the analysis of rock-mechanics problems, the method has received considerable attention in studying granular materials. In addition to the granular assemblies, DEM has been extended to simulating solid materials using bonded contact models. The early-day application of this method included several hundred particles; while in the last 20 years DEM has evolved to the capability of simulating a system of a few million particles. The application has been introduced in different fields and also in the asphalt mixture field (Meegoda and Chang, 1994; Meegoda and Chang, 1995; Papagiannakis et al., 2002; Kim et al., 2008; Kim and Buttlar, 2005; Gantt and Gatzke, 2005).

As with the growth of DEM, many codes have been developed for simulating granular and solid materials, including the early codes Ball (Cundall and Strack, 1979) and Trubal (Chen and Hung, 1991; Ng and Dobry, 1994; Washington and Meegoda, 2003), universal distinct element code (UDEC) (Dickens and Walker, 1996; Zhao et al., 2008), and discrete element code in three dimensions (3DEC) (Hart et al., 1988). In the AC area, the particle flow code (PFC) in 2D and 3D (PFC2D/3D, Itasca Consulting Group, 2005) has been widely used by researchers in AC (Buttlar and You, 2001; Dai and You, 2007; Liu and You, 2008; You and Buttlar, 2006; Abbas et al., 2005, 2006, 2007). The major reasons for PFC3D's wide use in AC is its commercial availability, flexibility for clumping mechanism, and viscoelastic contact models. This literature review focuses on collecting work done in simulating asphalt-based materials using the DEM method. Table 9.11 presents the major applications and their features. It should be noted that only essentially new applications were cited in Table 9.11.

Early in these research studies, asphalt mastics were modeled as an elastic material (Abbas et al., 2005; Buttlar and You, 2001; Collop et al., 2004, 2006, 2007; Dai and You, 2007; You and Buttlar, 2004, 2005, 2006). With these non-viscous models, however, the time-dependent properties such as relaxation, dynamic modulus, and phase angles cannot be predicted. Nevertheless, one of the earliest applications of DEM in AC (Meegod and Chang, 1994, 1995; Chang and Meegod, 1997, 1999) utilized a Maxwell element. Throughout the literature review, the Burger model has been used exclusively for modeling the viscoelasticity of asphalt binders or mastics in mixtures (Collop et al., 2006, 2007; Abbas et al., 2007; Liu et al., 2009). With these viscoelastic models, time-dependent behavior of asphalt mixtures was simulated, such as the dilation, dynamic modulus, and phase angles. In addition to the contact models above, there was the bilinear cohesive model developed by Kim and Buttlar for simulating fractures in asphalt mixtures, using the displacement-softening model provided in PFC 2D (Itasca Consulting Group, 2005; Kim and Buttlar, 2005; Kim et al., 2008). A large portion of the literature DEM simulations have been in stiffness prediction (Abbas et al., 2007; Dai and You, 2007; Liu and You, 2008; You and Buttlar, 2004, 2006), fatigue modeling

Author/Year	Problem Studied	Platform	2D/3D	Shape	Contact Model	Micro Parameter Calibration
Abas et al./ 2007	Prediction mixture dynamic modulus with neat binder and modified binder	PFC2D	2D	Yes	Burger's model	Assumed values
Collop et al./2006	Simulation of triaxial and uniaxial testing	PFC3D	3D	No	Burger's model, thin film effects considered	Literature data/ no experimentally calibrated data for micro parameters, film thickness model
Per Uliditlize /2001	Failure, the strength and fatigue	DEM	2D	Yes	Non-linear contact, strength	NA
Chang, Meegod/1994	Stress-strain relationship including softening	True Ball	3D	No	Viscoelastic	NA
You and Buttlar/2004	Mixture modulus prediction	PFC2D	2D	Yes	Linear	Measured mastic modulus
Dai and You/2008	Prediction of creep stiffness	PFC2D	2D	Yes	Linear	Measured mastics
Liu et al./ 2009	Prediction of creep stiffness	PFC2D	2D	Yes	Viscoelastic	NA

TABLE 9.11   Major research efforts on DEM applications to AC.

(Carmona et al., 2007), and fracture modeling (Kim and Buttlar, 2005; Kim et al., 2008), and failure (Ullidtz, 2001).

Abbas et al. (2007) built a viscoelastic model for mastic and asphalt mixture using the Burger model in PFC2D to simulate the dynamic modulus test and low temp fracture of indirect tensile test. The interaction among the discrete particles was defined using two contact stiffness models: the linear contact model and the viscoelastic Burger's contact model.

Similarly, using the PFC2D code, Liu et al. (2009) built a viscoelastic model of asphalt mixtures in which the viscoelastic behavior of asphalt mastics was represented by a Burger's model, while the aggregate particles were simulated with particles of irregular shapes consisting of balls bonded together. The asphalt mastic between aggregates was filled with balls bonded with the Burger's model. The entire micromechanical model was developed with four constitutive laws to represent the interactions at contacts of discrete elements within an aggregate, within mastic, between an aggregate and mastic, and between two adjacent aggregates. Three components, i.e., a stiffness model, a slip model, and a bonding model, were included in each constitutive law to capture the relationship between the contact force and relative displacement as well as to describe slipping and tensile strength at a particular contact (Liu et al., 2009). However, model parameter characterization presents a challenge.

In summary, DEM applications in AC include those using idealized microstructure (Collop et al., 2004, 2007, 2006); those that incorporated 2D micrsotructure (Buttlar and You, 2001; You and Buttlar, 2004, 2006); and those using a user-defined irregular particle model (Liu and You, 2008). In terms of contact models, the non-linear film-thickness-dependent contact model (Collop et al., 2006) deserves special attention. It was well recognized (Cheung and Cebon, 2007) that the stiffness of asphalt binder changes with the film thickness. This feature should be included for future DEM applications.

9.6   Equivalent Ellipsoid Approach

For irregular particles, the translational motion equation is the same as Equation 9-1. However, the rotational motion equations are quite different and can be written as in Equation 9-31.

$$\sum M_x = I_x \frac{d\omega_x}{dt} - (I_y - I_z)\omega_y\omega_z - I_{xy}(\frac{d\omega_y}{dt} - \omega_z\omega_x) - I_{yz}(\omega_y^2 - \omega_z^2) - I_{zx}(\frac{d\omega_z}{dt} + \omega_x\omega_y) \qquad (9\text{-}31a)$$

$$\sum M_y = I_y \frac{d\omega_y}{dt} - (I_z - I_x)\omega_x\omega_z - I_{yz}(\frac{d\omega_z}{dt} - \omega_x\omega_y) - I_{zx}(\omega_z^2 - \omega_x^2) - I_{xy}(\frac{d\omega_x}{dt} + \omega_z\omega_y) \qquad (9\text{-}31b)$$

$$\sum M_z = I_z \frac{d\omega_z}{dt} - (I_x - I_y)\omega_x\omega_y - I_{zx}(\frac{d\omega_x}{dt} - \omega_y\omega_z) - I_{xy}(\omega_x^2 - \omega_y^2) - I_{yz}(\frac{d\omega_y}{dt} + \omega_z\omega_x) \qquad (9\text{-}31c)$$

Where M_x, M_y, M_z are the external force momentums against x,y,z axes on the particle; ω_x, ω_y, ω_z are the rotational velocity of the particle; I_x, I_y, I_z, I_{xy}, I_{xz} and I_{yz} are the six mass momentums. These equations follow the general format of dynamics of rigid bodies (Hibbeler, 1974). Figure 9.22 presents real particles in contact. Compared to spheres in contact, real contact is much more complicated; the particle mass momentums are usually not isotropic. To completely abide by the irregular geometry of real particles is a challenging job, as it requires a full development of contact-detecting algorithms. The existing codes (Lin and Ng, 1997) prompt the possibilities or approximating irregular particles with ellipsoids with equal mass and mass momentums. The major reason for investigating the feasibility of this idea is the requirement of using a huge number of

a. Representation by Voxels

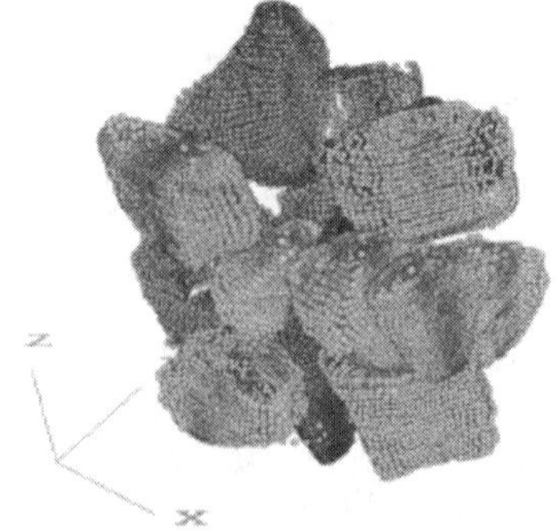

b. Representation by Clump

Figure 9.22   Representations of particles of irregular shape by using clump in PFC$^{3\text{-}D}$.

spheres to represent irregular particles even using the burn algorithms. The methods in Chapter 3 include one on the computation of the volume and mass center coordinates of irregular particles. For translational motion, the motion equation is the same as that of spheres. However, the mass momentum for irregular particles is quite different from that of spheres. The question is, for irregular particles, how different are the momentums from those of equivalent ellipsoids? The following presents the work by Wang et al. (2007) to answer this question.

9.6.1　Moment of Inertia Tensor

The momentum of inertia is a term used to describe the capacity of a cross-section to resist bending. It is a ma thematical property of a section concerning the shape and mass distribution about a set of reference axes. The reference axes usually run through the mass center of the objects. The momentum of inertia in a solid body with density $\rho(\mathbf{r})$ with respect to a given axis is expressed as I and defined by the volume integral:

$$I = \int \rho(r) r^2 dV \tag{9-32}$$

Where r is the distance from a volume element dV to the axis of rotation. The momentum of inertia of an object depends on its shape and distribution of mass within the object. It is directly related to rotational kinetic energy and angular momentum by the following relation:

$$K = \frac{1}{2} I \omega^2 \qquad L = I \omega \tag{9-33}$$

Where ω is the angular velocity; K is the rotational kinetic energy; and L is angular momentum.

For 3D motion, a general case, the "moments of inertia," about three rotation axes, are not enough to describe the shape and mass distribution. It generally requires six "products of inertia." The total of nine "moments of inertia" form a 3×3 matrix and is called moment of inertia tensor. It can be expressed as:

$$I_{jk} = \int_V \rho(r)(r^2 \delta_{jk} - x_j x_k) dV \tag{9-34}$$

Where r is the distance of a point to the axis; δ_{jk} is the Kronecker delta. Expanding this equation in terms of Cartesian axes gives the following equation:

$$I = \int_V \rho(x,y,z) \begin{bmatrix} y^2 + z^2 & -xy & -xz \\ -xy & z^2 + x^2 & -yz \\ -xz & -yz & x^2 + y^2 \end{bmatrix} dxdydz = \begin{bmatrix} I_{11} & I_{12} & I_{13} \\ I_{21} & I_{22} & I_{23} \\ I_{31} & I_{32} & I_{33} \end{bmatrix} \tag{9-35}$$

The diagonal elements of the tensor, I_{11}, I_{22}, and I_{33}, are just simple moments of inertia about x, y, and z axes. The other elements of the inertia tensor, such as I_{12}, are the "products of inertia," which make the moment of inertia tensor always symmetric. Using the boundary coordinates of the particle, the moment of inertia tensor for a solid body can be obtained numerically by summing the products of all the voxels contained in a particle. The moment of inertia tensor can be expressed as:

$$I = m_p s \sum \begin{bmatrix} y^2 + z^2 & -xy & -xz \\ -xy & z^2 + x^2 & -yz \\ -xz & -yz & x^2 + y^2 \end{bmatrix} \tag{9-36}$$

Where m_p is the mass for a volume element $1 \times 1 \times 1$ pixels and s is a number of pixels between two adjacent scanned slices. For example, the moment inertia tensor for the particle shown in Figure 9.23a can be obtained in mass•pixel^2 unit as:

$$I = \begin{bmatrix} 3.473 & 0.607 & -0.580 \\ 0.607 & 1.881 & -0.925 \\ -0.580 & -0.925 & 2.686 \end{bmatrix} \times 10^9$$

The crucial property of a tensor is that its components should transform under a rotation of coordinate axes in such a way as to keep its geometrical or physical meaning invariant. In a particular case of rotation of axes, the moment inertia tensor becomes a diagonal form where the products of inertia are null. This occurs when the three orthogonal axes of reference frame are parallel to the three "principal axes" of the object. In other words, a rigid body is dynamically balanced when its angular momentum is parallel to its angular velocity:

$$L = I\omega = \lambda \cdot \omega \tag{9-37}$$

Where λ is some (scalar) number. For this to be true, the angular velocity, ω, must point along a principal axis of the moment of inertia tensor. The corresponding value of I is called a principal moment of inertia. The principal moment of inertia of a rotating body is defined by finding values of I such that:

$$L = \begin{bmatrix} L_x \\ L_y \\ L_z \end{bmatrix} = \begin{bmatrix} I_{11} & I_{12} & I_{13} \\ I_{21} & I_{22} & I_{23} \\ I_{31} & I_{32} & I_{33} \end{bmatrix} \begin{bmatrix} \omega_x \\ \omega_y \\ \omega_z \end{bmatrix} = \lambda \begin{bmatrix} \omega_x \\ \omega_y \\ \omega_z \end{bmatrix} \tag{9-38}$$

Which is an eigenvalue problem. The principal moments of inertia are the eigenvalues of the moment of inertia tensor and the corresponding eigenvectors are the direction cosines of the principal axis. There will be three eigenvalues, which will be called I_1, I_2, and I_3 in order of decreasing magnitude and three corresponding principal axes defined as ω_1, ω_2, and ω_3. The principal axes, ω_1, ω_2, and ω_3, are unit vectors and are

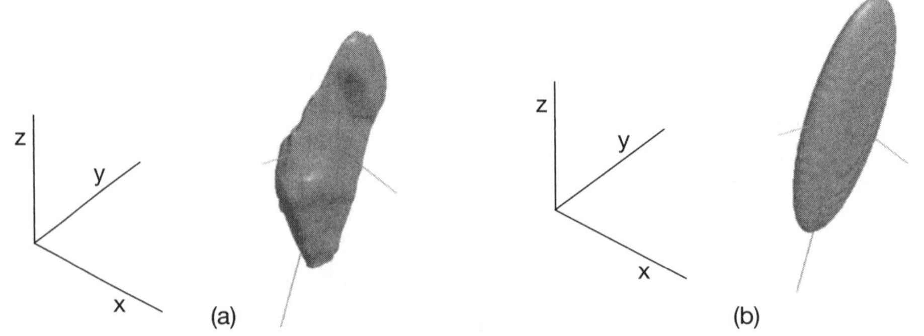

FIGURE 9.23   (a) Digitally reconstructed limestone particle (b) Simulated ellipsoid that has the same mass momentums and principal directions as those of the real particle.

orthogonal to each other. For the example aggregate particle, the principal moments of inertia and principal axes are obtained as:

$$I_1 = 3.819 \times 10^9, \quad I_2 = 3.266 \times 10^9, \quad I_3 = 1.023 \times 10^9$$

$$\omega_1 = \begin{bmatrix} 0.918 \\ 0.107 \\ -0.382 \end{bmatrix}, \quad \omega_2 = \begin{bmatrix} -0.239 \\ 0.619 \\ 0.748 \end{bmatrix}, \text{ and } \omega_3 = \begin{bmatrix} -0.317 \\ -0.778 \\ 0.543 \end{bmatrix}$$

9.6.2   Equivalent Ellipsoid

Due to the complexity and difficulty in reconstructing real particle shape, either sphere or ellipsoid is used in DEM or other computational models. The use of sphere to represent an aggregate particle is relatively simple and widely used in DEM simulation. However, a sphere does not reflect any shape accurately enough. An ellipsoid, however, presents much more potential to represent more complicated shapes with reasonable accuracy. The size and shape of an ellipsoid can be determined by its lengths of semi-axes a, b, and c in the ellipsoid Equation 9-39. The principal moments of inertia of the ellipsoid are determined by Equation 9-40 (Goldstein, 1950).

$$\frac{x^2}{a^2} + \frac{y^2}{b^2} + \frac{z^2}{c^2} = 1 \tag{9-39}$$

From Equation 9-40, the semi-axis lengths of the ellipsoid can be expressed as:

$$I_1 = \frac{1}{5}m(b^2 + c^2), \quad I_2 = \frac{1}{5}m(c^2 + a^2), \quad I_3 = \frac{1}{5}m(a^2 + b^2) \tag{9-40}$$

By substituting m with the mass converted from the scanned sectional images and I_1, I_2, and I_3 values obtained from the eigenvalue solutions, the semi-axis lengths of the ellipsoid can be conveniently determined. This means that the ellipsoid will have the same mass momentums as those of the real particle. This ellipsoid is named as the equivalent ellipsoid. Nevertheless, it should be noted that the volume of the ellipsoid might be different from that of the real particle. The magnitude of the deviation is related to the shape of the particle. By this approach, the semi-axis lengths of the ellipsoid for the example particle are obtained as:

$$a = 34.5, b = 63.2, \text{ and } c = 123.9 \text{ (pixels)}$$

Then the ellipsoid should be rotated based on the eigenvectors so that the orientation of the three principal axes of the ellipsoid is the same as those of the real particle. The eigenvector for the I_3, which is the smallest among the principal moments of inertia, will point to the direction of longest axis of the particle simply because the moment of inertia along the longest axis will be the smallest. The final ellipsoid configuration for the aggregate reconstructed from scanned images in Figure 9-23a is illustrated in Figure 9-23b. The blue line in both Figure 9-23a and Figure 9-23b is the direction of ω_3 that is corresponding to the longest axis. In addition, the red line and the green line are the directions of ω_1 and ω_2 that are corresponding to the shortest axis and the medium axis. From Figures 9-23a and 9-23b, it can be noted that all three axes also match well with the visually determined principal axes of the aggregate.

9.6.3 Example Applications

To demonstrate the soundness of the newly developed methodology, limestone and sandstone aggregate particles passing a 25-millimeter sieve size but retained on a 9.5-millimeter sieve have been studied by Wang et al. (2007). The particles were first placed in a film cartridge and X-ray scanned. The image size is 512 × 512 pixels and the in-plane resolution is 0.2 mm/pixel while the spacing between two adjacent sections is 0.4 mm. The gray images are then processed to obtain binary images. The method (Wang et al., 2004) is applied to reconstruct and analyze each particle to obtain the volume and mass momentums. The newly developed method is then used to obtain the principal directions of the particles and their corresponding equivalent ellipsoids. Figure 9.24 presents all the particles of the two specimens and their corresponding equivalent ellipsoids.

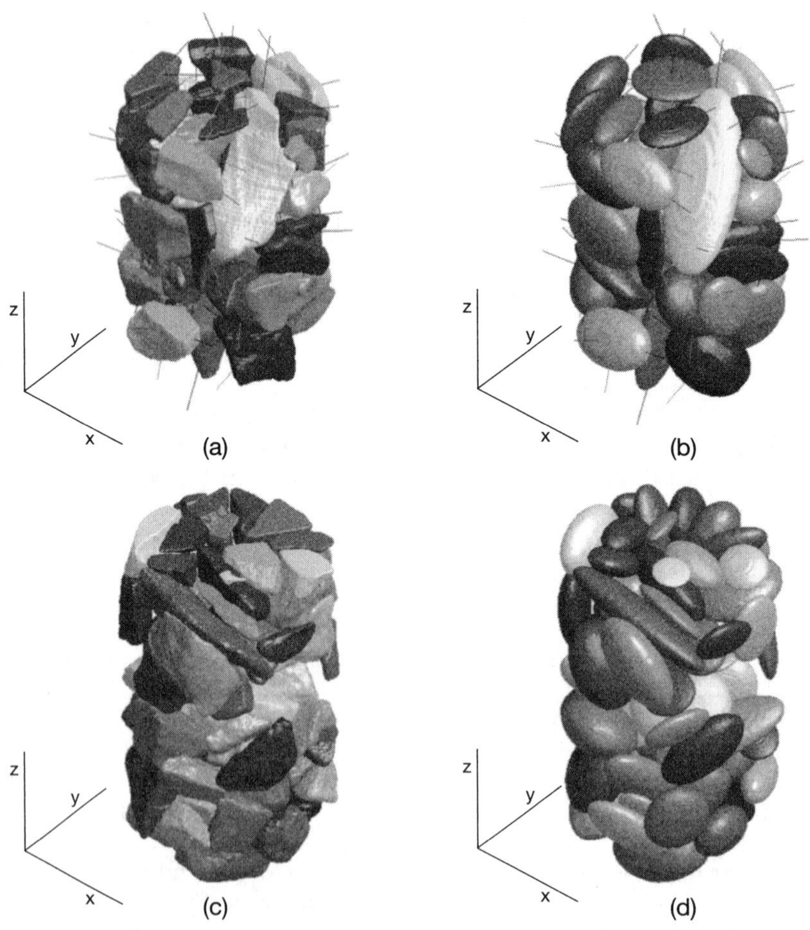

FIGURE 9.24 (a) Reconstructed sandstone specimen (b) Sandstone specimen of equivalent ellipsoids (c) Reconstructed limestone specimen (d) Limestone specimen of equivalent ellipsoids.

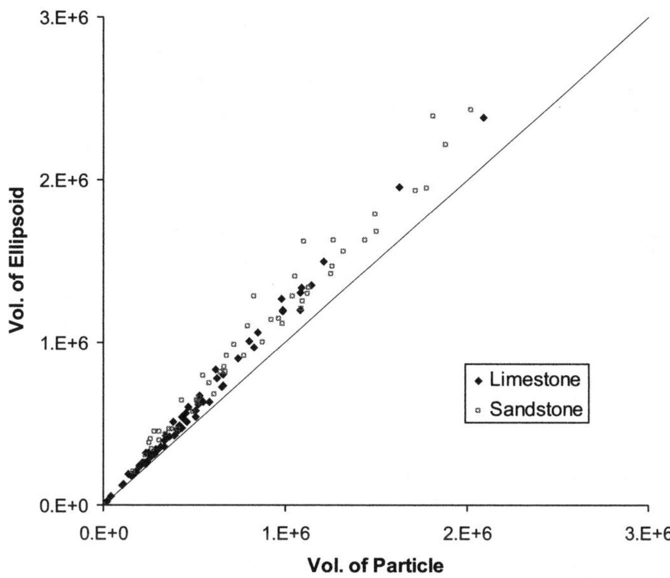

Figure 9.25   Relationship between the volume of real particles and the volume of simulated ellipsoids.

It should be noted that the equivalent ellipsoids usually have different volumes from their corresponding real particles. Figure 9.25 presents this correlation. The volumes of ellipsoids are about 15% larger than those of real particles in some cases. This can be adjusted by reducing the density of the particles.

9.6.4   Discoveries

The equivalent ellipsoid method has the same principal mass momentums as those of the real particle. The accuracy to approximate the particle of the equivalent ellipsoid is smaller than that of the clustering method, but reduces the number of particles to one. The representation methods present potential to better simulate real particulate systems.

9.7   Combined FEM and DEM

The flexible container problem (Munjiza, 2005) is a good example of the combined continua-discontinua problem. A flexible container is made with particles of soft rubber. In addition to interacting with one another, the soft balls deform as well. The walls of the container also deform. The deformation of the container and deformation of individual particles significantly influences the way particles move inside the container. Each individual particle deforms under external forces and interacts with other particles and walls. Change in the shape and size of individual particles is a problem of finite strain elasticity. The deformability of individual particles is well presented by a continuum-based model. Interaction among individual particles and interaction between particles and the container is well presented by a discontinua-based model. This flexible container problem involves both continua and discontinua. The combined FEM/DEM method is an advanced tool to solve such combined continua-discontinua problems.

DEMs are numerical techniques developed to simulate systems comprised of multiple distinct bodies that interact with one another through contact forces. However, in some cases, it is useful to consider the deformation of the distinct bodies in order to evaluate their stress and strain distributions (e.g., the mastics in asphalt concrete). The combined finite-discrete element method is just a numerical method that could combine aspects of both finite elements and discrete elements together. Starting from the finite element method, a solid domain is discretized into finite elements; if this solid domain fails or fractures, it would be transformed into several subsolid domains interacting with one another and each of them could also be discretized into its own finite element mesh. During this process, a transition from continua to discontinua algorithms is involved, which is the main problem solved by the combined discrete-finite element method.

The contact between these sub-solid domains is still considered the same as the contact in the discrete element method. The difference is each of them could be discretized into finite elements, so the contact solutions of the finite element method could also be utilized for both contact detection and contact interaction. The purpose of the contact detection is to locate those domains that are close enough to be considered contacting one another. The main challenge of this is CPU efficiency. The purpose of the contact interaction is to evaluate the contact forces between interacting domains and the major difficulty is the solution of contact kinematics, which is very complex, especially in three dimensions. The existing FEM could help in solving the contact kinematics and a potential contact force concept is introduced to solve the contact interaction problem.

The combined FEM/DEM method is becoming a wildly recognized tool utilized by many researchers. Munjiza and his research group made great strides in this area. Munjiza began research involving a large number of separate bodies that interact with one another in 1999, which is a main problem in the large-scale discrete element simulations or combined finite-discrete element simulations. In this work (Munjiza and John, 2002) a fracture algorithm applicable to simultaneous multiple fracture of a large number of bodies is developed and a fracture model is used in the combined discrete-finite element method for both initiation and propagation of Mode I loaded cracks in concrete. The algorithm is based on the approximation of experimental stress-strain curves for concrete in tension. The standard finite element formulation for the hardening part of the constitutive law is combined with the single-crack model for the softening part of the stress-strain curve. Finite elements are used to model the behavior of the material until the ultimate tensile strength and a discrete crack model is implemented through crack openings and separation along edges of finite elements. Based on the conclusion, the major advantages of this model are its ability to model both crack propagation and crack initiation of multiple cracks allowing creation of a large number of distinct interacting fragments without considerable additional CPU requirements.

Munjiza et al. (2000) published another piece of work utilizing the combined finite-discrete element method to simulate the fracture of solid considering the effect of expansion of detonation gas to cracks. The proposed equation enables gas pressure to be obtained in a closed form for both reversible and irreversible adiabatic expansion, while the gas flow model proposed considers only 1D compressible flow through cracks. When coupled with finite-discrete element algorithms for solid fracture and fragmentation, the model enables gas pressure to be predicted and energy balance to be preserved.

Munjiza et al. (2004) reported the work to use combined finite-discrete element model to investigate the failure and collapse of reinforced concrete beam and column. Algorithms specific to reinforced concrete structures and beam type structural elements were developed. These solutions have two main characteristics: 1) finite element discretizations are able to capture pre-failure behavior accurately; and 2) after the failure and collapse have occurred, the same finite element discretizations are able to capture interaction between failing and collapsing structural elements. A simple two-node finite element was proposed together with numerical integration, which enables the nonlinear behavior of both concrete and reinforcement to be captured. The proposed numerical solution was compared with an analytical solution for linear elastic behavior. Experimental verification of the proposed numerical solution for non-linear behavior and collapse was conducted using existing experimental results. Based on the conclusion of this work, the results obtained using the combined schemes compare well with analytical and experimental results.

Bangash and Munjiza (2003) conducted a comparison between an experimental result, undertaken at the Swiss Federal Institute, and a combined finite discrete element simulation to model the pre-failure and post-failure transient dynamics of reinforced concrete structures. In this work, a novel beam element is introduced in order to increase CPU and RAM efficiency. The accuracy and reliability of this element is assessed when used in dynamic loading conditions. Based on the conclusion, the element introduced is capable of accurately modeling inertia and contact effects in pre- and post-failure dynamics. Komodromos and Williams (2002) used a combined DEM and FEM to simulate deformable multi-body systems. In this work, an updated Lagrangian finite element formulation and an explicit time integration scheme were used together with some simplifying assumptions to liberalize this highly nonlinear contact problem. In particular, the DEM is employed to identify, at each simulation step, the bodies in contact and determine the contact forces. Then, either an FE or a DE formulation, depending on whether the body under consideration is deformable or rigid, respectively, is used for the individual body to describe the equations of motion. In case of a deformable body, the strains are assumed to be sufficiently small to permit a small strain analysis. The deformability of individual bodies is considered using a displacement-based updated-Lagrangian FE formulation.

Finally, an explicit time-integration method, specifically the central difference method, is used to perform the numerical direct integration of the equations of motion and determine new displacements, as well as the deformations and stresses wherever needed, of each body. After computation of the motion for each discrete body at a new time step, the positions of all discrete bodies are updated and a new contact detection process determines the new contacts and evaluates the corresponding contact forces, which are then used in the following time step of the simulation. A wave propagation simulation is presented as a very simple example that demonstrates a software implementation of the combined DEM and FEM to simulate deformable bodies using Java and database technologies. An extendable, object-oriented, and portable computational tool, which enables numerical simulations of multiple distinct bodies that interact through contact forces while allowing selected bodies to be deformable, is used for this verification example.

Lewis et al. (2005) used the combined finite-discrete element method to simulate the pharmaceutical powder and tableting process and contact dynamics for irregular-shaped particles. The main achievement is that a particle-scale formulation and two-stage contact detection algorithm were developed, which enhances the overall calculation efficiency for particle interaction characteristics. In this work, the irregular particle shapes and random sizes are represented as a pseudo-particle assembly having a scaled-up geometry, but based on variations of real powder particles. The simulation results show that particle size, shapes, and material properties have a significant influence on the behavior of compaction and deformation. Gethin et al. (2006) utilized the combined finite-discrete element method to simulate the flow and compaction of irregular randomly packed particles to form a tabletted product. In this work, the techniques which have been adopted to achieve computational efficiency are described, which include contact detection, particle-level deformation analysis, and a homogenization strategy. The computational scheme is validated using published data and is shown to be capable of simulating effects of particle shape, size, and friction on the flow rate. The advantage of the scheme is the flexibility that it offers to capture a mixture of material properties and particle shape and that no restrictions are necessary on the contact models since these are integral in the calculation procedures. Mamaghani (2006) used the discrete finite element method (DFEM) to model masonry bridges, which are composed of a finite number of distinct interacting blocks that have a length scale relatively comparable to the structure of interest. The developed DFEM is based on the principles of the FEM that incorporate contact elements. DFEM considers blocks as sub-domains and represents them by solid elements. Contact elements, which are superior to joint or interface elements, are used to model the block interactions, such as sliding or separation. In this work, some typical examples are illustrated in order to analyze the applicability of the DFEM. Based on their conclusion, the DFEM could become a useful tool for researchers in designing, analyzing, and studying the behavior of masonry bridges under static and dynamic loading.

9.8   Similarities between DEM and Molecular Dynamics

An analysis of the similarity between the DEM approach and the molecular dynamics (MD) approach would be interesting in that these two techniques may be combined for multiscale modeling. Both methods use the finite difference method to simulate the multi-body interactions. MD uses potential and derives forces from potentials. It can consider not only the nearest neighbor interactions, but also interactions with molecules some distances away. However, the DEM approach can only consider the interactions of particles in contact. A combination of the two methods may present potential for multiscale modeling.

References

Abbas, A., Masad, E., Papagiannakis, T. and Shenoy, A. (2005). Modeling asphalt mastic stiffness using discrete element analysis and micromechanics-based models. *International Journal of Pavement Engineering*, Vol.6, No.2, pp.137–146.

Abbas, A., Masad, E., Paapagiannakis, T. and Harman, T. (2007). Micromechanical modeling of the viscoelastic behavior of asphalt mixtures using the discrete-element method. *International Journal of Geomechanics*, Vol.7, No.2, pp.131–139.

Abbas, A. R., Papagiannakis, A. T. and Masad, E. A. (2006). Micromechanical simulation of asphaltic materials using the discrete element method. *Geotechnical Special Publication*, No.146, pp.1–11.

Bangash, T. and Munjiza, A. (2003). Experimental validation of a computationally efficient beam for combined finite-discrete element modeling of structures in distress. *Computational Mechanics*, Vol.30, No.5/6, pp.366–73.

Buttlar, W. G. and You, Z. (2001). Discrete element modeling of asphalt concrete: microfabric approach. *Transportation Research Record,* No.1757, pp.111–118.

Carmona, H.A., Kun, F., Andrade, J. S. and Herrmann, H. J. (2007). Computer simulation of fatigue under diametrical compression. *Physical Review E,* Vol.75, No.4, pp.1–8.

Chang, G.K., and Meegoda, J.N. (1997). Micromechanical simulation of hot mixture asphalt." *Journal of Engineering Mechanics,* Vol.123, No.5, pp.495–503.

Chang, K.G. and Meegoda, J.N. (1999). Micromechanical Model for Temperature Effects of Hot-Mix Asphalt Concrete. Transportation Research Record, No.1687, 95-103.

Chen, Y.-C. and Hung, H.-Y. (1991). Evolution of shear modulus and fabric during shear deformation. *Soils and Foundations,* Vol.31, No.4, pp.148–160.

Cheung, C.Y. and Cebon, D. (1997). Thin Film Deformation Behaviour of Power Law Creeping Materials. *Journal of Engineering Mechanics,* Vol.123, No.11, pp. 1138–1152.

Collop, A. C., McDowell, G. R. and Lee, Y. W. (2004). Use of the Distinct Element Method to Model the Deformation Behavior of an Idealized Asphalt Mixture. *Taylor and Francis,* London.

Collop, A. C., McDowell, G. R. and Lee, Y. W. (2006). Modeling dilation in an idealized asphalt mixture using discrete element modeling. *Granular Matter,* Vol.8, No.3–4, pp.175–184.

Collop, A. C., McDowell, G. R. and Lee, Y.W. (2007). On the use of discrete element modeling to simulate the viscoelastic deformation behavior of an idealized asphalt mixture. *Geomechanics and Geoengineering,* Vol.2, No.2, pp.77–86.

Cundall, P. A. (1971). A computer model for simulating progressive large scale movements in blocky rock systems. *Proc., Symposium of the International Society of Rock Mechanics,* Nancy, France, Vol.1, Paper No.II-8, pp.129–136.

Cundall, P. A., and Strack, O. D. L. (1979). A discrete numerical model for granular assemblies. *Geotechnique,* Vol.29, No.1, pp.47–65.

Cundall, P.A., Drescher, A. and Strack, O.D.L. (1982). Numerical experiments on granular assemblies: measurements and observations. *IUTAM Conference on Deformation and failure of granular materials* (Eds: Vermeer and Luger), pp.355–370, Balkema.

Cundall, P.A. and Hart, R.D. (1992). Numerical modeling of discontinua. *Engineering Computations,* Vol.9, pp.101–113.

Dai, Q.L, and You, Z. (2007). Prediction of creep stiffness of asphalt mixture with micromechanical finite element and discrete element models. *Journal of Engineering Mechanics,* Vol.133, No.2, pp.163–173.

Dickens, J.G., and Walker, P.J. (1996). Use of distinct element model to simulate behavior of drystone walls. *Structural Engineering Review,* Vol.8, No.2–3, pp.187–199.

Fu, Y. (2005). DEM Simulation and experimental validation of micro-macro behaviors of granular materials using X-ray tomography imaging. *Ph.D. Dissertation,* Louisiana State University.

Fu, Y., Wang, L. and Tumay, M.T. (2007). Experimental quantification and DEM simulation of irregular particle kinematics and local Strains. *Journal of Mechanics and Materials,* Vol.134, No.2, pp.143–154.

Fu, Y.R., Wang, L.B. and Zhou, C.B. (2010). 3D clustering dem simulation and non-invasive experimental verification of shear localization in irregular particle assemblies. *A Special Issue of IJPE on Multiscale Characterization and Modeling of Stone-based Infrastructure Materials.* Guest Editors: Linbing Wang and Erol Tutumluer. In Press.

Gantt, J.A. and Gatzke, E.P. (2005). Discrete element method application for verification of kinetic theory of granular flow in a high shear mixer. Cincinnati, OH, United States: *American Institute of Chemical Engineers,* New York, NY 10016-5991.

Gethin, D.T., Yang, X.S. and Lewis R.W. (2006). A two dimensional combined discrete and finite element scheme for simulating the flow and compaction of systems comprising irregular particulates. *Computer Methods in Applied Mechanics and Engineering,* Vol.195, pp.5552–5565.

Ghaboussi, J. (1990). Three-dimensional discrete element method for granular materials. *International Journal for Numerical and Analytical Methods in Geomechanics*, Vol.14, No.7, pp.451–472.

Golchert, D., Moreno, R., Ghadiri, M. and Litster, J. (2004). Effect of granule morphology on breakage behaviors during compression. *Powder Technology*, Vol.143–144, pp.84–96.

Goldstein, H. (1950). Classical Mechanics. *Addison-Wesley*, Reading, MA.

Hart, R., Cundall, P.A. and Lemos, J. (1988). Formulation of a three-dimensional distinct element model—part II. Mechanical calculations for motion and interaction of a system composed of many polyhedral blocks. *International Journal of Rock Mechanics*, Vol.25, No.3, pp.117–125.

Hibbeler, R.C. (1974). Engineering Mechanics: Dynamics. *Macmillan Publishing Co., Inc.*, New York.

Itasca (2005). PFC Manuals, *Itasca Consulting Group, Inc.*

Iwashita, K. and Oda, M. (1998). Rotational resistance at contacts in the simulation of shear band development by DEM. *Journal of Engineering Mechanics*, Vol.124, No.3, pp.285–292.

Jensen, R.P., Edil, T.B., Bosscher, P.J., Plesha, M.E. and Kahla, N.B. (2001a). Effect of particle shape on interface behavior of DEM-simulated granular materials. *The International Journal of Geomechanics*, Vol.1, No.1, pp.1–19.

Jensen, R., Plesha, M.E., Edil, T.B., Bosscher, P.J. and Kahla, N.B. (2001b). DEM simulation of particle damage in granular media-structure interface. The *International Journal of Geomechanics*, Vol.1, No.1, pp.21–39.

Kim, H. and Buttlar, W.G. (2005). Micro mechanical fracture modeling of asphalt mixture using the discrete element method. *Proceeding of the GeoFrontier Conference*, Advances in Pavement Engineering (GSP 130), Austin, TX, United States.

Kim, Y.R., Allen, D.H. and Little, D.N. (2005). Damage-induced modeling of asphalt mixtures through computational micromechanics and cohesive zone fracture. *Journal of Materials in Civil Engineering*, Vol. 17, No. 5, pp.477–484.

Kim, H., Wagoner, W.P. and Buttlar, W.G. (2008). Simulation of fracture behavior in asphalt concrete using a heterogeneous cohesive zone discrete element model. *Journal of Materials in Civil Engineering*, Vol.20, No.8, pp.552–563.

Komodromos, P.I. and Williams, J.R. (2002). On the simulation of deformable bodies using combined discrete and finite element methods. *Geotechnical Special Publication*, No.117, pp.138–44.

Lewis, R.W., Gethin, D.T., Yang, X.S. and Rowe, R.C. (2005). A combined finite-discrete element method for simulating pharmaceutical powder tableting, *International Journal for Numerical Methods in Engineering*, Vol.62, pp.853–869.

Lin, X. and Ng, T. (1997). A three-dimensional discrete element model using arrays of ellipsoids. *Geotechnique*, Vol.47, pp.219–329.

Liu, Y., and You, Z. (2008). Simulation of cyclic loading tests for asphalt mixtures using user defined models within discrete element method. *Proc., GeoCongress 2008, The Challenge of Sustainability in the Geoenvironment*, ASCE, Reston, Va., pp.742–749.

Liu, Y., Dai, Q., and You, Z. (2009). Viscoelastic model for discrete element simulation of asphalt mixtures. *Journal of Engineering Mechanics*, Vol.135, No.4, pp.324–333.

Mamaghani, I.H.P. (2006). Analysis of masonry bridges: discrete finite element method, *Transportation Research Record*, No.1976, pp.13–19.

Meegoda, J.N. and Chang, K.G. (1994). Modeling viscoelastic behavior of hot mix asphalt (HMA) using discrete element methods. *Proceeding of 3rd ASCE Materials Engineering Conference—Infrastructure: New Materials and Methods of Repair*, San Diego, pp.804–811.

Meegoda, J.N. and Chang, K.G. (1995). Fundamental study on the discontinuities and heterogeneities of asphalt concrete. *Proceedings of 10th Conference*, ASCE, Boulder, CO, USA.

Mirghasemi, A.A., Rothenburg, L. and Matyas, E.L. (2002). Influence of particle shape on engineering properties of assemblies of two-dimensional polygon-shaped particles. *Geotechnique*, Vol.52, No.3, pp.209–217.

Munjiza, A. (2005). The Combined Finite-Discrete Element Method. *John Wiley & Sons, Ltd.*

Munjiza, A., Bangash, T. and John, N.W.M. (2004). The combined finite-discrete element method for structural failure and collapse. *Engineering Fracture Mechanics*, Vol.71, No.4–6, pp.469–483.

Munjiza, A. and John, N.W.M. (2002). Mesh size sensitivity of the combined FEM/DEM fracture and fragmentation algorithms. *Engineering Fracture Mechanics*, Vol.69, No.2, pp.281–295.

Munjiza, A., Latham, J.P. and Andrews, K.R.F. (2000) Detonation gas model for combined finite-discrete element modelling of fracture and fragmentation. *International Journal of Numerical Methods in Engineering*, Vol.49, pp.1495–1520.

Ng, T.T. and Dobry, R. (1994). Numerical simulations of monotonic and cyclic loading of granular soil. *Journal of Geotechnical Engineering*, Vol.120, No.2, pp.388–403.

Ni, Q., Powrie, W., Zhang, X. and Harkness, R. (2000). Effect of particle properties on soil behavior 3-D numerical modeling of shear box tests. *ASCE Geotechnical Special Publications, Numerical Methods in Geotechnical Engineering*, No.96, pp.58–70.

Oda, M. and Kazama, H. (1998). Micro-structure of shear band and its relation to the mechanisms of dilatancy and failure of dense granular soils. *Geoechnique*, Vol.44, No.1, pp.1–19.

Oda, M. and Iwashita, K. (Editors) (1999). Mechanics of Granular Materials: An Introduction. *Elsevier*.

Papagiannakis, A.T., Abbas, A. and Masad, E. (2002). Micromechanical analysis of viscoelastic properties of asphalt concretes. *Transportation Research Record*, No.1789, pp.113–120.

Rothenburg, L. and Bathurst, R.J. (1989). Analytical study of induced anisotropy in idealized granular materials. *Geotechnique*, Vol.39, No.4, pp.601–614.

Rothenburg, L. and Barthurst, R.J. (1992). Micromechical features of granular assemblies with planar elliptical particles. *Geotechnique*, Vol.42, pp.79–95.

Thornton, C. (1992). Applications of DEM to process engineering problems. *Engineering Computations*, Vol.9, No.2, pp.289–297.

Ullidtz, P. (2001). Distinct element method for study of failure in cohesive particulate media. *Transportation Research Record*, No.1757, pp.127–133.

Walton, O.R. and Braun, R.L. (1993). Simulation of rotary-drum and repose tests for frictional spheres and rigid sphere clusters. *Joint DOE/NSF Workshop on Flow of Particulates and Fluids*, Ithaca, NY, Vol.29, pp.1–18.

Wang, L.B., Frost, J.D. and Lai, J.S. (2004). 3D digital representation of the microstructure of granular materials from x-ray tomography imaging. *Journal of Computing in Civil Engineering*, Vol.18, No.1, pp.28–35.

Wang, L.B., Park, J.Y. and Fu,Y.R. (2007). Representation of real particles for dem simulation using x-ray tomography. *Journal of Construction and Building Materials*, Vol.21, No.2, pp.338–348.

Washington, D.W. and Meegoda, J.N. (2003). Micro-mechanical simulation of geotechnical problems using massively parallel computers. *International Journal of Numerical and Analytical Methods in Geomechanics*, Vol.27, No.14, pp.1227–1234.

Williams, J.R. and Pentland, A.P. (1991). Superquadrics and modal dynamics for discrete elements in concurrent design. *Intelligent Engineering Systems Laboratory, Technical Report, Order No. IESL 91-12*, Massachusetts Institute of Technology.

Yoshida, T., Tatsuoka, F., Siddiquee, M.S.A., Kamegai, Y. and Park, C.S. (1994). Shear banding in sands observed in plane strain compression. *Symposium on Localization and Bifurcation Theory for Soils and Rocks (Eds: R. Chambon, J. Desrues and I. Vardoulakis)*, pp.165–179, Balkema.

You, Z. and Buttlar, W.G. (2004). Discrete element modeling to predict the modulus of asphalt concrete mixtures. *Journal of Materials in Civil Engineering,* Vol.16, No.2, pp.140–146.

You, Z. and Buttlar, W.G. (2005). Application of discrete element modeling techniques to predict the complex modulus measurements of asphalt-aggregate hollow cylinders subjected to internal pressure. *Transportation Research Record,* No.1929, pp.218–226.

You, Z. and Buttlar, W.G. (2006). Micromechanical modeling approach to predict compressive dynamic moduli of asphalt mixture using the distinct element method. *Transportation Research Record,* No.1970, pp.73–83.

Zhao, X.B., Zhao, J., Cai, J.G. and Hefny, A.M. (2008). Modeling on wave propagation across fractured rock masses. *Computers and Geotechnics,* Vol.35, No.1, pp.97–104.

CHAPTER **10**

Digital Specimen and Digital Test-Integration of Microstructure into Simulation

10.1 Introduction

X-ray computerized tomography (XCT) imaging and computational simulation has made it possible to characterize the properties of asphalt concrete (AC) through reconstructing its three-dimensional (3D) microstructure and computational simulation based on the 3D microstructure. The digital representation of the real 3D microstructure (not a simulated 3D or 2D microstructure) of a physical specimen is called a digital specimen. It is the digital counterpart of the physical specimen in every required detail. A digital specimen is usually represented as a 3D digital image and naturally includes scientific visualization of the initial/evolved microstructures and behavior related to rutting, fatigue cracking, and thermal cracking. Figure 10.1a presents one digital specimen. It is a stack of 81 slices of a physical specimen non-destructively acquired. The required details at particle level for mechanical modeling are available from the digital specimen. Figure 10.1b represents a visualization of the void structure of the digital specimen presented in Figure 10.1a. It is also a digital specimen if the study is on the permeability and moisture damage of the asphalt mix. A digital specimen can be transferred through the Internet and shared by many researchers for use in multi-purpose studies such as model verification and behavior simulation. With detailed representation of the microstructure of AC, modeling and simulation can be based on the real microstructure of the material; experimental observation of the microstructure evolution can be compared with computational simulation to validate the models and the understanding of the fundamental mechanisms of strength and deformation of AC.

Computational simulation of a mechanical test, which is based on digital specimens and considers every required detail of the microstructure and its evolution, is called a digital test. It is different from conventional computational simulations that assume either a continuum model (Collop et al., 2003) or a simulated microstructure (Chang and Meegoda, 1997, 1999; Buttlar et al., 1999; Buttlar and You, 2001; Bjorn et al., 2002; Sadd et al., 2004). For example, Figure 10.1c presents a simulation of a triaxial compression test (digital test) on the digital specimen, while Figure 10.1d presents a simulation

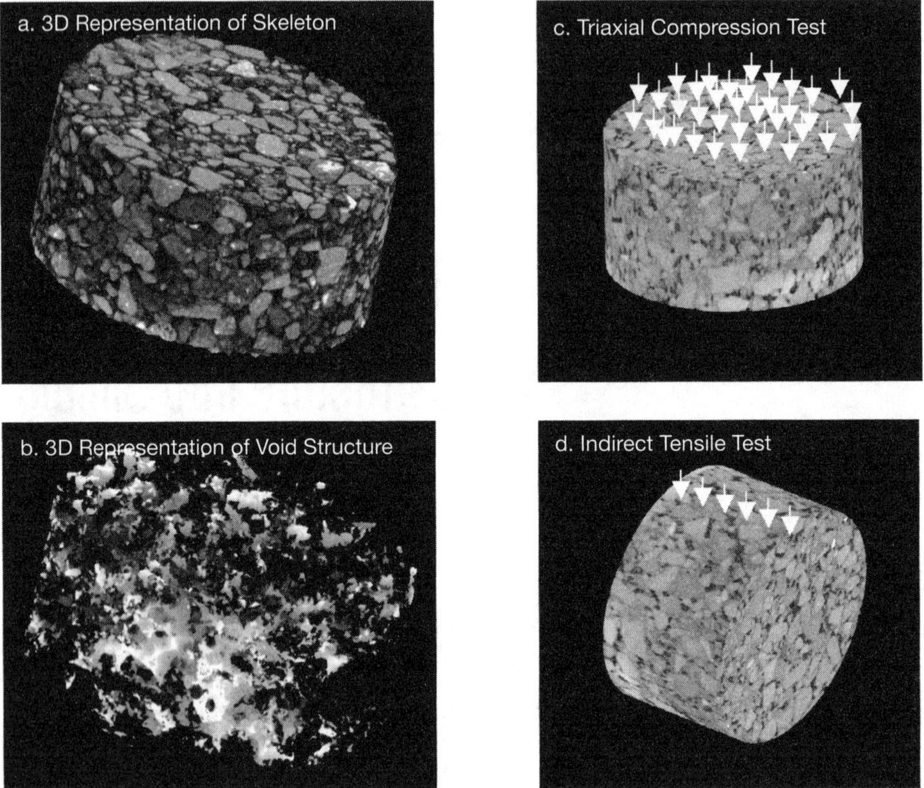

Figure 10.1 Illustration of the concepts of digital specimen and digital test.

of an indirect tensile test (digital test) on the same specimen. One of the advantages of digital specimens and digital testing lies in its capability to conduct multiple tests on the same specimen, saving different sets of equipment. Another advantage is that each constituent can be considered as homogeneous. For example, asphalt binder can be well modeled as a viscoelastic material, while aggregates can be modeled as elastic, so material characterization can be more accurate.

Digital specimen and test techniques permit investigation of the strength and deformation mechanisms of AC in a microscopic view that integrates mechanism identification, numerical simulation, and experimental observations. It represents the trend for the future mix design (Roberts et al., 2003). In addition, the generic approach to integrate microstructure into modeling and simulation can be extended into other materials such as unbound aggregates, cement concrete, and other materials with significant heterogeneity in material structure.

10.2 Digital Simple Performance Test of AC

Simple performance tests are used to characterize the viscoelastic, viscoplastic, and continuum damage mechanics (CDM) material properties in pavement design. Indirect tensile tests and dynamic modulus tests are widely used to predict the mechanical properties of AC in terms of resilient modulus, phase angle, and dynamic modulus. In practice, the interpretation of the test is simplified using elasticity or viscoelasticity. For an indirect tensile test, the simplified theoretical solutions for the plane stress condition along the horizontal and vertical diameters (Hondros, 1959) are usually used to obtain stresses. These stresses, in conjunction with measured strains, are used to calculate the modulus. However, due to the heterogeneity, the stresses are not uniform as the theoretical predictions of elasticity or viscoelasticity. Therefore, the modulus obtained often demonstrated large scattering. Accounting for the heterogeneity is necessary.

AC mixture shows temperature and time dependency under loading. Viscoelastic (Schapery, 1984, 1990; Park et al., 1996; Zhang et al., 1997; Lee and Kim, 1998) and viscoplastic (Perzyna, 1966; Seibi et al. 2001; Schwartz et al., 2002; Collop et al., 2003; Tashman et al., 2005) material models were introduced into the study of asphalt concrete mixture. These models are based on additive strain decomposition. Kichenin et al. (1996) proposed a model with two-dissipative mechanisms, associating an elastic-viscous and an elastoplastic model in parallel. This stress-overlay-based model reflects the stress transfer between the two mechanisms and is suitable to describe materials consisting of multiple constituents. However, these continuum mechanics models have difficulties capturing the material behavior due to the microstructure of different mixes.

X-ray-imaging-based digital testing was developed to take into account this material heterogeneity and stress nonuniformity (Zhang, et al., 2006). In order to study the viscoplastic material characteristics of asphalt mixture, the two-layer viscoplastic model was used to characterize the mixture behavior, serving as the basis for the numerical performance test. This chapter will first present a macroscopic parametric study to evaluate the sensitivity of model parameters to the deformation response of the sample. It then presents microscopic models for indirect tensile testing and dynamic modulus testing considering the viscous behavior of the asphalt binder or mastics and the phase configuration of the AC mixture to build a realistic digital specimen and digital tests that enable the characterization of mixture properties as well as properties of individual materials. Finally, it presents the digital APA test and a perspective on digital mix design method currently funded by the National Science Foundation.

10.2.1   Digital Indirect Tensile Test

Indirect tensile testing is widely used in evaluating the rutting and fatigue properties of AC. This section explains the development of the digital indirect tensile test. As described in the previous chapters and sections, a realistic material model for asphalt binder should be a viscoplasticity model to represent the contribution of the viscoplasticity of asphalt mixture. From a micromechanics point of view, asphalt mixture can be considered a composite including cavity and aggregate inclusions in the binder matrix. As aggregates are elastic (under normal traffic loading, but not at the contact points), the viscoplasticity of the mixture is actually from the binder. Therefore, a viscoplasticity model of the same structure can be used for both the binder (mastics) and the mixture, with different material constants. In the following sub-section, a viscoplasticity model is first tested on an asphalt mixture at the macroscopic scale (homogeneous continuum). After its verification, the same model is then used for representing the constitutive behavior of the mastics of binder in establishing the digital specimen and digital tests.

10.2.1.1   Finite Element (FE) Model

An FEM indirect tensile test was built to reflect the actual displacement and traction boundary of the test. Due to the symmetrical geometry of the specimen, only a quarter of the sample was modeled (Figure 10.2a, the whole geometry must be used for the digital model due to non-symmetric microstructure). Distributed traction force was used to represent the loading in order to avoid contact problems. Roller supports were applied to each node at the symmetric axis. The location of the displacement measurement point was determined according to actual testing configurations. The loads are modeled as a repeated haversine pulse. The amplitude and frequency were set to be the same as those in the actual testing (Figure 10.2c).

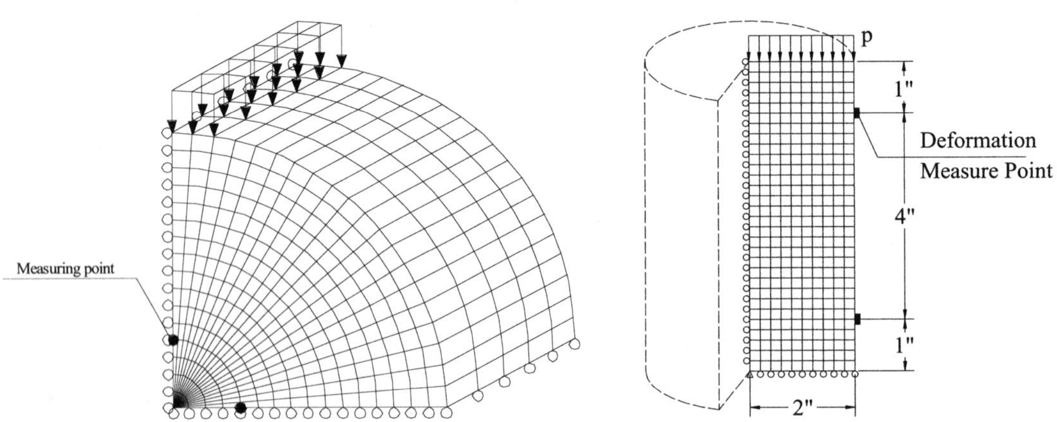

a. FE Model Indirect Tensile Test   b. FE Model Dynamic Modulus Test

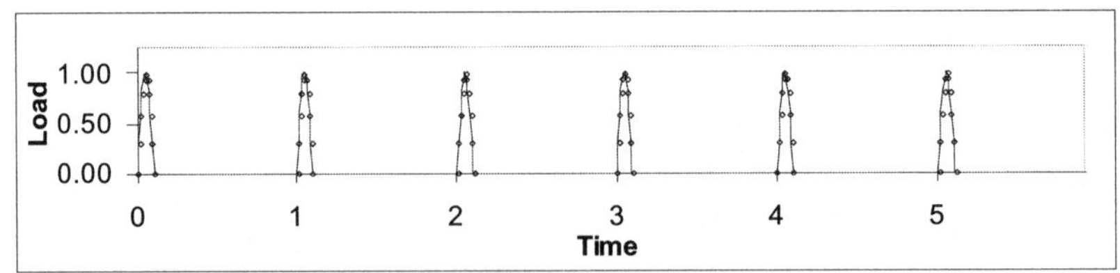

c. Indirect Tensile Test Pulse Loading

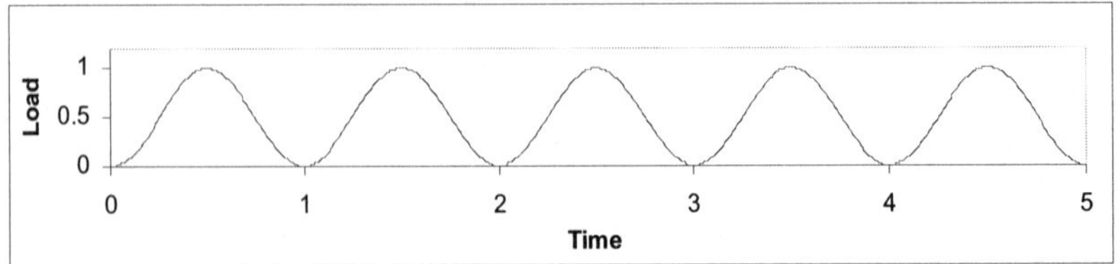

d. Dynamic Modulus Test Sinusoidal Loading

FIGURE 10.2   Model components for indirect tensile test and the dynamic modulus test.

The simulation model of dynamic modulus test is shown in Figure 10.2b. Due to the axisymmetric configuration of the macroscopic model, a four-node bilinear axisymmetric solid element with reduced integration and hourglass control was used. The loads are modeled as distributed boundary traction with sinusoidal repetition (Figure 10.2d) on the top edge of the sample.

10.2.1.2 Material Model

The stress-overlay-based two-layer viscoplastic model (ABAQUS 1995) was used to describe the constitutive relationship (Figure 10.3) of the AC mixture in the macroscopic study, while for asphalt binder only in the microscopic study.

The total stress and strain in the networks is expressed in Equation 10-1.

$$\sigma = f\sigma_{VE} + (1-f)\sigma_{EP} \qquad \varepsilon = \varepsilon_{EP} = \varepsilon_{VE} \tag{10-1}$$

Where subscript $_{VE}$ represents the viscoelastic network and $_{EP}$ represents the elastoplastic network, f is the stiffness ratio of the viscoplastic network and is expressed in the following equation.

$$f = \frac{K_{VE}}{K_{EP} + K_{VE}}$$

Where K is the instantaneous modulus. The von Mises type plasticity was used in the elastoplastic component as in Equation 10-2.

$$\dot{\varepsilon}'_{ij} = \frac{1}{2\mu} S_{ij} + \frac{1-2v}{E}\sigma_{ij}\delta_{ij}, \qquad \dot{\varepsilon}''_{ij} = \dot{\bar{\varepsilon}}^{pl}\cdot\frac{3}{2}\frac{S_{ij}}{q} \tag{10-2}$$

The strain creep law, as shown in Equation 10-3, was used.

$$\dot{\varepsilon}''_{ij} = \dot{\bar{\varepsilon}}^{cr}\cdot\frac{3}{2}\frac{S_{ij}}{q}, \qquad \dot{\bar{\varepsilon}}^{cr} = \left(Aq^{n}\left[(m+1)\bar{\varepsilon}^{cr}\right]^{m}\right)^{\frac{1}{m+1}} \tag{10-3}$$

The hardening follows Equation 10-4.

$$Y = \sigma_{yp} + Be^{C\varepsilon_{EP}} \tag{10-4}$$

Where $\dot{\bar{\varepsilon}}^{cr}$ is the uniaxial equivalent creep strain rate, $\bar{\varepsilon}^{cr}$ is the uniaxial equivalent creep strain, S_{ij} is the deviatoric stress tensor, q is the uniaxial equivalent deviatoric stress, and A, n, m are material constants. There are seven parameters that need to be calibrated: elastic modulus E, Poisson's ratio μ, modulus ratio f, viscous parameters A, m, n, the initial yield stress σ_{yp} and hardening parameters B and C.

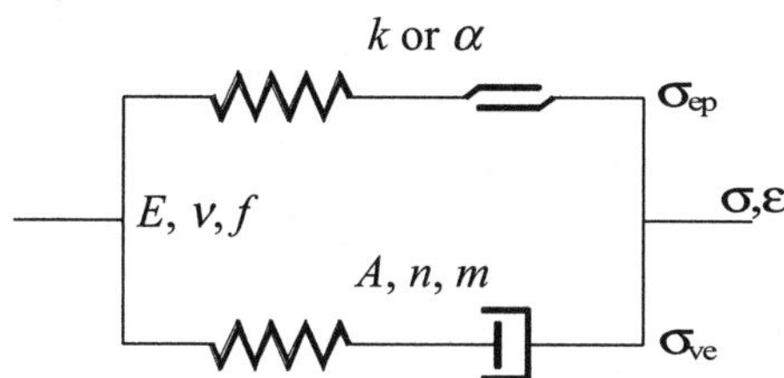

FIGURE 10.3 Two-layer Viscoplastic model (ABAQUS, 1995).

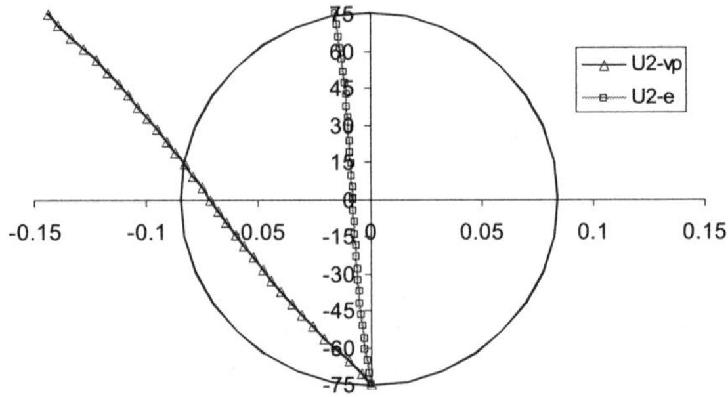

Figure 10.4 Displacement for elastic and viscoplastic solutions.

10.2.1.3   Numerical Experiment

The simulation was conducted on the Inferno2 parallel computing facility at Virginia Tech. The stress distributions along diameters are almost the same as the elastic solution. However, the total displacement is apparently different. Figure 10.4 shows the vertical displacement along the vertical diameters for the elastic solution and the viscoplastic simulation.

10.2.1.4   Physical Experiment

Some physical indirect tensile tests and dynamic modulus tests (see Figure 10.5) conducted at the Virginia Tech Transportation Institute (VTTI) were used to compare with the simulation results. In the indirect tensile test, only the final part of the cumulative data was recorded. Consequently, only the results of the dynamic modulus test were shown in this study. Test results at 5°C and 40°C were used to back calculate material parameters for 3D stress conditions.

10.2.1.5   Parameter Back Calculation

The concept of the back calculation is to obtain a set of parameters that renders the closest deformation profile to the testing results with the objective function of Equation 10-5.

$$D(u) = \sum_{i=1}^{N} (\delta_i^m - \delta_i^p)^2 \tag{10-5}$$

Where N is the number of time points for the deformation history, δ_i^m, δ_i^p are measured and predicted deflections, respectively. The instantaneous response was used to estimate the elastic modulus and the stiffness ratio f between the two mechanisms.

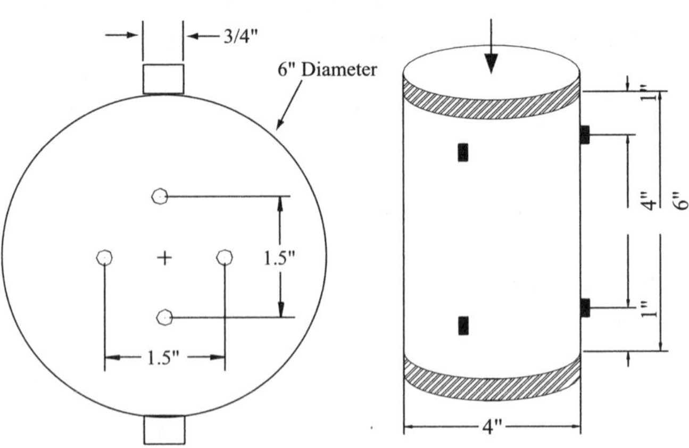

Figure 10.5 Illustration of setups for indirect tensile test and dynamic modulus test.

First, an initial simulation profile was obtained with a set of parameters that are typical values for AC. Second, sensitivity analysis of each parameter was conducted to find the effect of each parameter to the simulation results. Finally, a series of simulations were carried out to obtain the set of parameters leading to the minimization of the objective function. More details on back calculation are presented in Section 10.3.3.

In order to simplify the procedure, Poisson's ratio is taken as 0.3 for both networks and linear hardening was assumed for the elastoplastic network. The effect of each parameter on the final deformation is plotted in Figure 10.6 and the percentage change of the deformation caused by a 10% change of each individual parameter is listed in Table 10.1. The parameters A, m, and σ_{yp} have major effects on the permanent deformation profile.

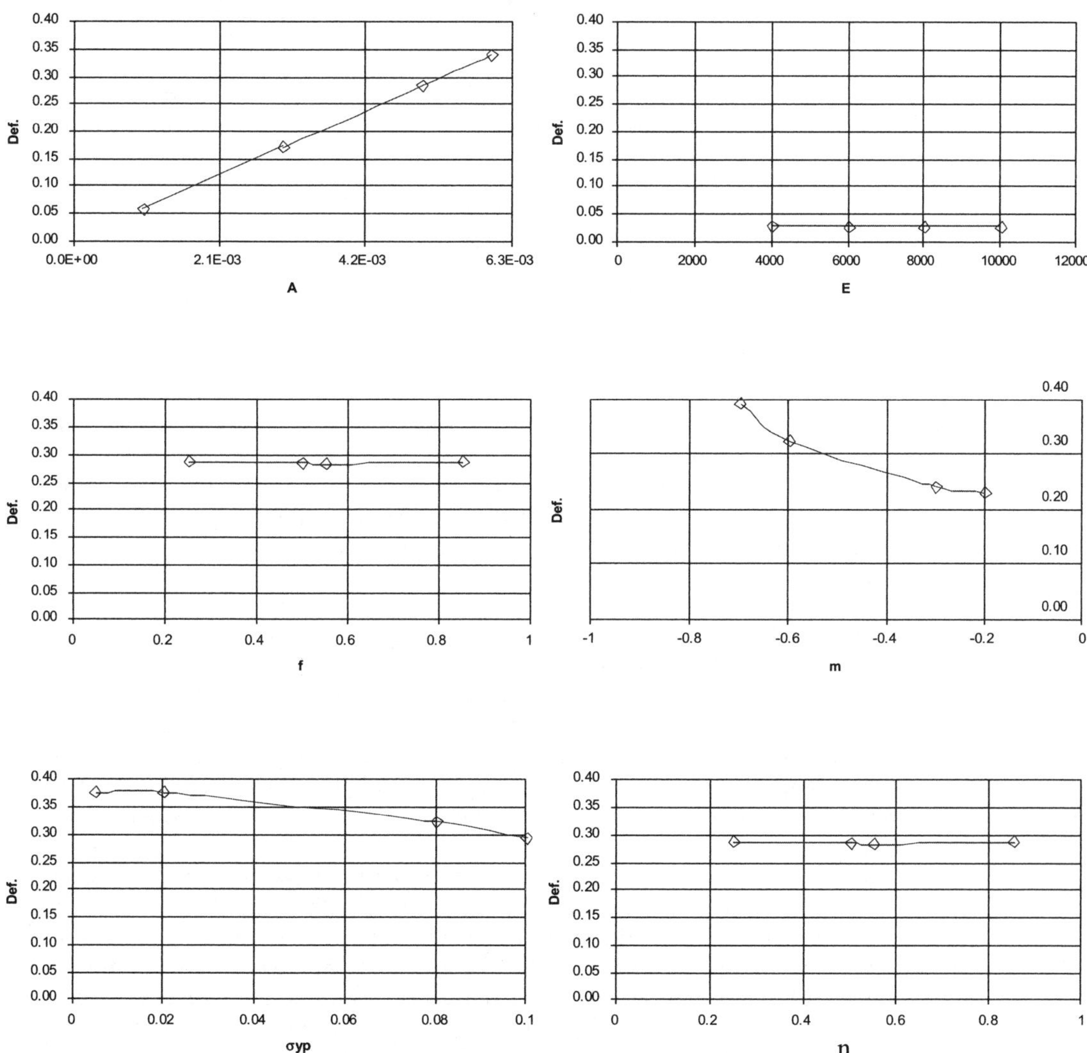

FIGURE 10.6 Parameter sensitivity to deformation profile.

	Initial Parameter	Average Percentage Change
E	6000	1.06%
v	0.35	
σ_{yp}	0.08	0.04%
A	0.005	9.76%
f	0.55	0.28%
m	−0.2	3.41%
n	0.6	3.60%

TABLE 10.1   Percentage change of deformation caused by a 10% change in parameters.

Parameters		5°C	40°C
E	(N/mm²)	30,000	4,100
v		0.3	0.35
σ_{yp}	(N/mm²)	0.08	0.08
B	(N/mm²)	0.8	0.8
C		0.6	0.6
A		1.0E-06	4.0E-05
n		0.8	0.8
m		−0.2	−0.2
f		0.85	0.85

TABLE 10.2   Optimized material parameters using simulation.

By using optimized parameters (Table 10.2) in the material model for 5°C and 40°C, respectively, the simulation results and the testing results are plotted against time in Figure 10.7. From the loading and deformation profiles, it is possible to get the dynamic modulus and the phase angle for each testing frequency. Figure 10.8 shows the effect of the stiffness ratio parameter f, on the lag of strain and the magnitude of the response.

10.2.1.6   Microscopic Study

3D microscopic models can be built for indirect tensile testing and dynamic modulus testing considering phase variation of different mixtures to achieve the digital specimen and the digital test functionality. The microstructural information is from the X-ray scanning of a real sample. A linear elastic material model is used for aggregates, while the two-layer viscoplastic material model is used for asphalt binder. The displacement of the loading point is monitored and the displacement history is recorded along with the loading history.

Due to computing time and the limitation of the computer memory and disk space, especially for repeated loading simulation, all the images with original 512 × 512 resolution were re-digitalized before the building of the finite element model. This is done by maintaining the volume fractions of both the voids and aggregates during the conversion. A FORTRAN program was developed to carry out the conversion of the microstructure and the generation of the finite element model (Figure 10.9).

In order to validate these concepts, mixtures from the WesTrack project were numerically tested following the procedure above. Three mixtures were subjected to the

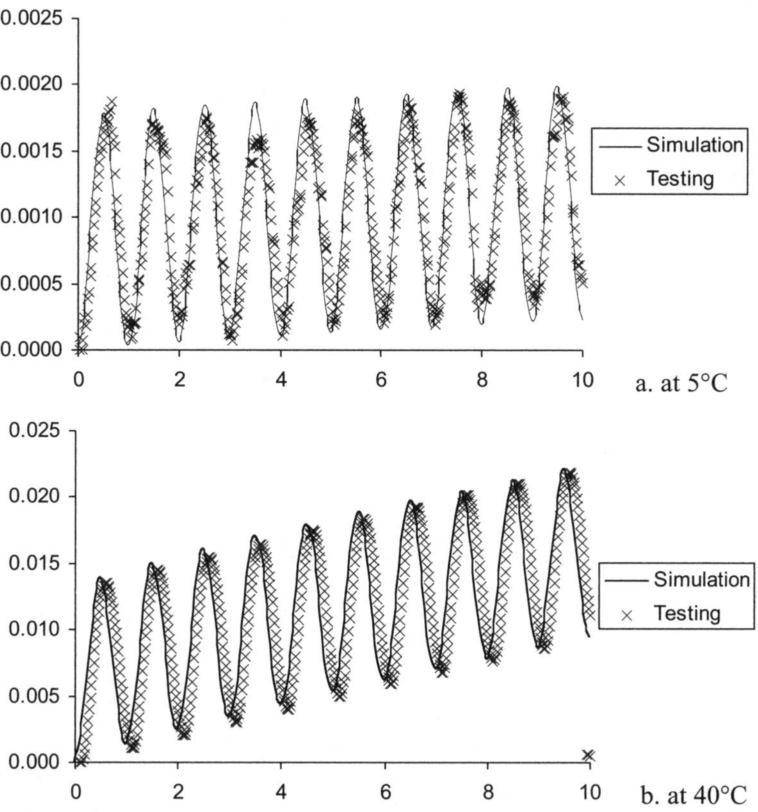

Figure 10.7   Simulation and experimental results at 5°C and 40°C for macroscopic dynamic modulus testing, an actual test with 1Hz frequency was used.

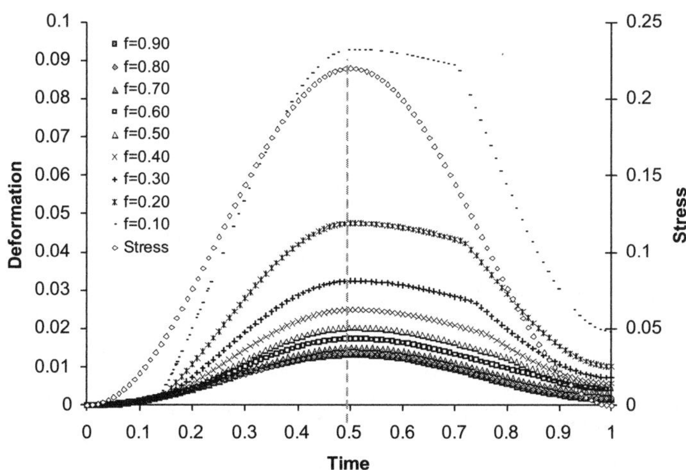

Figure 10.8   Deformation variation for parameter f ($0 < f < 1.0$), the plot shows the effect of the parameter f on the phase angle.

FIGURE 10.9 Microscopic FE model for the indirect tensile test.

same loading pattern and magnitude so that comparable deformation responses for different mixtures could be obtained. The results plotted in Figure 10.10 indicate the different strain responses for different mixtures with the same material properties for each component. The fine-plus mix, as expected, demonstrated larger deformation than the fine mix. However, the coarse mix experienced less deformation than the fine plus mixture. This may be due to the selection of material parameters and the interpretation of the phases from scanning images based on volume fractions. Meanwhile, three levels of re-digitalization are conducted and numerically tested with the same set of material parameters. The simulation of five loading cycles with the same sample size and loading pattern (Figure 10.11) shows that the loss of accuracy when converting images from high resolution to low resolution was minimized by keeping consistent phase fractions.

The parameters can be calibrated using the back-calculation method following the same approach as in the macroscopic study illustrated previously. As for the time history, the difference of the deformation for these three mixes increases with the increase

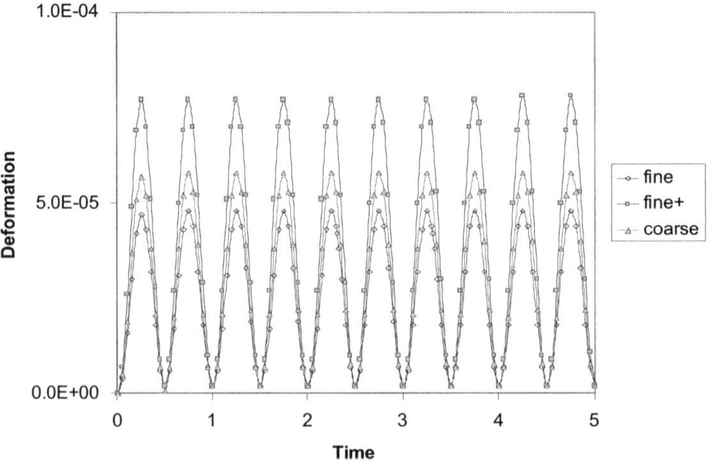

FIGURE 10.10 Microscopic numerical testing of samples from WesTrack mixtures.

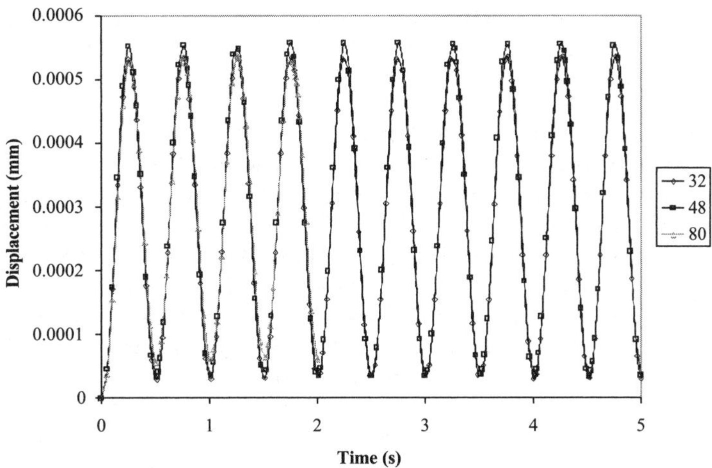

FIGURE 10.11   Displacement profile for mesh sizes 5, 3, and 2 mm.

of the loading cycles. This shows the importance of the sample microstructure when dealing with a large number of cycles of repeated loading, which is the real situation of asphalt pavement.

The simulation results actually indicate that the rate dependent material model for asphalt binder has rendered realistic simulation of the response for the asphalt mixture in comparison with the experimental results. In the sense of engineering estimation, the resolution change may not cause significant effects on the material parameter estimation as long as the major configurations and the overall microstructure (void's volume fraction and aggregate's volume fraction) of the sample are kept at the same level.

10.2.2   Digital Compression Test

The above algorithms and procedures can be further verified with laboratory testing and microscopic-level analyses. Small AC samples that can be scanned with a microtomographic scanner were prepared with different aggregate volume fractions. The purpose of the experiment is to study the effect of aggregate volume fractions on the behavior of the sample and to further develop the concept and verify the algorithm and method. An X-ray microtomographic scanner was used in the study to obtain the microstructure of small asphalt samples and the incorporated testing stage was used to conduct the uniaxial compressive test at the same time when the specimen was scanned.

10.2.2.1   Testing Samples

Eight samples were prepared with four levels of aggregate volume fraction: 0%, 5%, 25%, and 50% (corresponding mass ratios are 0:1, 0.13:1, 0.87:1, and 2.65:1). All samples are cylinders with 6 mm diameters and 9 mm heights. The actual aggregate volume fractions may be slightly different from these design values due to the presence of voids, which are difficult to control for the small samples. The actual volume fractions of aggregates and voids can be quantified through 3D image analysis. The size of aggregates is the same for all the samples. Aggregates passing through a No.16 sieve and retained on a No.30 sieve (i.e., 0.6 ~ 1.18 mm) were used. An aluminium mould was designed to make samples in batches. Mixtures were made by controlling the weight of each component proportionally. The weight of each component was used to calculate the aggregate volume fractions.

10.2.2.2 Sample Scanning and Testing

The SkyScan microtomographic system and testing stage (SKYSCAN 2006) were used in this study to obtain the microstructure of small samples and to conduct the uniaxial compressive test. The X-ray source is 20 – 50 kV with maximum power of 40 W. The pixel size is from 6 to 30 µm. The material testing stage can perform compression, tension, and torsion tests. The testing stage applies displacement to the top and bottom of a sample. Samples with aggregates were scanned before testing. For these small samples, the whole sample was scanned with one rotation using the cone beam mode. The resolution of the reconstructed images was 512×512 pixels. All samples were tested at room temperature under uniaxial compression. The testing stage is displacement controlled and the speed of the displacement was set at 15 µm/s. When the resistant force reached the maximum allowable value of the testing stage, the application of displacement was stopped. For the small sample used in this study, the application of displacement was manually stopped at a stress of 0.8 MPa, i.e., 22.64N force on the sample. Prior to each test, the testing stage was calibrated.

10.2.2.3 Sample Reconstruction, Test Simulation, and Parameter Back Calculation

After scanning, sectional images of the small sample were reconstructed. The reconstructed sectional images were used to conduct 3D analysis, which determines volume fractions for aggregates, binder, and voids. They were used to obtain the threshold value that will be used in the generation of the FE model. 3D visualization was carried out to generate the digital representations of these physical samples.

Reconstructed sectional images were read with programs developed in Interchange Data Language (IDL) and transferred into a data file read by a FORTRAN program to generate the required inputs for the FEM. In this study, adaptive meshes were used for asphalt binder to account for large distortions that may occur within thin asphalt layers between two aggregates.

The simulation of the uniaxial compression test was conducted as a small-scale dynamic modulus test. The two-layer viscoplastic material model was used for asphalt binder and an elasticity model was used for aggregates. Loading plates were modeled as rigid surfaces. The movement of the top rigid surface was controlled with a downward displacement velocity of 15 µm/s, which is the same as that in the physical test.

Parameters in the material model were estimated with back calculation based on a parameter sensitivity study. Material parameters were calibrated from one set of samples. Then, the same set of material parameters was used in the simulation of other samples with different aggregate volume fractions.

Figure 10.12 shows samples being scanned within the testing stage and a reconstructed 3D image. The height of the sample can be measured on the screen. In the 3D image, the black phase is aggregates.

Through reconstruction of the 3D microstructure and inclusion of the microstructure into FEM simulation (the digital test, Figure 10.13), the compression test can be modeled. In this model, the asphalt binder is modeled as viscoplastic (Kichenin et al., 1996); aggregates are modeled as elastic. In this study, specimens of the pure binder, 5% aggregates (by volume of aggregate binder mixture), 25% aggregates, and 50% aggregates were tested. In each case the mastic microstructure is scanned and represented in

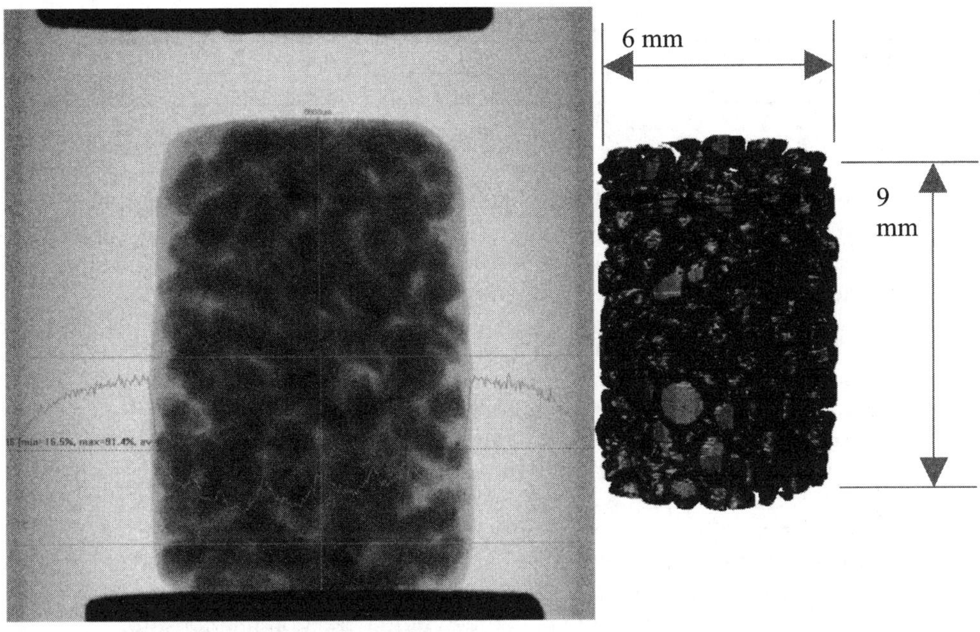

Figure 10.12 Samples during scanning and reconstructed 3D visualization.

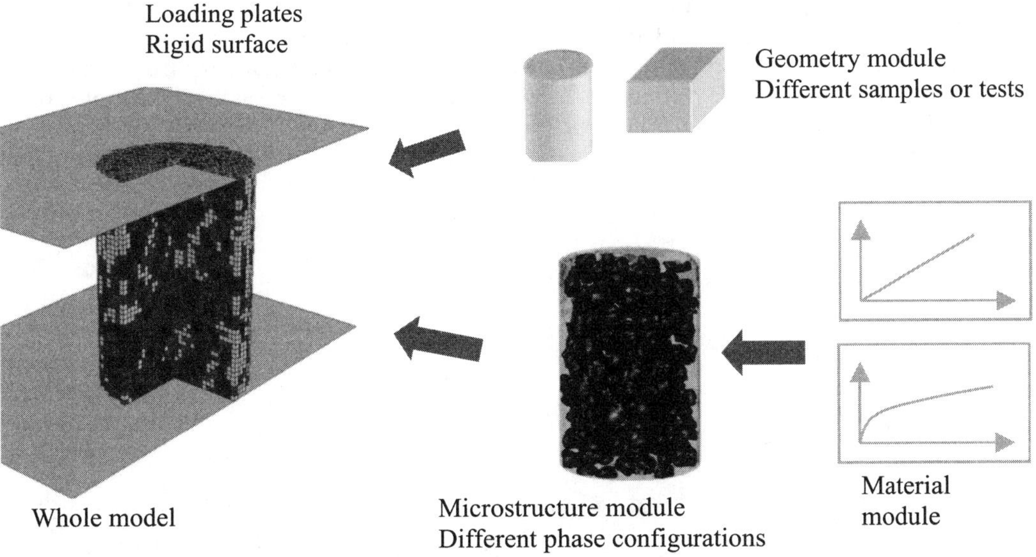

Figure 10.13 Illustration the digital test technique.

Sample ID		2a	2b	3a	3b	4a	4b
Volume fraction	Aggregates	5.87	4.04	22.42	23.49	50.52	49.91
	Voids	3.48	1.18	1.77	2.46	12.97	11.26

TABLE 10.3   Volume fraction (%) of aggregates and voids.

the FEM simulation. The pure binder specimen and the 5% aggregate specimen were tested for back-calculating the material constants of the elastic and viscoplasticity models for aggregates and the binder, respectively. Then these material constants were used for modeling the other two specimens.

10.2.2.4 Volume Fraction of Aggregates and Voids

The actual volume fractions of aggregates and voids were obtained through 3D analysis of reconstructed images. The volume fractions of aggregates and voids for each sample are listed in Table 10.3. These values are important for the generation of the FEM.

For these small samples, the void volume fraction may be related to the aggregate content. If more samples were made, a more accurate trend between the volume fractions of the two phases might be obtained. This may imply that the void content can be controlled by the aggregate content based on some statistic study results from 3D image analysis. However, it should be pointed out that other factors such as the shape and gradation of the aggregate will affect the volume fraction relationship.

10.2.2.5 Sample Testing

The maximum displacement recorded varies from 1 to 3 mm. However, only the beginning part of the results was used to compare with finite element simulation results and to back-calculate material parameters. Due to the lack of a damage or softening component, simulation at large displacement may not be reasonable.

According to the force-displacement relationship of the pure binder sample, shown in Figure 10.14 (left), the stress-strain relationship for the pure binder sample can be established, from which the elastic modulus, yielding stress, and hardening property for the binder can be estimated. They were used as initial parameters in the back calculation. After adding aggregates into the asphalt binder, the behavior of the composite material is significantly different from that of pure binder, especially for samples with higher aggregate volume fractions. However, samples with 5% aggregate are only slightly different from pure binder samples in terms of force-displacement response.

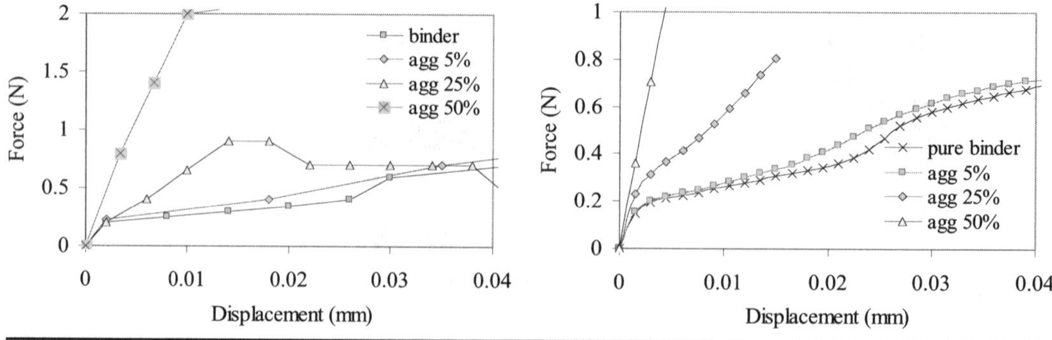

FIGURE 10.14   Force-displacement curves from test (left) and simulation (right).

FIGURE 10.15   3D visualization of small samples with different aggregate volume fractions.

10.2.2.6   3D Visualization and Test Simulation

The visualization of reconstructed 3D digital samples representing different aggregate volume fractions is shown in Figure 10.15. In the visualization, voids and binder were set to be partially transparent. Similarly, individual phases can be visualized separately. This is particularly useful in the study of void structure or connectivity.

The simulation of the test was conducted at Virginia Tech's high performance computing facility which utilizes parallel computing and provides memory and disk space for solving large problems. The model has approximately 30,000 elements. The mesh size is small enough to capture the aggregates and voids and to reflect their effects on the deformation response. The simulative finite element models with sample microstructures are shown in Figure 10.16.

The whole simulation model includes the deformable sample and the rigid loading surfaces. When comparing the aggregate configurations before and after loading from the displacement contour plot (Figure 10.17), one can observe some areas with much larger deformations associated with aggregate rotation. They are usually close to places with a high concentration of voids.

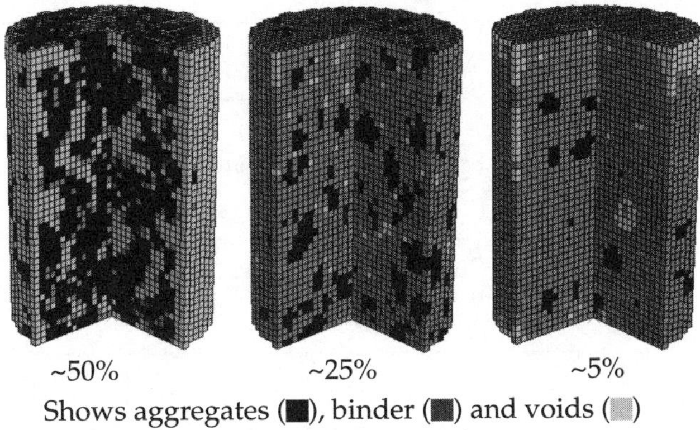

FIGURE 10.16   FE model for small samples, one quarter of the digital sample is taken out to show the internal microstructure.

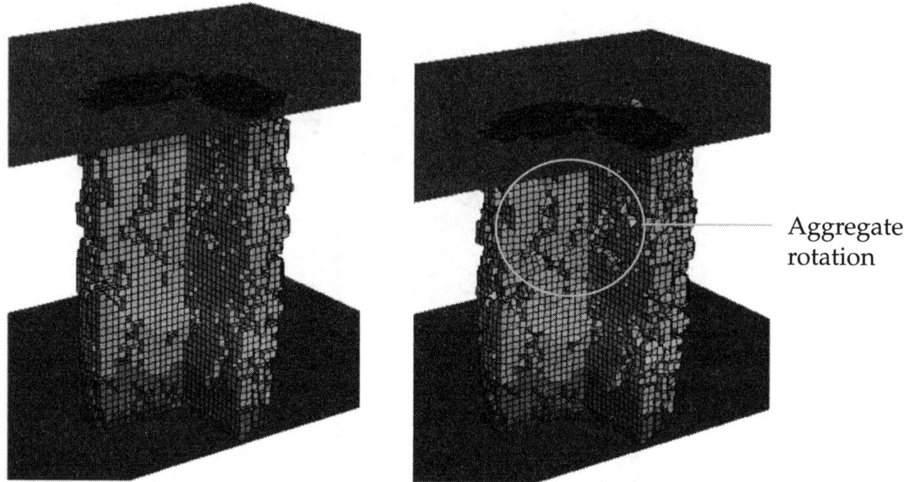

FIGURE 10.17   Finite element simulation model for the uniaxial compressive test of small asphalt samples, (left) displacement before loading, (right) displacement after loading.

From the stress-strain relationship established in the test of the pure binder sample, one can obtain the initial material parameters for the simulation. Based on the simulation and test of the pure binder sample, material parameters were optimized by minimizing the objective function. Material parameters obtained through back calculation are listed in Table 10.4. They were used in all the simulations for samples with different aggregate volume fractions. Parameters for the elastic model used for aggregates were obtained by adjusting the parameters of the elastic model only in the simulation of the sample with a 5% aggregate volume fraction. The back-calculated parameters for aggregates are also listed in Table 10.4. However, the viscoplastic material parameters remained unchanged during the optimization of the elastic part. This made the calibration relatively easier. The back-calculated parameters listed in Table 10.4 were used in the remaining simulations, i.e., for samples with aggregate volume fraction of 25% and

Parameters			**Back-calculated Values**
Asphalt binder	E	(N/mm²)	31.83
	μ		0.45
	σ_{yp}	(N/mm²)	0.0070736
	B	(N/mm²)	0.028294
	C		0.004
	A		1.0E-6
	n		0.8
	m		−0.18
	f		0.85
Aggregates	E	(N/mm²)	210
	μ		0.167

TABLE 10.4   Material parameters from back calculation.

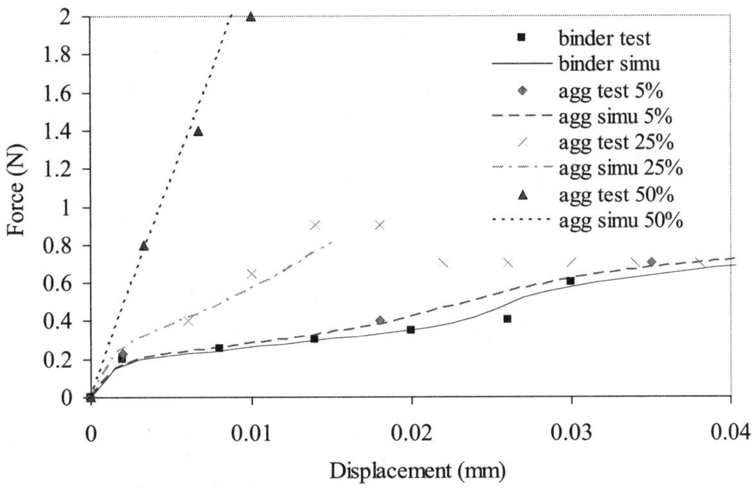

FIGURE 10.18 Simulation results with calibrated material parameters versus testing results.

50%. The force-displacement curve for the simulation with these parameters is presented in Figure 10.14 (right).

By putting the results of physical tests together with those of numerical simulation, the force-displacement processes were compared in Figure 10.18. The consistency indicates the validity of the method. It is clear that the results between simulation and test are close, especially for samples with lower aggregate content. The material parameters obtained from tests of the pure binder sample and the sample with 5% aggregate are able to be used in the simulation of other samples, which have different aggregate contents but the same material properties. It may also imply that material parameters obtained from individual material tests could be used in the simulation of composite material with its real microstructure. In this case, the real microstructural configuration is

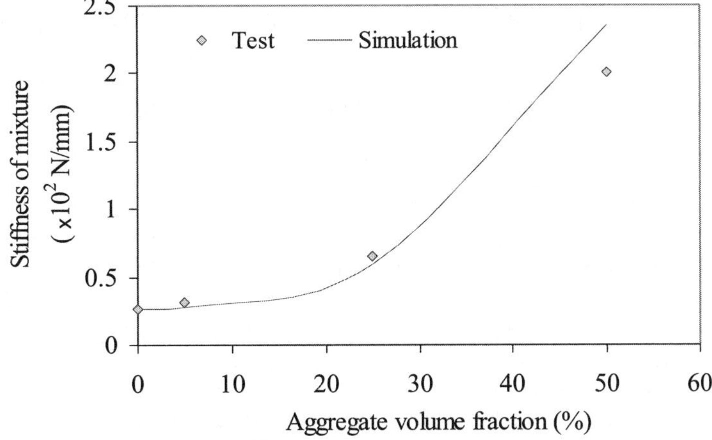

FIGURE 10.19 Stiffness versus aggregate volume fraction for small samples at the deformation of 0.01 mm.

very important in the simulation of composite materials besides the properties of the individual component.

In both the physical and simulation of the test, the stiffness of the asphalt binder is fairly small compared with samples with aggregates. The stiffness of each sample was calculated at a displacement of 0.01 mm from both physical and simulative test results. The relationship between stiffness and aggregate volume fraction, shown in Figure 10.19, demonstrated a trend of stiffening with aggregate content. It also indicates an increased rate of stiffness enhancement with the increase of aggregate volume fraction.

10.3   Digital Test for Simulative Test (APA Test)

Following the same philosophy, the digital specimen and digital test techniques can be applied to simulative tests such as the Asphalt Pavement Analyzer (APA) test. The simulative tests mean applying a repeated scaled (could be full scale) load on a scaled wheel (could be full scale) onto a scaled pavement section (could be full scale) to induce pavement responses and to evaluate its performance. The simulative tests include the full-scale accelerated pavement test, model mobile load simulator, the Hamburg test and the APA test. In this section, the APA test is used as an example for developing its digital counterpart, the digital specimen, and digital test. Figures 10.20a, b, and c illustrate the three sequent steps in building the digital specimens, applying boundary conditions and loading conditions. Figure 10.20d illustrates how a digital specimen is taken from a 3D image.

In this application, a slightly different method is adopted. Test results are compared statistically. Two specimens (150 mm diameter) for each of the three mixes were cored from the WesTrack project (Wang et al., 2001). The six specimens were then scanned using XCT. During the construction process, specimens were manufactured and then tested in an APA. The digital tests are then based on the digital specimens from the six scanned cores. During the test, a rolling wheel applied load to the specimens through a pressurized hose placed on the specimens, and the specimens with the holding mold were fixed in a temperature-controlled chamber, so that the specimens could be tested under the specified temperature condition. For this study, the applied load was 100 lbs. The pressure in the hose was 100 psi, and the testing temperature was 40°C. For a standard APA test, a typical 8000 cycles of loading are applied to each specimen. Five digital specimens were cored from different locations of the same scanned specimens as illustrated in Figure 10.20d. The resolution of the images was 0.29 mm/pixel. The spacing between two images was 1 mm. Based on the resolution of the image, the particle size smaller than 0.29 mm is invisible.

10.3.1   Computational Efficiency

The objective of the image processing and analysis step is to decrease the image size to improve the computational efficiency. The size of the original images is 512×512 pixels. If the original images are used for simulation with 1 pixel (voxel) as one element, the computation time will be too expensive. Figure 10.21 shows the relationship between the computation time and the number of elements. The computation time drastically increases with the increase of the number of elements. Therefore, the re-digitized images with reduced pixel numbers were used in the simulations. Image analysis software, Image-Pro Plus, was used to process the images. An analysis was conducted to

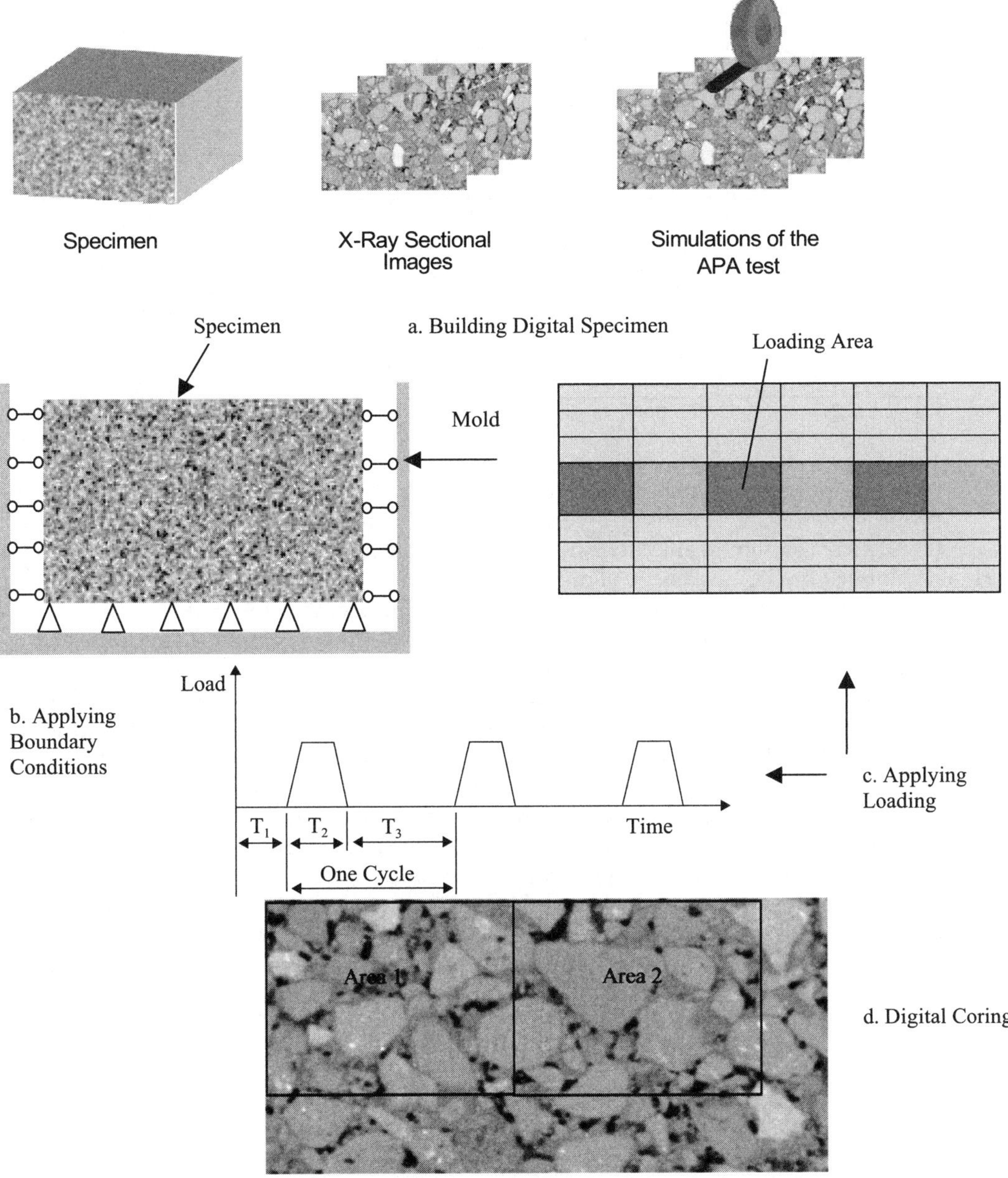

FIGURE 10.20   Process of building the digital specimen and test.

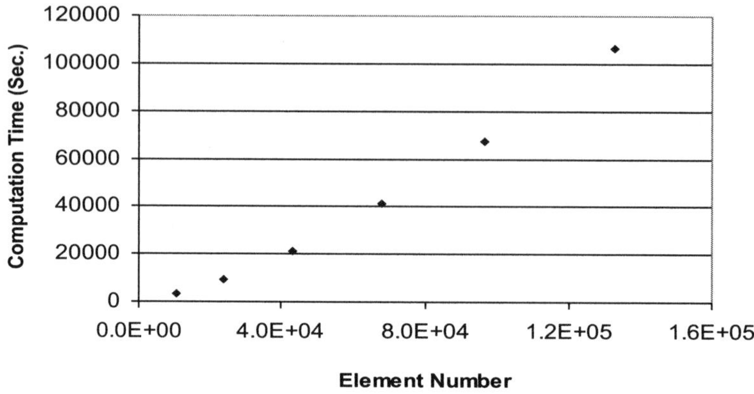

FIGURE 10.21   Computation time versus number of elements.

compare the different digitized ratios on the influence of overall deformation. An appropriate ratio was selected to reduce the element number without loss of computational accuracy significantly. The deformations obtained from the images with different resolutions are compared in Figure 10.22. After the comparison, the image with the resolution of 0.97 mm/pixel (30% of original image size) was selected.

Through the above process, the number of elements was drastically reduced. However, to simulate the APA test with many loading cycles, the computation time is still very expensive. To solve the difficulty, an analysis was conducted to study the affecting area under the loading application. From St. Venant's principle, a load applied to the material will only have local effects around the location where it is applied, and for the area that is far away, the responses of the material are not significantly affected. Based on this principle, simulations with certain loading cycles were conducted to determine the affecting area. Figures 10.23 and 10.24 show the affecting areas after the application of load in vertical and horizontal directions, respectively.

The affecting area was determined based on comparison of the deformations with different coring areas. At first, the deformations for the coring area with different widths and original height of the specimen were compared in Figure 10.25 to determine the

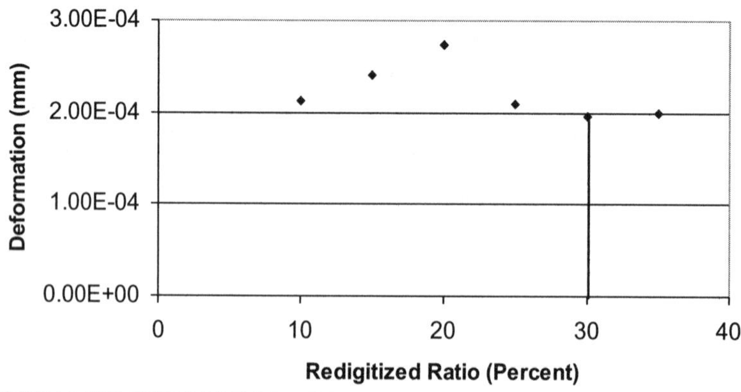

FIGURE 10.22   Deformations versus digitized ratio.

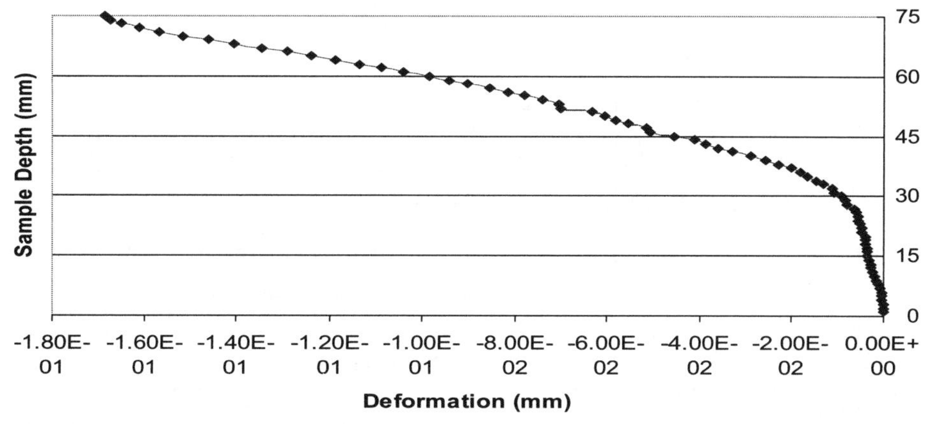

FIGURE 10.23   Affecting area in vertical direction.

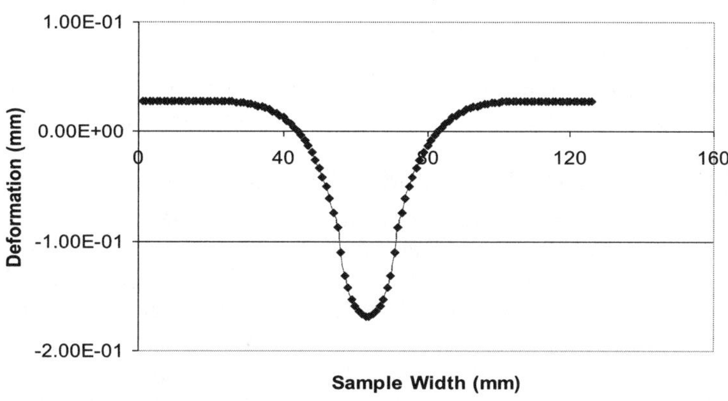

FIGURE 10.24   Affecting area in horizontal direction.

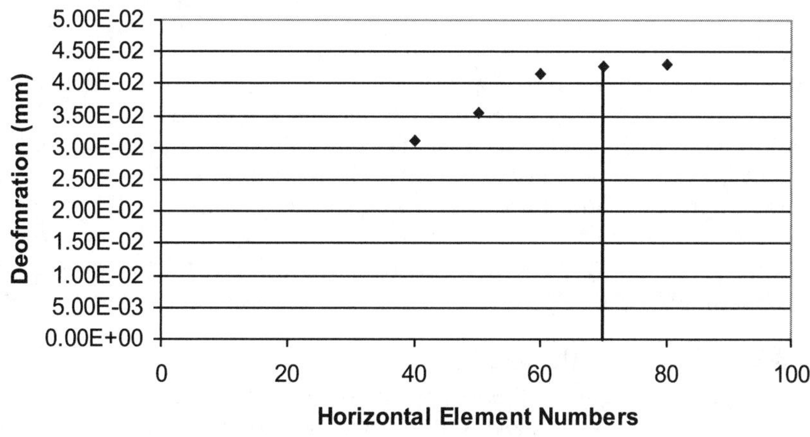

FIGURE 10.25   Deformations versus horizontal element numbers.

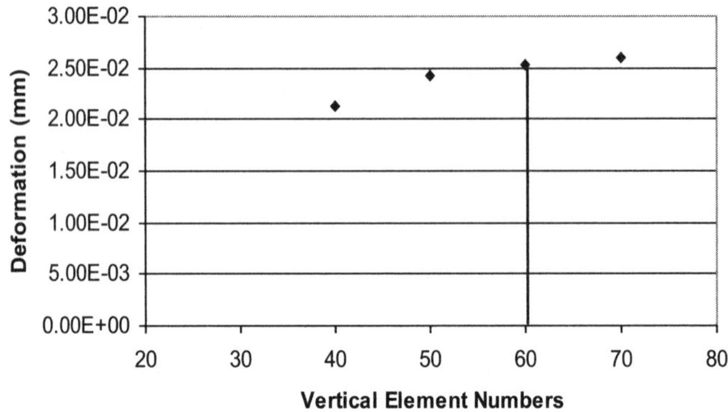

FIGURE 10.26   Deformations versus vertical element numbers.

coring width. After the comparison, the width is selected as 70 elements. Then, the comparison is conducted with the selected width and different depths as shown in Figure 10.26. The depth of the coring area is selected as 60 elements. After the area is determined, the small portion can be cored from the re-digitized images for the simulation purpose. While the reduced sizes may affect simulation results, it may not be so significant for permanent deformation which is usually very localized.

10.3.2   Phase Separation and Model Building

The obtained images contain the microstructure information of the scanned specimen. Each image is an assembly of pixel arrays that have a certain value of intensities. For each pixel, its coordinates can be recorded. For example, if black represents voids, the pixels that have pixel values of zero are voids and their location can be determined. The purpose of image analysis is to determine the threshold value for separating the three phases of AC. The threshold values are determined through analyzing the volume fractions of each phase in the mixture. The grouping of three phases is conducted by assigning different threshold values to different phases.

Pixel values and their geometric information are imported into a self-developed FORTRAN code to build the FEM geometry model. It should be noted that the pixel in a 2D space is represented by volume elements — voxels in a 3D space. The voxels are not cubic elements in this study. They are rectangular with x, y cross-sections represented as pixels (squares). The z dimension takes the spacing between the slices.

After the above procedure, the materials are separated into three element groups: mastic (binder with mineral filler smaller than 0.29 mm), aggregate, and air void. Each element group was assigned different material properties. The element group representing aggregates and mastics was assigned with elastic and viscoplastic material properties, respectively, while the group representing air voids was removed during the loading steps. The model with aggregates, mastics, and air voids is represented in Figure 10.27, where different gray levels represent aggregates, mastics, and voids.

10.3.3   Simulation Results

As mentioned above, a small portion was cored from the specimens for simulation purposes. Two portions can be cored from each image based on the affecting area determined as shown in Figure 10.20d. Each cored area forms a block with the areas cored from the

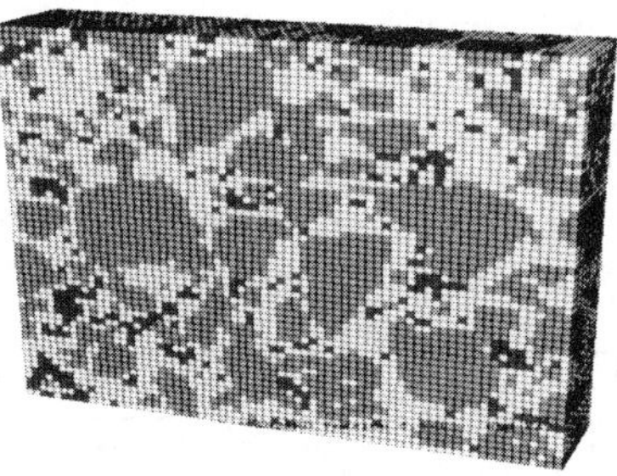

FIGURE 10.27   FEM geometrical model with three phases.

image slices adjacent to the current one. There are a total of seven layers for each block. For each block, the number of elements is 29,400, and total nodes are 34,648.

A cyclic loading with a loading magnitude of 100 psi was applied to the above block. A series of deformation profiles was obtained from the two-cored areas.

For the three mixes of the WesTrack project, three cored samples were taken for each of them. At first, simulations were conducted for the fine mix to optimize the model parameters. Figure 10.28 shows the experimental results for three specimens from the fine mix, and their average deformation. The average deformation was used to compare with the simulation results to optimize the model parameters. After the parameter optimization, the simulated results for different cores are presented in Figure 10.29. The

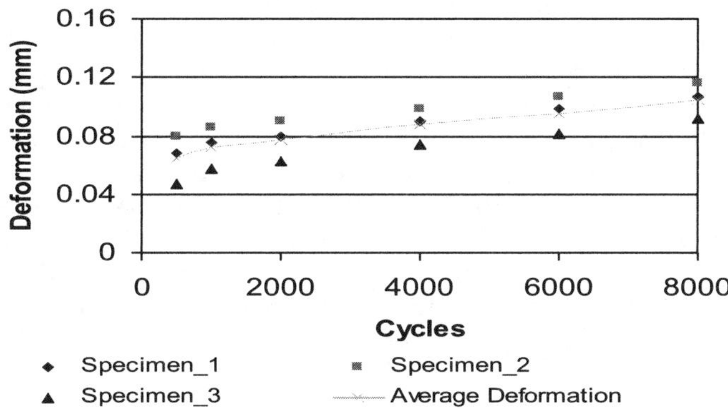

FIGURE 10.28   Experimental results and their average deformation for fine mix.

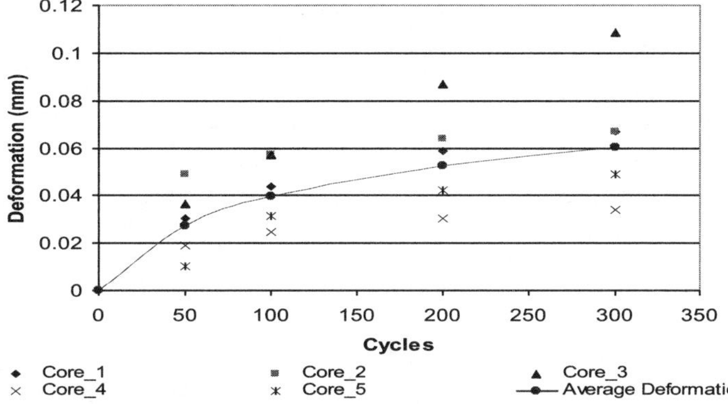

FIGURE 10.29   Simulated results and their average deformation for fine mix.

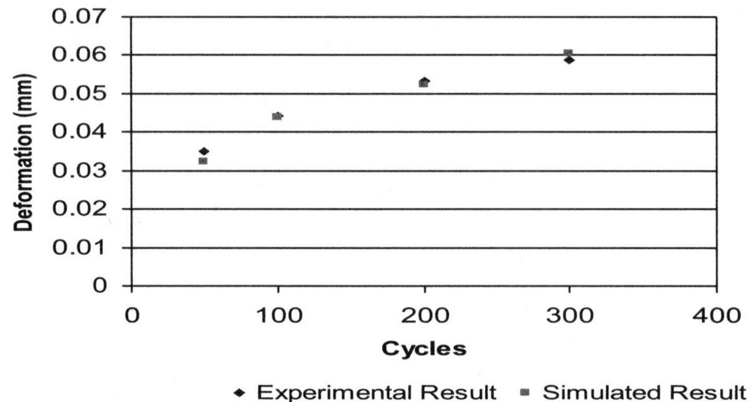

FIGURE 10.30   Average experimental result versus simulated result for fine mix.

simulation results after the parameter optimization and experimental results are compared in Figure 10.30.

After obtaining the model parameters from the fine mix, the parameters were used in the simulations for the fine-plus mix and the coarse mix. For the fine-plus mix, the experimental results are shown in Figure 10.31. The simulation results for different cores are shown in Figure 10.32. The simulation results are compared with the experimental results in Figure 10.33. For the coarse mix, the experimental results are shown in Figure 10.34. The simulation results for different cores are shown in Figure 10.35. The simulation results are compared with the experimental results in Figure 10.36.

The comparisons indicate that plastic deformation is very much localized and for simulation purposes, a small contact area on even a smaller size of the specimen may be adequate in obtaining realistic deformation history. The model can capture the deformation history more accurately when the presence of both aggregates and air voids is considered.

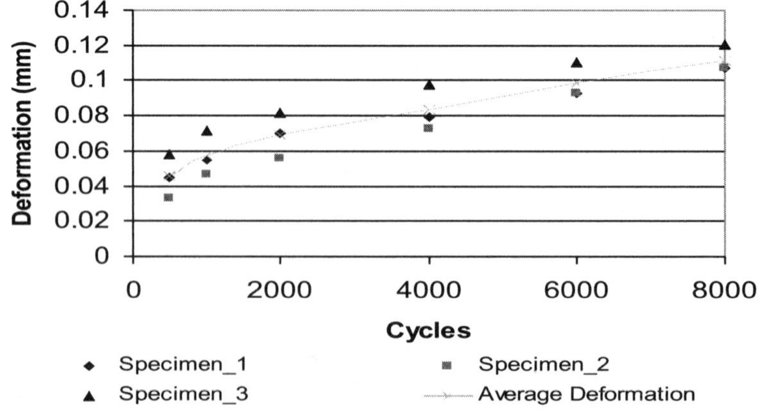

FIGURE 10.31   Experimental results and their average deformation for fine-plus mix.

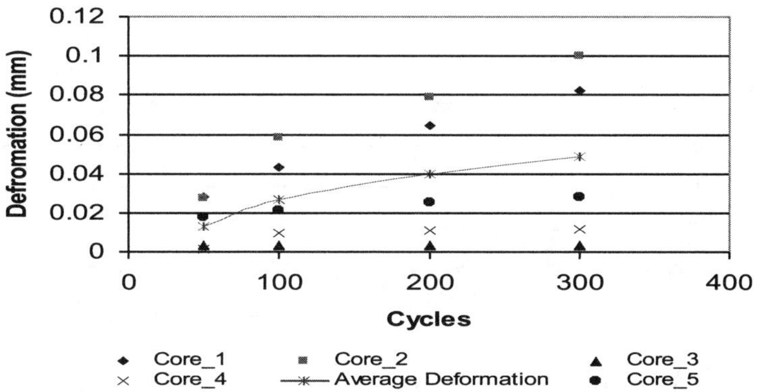

Figure 10.32 Simulated results and their average deformation for fine-plus mix.

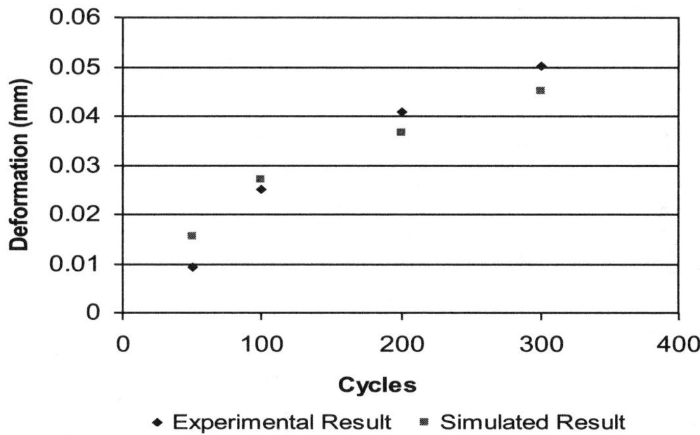

Figure 10.33 Average experimental results versus simulated results for fine-plus mix.

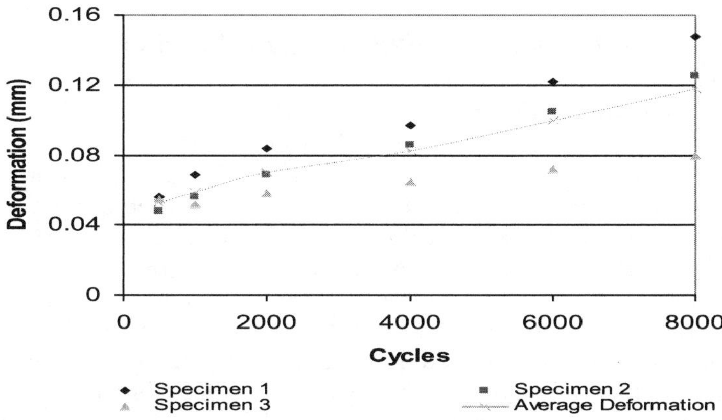

Figure 10.34 Experimental results and their average deformation for coarse mix.

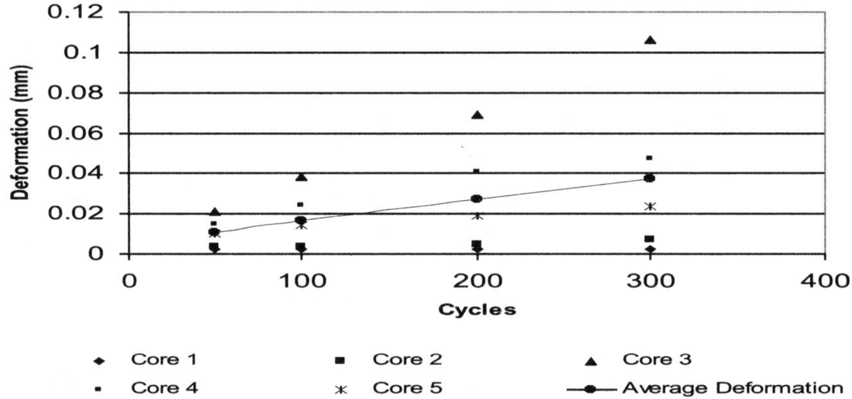

Figure 10.35 Simulated results and their average deformation for coarse mix.

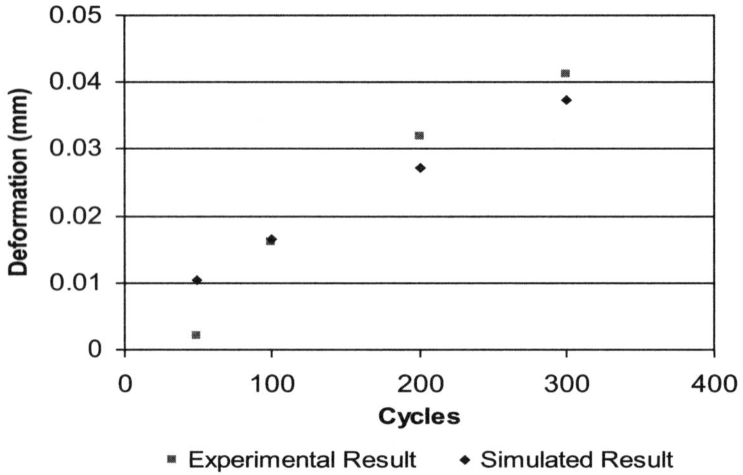

Figure 10.36 Average experimental result versus simulated result for coarse mix.

10.3.4 Inverse Approach for Material Characterization

Inverse analysis has been widely used to characterize the material parameters in complicated models. In this section, more details are described because this approach is viable and flexible in model calibration, especially using simulative tests where boundary conditions are complicated. In the inverse analysis, with the help of numerical simulation techniques, the unknown parameters are obtained from the actual results of the measurements to infer the optimum values of the model parameters. The goal of the back analysis is to find the values of the model to best fit the measured data with the simulated data. The advantage of inverse analysis is that the parameters that need to be identified refer directly to the numerical model used for the predictions and optimization studies. To optimize the model parameters, an objective function is defined. The minimization of the objective function is performed on the basis of a variety of optimization schemes. The result of the analysis depends on the selection of the algorithms of optimization. The trial- and error-method, least-square method, and other optimization methods are usually used for minimizing the objective function. In this section, an improved trial-and-error optimization algorithm was selected to obtain the model parameters. For a model with a few parameters, the trial-and-error method is usually very tedious and time-consuming. However, for the developed method, this disadvantage can be overcome through conducting the sensitivity analysis to the model parameters. The improved method needs no more complicated mathematics for solving the inverse problems, and avoids the tedious iteration process of the trial-and-error method. The sensitivity analysis determines the moving directions and amounts of changes for the model parameters. Therefore, it saves numbers of iterations for seeking the minima of the objective function. The procedure of the parameter optimization is related with a series of calls between ABAQUS and a developed program, Param Opti, to minimize the defined objective function, which is the function of measured and simulated deformations taken from the experimental measurements and FEM simulations with different loading cycles. The function defined in Equation 10-5 is used for simulative tests as well. If u represents a function of the vector of model parameters, D is the function of vector u.

The following is the optimization procedure:

Step 1. Guess an initial set of parameters. Do simulations to obtain an initial deformed profile with different loading cycles.

Step 2. Perform sensitivity analysis of the parameters.

Step 3. Input the deformed profile to the Param Opti program to calculate the value of the objective function. If the value of the objective function does not satisfy the selected convergence criteria, the program will generate a new set of parameters.

Step 4. Input the new set of parameters into ABAQUS input file to do simulations to obtain the deformed profiles corresponding to the new set of parameters.

Step 5. Repeat steps 3 and 4 until the selected criterion is satisfied.

Step 6. Output the parameter set.

The objective of parameter optimization is to find a set of parameters to minimize the objective function. For the two-layer elasto-viscoplastic model, there are 10 parameters: E, v, A, n, m, f, Y_0, B, C, D, that need to be characterized. According to Huang (2004), the typical value of the Poisson's ratio for AC was given as 0.35. In this study, this value was used in the simulations. From the studies conducted by Perl et al. (1983) and Huang (2004), the parameter n is related to the contact pressure between the pressurized hose and the specimen in the APA test. In Huang (2004), the value 0.8 for n was used under the contact pressure of 95 psi. In this study, the contact pressure is 100 psi. Therefore, the same value will be used. After fixing two parameters, there are eight parameters for the optimization.

To minimize the number of trials in the optimization process, a sensitivity analysis needs to be carried out before the optimization. The purpose of the sensitivity analysis is to study the influence of the change of the parameters on the deformation. Through the sensitivity analysis, the percent of change for each parameter in every iteration call between the ABAQUS and the Param Opti program can be determined. The values will be input into the Param Opti program. After obtaining the simulation result using the parameter set, if the convergence criterion of the objective function is not satisfied, the program will generate a new set of parameters to obtain new deformed profiles. Otherwise, the program will output the parameter set. The following are steps for the sensitivity analysis. The result of the sensitivity analysis is shown in Figure 10.37.

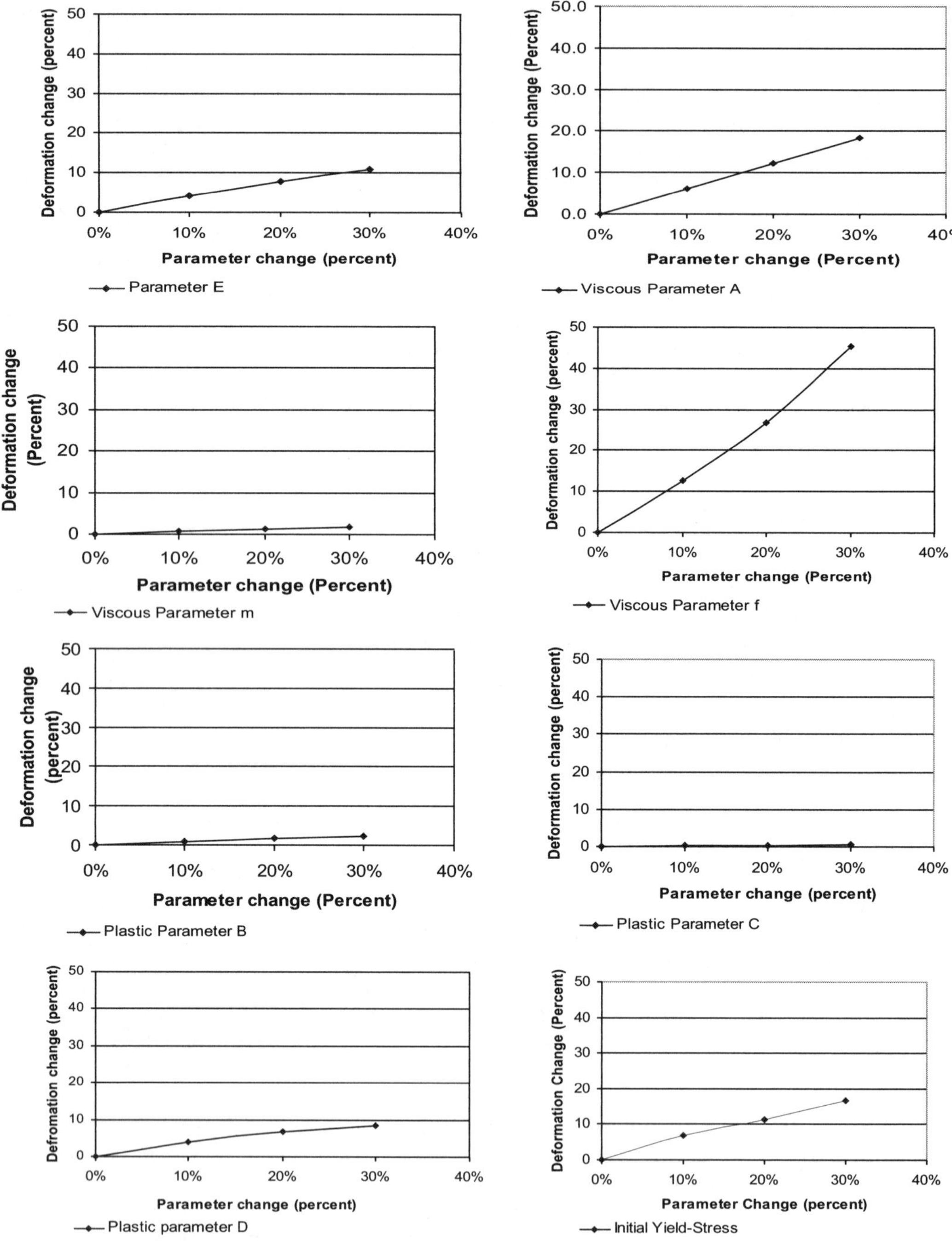

FIGURE 10.37 Parameter sensitivity analysis.

Step 1. Guess an initial set of parameters based on the previous studies and apply it to the model in order to obtain the original deformed profiles for the testing sample.

Step 2. Change one parameter by 10% from its initial value and fix the other parameters. Then put this new set of parameters into the model and do numerical simulation to obtain a new deformed profile.

Step 3. Compare the new deformed profile with the original profile to study the sensitivity of the parameter on the deformation.

Repeat steps 2 and 3 for other parameters.

From the sensitivity analysis it can be seen that the total deformation is more sensitive to the viscous parameters than the plastic parameters, and the most sensitive parameters are the viscous parameter f and the plastic parameter, initial yield stress Y_0.

The initial yield stress Y_0 controls the increment of the deformation corresponding to the increment of the loading cycles. Figure 10.38 shows the influence of the initial yield stress on the deformation. When an increase or decrease in initial yield stress from the optimized initial yield stress occurs, the deformation increment between cycles will decrease or increase from the optimized deformation result. The optimized result for the initial yield stress matches the deformation increment from the experimental measurement.

The other most sensitive parameter, parameter f, determines the amount of the permanent (plastic) deformation. Figure 10.39 shows the influence of the parameter f on permanent deformation. It can be seen from the figure that plastic deformation increases with the decrease of the parameter f.

Parameter f (Equation 10-2) is the ratio between modulus in the elastic-viscous network and the sum of the modulus in the networks. Transforming Equation 10-2 into the following Equation 10-6, it can be seen that when the stiffness in the elastic-viscous network is very large compared to the stiffness in the elastic-plastic network, i.e., $K_v \gg K_p$, the parameter f will approach 1. Under this condition, the elastic-viscous branch will take most of the deformation under the applied load as shown in Figure 10.40.

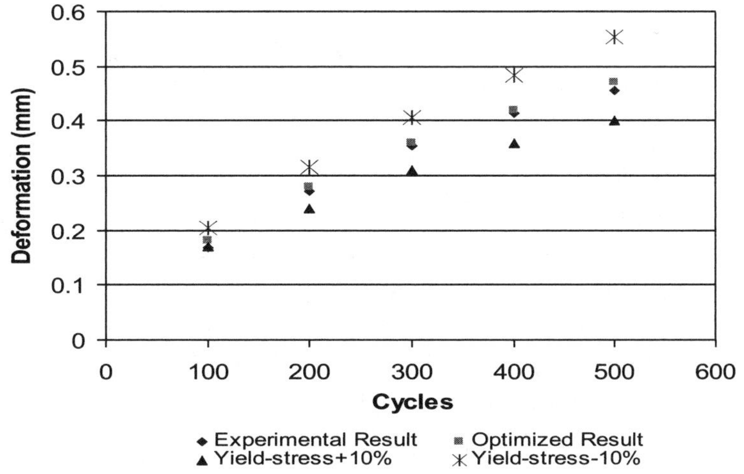

FIGURE 10.38 Influence of initial yield stress Y_0 on the deformation.

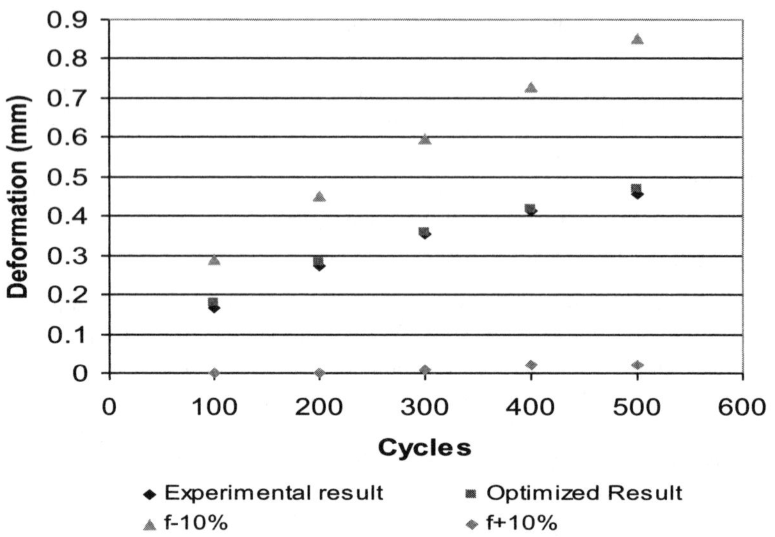

FIGURE 10.39 Influence of the parameter *f* on the permanent deformation.

Figure 10.40 shows the situation when the parameter $f = 0.99$. It has been shown that the plastic deformation equals almost to zero, and the viscous deformation is very close to the total deformation.

$$f = \frac{K_V}{K_P + K_V} = \frac{1}{1 + \dfrac{K_p}{K_v}} \tag{10-6}$$

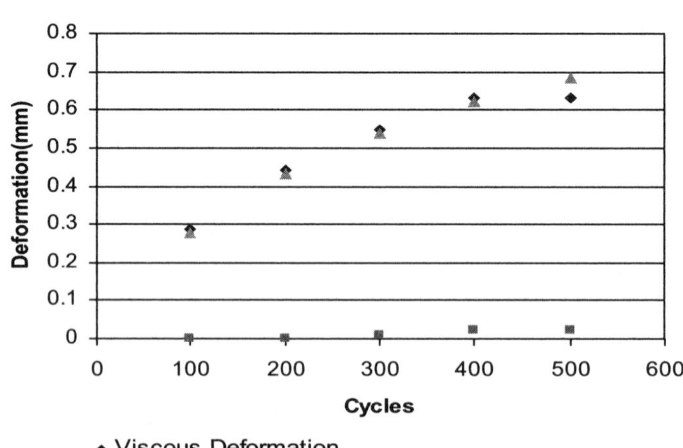

FIGURE 10.40 Comparison of separated deformation and total deformation.

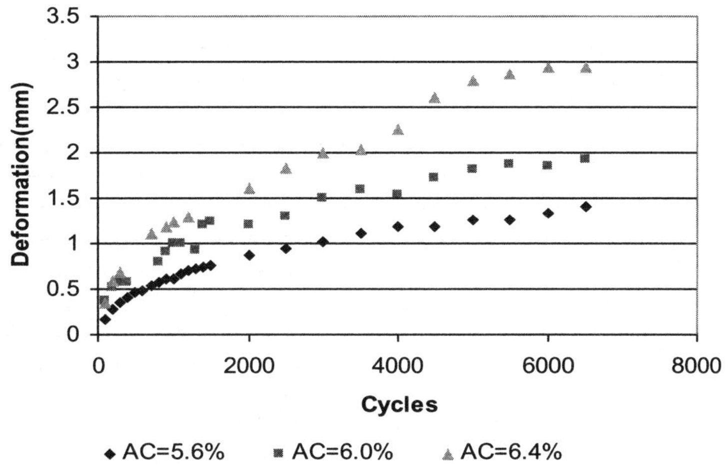

FIGURE 10.41   Permanent deformations of three mixes versus loading cycles.

10.3.5   Model Rationality Analysis

The proposed model rationality was analyzed by comparing the simulated results with the experimental results obtained through the APA tests at the Virginia Transportation Research Council (VTRC) (Wang et al., 2008). The tests were carried out under the same temperature (49°C), loading frequency and loading level (0.8 Mpa). Three mixes with the same asphalt binder (PG76-22), different AC contents, and the same aggregate gradation were tested. The experimental results are presented in Figure 10.41. This figure shows that the AC mixture with higher AC content has larger permanent deformation. For AC, most of the deformation occurred under 500 loading cycles. The numerical simulation results, compared with the experimental results for up to 800 loading cycles, were presented in Figures 10.42 through 10.44. After the parameter optimization, a good correlation exists between the experimental results and the numerical results. The parameters f

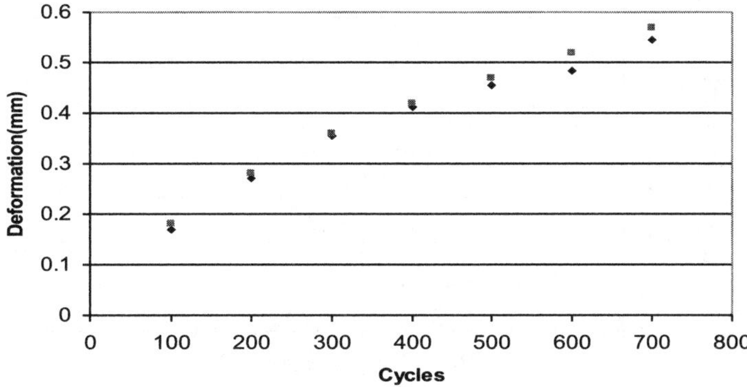

FIGURE 10.42   Experimental result versus simulated result (AC = 5.6%).

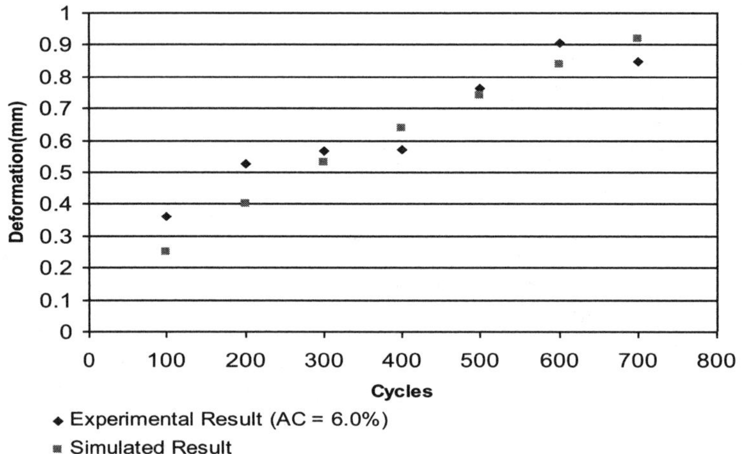

Figure 10.43 Experimental result versus simulated result (AC = 6.0%).

and Y_0 for the three mixes are presented in Table 10.5. Except for the parameter f and initial yielding stress Y_0, other parameters for the three mixes are the same as shown in Table 10.6. Table 10.5 shows that the initial yield stress Y_0 decreased with the increase of the asphalt contents in the mixtures. Compared with aggregates, asphalt binder is a weak material. Under the applied load, it is easier to yield. Therefore, the initial yield stress decreases with the increase of the asphalt content. On the other hand, the parameter f also decreases with the increase of the asphalt contents. From Equation 10-6, if parameter f decreases, the ratio $\dfrac{K_p}{K_v}$ must be increased. If K_p is fixed, the stiffness of the viscous components will decrease, which is consistent with the intuitive understanding (more asphalt, more viscous, less stiff). Assuming the plastic deformation is related to the aggregate skeleton, the permanent deformation will increase.

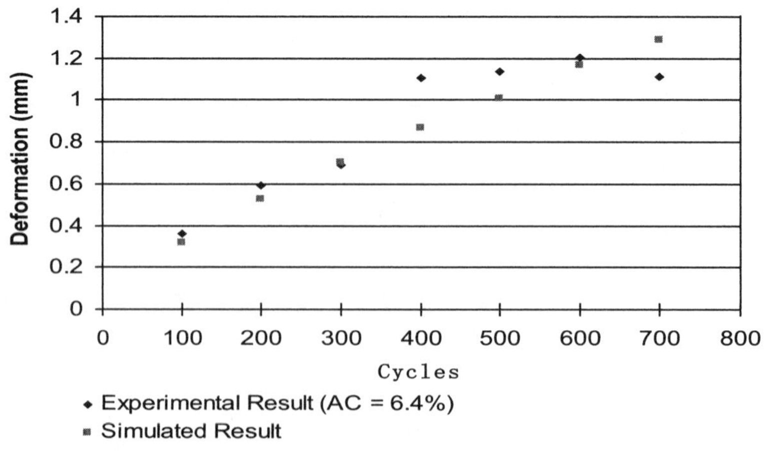

Figure 10.44 Experimental result versus simulated result (AC = 6.4%).

Mixes	Parameter *f*	Initial Yield Stress Y_0 (Mpa)
AC = 5.6%	0.9	0.037
AC = 6.0%	0.85	0.032
AC = 6.4%	0.8	0.028

TABLE 10.5   Parameter *f* and initial yield stress Y_0.

A	n	m	B	C	D
2.3E-3	0.8	−0.24	1.0E-2	3	4.5

TABLE 10.6   Other parameters.

10.4   Other Approaches to Incorporate Microstructure into Simulation

10.4.1   Micromechanical Finite Element Method

Dai et al. (2004) presented a micro-finite element method (MFEM). Each contact pair is modeled as an element between two aggregate particles with a binder layer in between. Figure 10.45 presents the idealization and the element analysis. Through element analysis, the element stiffness matrix can be established.

$$[K] = \begin{bmatrix} K_{nn} & 0 & K_{nn}e & -K_{nn} & 0 & -K_{nn}e \\ & K_{tt} & K_u r_1 & 0 & -K_{tt} & K_u r_2 \\ & & K_u r_1^2 + \dfrac{K_{nn}}{3}(w_2^2 - w_1 w_2 + w_1^2) & -K_{nn}e & -K_u r_1 & K_u r_1 r_2 - \dfrac{K_{nn}}{3}(w_2^2 - w_1 w_2 + w_1^2) \\ & & & K_{nn} & 0 & K_{nn}e \\ & & & & K_{tt} & -K_{tt} r_2 \\ & & & & & K_{tt} r_2^2 - \dfrac{K_{nn}}{3}(w_2^2 - w_1 w_2 + w_1^2) \end{bmatrix}$$

$$(10\text{-}7)$$

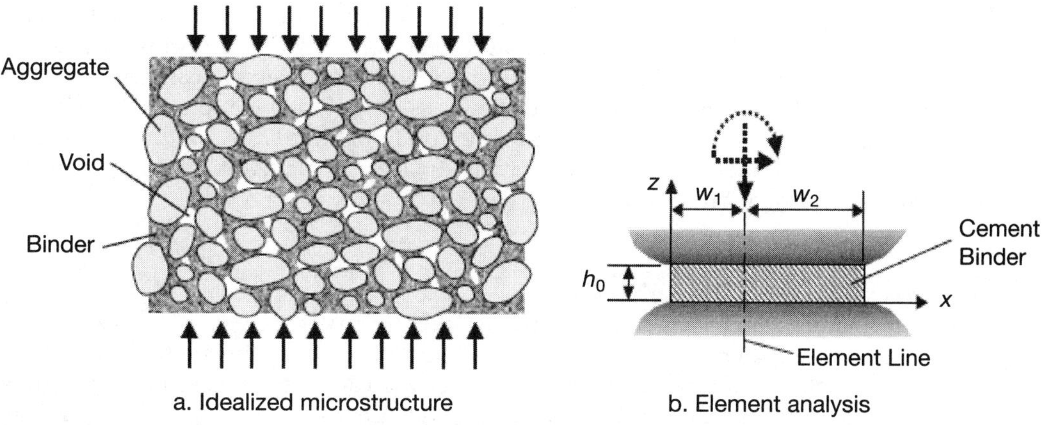

FIGURE 10.45   Illustration of the MFEM (Courtesy Qingli Dai and Sadd).

Dimension	Case 1		Case 2		Case 3	
	2D	**2D**	**2D**	**2D**	**3D**	**3D**
Total element number	10000	10000	10000	10000	5000	5000
Rigid element number	0	80	0	80	0	Defined by image, elements representing aggregate of mixture
Element type	4-node bilinear, reduced integration	4-node bilinear, reduced integration	4-node bilinear, reduced integration	4-node bilinear, reduced integration	8-node linear brick	8-node linear brick
Constitutive model	Elastic	Elastic	Damage concrete plasticity	Damage concrete plasticity	Elastic (different for each component of the mixture)	Elastic (different for each component of the mixture)
Loading application	Top pressure	Top pressure	Top pressure	Top pressure	Top pressure	Top pressure
Calculation time (s)	4s	3s	8s	4s	8s	3s

TABLE 10.7 Computational efficiency using rigid element.

10.4.2 Rigid Element Method

Considering the significant difference in stiffness between aggregates and asphalt binder at relatively high temperatures, the aggregates may be treated as rigid. The major advantage of representing aggregates as rigid bodies rather than deformable finite elements is computational efficiency. Element-level calculations are not performed for elements that are part of a rigid body. Although some computational effort is required to update the motion of the nodes of the rigid body and to assemble concentrated and distributed loads, the motion of the rigid body is determined completely by a maximum of six degrees of freedom at the reference node. Table 10.7 (Wang, 2010) presents a preliminary estimation of the computation efficiency when ABAQUS is run on the System X platform at Virginia Tech.

10.5 Extension to Other Materials

These techniques can be extended to various other materials including composite metals, soils, cement concrete, wood composite (Figure 10.46), and other composite materials at different length scales. The digital specimens can also be used to study other phenomena including transport, conductivity, and stress-strain relation. They are fundamental tools for many other materials.

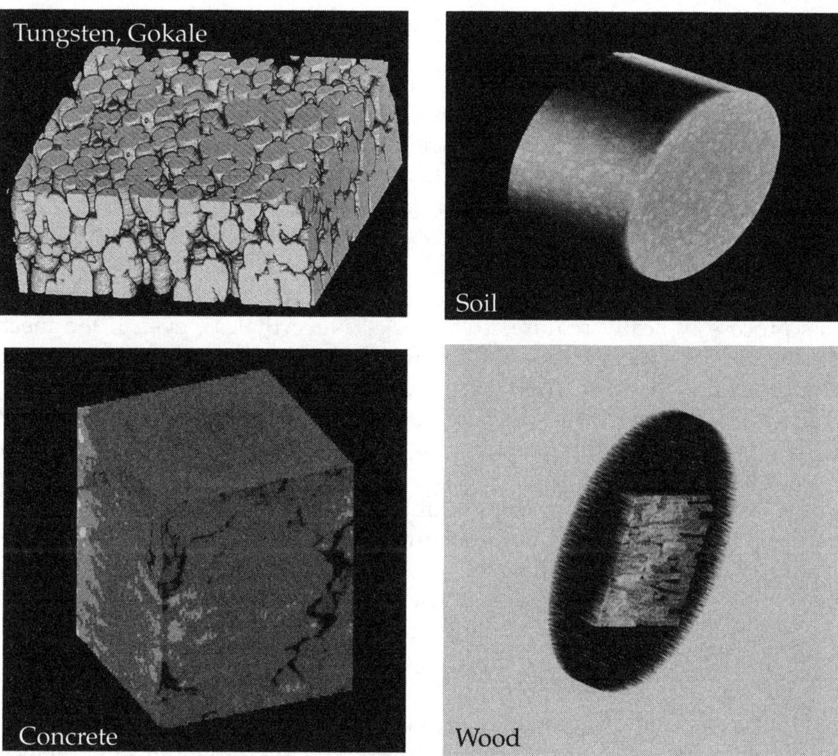

FIGURE 10.46 Application of the digital specimen and digital tester techniques to other materials.

10.6 Perspective for Digital Mix Design

Mix design for asphalt concrete is one of the most important steps in determining the performance of constructed pavements. The principles for guiding either the Marshall mix design method or the SuperPave mix design method are largely empirical. Although empirical methods may not be eliminated due to the tremendous experience accumulated historically, significant improvements over current mix design methods may be made using modern tomography and computational techniques. Current mix design methods have limitations including: 1) characteristics of the constituents (asphalt binder and aggregates) are not directly revealed in the mix design or linked to performance; 2) there is a lack of mechanism on how aggregate properties such as aggregate shape, angularity and texture, and binder rheology properties are compatible with each other and contribute to the performance of mixes; 3) the mix performances against rutting and fatigue cracking are not balanced; for example, a mix may have too strong a resistance against rutting but not enough resistance against fatigue cracking; 4) there is a lack of a true "performance test" to characterize the "strength" and "deformation" properties; 5) mechanisms to link lab performance with field performance cannot be thoroughly understood due to the difficulties in developing rational models; 6) the coupled phenomena of deformation and cracking at microscopic scale are not well explored both fundamentally and experimentally; and 7) the expenses for purchasing different testing equipment, training technicians, and performing tests are very high.

In some cases it is even worse as different testing devices or procedures may yield conflicting results.

Mix design methods typically include the following steps: 1) selection of binder and aggregate; 2) characterization of binder and aggregate; 3) selection of a gradation; 4) determination of the optimum AC content and volumetric characteristics; and 5) evaluation of the performance of the mixtures. Once the materials (asphalt binder and aggregate) are selected, the most important steps are the determination of the gradation, the optimum asphalt content, and evaluation of the performance through mechanical testing such as the simple performance test (SPT) or the simulative test (ST). If results of the mechanical tests do not meet the specifications, another gradation shall be tried. This process typically requires two weeks. Nevertheless, even if the mechanical test results meet the specification requirements, the mixes may not perform satisfactorily when placed in the field. This is a very challenging problem: is there any way by which a mix can be designed with predicted performances of reasonable accuracy?

Before answering this question, the important performance criteria should be analyzed. Historically, performance has been measured in terms of PSI (Present Service Index), roughness index, and individual distresses such as rutting, cracking (thermal cracking, fatigue cracking), and moisture damage. These distresses have been demonstrated. However, it was still not clear what had happened inside the mixture even after the SuperPave mix design was developed. In recent years, several research projects have focused on the understanding of the fundamental mechanisms of the development of these distresses. A common feature of these projects is to investigate what factors truly control the development of these distresses. As AC is a highly heterogeneous material composed of aggregates, asphalt binder, and air voids, stresses and strains are significantly localized in different constituents. Therefore, an understanding of how each constituent and/or its interaction affect the mixture performance would help develop a mix that performs well.

With recent development in X-ray computed tomography (XCT) and computational simulation (the digital specimen and digital test technique, for example), mix design can be significantly enhanced and even replaced to some degree with a computation-based approach. Digital mix design may take the following four steps.

Step 1: Representing Aggregate Particles

Recent efforts at Virginia Tech and the National Institute of Standards and Technology are developing databases for 3D aggregate images acquired through XCT and laser scanning. These images are not simple pictures of the particles. They are computer visualizations, including all the detailed information such as the surface coordinates of these particles. The particle shape, angularity, and texture information can be abstracted from these images. In other words, aggregate evaluations can be achieved through X-ray tomography imaging and image analysis at the same time they are represented for packing. The details of the methods on how to represent the particles can be found in Wang et al., (2004).

Step 2: Packing Particles to Form a Gradation and Gradation Optimizing

Once individual particles are represented in 3D, a set (or a gradation) of particles can be assembled together to form a skeleton structure. In this structure, particle-particle contacts can be assessed (Figure 10.47). By manipulating the particle locations and orientations and placing smaller particles in the voids among larger particles, an optimum gradation of the aggregates can be achieved through maximizing the aggregate-

a. Step 1-Representing Individual Particles

b. Step 2- Particle Packing to Form Gradation

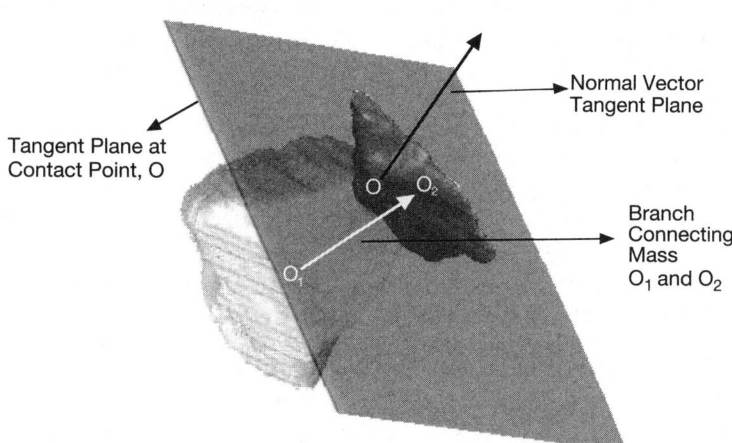

c. Step 3- Determining Particle-Particle Contact for Gradation Optimization and Selection

FIGURE 10.47   Step 1 and Step 2 for digital mix design.

aggregate contacts, forming the most stable structure under various loading. This is actually the step in parallel with selecting a gradation in the traditional mix design. Differing from the traditional approach, the selection of gradation in this method directly utilizes the criteria for mechanical properties or performance of the mixture, such as aggregate-aggregate contacts and local stability, etc.

Step 3: Determining the Asphalt Content

Once a gradation is selected, the voids in mineral aggregate (VMA) can be determined. For a certain air void content, such as 4%, the mastics volume, the binder volume, or the asphalt content can be determined. Compared to the traditional approach, this method also determines the surface area of all the aggregates and therefore can determine the film thickness. The asphalt content can also be determined using an alternative approach that is based on modeling the particle-particle contact with a binder layer. As the binder layer will affect the plastic/permanent deformation of particles in contact, this layer thickness will be directly linked to the rutting performance of the mixture. On the other hand, when the asphalt film thickness is too small, wearing of the asphalt binder will reduce its fatigue properties.

Step 4: Performing a Digital Test to Predict the Performance

Once the aggregate gradation and asphalt content are determined, the mechanical behavior of the mixture can be evaluated with, for example, the dynamic modulus test or the indirect tensile test using the discrete element method (DEM) and the finite element method (FEM), or the digital specimen and digital test techniques. In this approach, the characterized binder properties (needed in modeling the contact and the heterogeneous mixture), aggregate shape (contact curvature), and contact asperity can be directly incorporated in the tests.

References

ABAQUS (1995). User's Manual. Hibbitt, Karlsson & Sorensen, Inc.,Pawtucket, R.I.

Bjorn, B., Boonchai, S. and Reynaldo, R. (2002). Prediction of the viscoelastic response and crack growth in asphalt mixtures using the boundary element method. *Transportation Research Record*, No.1789, pp.129–135.

Buttlar, W.G., Al-Khateeb, G.G. and Bozkurt, D. (1999). Development of a hollow cylinder tensile tester to obtain mechanical properties of bituminous paving mixtures. *Journal of the Association of Asphalt Paving Technologists*. Vol. 68, pp.369–403.

Buttlar, W.G. and You, Z. (2001). Discrete element modeling of asphalt concrete: microfabric approach. *Transportation Research Record*, Vol.1757, pp.111–118.

Chang, G.K. and Meegoda, J.N. (1997). Micromechanical simulation of hot mixture asphalt. *Journal of Engineering Mechanics*, Vol. 123, No. 5, pp.495–503.

Chang, G.K. and Meegoda, J.N. (1999). Micro-mechanic model for temperature effects of hot mixture asphalt concrete. *Transportation Research Record*, No.1687, pp.95–103.

Collop, A.C., Scarpas, A.T., Kasbergen, C. and Bondt, A.D. (2003). Development and finite element implementation of a stress dependent elasto-visco-plastic constitutive model with damage for asphalt. *Transportation Research Record*. No.1832, pp.96–104.

Dai, Q. and Sadd, M.H. (2004). Parametric model study of microstructure effects on damage behavior of asphalt samples. *International Journal of Pavement Engineering*. Vol.5, No.1, pp.19–30.

Hondros, G. (1959). Evaluation of Poisson's ratio and the modulus of materials of a low tensile resistance by the Brazilian (indirect tensile) test with particular reference to concrete. *Australia Journal of Applied Science*, Vol.10, No.3, pp.243–268.

Huang, B.S, Mohammad, L.N. and Wathugala, G.W. (2004). Application of a temperature dependent viscoplastic hierarchical single surface model for asphalt mixtures. *ASCE Journal of Materials in Civil Engineering*. Vol.16, No.2, pp.147–154.

Kichenin, J., Van, K.D. and Boytard, K. (1996). Finite-element simulation of a new two-dissipative mechanisms model for bulk medium-density polyethylene. *Journal of Materials Science*, Vol.31, No.6, pp.1653–1661.

Lee, H.J. and Kim, Y.R. (1998). Viscoelastic constitutive model for asphalt concrete under cyclic loading. *Journal of Engineering Mechanics*, Vol.124, No.1, pp.32–40.

Park, S.W., Kim, Y. R. and Schapery, R.A. (1996). A viscoelastic continuum damage model and its application to uniaxial behavior of asphalt concrete. *Mechanics of Materials*, Vol.24, No.4, pp.241–257.

Perzyna, P. (1966). Fundamental problems in viscoplasticity. *Advances in Applied Mechanics*. Vol. 9, pp.244–368.

Roberts, F.L., Mohammad, L.N. and Wang, L.B. (2002). History of hot mix asphalt mixture design in the united states. ASCE 150th anniversary civil engineers paper. *Journal of Materials in Civil Engineering*, Vol.14, No.4, pp.279–293.

Sadd, M.H., Dai, Q. and Parameswaran, V. (2004). Microstructural simulation of asphalt materials: modeling and experimental studies. *Journal of Materials in Civil Engineering*. Vol. 16, No.2, pp.107–115.

Schapery, R.A. (1984). Correspondence principles and a generalized J-integral for large deformation and fracture analysis of viscoelastic media. *International Journal of Fracture*, Vol. 25, No.3, pp.195–223.

Schapery, R.A. (1990). A theory of mechanical behavior of elastic media with growing damage and other changes in structure. *Journal of the Mechanics and Physics of Solids*, Vol. 38, pp.215–253.

Schwartz, C.W., Gibson, N.H., Schapery, R.A. and Witczak, M.W. (2002). Viscoplasticity modeling of asphalt concrete behavior. *ASCE Geotechnical Special Publication*, No.123, pp.144–159.

Seibi, A.C., Sharma, M.G., Ali, G.A. and Kenis, W. J. (2001). Constitutive relations for asphalt concrete under high rates of loading. *Transportation Research Record*, No.1767, pp.111–119.

SKYSCAN (2006). SkyScan Brochure 2006, *SkyScan Microtomography, Micro-CT Manual*, SkyScan.

Tashman, L., Masad, E., Zbib, H., Little, D. and Kaloush, K. (2005). Microstructural viscoplastic continuum model for permanent deformation in asphalt pavements. *Journal of Engineering Mechanics*, Vol. 131, No.1, pp.48–57.

Wang, D. (2010). Integration of microstructure into modeling for stone-based infrastructure materials, *Internal Report*, Virginia Tech.

Wang L.B., Frost, J.D. and Lai, J.S. (2004). 3D digital representation of the microstructure of granular materials from x-ray tomography imaging. *ASCE Journal of Computing in Civil Engineering*, Vol.18, No.1, pp.28–35.

Wang, L.B., Frost, J.D. and Shashidhar, N. (2001). Microstructure study of westrack mixes from x-ray tomography images. *Transportation Research Record*, No.1767, pp.85–94.

Wang, Y.P., Wang, L.B. and Maupin, B. (2008). An inverse approach for evaluating the properties of asphalt concrete using the APA test. *Journal of Road Materials and Pavement Design*, Vol.9, Special Issue, pp.201–218.

Zhang, B., Wang, L. and Tumay, M.T. (2006). An evaluation of the stress non-uniformity due to the heterogeneity of ac in the indirect tensile test. *ASCE Geotechnical Special Publication*, No.146, pp.29–43.

Zhang, W., Drescher, A. and Newcomb, D.E. (1997). Viscoelastic analysis of diametral compression of asphalt concrete. *Journal of Engineering Mechanics*, Vol.123, No.6, pp.596–603.

CHAPTER 11

Simulation of Asphalt Compaction

11.1 Introduction

Compaction of asphalt concrete (AC) is one of the most important steps in controlling the quality of pavement construction. More importantly, many problems arise during compaction: some mixes may not be easy to compact compared to others (perhaps indicating a good mix); too thin a layer thickness may prevent effective compaction; a weaker mastic may make compaction easy but it does not indicate good quality; a soft base or sub-base may cause difficulties for compaction (compaction energy is absorbed by the base or sub-base); the non-uniformity of the base or sub-base support causes non-uniform compaction if operation parameters do not vary correspondingly; and a change of environmental conditions such as temperature and wind speed results in a change of viscosity and therefore compaction effectiveness. An understanding of how these, and other factors, affect compaction in the point of view of fundamental mechanics will help achieve better compaction and optimize layer thickness.

Due to the many influencing factors, it is difficult to develop simple mathematic compaction models to relate these influencing factors with the end product properties such as density, modulus, and performance. In addition, due to the significant differences between field conditions and laboratory conditions in loading, boundary, and environmental constraints, it is challenging to develop the relationship between lab compaction and field compaction. Research on compaction modeling may help: 1) relate laboratory to field compaction; 2) identify the controlling factors that affect density and performance; and 3) provide guidance on construction process control for better quality AC.

A tremendous amount of research has been conducted in recent years relating to lab and field compaction (Huerne 2004; Masad et al., 2009; Wang et al., 2007; Partl, 2007a, 2007b). The following presents the modeling of AC in two categories—the lab compaction and the field compaction.

11.2 Lab Compaction

11.2.1 Empirical Method-Soil Consolidation Analogy

Figure 11.1 presents a typical idealized gyratory compaction curve. Figure 11.2 presents the same curve in the air-void reduction view. It should be noted that the gyratory compaction is similar to a one-dimensional (1D) consolidation. If the volumetric deformation of asphalt binder and aggregates is neglected, the change of the air-void content is actually the incremental volumetric strain of the sample. In the 1D situation, the

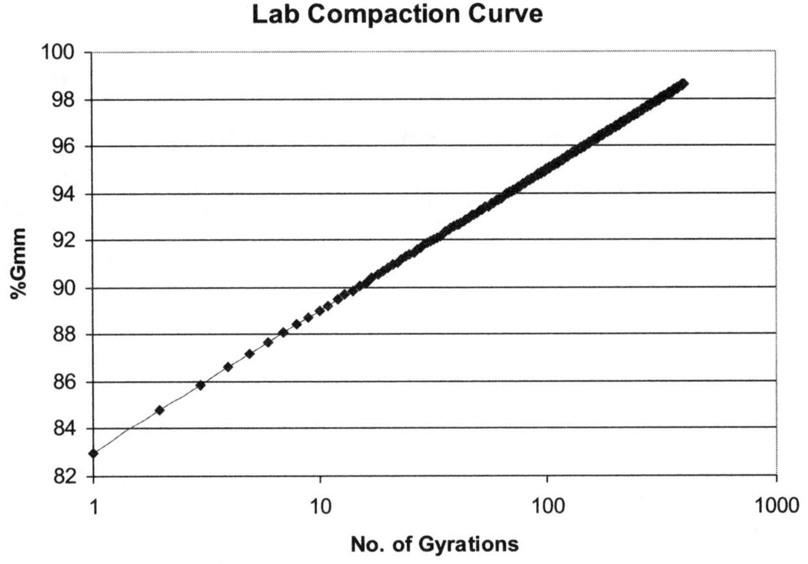

Figure 11.1 Idealized typical gyratory compaction curve.

incremental volumetric strain is equal to the incremental vertical strain $\Delta\varepsilon_z$. This view is very important in that it relates the change of a physical parameter (air void content) to the mechanical parameter of volumetric strain and 1D strain, which allows researchers to further associate it to dissipated strain energy, constitutive modeling, and field compaction on the basis of rigorous mechanics principles.

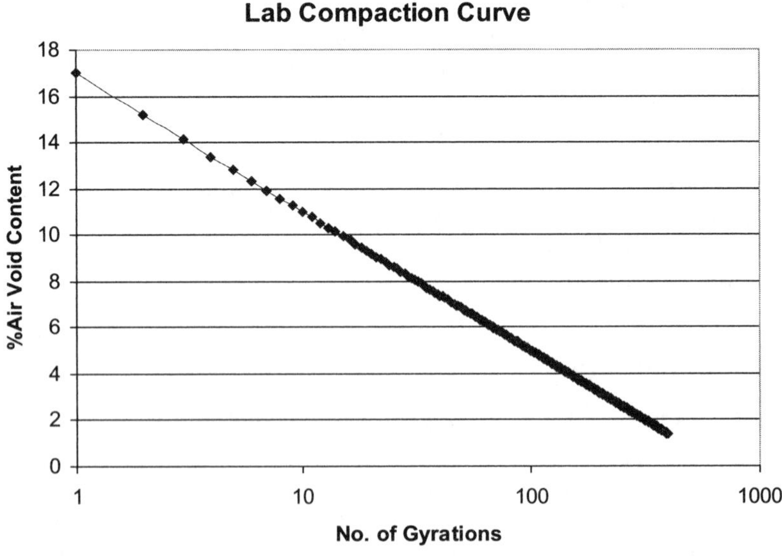

Figure 11.2 Air void reduction in the compaction process.

If the compaction process is viewed as a process where air-voids are reduced and particles are re-oriented, the mathematical representation of the air-void reduction process (Equation 11-1) can be conveniently derived from Figure 11.2.

$$A_v = A_{v0} - K^l \bullet LogN \qquad (11\text{-}1)$$

where A_{v0} = air-void content after the first gyration;

$\qquad A_v$ = air-void content after the Nth gyration;

$\qquad K^l$ = slope of lab compaction, or the rate of air-void content reduction.

$(A_v - A_{v0})$ corresponds to the accumulative volumetric strain. In a 1D case it is also the vertical strain.

In other words, if density is the only parameter of interest (density may not be the only parameter for performance), Equation 11-1 is a mathematical model. It is necessary to associate the model parameters A_{v0} and K^l with other material properties such as aggregate characteristics, gradation, binder content, and binder properties (temperature dependent). However, it should be noted that the current gyratory compaction procedure uses a constant pressure. How compaction varies with the magnitude of the pressure may need further exploration in order to associate lab compaction to field compaction because field compaction has a varying pressure to different volume elements (the AC in a layer feels a varying pressure when the compactor moves). Figure 11.3 illustrates this difference, where the three volume elements may be subjected to different levels of stress or compactive efforts. If compaction curves at different compactive efforts can be acquired in the lab, through calculating the compactive energy transferred to the different elements in the field, the compaction levels at different locations can be evaluated. This presents a logical reasoning to tackle the problem, although the mathematic models may take different formats.

The $A_v \sim LogN$ consolidation relationship is not unique to AC. Figure 11.4 presents the consolidation process of a soil, which is well known to civil engineers. Obviously, soil consolidation (*e-log p*) is very similar to the consolidation/compaction of AC. A further interpretation of the soil consolidation curve was provided by Wang and Frost (2004) to transfer this curve (*e-log p*) space to the $E_d\text{-}p$ (dissipated strain energy-effective

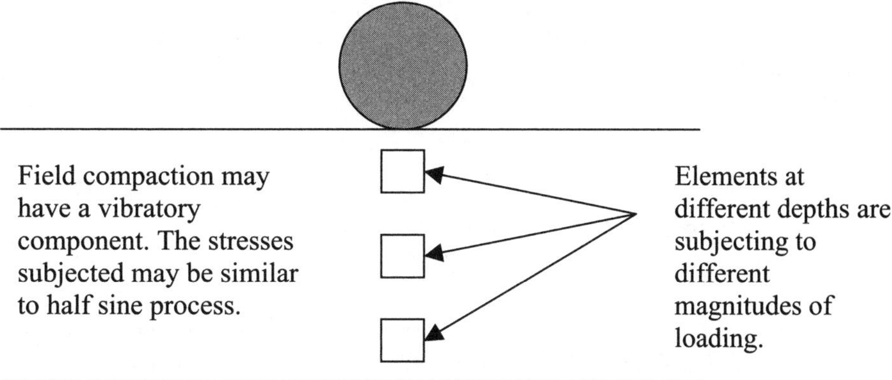

Figure 11.3   Illustration of the loading difference of volume elements in field compaction.

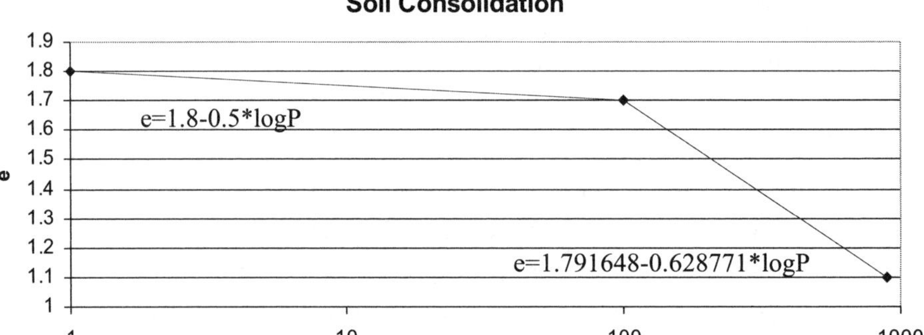

FIGURE 11.4   An analogy to soil consolidation.

consolidation stress) space. The same transformation can be performed on the asphalt compaction curve.

As mentioned before, $(A_v - A_{v0})$ is actually the accumulative volumetric strain, and the vertical strain in the 1D case. The term ΔA_v corresponds to the incremental volumetric strain or incremental vertical strain in the 1D case at the nth gyration. If the pressure during the compaction process is kept as a constant (P), then the dissipated work (assuming the deformation during the compaction process is irrecoverable) on a unit volume of the sample is equal to $P\Delta A_v$. Therefore, the accumulative dissipated work (dissipated strain energy in terms of the material) is:

$$E_d = \sum_{N0}^{N} P\Delta A_v \tag{11-2}$$

An integral format of this formula would be:

$$E_d = \int_{A_{v0}}^{A_v} PdA_v \int_{N_0}^{N} PK^l / NdN = PK^l Log(N / N_0) \tag{11-3}$$

Therefore, the slope K^l is also a measurement of the dissipated strain energy rate required to consolidate AC. For different mixes, such as mix A and Mix B in Figure 11.5, the slopes are different. Many research projects have indicated that this slope is related to the gradation, aggregate shape, angularity and texture, and asphalt binder rheology properties (temperature dependent as well). In other words, K^l may be decomposed into three components:

$$K^l = K_g K_m K_b \tag{11-4}$$

where K_g = gradation contribution;

$\quad K_m$ = aggregate morphological contribution;

$\quad K_b$ = binder rheology contribution.

It should be noted that K_m is related to aggregate shape, angularity, and texture. It is a function of the shape, angularity, and texture factors.

While the variation of K^l with the above parameters as lump sum is available in many laboratory testing programs, the decomposition in Equation 11-4 format is not available.

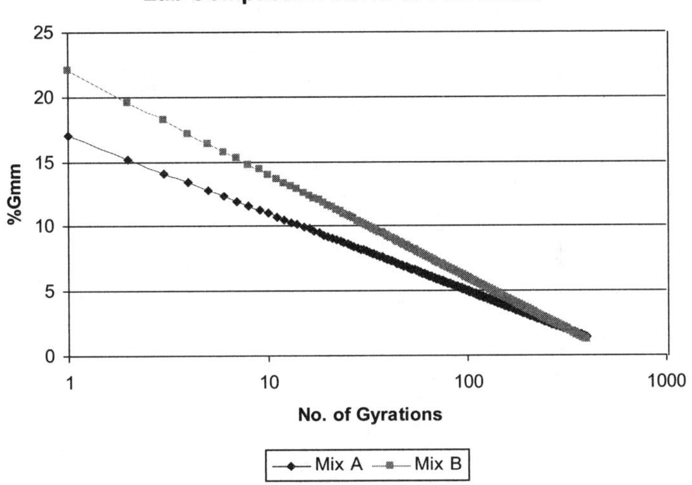

FIGURE 11.5   Compaction of two different mixes.

Future research on the characterization of K_g, K_m, and K_b through analyzing experimental data, micromechanics models, and computational simulation is a viable topic.

A third view of looking at the compaction curve is in terms of the stiffening of the mixture due to the reduction in air-void content and the particle orientation. Figure 11.6 shows some experimental correlations (NCHRP 478), where the shear modulus and the slope K^l are correlated. This makes sense in that by first degree mixture theories, the shear modulus is a function of the air-void content; the larger the slope, the easier the mixture is compacted to a dense configuration. More complicated micromechanics theories not only incorporate volumetric compositions and modulus of the components, but also the shape of inclusions. This leads to the micromechanics approach in the next

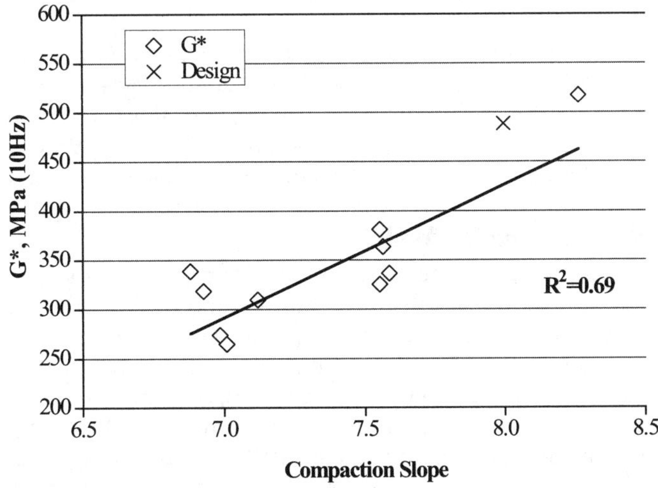

FIGURE 11.6   Stiffness and the slope relationship.

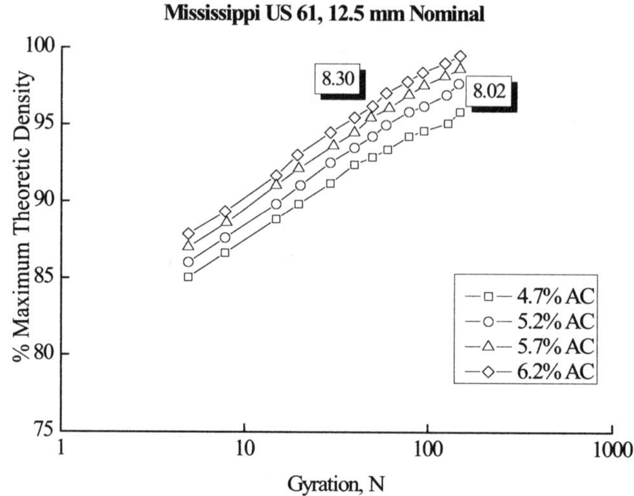

FIGURE 11.7 Lab compaction characteristics at different binder contents.

section, which places the predictions of the slope in a more rigorous mechanics framework. Currently, most intelligent compaction models are based on the stiffening concept that is when air void content is reduced, it makes the mixture stiff.

There are also experimental observations to relate asphalt binder content with compaction (Figure 11.7, NCHRP 478). It is shown that the compaction curve will shift the initial void content (the percent Gmm, the theoretical maximum density) at the end of cycle 1. The curves are almost parallel. It makes sense in that, for the same type of binder, its rheology properties are the same, as is the slope (the same gradation and the same aggregates). It also affects the final maximum achievable density. Different asphalt contents, however, will make a difference for the initial and final density. In other words, this observation indicates that both K^l and A_{v0} are affected by asphalt binder content. The term A_{v0} may also be very much related to the air void content of un-compacted aggregates.

It should also be noted that although there are fewer datasets available on the field compaction of AC, quite a few of those datasets demonstrate similarities between field compaction and lab compaction; for example, the density increase versus the shear modulus relationship presented in Figure 11.8 (from Sakai). A modeling technique also provides a similar relationship for field compaction by Wang et al. (2007).

11.2.2 Micromechanics Models

Both mechanics principles and experimental data point to the fact that the slope K is directly related to the mixture shear modulus and thus, bulk modulus and dynamic modulus (assuming the relationship among the shear modulus, Young's modulus, and the Poisson's ratio exists). Therefore, finding the slope and the modulus relation, and the modulus-component properties relationship, becomes essential.

Micromechanics models to predict the mixture modulus have been developed by Voigt (1889), Einstein (1911), and Reuss (1929). Buttlar and Dave (2005) presented a comprehensive summary on the use of models with both non-interacting and interacting particles. They have shown that existing micromechanical models, such as the

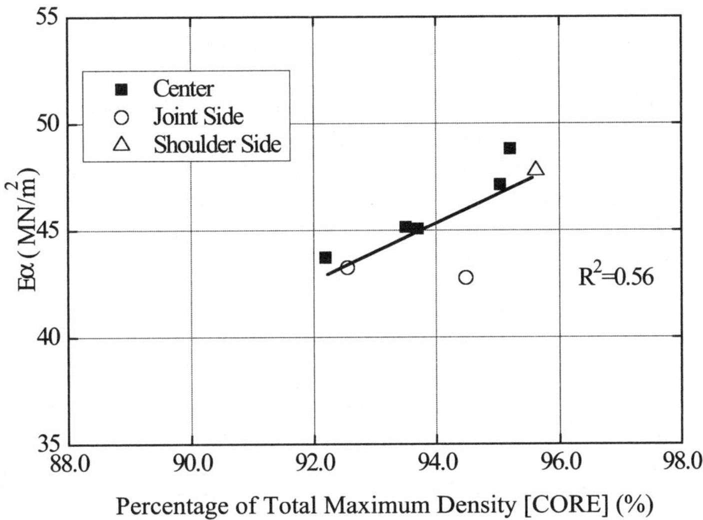

Figure 11.8   Stiffness versus density during breakdown rolling (Sakai).

composite spheres model and the arbitrary phase geometry model, do not adequately describe the complex microstructure of asphalt mixture. Most micromechanical models (Li et al. 1999; Papagiannakis et al., 2002; Li and Metcalf, 2005) may either over- or under-predict the stiffness (or modulus) of asphalt mixtures (Buttlar and You, 2001; Buttlar and Dave, 2005). This is primarily attributed to the inability of the models to account for the contribution of the aggregate interlock (i.e., lack of microstructure in the model) to the overall response of the mixture (You and Buttlar, 2004, 2005). Nevertheless, these models may serve as the basis formula on which modifications may be developed to include experimental calibrations.

Evaluations of the mixture modulus from the modulus of the components and their volume fractions are quite often used in AC with the use of experimental calibrations. Tremendous work has been done in this area. From the rule of mixture, to Eshelby tensor, to computational assessments, these micromechanics models present tools to predict the mixture properties for guiding mix design. A model that is popularly used in the AC area is the Hirsch model (Christensen et al., 2003; Chapter 5), which is actually the application of the rule of mixture. To avoid complicated formulas, this model basically shows that the mixture modulus is a function of the aggregate modulus, binder modulus (or mastic modulus), their corresponding volume fractions, and voids in mineral aggregate (*VMA*) and voids filled with asphalt (*VFA*).

$$E_c = f(E_a, E_b, VMA, VFA) \tag{11-5}$$

E_c, E_a and E_b are the mixture, aggregate, and binder modulus, respectively. It should be noted that VMA, VFA, $V_{\text{beffective}}$ (effective volume of binder), and Va (air-void content) are related.

Therefore, the mixture modulus is related to air-void content. When air-void content decreases, the mixture modulus increases. To a certain degree this model shows that the consolidation process will stiffen the mixture. It should be noted that E_b, the binder modulus, is a function of temperature and loading frequency. In applying this approach, the binder master curve or mastic master curve and the time-temperature superposition principle may be used.

There are many other analytical micromechanics formulas (upper or lower bounds) to estimate the mixture modulus from the components' moduli and their volume fractions. A simple one, the Mori-Tanaka method, may have its advantages due to it simplicity (Wang and Lai, 1997). Typically, these methods predict only the isotropic modulus. Formulas based on the Eshelby mechanics (Chapter 5), on the other hand, can account for the anisotropic modulus resulted from ellipsoid inclusions. This approach translates the predictions as:

$$E_{ijkl} = f(E^a_{ijkl}, E^b_{ijkl}, v_a, v_b, a/b, b/c) \tag{11-6}$$

E_{ijkl}, E^a_{ijkl} and E^b_{ijkl} are the elasticity tensor for the mixture and component a and component b, respectively; v_a and v_b are the volume fractions of the two components; a, b, and c are axial lengths of the ellipsoid inclusions. By this formula, the moduli in different directions may not be the same when the particle shape and orientations are different. Even with the same particles, when the particle orientations change, the anisotropy will change as well. During the compaction, aggregates of different shapes may change their orientations, resulting in the evolutions of the anisotropy. In summary, through this set of formulas, volume fractions, aggregate shapes and their orientations, and interlocking are important to the stiffening of the mixtures (e.g., the vertical modulus).

As particle shapes and their orientations are complicated in estimating the composite modulus, a powerful approach is the computational mechanics approach to calculate the effective modulus of a mixture by including its microstructures in FEM simulations (Zhang and Wang, 2007).

A final note is that the micromechanics method has been widely applied in pavement design since 1978 (Shell). The Shell normograph, although based on tremendous experimental data, is actually a micromechanics approach. It estimates the mixture stiffness from the binder modulus, penetration grade, temperature, volume fractions of binder and aggregates, and vehicle speeds.

11.2.3   Macro Thermodynamics Modeling

Krishnan and Rao (2000) developed a mixture-theory-based constitutive model (see also Chapter 5) and performed simulation of the air-void reduction in the compaction process. Krishnan and Rajgopal (2004) further developed a constitutive model that is based on fundamental thermodynamics principles.

Koneru et al. (2008) developed a thermodynamics framework for a constitutive model of AC and applied it to solving the gyratory compaction of asphalt concrete. An analytical solution was derived for the 1D case (gyratory compaction). The work was then extended to the 3D case using FEM code ABAQUS with UMAT implementation.

11.2.4   DEM Simulations

Analytical micromechanics estimations are usually only the upper or lower bounds of the modulus and are static. Actual solutions typically require incorporating the microstructure and computational methods such as FEM. Considering the special properties of AC in compaction, when binder is very soft and aggregate particles almost translate and orientate as rigid bodies, the DEM approach provides a viable tool. In recent years, Fu (2005) and Fu et al. (2007) developed a clustering DEM approach, to include real particle shapes in the DEM simulations of true 3D behavior of granular materials for both compression and shearing tests. You and Buttlar (2004, 2005) developed a micro-fabric DEM approach for 2D simulations. These approaches show great advantages in handling particle movements in large displacements and rotations. These developments, in conjunction with X-ray computed tomography, allow for accurately modeling the compaction process and experimentally validating the models. Wang et al. (1999, 2007) also developed methods to trace particle movements and compute the effective strains. This approach can be used to assess the particle shape, angularity, and texture properties, and also provide information for model verification. The quantification methods (Wang et al., 2004) can be used to calculate the shape, angularity, and texture properties of aggregates for more accurate contact modeling. The rough contacts, according to contact mechanics, make contact stiffer, and therefore qualitatively agree with the experimental observations that rough aggregates yield stiff mixes.

Another advantage of the DEM approach is its ability to handle large deformations. This is especially useful in modeling the field compaction, where the layer thickness might be 2.5 to 3.5 times the maximum size of aggregate particles. In that case, individual particle movements may significantly affect the compaction effectiveness and uniformity. The number of particles may not be so large so the computation power may not be that demanding.

For the gyratory compaction, the digital test and digital specimen (Chapter 10) approach can also be utilized to model the compaction process. This approach, in conjunction with the particle identification techniques (Fu, 2005; Fu et al., 2007; Wang et al., 2007b) presents a very powerful tool to experimentally observe how aggregate particles move during the compaction process.

More recently, a viscoleasticity contact model has been developed and validated through X-ray images (Wang, 2007). Figure 11.9 illustrates the experimental validation of the contact model through X-ray tomography imaging, where a binder film between two aggregate particles under compression load were modeled and tested.

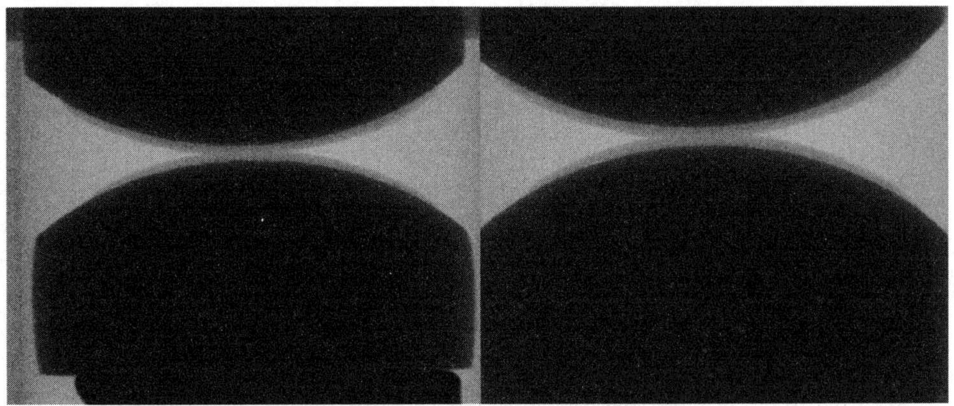

FIGURE 11.9 Validation of a viscoelasticity contact model.

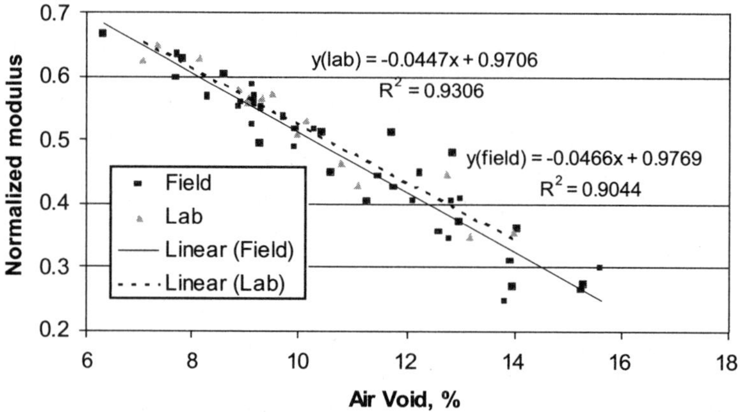

Figure 11.10 Normalized modulus versus air-void levels on DEM models from laboratory and field specimens (courtesy Zhanping You).

You recently used a relatively simple model to obtain consistent results with experimental observations (Figure 11.10). Through incorporation of the shape and visco-elasticity contact models as described before, more realistic results can be obtained. Future research in this area can help better understand the compaction mechanism.

Through modeling the compaction process using the DEM approach, aggregate shapes and particle kinematics during the compaction can be modeled. The differences in slopes of several specially designed mixes (designed to evaluate the binder stiffness and temperature effects, aggregate gradation effects, and aggregate shape, angularity, and texture effects) can provide the calibrations of the K_g, K_m, and K_b.

11.3   Field Compaction

11.3.1   Macro Constitutive Models—Porous Viscoplasticity Model

There are few studies in the continuum regime for investigating the compaction mechanisms. The use of mixture theory, to a certain degree, explains void reduction during the compaction process. One of the more popular methods is using a viscoplasticity model in conjunction with FEM numerical simulations. Huerne (2004) investigated the simulation of the compaction process with an analogy between hot asphalt mixtures and wet soils. A critical state soil mechanics model (Wood, 1990) was used as a fundamental model to describe the behavior of asphalt mixture in his study. He analyzed the influence of the compaction rate on mechanical properties of asphalt mixtures with the help of FEM simulation. A non-standard FEM approach, the arbitrary Lagrangian Eulerian (ALE) method, was used to simulate the behavior of mixture during compaction considering the fact that the behavior of the material during this stage is between a solid and a liquid (Huerne, 2004).

The compaction process involves large irrecoverable deformations. It is also rate dependent. Therefore, it requires typically the viscoplasticity models. Quite a few viscoplascity models have been proposed in recent years to model the properties of AC (see the references in Chapter 7 and Chapter 10). The parallel layer model (Kichenin et al., 1996) has proven to be quite accurate in predicting the permanent deformations of AC. This model has also been used by Scarpas and many others. Zhang and Wang (2007) used this model to represent the mastic properties and successfully applied it to model the static compression of AC. Model predictions and experimental observations agree very well. Zhang and Wang's work falls into the theme topic of digital specimen and digital test techniques where the microstructure of asphalt concrete is considered.

Another model is the porous viscoplasticity model. This viscoplasticity model takes the void content as an internal variable and has been successfully used by Wang et al., (2007a) to model the field compaction process. This model is described here in more detail.

The model is based on the theory proposed by Gurson (1977) and modified by Tvergaard (1981). Rather than assuming plastic incompressibility, the porous plastic model considers hydrostatic components of stresses and strains. It takes the effect of void nucleation and growth into account. It shows the importance of plastic dilatation. Guler *et al.* (2002) used this Gurson-Tvergaard model in compaction simulation trying to calibrate the material parameters using experimental results.

The material model includes the following components:

- Linear elasticity:

$$\varepsilon_{ij}^{el} = C_{ijkl}^{el}\sigma_{kl} \tag{11-7}$$

- Yield function dependent on confining stress:

$$F = \left(\frac{q}{\sigma_y}\right) + 2q_1 f \cosh\left(-q_2\frac{3p}{2\sigma_y}\right) - \left(1 + q_3 f^2\right) = 0 \tag{11-8}$$

- Associated plastic flow:

$$\dot{\varepsilon}_{ij}^{pl} = \dot{\lambda}\frac{\partial F}{\partial \sigma_{ij}} \tag{11-9}$$

- Isotropic strain hardening:

$$\sigma_y = \sigma_y\left(\bar{\varepsilon}_m^{pl}\right) \tag{11-10}$$

The evolution of the equivalent plastic strain:

$$\left(1 - f\right)\sigma_y\dot{\bar{\varepsilon}}_m^{pl} = \sigma_{ij}\dot{\varepsilon}_{ij}^{pl} \tag{11-11}$$

The evolution of the volume fraction f:

$$\dot{f} = \left(1 - f\right)\dot{\varepsilon}_{ii}^{pl} + A\dot{\bar{\varepsilon}}_m^{pl} \tag{11-12}$$

Rate dependent yielding can be modeled in ABAQUS and the failure can be modeled in ABAQUS explicit. Temperature-dependent material parameters can be defined as a tabular function of temperature.

Notation

σ_{ij}, ε_{ij}	The total stress and strain tensor
ε_{ij}^{el}	The elastic strain stress tensor
D_{ijkl}	The elasticity tensor
p	The hydrostatic pressure, $p = -\dfrac{1}{3}\sigma_{ij}\delta_{ij}$
S_{ij}	The deviatoric stress tensor, $S_{ij} = \sigma_{ij} + p\delta_{ij}$
q	The effective Mises stress, $q = \sqrt{\dfrac{3}{2}S_{ij}S_{ij}}$
σ_y	The yield stress of the fully dense material ($f = 0$), $\sigma_y = \sigma_y\left(\bar{\varepsilon}_m^{pl}\right)$
$\bar{\varepsilon}_m^{pl}$	The equivalent plastic strain in matrix $\bar{\varepsilon}_m^{pl} = \sqrt{\dfrac{2}{3}\varepsilon_{ij}^{pl}\varepsilon_{ij}^{pl}}$
f	The void volume fraction
q_1, q_2, q_3	Material parameters
A	$A = \dfrac{f_N}{s_N\sqrt{2\pi}}\exp\left[-\dfrac{1}{2}\left(\dfrac{\bar{\varepsilon}_m^{pl} - \varepsilon_N}{s_N}\right)^2\right]$
ε_N	The mean value of nucleation strain
s_N	The standard deviation of nucleation strain
f_N	The volume fraction of the nucleated voids (in tension)
EVOL	Element volume
VVF	Void volume fraction

The application of this model to the simulation of the field compaction will be presented following a discussion about the field compaction process and influencing factors.

Considering the field conditions, more properties need to be considered. According to Wang et al. (2007a), factors that affect compaction can be generally classified into three categories: properties of the materials, environmental variables and boundary conditions, and compaction equipment and operation control. The properties of the materials include the gradation, shape, angularity, and texture of aggregates, the grade, and the amount and the temperature sensitivity of the binder. The environmental variables and boundary conditions include air and base temperatures, humidity, sunshine or cloudiness and wind velocity, layer thickness, stiffness of the underlying layer, and mix lay-down temperature. Layer thickness is one of the most important variables that affect other influencing factors such as the rate of cooling of the asphalt mixtures, and the particle movement during compaction. Compaction equipment and operation control include types of roller, rolling patterns, number of passes, and rolling speed. These factors can be adjusted during the compaction. An understanding of the compaction mechanism plus the intelligent compactor technique (Briaud, 2003) will certainly enhance the compaction quality significantly (intelligent compaction requires a fundamental understanding of the compaction mechanism to make it wise; wrong understanding would make it unwise). The factors in the three categories are listed as follows (see Figure 11.11).

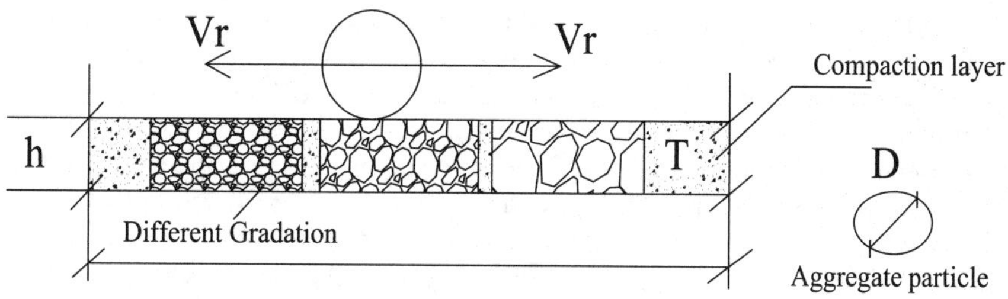

FIGURE 11.11   Illustration of the compaction model.

Properties of the Materials

Aggregate gradation (filler percentage as well)
Maximum size of aggregate (or the dominant aggregate size) D
Aggregate shape, angularity, and texture
Asphalt content
Asphalt type and temperature sensitivity

Environmental Variables and Boundary Conditions

Mixture temperature when laid T_l
Air temperature T_a
Base temperature T_b
Wind velocity V_w
Other weather conditions (sunshine, cloudiness, humidity)
Layer thickness h
Stiffness of the lower course S
Uniformity of the above parameters

Compaction Equipment and Operation Control

Type of compactor
Rolling pattern
Number of roller passes N_r
Rolling speed V_r
Compaction time t

Obviously, a thorough investigation to explore how these factors affect compaction independently or in a coupled manner would require a very complicated model or a huge experimental factorial and tremendous costs. Identification of the critical factors would help reduce the efforts, while still enabling solving the essential problems in achieving effective compaction. A literature review indicates that critical factors include layer thickness, gradation, dominant aggregate particle size (the size that constitutes the largest percentage in the gradation), aggregate shape and angularity, temperature (binder viscosity), base/sub-base support, and the number of passes (the compaction system is assumed to be fixed). In addition, environmental conditions are much more difficult to control and fundamental mechanisms at the materials scale (material properties are sensitive to the environmental conditions) cannot be properly investigated if an experimental method is adopted because of the difficulties in controlling the environmental conditions. Therefore, an advanced modeling and computational simulation approach is advantageous. The DEM has the advantage of dealing with the components' properties separately (aggregate, binder, and void) while the FEM, with a continuum type of modeling using porosity as an internal variable, can conveniently model the reduction of air voids in the compaction process. Investigations using both methods will be presented in this section.

While the described simulation procedure can be used for simulation of a compaction process, industrial applications may seek empirical relationships developed from either experiments or numerical simulations. It is anticipated that the void reduction process in field compaction can be similarly represented as a function similar to the lab compaction with:

$$K^f = K_g K_m K_b K_h K_l K_s \qquad (11\text{-}13)$$

Where additional layer thickness factor K_h, load magnitude factor K_l, and soil support factor K_s are introduced. Other factors may need to be introduced, such as the loading speed factor and temperature drop factor. It should be noted that the aggregate morphology factor is actually the combined shape, angularity, and texture factor; the temperature factor may be included in the binder factor. It should also be noted that K_g, K_s, K_b may not be the same as those for laboratory compaction.

The following presents the simulation of field compaction conducted by Wang et al. (2007a) to provide a guide in identifying critical compaction factors. Wang et al. (2007a) used the porous viscoplasticity (Gurson-Tvergaard) model to study the effects of a few influencing factors. The purpose of that study was to determine dominating factors and to show whether modeling techniques can capture the essential compaction features. In this compaction simulation model (Figure 11.12), the roller was assumed to be rigid and the porous plasticity model was used for AC. Besides the constant loading applied on the roller, a vibration force with a smaller magnitude was also applied on the roller. The simulation was conducted using ABAQUS.

Using this model the effects due to compaction force, thickness, and soil support factors were evaluated. The variables used for the simulations include:

Forces applied on the roller: 1F, 2F, 3F, 4F
Vibration of the roller: 0.02F*sin (62.8T)
Thickness of the AC section: 1.0t, 1.5t, 2.0t, 2.5t
Where $F = 5000$N, $t = 30$mm, and T is the loading time.

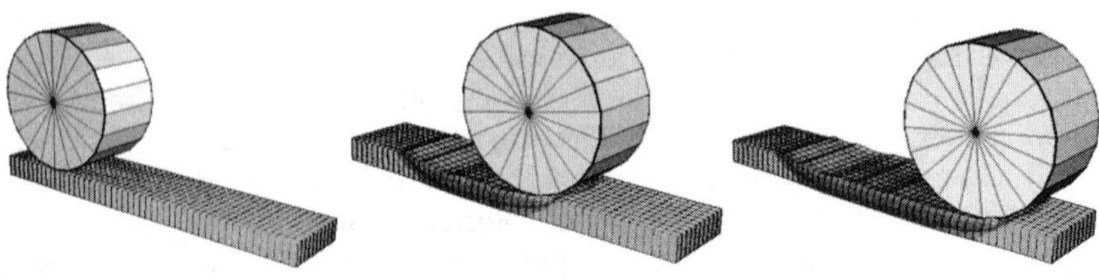

Figure 11.12   Finite element model for compaction simulation.

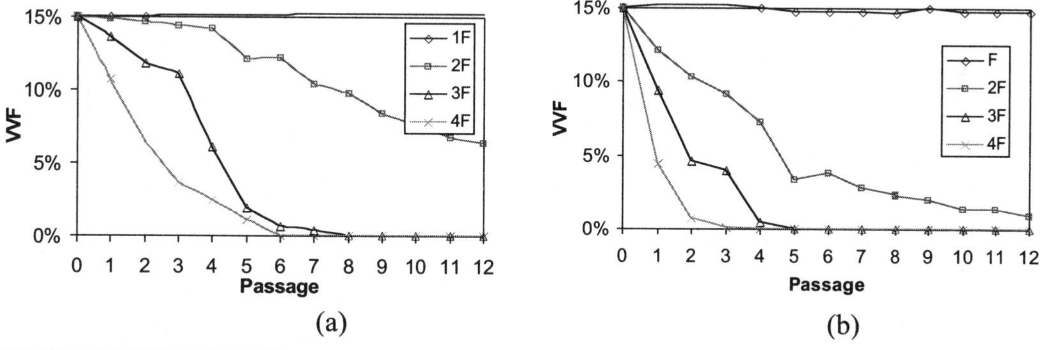

(a)　　　　　　　　　　　　　(b)

Figure 11.13 Change of VVF under different roller pressures at (a) bottom and (b) mid height of a typical section.

The changes of VVF (void volume fraction, or air-void content) at the bottom and mid-height of the AC section against the roller passage are plotted in Figure 11.13. It can be seen that the targeted compaction air-void content (or density) can be achieved much earlier (four to six passages) for forces $3F$ and/or larger in this case. Figure 11.13 also indicates that the bottom layer is less effectively compacted.

To further explore how compaction is achieved at different locations of a layer, responses of sub-layers corresponding to different locations are presented in Figure 11.14. It can be noted that the layer at the bottom is typically less compacted. With larger compaction forces, the compaction rates increase.

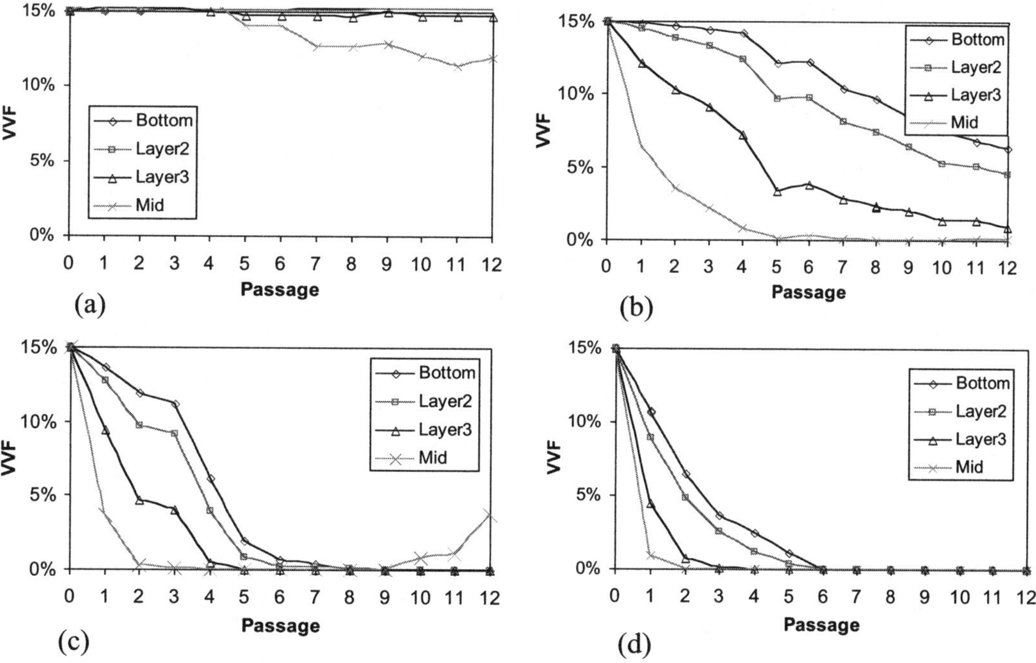

(a)　　　　　　　　　　　　　(b)

(c)　　　　　　　　　　　　　(d)

Figure 11.14 Change of VVF at different locations under different roller forces: (a) 1F, (b) 2F, (c) 3F, and (d) 4F.

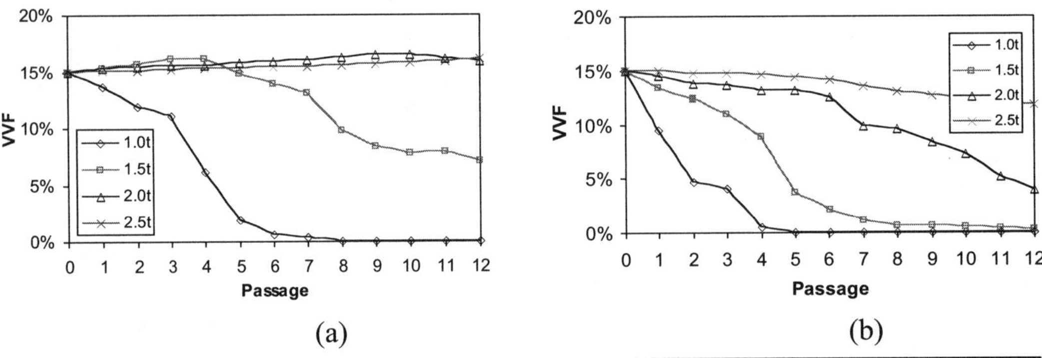

FIGURE 11.15 Change of VVF for AC with different thickness at (a) bottom and (b) mid height of the compaction section.

The VVF was obtained at the bottom, the mid-height, and other two points (Layer 2 and Layer 3) between them.

Figure 11.15 presents another view of looking at the compaction effectiveness. It presents the compaction process when the same compaction force is applied to different layer thicknesses. It can be noted that when the layer thickness increases, the same compactive efforts will achieve less compaction.

Figure 11.16 shows that the thickness and loading magnitude effects on maximum compaction deformation are not linear. When the thickness is too small it may not achieve compaction effectively. This is consistent with field compaction observations.

Figure 11.17 shows the effect of soil support. It indicates that soft soil will reduce compaction effectiveness. This is rational as more compactive energy may be transferred to soft soil.

The simulation results indicate:

- Optimal compaction results depend on both compaction pressure and the layer thickness of asphalt concrete. An optimum interrelationship can be determined from the simulation.

- The number of passages required for compaction could be determined from the EVOL and VVF evolution curves.

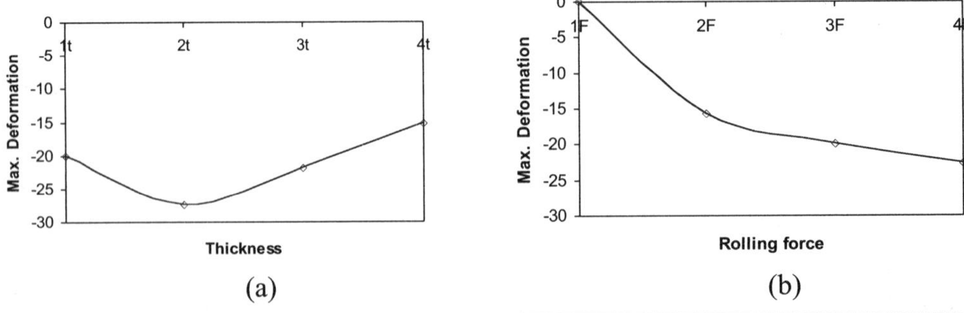

FIGURE 11.16 Center maximum deformations versus (a) thickness and (b) compaction pressure.

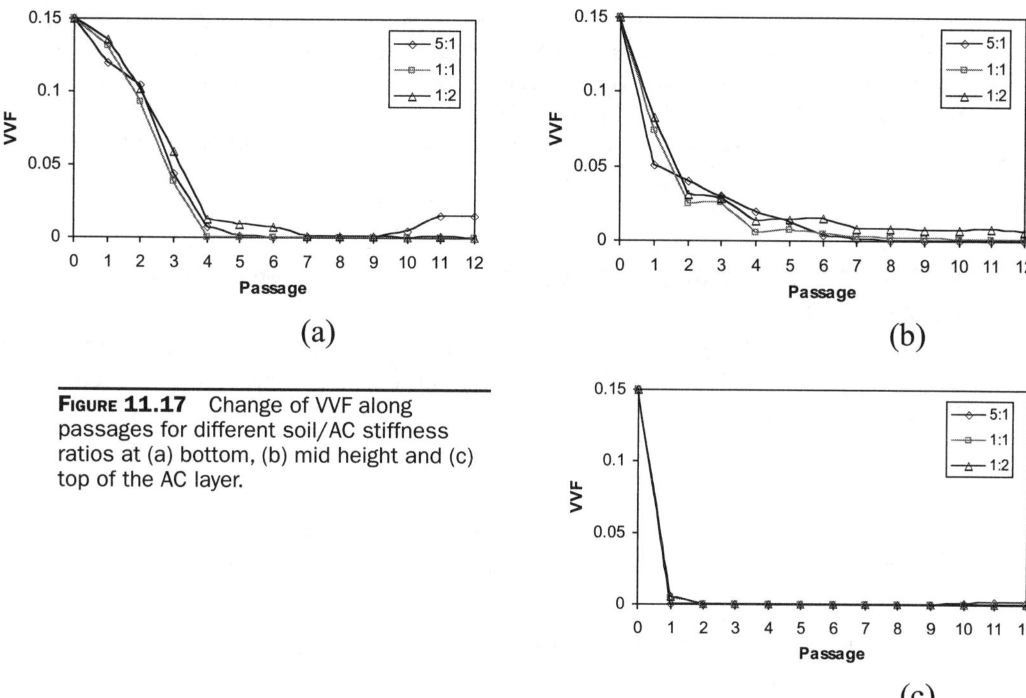

FIGURE 11.17 Change of VVF along passages for different soil/AC stiffness ratios at (a) bottom, (b) mid height and (c) top of the AC layer.

- The optimum compaction passages, compaction pressure, or the thickness of the compaction layer could be determined based on compaction simulations.

- Stiffer sub-base material will lead to more effective compaction of the AC layer above.

- The effect of the stiffness of the base/sub-base on the compaction is greater at the bottom of the AC layer than at the top.

- If the stiffness of the base/sub-base material is smaller than the stiffness of the AC layer during compaction, it may cause compaction problems.

These results indicate that the model and FEM simulation of compaction can be used to evaluate factors affecting compaction. With good calibration, the model may capture the compaction characteristics well.

Other unique models that can be used for modeling compaction include a model developed by Tashman et al., (2007). This model introduces one aggregate orientation parameter and the modified effective stress concept. However, this particle orientation parameter can be only a constant or static in that model. There is no evolution law to govern the change of this parameter during compaction. The modified effective stress is adopted from granular mechanics where there is no other constraint except particle-particle contact. The model developed by Wang et al. (2003) is a mixture-theory-based model. Although it includes void as a component, it does not involve repetitive loading. Therefore, it can be used in modeling static loading only.

The above evidence indicates that viscoplasticity models can be used to model the compaction process. The model parameters such as the elasticity modulus, the yielding stress, and other parameters are related to the binder stiffness, aggregate gradation, shape, angularity, and texture. Through calibrating the model constants of a limited number of selected mixes (designed to evaluate the binder stiffness and temperature effects, aggregate gradation effects, and aggregate shape, angularity, and texture effects), the model predicted compaction process will yield differences in slope and the initial void content; these will present the calibrations of both laboratory and field compaction slopes. The ratio between the field slope and the lab slope presents a measure of the shift factor.

11.3.2 Macro-Constitutive Models-Critical State Soil Mechanics (CSSM) Model

Huerne (2004) made an analogy between soil and AC and applied the CSSM (Wood, 1990) model to simulate asphalt compaction. Huerne's theory is summarized as follows:

1. *p-q* space where:

$$p = \frac{\sigma_{xx} + \sigma_{yy} + \sigma_{zz}}{3}, \quad q = \left[\frac{(\sigma_{xx} - \sigma_{yy})^2 + (\sigma_{yy} - \sigma_{zz})^2 + (\sigma_{zz} - \sigma_{xx})^2}{2} + 3(\tau_{xy}^2 + \tau_{yz}^2 + \tau_{zx}^2) \right]^{1/2} \quad (11\text{-}14)$$

2. Mohr-Coulomb yield surface and critical state

3. Associated flow

4. VMA and void ratio equivalency:

Recoverable behavior: $VMA = VMA_k - k\log(p)$

Irrecoverable behavior: $VMA = VMA_\lambda - \lambda\log(p)$

Where VMA_k, k, VMA_λ, and λ are material constants related to aggregate gradation characteristics, and binder contents.

Huerne also introduced the modeling of cooling through the use of Fourier's law and the empirical relationship to temperature drop. The FEM simulations of experimental sections indicate that the model and the FEM procedure simulate the compaction process nicely.

11.3.3 DEM Simulation of Field Compaction

In investigating how important factors influence the field compaction of pavement materials, the 3D particle flow code (PFC3D, Itasca, 2003), which is based on DEM, is used as the simulation tool (Figure 11.18). The investigation includes:

- Particle shape effect;
- Particle contact property effect; and
- Temperature effect.

11.3.3.1 Particle Shape Effect

The DEM view can look at how aggregates move during the compaction process. The particle shape effect is evaluated through clustering small balls together. Case 1 (Figure 11.19) is the simplest case for balls only. Case 2 clusters two balls into a rectangular

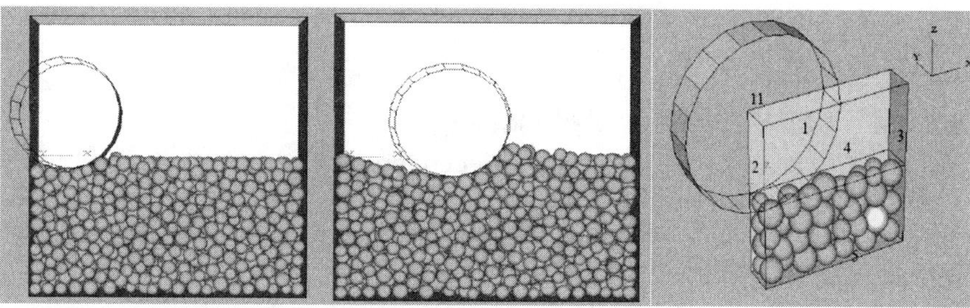

Figure 11.18   Illustration of the DEM model.

shape, while Case 3 clusters three balls into a triangle. If the ratio between the largest dimension and the smallest dimension is considered as a shape factor, then Case 1, Case 2, and Case 3 would have a shape factor equal to 1, 2, and 1.07, respectively. During compaction, the particle rotation is one of the important kinematics parameters that affect the compaction of the mixture. It may be intuitively interpreted that the more easily a particle can be rotated, the less difficult to move the particle into a denser configuration. The results indicate that the ball is easiest to rotate, followed by the triangle and the rectangular. If this simulation observation can be extended, one of the rational guesses is that particles of cubical shapes are easier to compact than flat or elongated particles (these particles may break, however). This is consistent with field observations.

11.3.3.2 Particle Contact Property Effect

For two particles of the same materials, it is understood that the contact normal stiffness of the two particles would be larger for rougher particles. The contact normal stiffness could also be larger if the mastics between the two particles are stiffer. By assigning a larger normal stiffness to the contact pair, the investigations by Wang et al. (2007) indicated that rougher aggregates or stiffer mastics/binder would make asphalt mixture more difficult to compact.

11.3.3.3 Temperature Effect

When the mixture temperature drops during transportation, the stiffness of the binder/mastics will increase, resulting in an increase of the normal contact stiffness. The simulation indicates that the lower the temperature, the more difficult to compact. It may also be deduced that when the surface temperature drops during compaction, forming a relatively stiffer layer at the surface, the mixture would be more difficult to compact. It may be summarized that:

- Mixtures with aggregates of elongated and flat particles, larger roughness, and coarser textures are more difficult to compact.

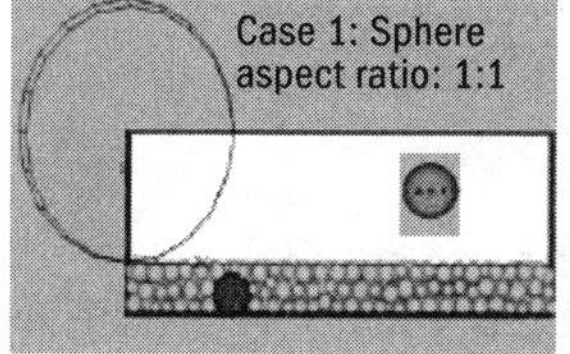

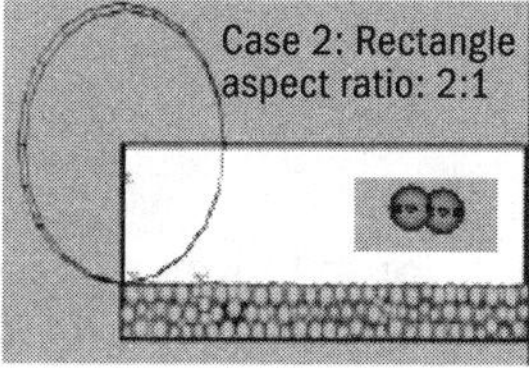

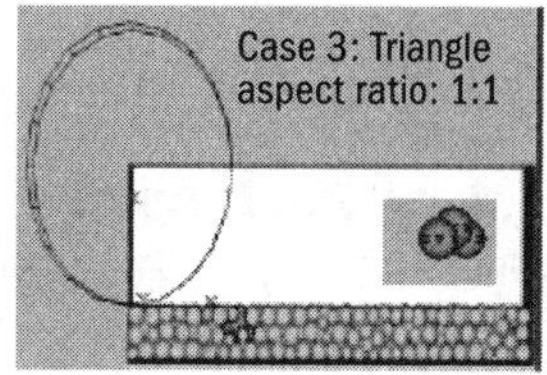

Figure 11.19   Particle shape effect.

Temperature drop will make compaction more difficult due to the larger stiffness for particle contacts and/or a stiffer top layer.

11.3.4 CAP Viscoplasticity Modeling

Xia and Chi (2008) adapted a rate-independent crushable foam plasticity model into a viscoplasticity model. They developed a user's material subroutine for ABAQUS and simulated the roller compaction process. The essential features of the model include non-associated flow represented by the yielding function F and the potential function g.

$$F(\sigma) - \sqrt{\frac{3}{2}\|s\|^2 + \alpha^2(p - p_0)^2} + B - 0 \quad \text{and} \quad g(\sigma) = \sqrt{\frac{3}{2}\|s\|^2 + \beta^2 p^2} \qquad (11\text{-}15)$$

Where p is the first stress invariant, and S is the deviatoric stress tensor. It is an extension of the CAP model. They used the additive strain decomposition and the Perzyna (1966) viscoplastic strain versus potential gradient relation, Equations 11-16 and 11-17.

$$\dot{\varepsilon} = \dot{\varepsilon}^e + \dot{\varepsilon}^{vp} \qquad (11\text{-}16)$$

$$\dot{\varepsilon}^{vp} = \eta \left\langle \Phi\left(\frac{F}{\sigma_y}\right)\right\rangle \frac{\partial F}{\partial \sigma} \qquad (11\text{-}17)$$

No comparison between simulation results and experimental observations is available.

11.4 Empirical Measurements of Compactibility

While the previous sections discussed the theories and backgrounds of compaction, this section summarizes some literature on the historical development of empirical measurements of compactibility. Although most of these measurements do not fall into the rigorous scheme of mechanics, they may help guide developing simple mechanics models.

11.4.1 The *C* Value Approach

Kezdi (1969) developed an empirical formulation that associates the densities of the material at different compaction stages with the accumulative compaction energies.

$$\rho_A(S) = \rho_{A\bullet} - (\rho_{A\bullet} - \rho_{A0})e^{\frac{S}{C}} \ [\text{kg/m}^3] \qquad (11\text{-}18)$$

Where S is the compaction energy, $\rho_{A\bullet}$ and ρ_{A0} are the final and initial densities, and C is a measurement of the compactibility. For $C = 10$, the material is easy to compact, while $C = 30$ is difficult to compact. Renken (1980) and Arand (1985, 1987) have also employed this approach.

11.4.2 The R_f Approach

Nijboer (1948) used Mohr-Coulomb failure criterion and triaxial tests to characterize the compactibility of asphalt concrete. He assumed that the internal friction angle, the "cohesion" resistance, and the viscosity of the mixture can be used to measure the compactibility of asphalt concrete and developed the R_f factor as defined in the following equation.

$$R_f = \frac{\frac{P}{LD} - R_t \tau_{cb}}{\eta_m} n \left(\frac{h}{v} \right)^{0.4} \quad \text{[N.mm.sec]} \tag{11-19}$$

where P = weight of the roller (N)

   L = width of the roller drum (mm)
   D = diameter of the roller drum (mm)
   R_t = roller type factor (dimensionless)
   τ_{cb} = shear cohesion of bituminous mixtures (N/mm^2)
   η_m = viscosity of the compacted mixture (poise)
   n = number of roller passes
   h = thickness of asphalt layer (mm)
   v = speed of the compactor (mm/s)

Vizi (1981) also used this approach to evaluate compactibility.

11.5 Microscopic Study of Lab Compaction Process

11.5.1 Void Distribution and Variation Using XCT

A more recent systematic study at the microscopic level on lab compaction methods including Marshall, Gyratory, and LCPC rolling-wheel-compactor (RWC) has been achieved by Partl et al. (2007a, 2007b). In this study, specimens of the same mix were compacted using the above compaction methods at the initial, medium, and final compaction. The void distribution in these mixes was evaluated using computerized tomography. Its discoveries include: 1) the void reduction in the Marshall compaction is almost linear; LCPC RWC is non-linear and the gyratory compaction is somewhere between; 2) the void distribution within the specimen is non-uniform; and 3) there is significant lateral movement of the mixtures during the RWC compaction. The structurally different specimens produced in each of the compaction methods indicate that these methods may not be able to produce specimens whose structure is similar to those produced in the field.

11.5.2 Other Experimental Studies

Masad et al. (2002) studied the gyratory compaction process using XCT. They analyzed the void distributions of the gyratory specimens and discovered that void contents at both the bottom and top of the specimens are much higher than the void content at the middle. In that investigation, compaction energy and the dissipation process were also investigated.

11.6 Perspective on Gradation Effect Representation

In applied research, empirical representation of compatibility is highly practical to material and paving engineers. Development of these formulations will help practicing engineers to assess the compatibility of a mix conveniently. While Section 11.4 presents some of these empirical relationship, one parameter on the gradation is difficult to be carried in these formulations. Of the inputs on the material sides, representing a gradation is challenging.

Taking the formulation for the compaction slope as an example, a reference case may be specified, which might be a uniform size of marble spheres of 0.5 in as the reference gradation and aggregates. For this material, the gradation factor is defined as 1.0, the combined aggregate shape, angularity, and texture contribution factor is also set to 1.0 approximately. In order to calibrate it, three different aggregates (including the uni-size marble spheres) shall be characterized for the shape, angularity, and texture factors. For this case, uniform size aggregates, asphalt binder content, and asphalt rheology contributions can be well calibrated through mixing and compacting the mixture at different temperatures. For this reference case, through X-ray scanning of the aggregate locations, the compaction process can be accurately modeled.

Aggregate gradation design, as part of the SuperPave mixture design method, is a process of selecting aggregate blends to ensure appropriate volumetric properties such as VMA for the mix stability and durability; suitable maximum aggregate size for compaction; and adequate aggregate skeleton for strength. The current SuperPave mix design method, however, does not provide specific guidance in the selection of gradation, and the gradation design is still a trial and error process. In addition, the traditional gradation design method requires a series of percentages of aggregate particles passing a certain number of sieve size to describe a complete gradation blend. There is no single parameter or parameters that can be used for evaluating a gradation and linking the gradation to other performance parameters. Representing a gradation is still a challenging problem.

11.7 Intelligent Compaction

One of the important applications of modeling and simulation of the compaction process is interpretation of the compaction data collected automatically during the compaction. Intelligent compaction (IC) is supposed to provide oversight and better control of the compaction process and results in improved quality and productivity of roadway pavements.

IC originated in the late 1970s with the work of the European equipment makers BOMAG, AMMANN, and Geodynamik that still dominate the market today. IC technology is currently used in Europe and Asia, and has recently been introduced in the United States. IC refers to the compaction of subgrade soils, unbound aggregate base/sub-bases, or AC using vibratory rollers equipped with an in-situ measurement and feedback system. Global positioning system (GPS) based mapping is often used to automate the documentation of results. By integrating measurement, documentation, and control systems, the use of IC rollers allows for real-time corrections in the compaction process.

The most common quality control test method in road construction is the spot test, including the static plate load test and the falling weight deflectometer (FWD) test for modulus; and the nuclear gauge, the water balloon, and the sand replacement tests for density. In asphalt compaction, the drilled cores, radiometric sounds (modulus), and pavement quality indicator (PQI) tests are used.

IC technology combines the continuous measurement of absolute values (density and/or stiffness) of the achieved compaction with a regulation system that uses the measured information to adapt the equipment performance continuously to the required conditions. This system controls the different compaction parameters of the roller: amplitude, frequency, and working speed (impact distance). The modulus measurements are made by the instrumented roller itself. The system is pre-conditioned with the range of acceptable modulus values (target value) and automatically adjusts the roller settings to achieve the target modulus if the readings are not within tolerance.

A dynamic compactor produces nonlinear oscillations whose characteristic properties can be described analytically using a theoretical model of the interaction between a vibratory roller and the ground. The soil-drum-interaction-force (F_B) is defined as follows in Equation 11-20 (Anderegg, 2000).

$$F_B \cong -m_d \ddot{x}_d + m_u r_u \Omega^2 \cdot \cos(\Omega t) + (m_t + m_d) \cdot g \tag{11-20}$$

In Equation 11-20, m_d = mass of the drum (kg), x_d = vertical displacement of drum (m), $\ddot{x}_d$ = acceleration of drum (m/s^2), m_f = mass of the frame (kg), m_u = unbalanced mass (kg), r_u = radial distance at which m_u is attached (m), $m_u r_u$ = static moment of the rotating shaft (kg·m), $\Omega = 2\pi \cdot f$, t = time elapsed (s), g = acceleration due to gravity (m/s^2), and f = frequency of the rotating shaft (Hz). The acceleration of the drum and the phase angle between excitation and oscillation can be measured. With this information, it is possible to calculate F_B.

If the underlying layer is described as a spring and dashpot system, the equation for the soil/asphalt-drum interaction force is given in Equation 11-21, where k_B = stiffness of soil/asphalt (kN/m), d_B = damping coefficient (kN·s/m) (a damping ratio of 0.2 is usually assumed, Anderegg, 2000), and $\dot{x}_d$ = velocity of drum (m/s).

$$F_B \cong k_B x_d + d_B \cdot \dot{x}_d \tag{11-21}$$

Combining Equation 11-20 and Equation 11-21 leads to the solution of the soil stiffness k_B, assuming the damping ratio to be equal to 20%.

Since the soil stiffness k_B is not an independent soil parameter, the soil modulus E as a more independent soil parameter can be obtained from the measured stiffness k_B, using the Lundberg solution (1939) developed based on previous work done by Hertz (1895). The Lundberg solution provides the relationship between the load on a roller and the imprint area created by the roller on an elastic half space. The solution can be used to solve for the modulus E of the material below the roller using the stiffness k_B [MN/m], as formulated in Equation 11-22, where L is the drum width, v is Poisson's ratio, m_f and m_d are the masses contributed by the frame and the drum of the roller, R is the radius of the drum, and g is the acceleration due to gravity.

$$k_B = \frac{E \cdot L \cdot \pi}{2 \cdot (1 - v^2) \cdot \left(2.14 + \frac{1}{2} \cdot \ln \left[\frac{\pi \cdot L^3 \cdot E}{(1 - v^2) \cdot 16 \cdot (m_f + m_d) \cdot R \cdot g} \right] \right)} \tag{11-22}$$

While it might be helpful to use the estimates for soil or AC modulus as compacted, direct application of the method may not be so valid because of the viscoplastic behavior of asphalt mixtures. FEM modeling of AC using advanced viscoplasticity models may present better interpretation of the compaction process.

11.8   Compaction Simulations in Other Fields

Compaction is not a problem that is unique to asphalt concrete. It is also a problem studied in soil, granular materials, and cement concrete. While these materials are significantly different, the compaction process may demonstrate similarities, for example, the log-scale law (e-logp, e-logN) relationships. At microscopic scales, these materials (granular soil, asphalt concrete, and cement concrete) have a similar aggregate (particle) skeleton with different binding media. Collaboration research among different disciplines is beneficial.

References

ABAQUS (1995). User's Manual. *Hibbitt, Karlsson & Sorensen, Inc.,*Pawtucket, R.I.

Anderegg, R. (2000). ACE Ammann compaction expert - automatic control of the compaction. *Workshop on Compaction of Soils and Granular Materials, Modeling of Compacted Materials, Compaction Management and Continuous Control, International Society of Soil Mechanics and Geotechnical Engineering (European Technical Committee).* pp.229–236, Paris.

Arand, W. (1985). Compaction of classifying characteristics for asphalt mixtures. *Procedings of Third Eurobitume symposium.* The Hague, pp.11–13.

Arand, W. (1987). Arbeitsanlagen fur die bestimmung de verdichtbarkeit von walzasphalt mit hilfe des marshallverfahrens, (Operational instructions for deduction of the compactibility of asphalt by means of the Marshall compactor). *Forschungsgesellschaft fur Strassen und verkehrswesen.*

Briaud, J.L. and Seo, J. (2003). Intelligent compaction: overview and research needs, *Texas A&M University.*

Buttlar, W.G. and You, Z. (2001) Discrete element modeling of asphalt concrete: a micro-fabric approach. *Journal of the Transportation Board*, National Research Council, Washington, D.C., No. 1757, pp.111–118.

Buttlar, W. G. and Dave, E. V. (2005) A micromechanics-based approach for determining presence and amount of recycled asphalt pavement material in asphalt concrete. *Long Beach, CA, United States: Association of Asphalt Paving Technologist*, pp.829–883.

Christensen, D. W., Pellien, T. and Bonaquist, R. F. (2003). Hirsch models for estimating the modulus of asphalt concrete. *Association of Asphalt Paving Technologist*, Vol. 72, pp.97–121.

Einstein, A. (1911). *Eine neue Bestimmung der Molekuldimensionen*, Vol.34(3), pp.591–592.

Fu Y. R. (2005). Experimental quantification and dem simulation of micro-macro behavior of granular materials using X-ray tomography imaging. *Ph.D. Dissertation*, Louisiana State University.

Fu, Y.R., Wang L.B. and Tumay, M. (2007). Quantification and simulation of particle kinematics and local strains in granular materials using x-ray tomography imaging and discrete element method. *Journal of Engineering Mechanics*, Vol.134, No.2, pp.143–154.

Guler, M., Bosscher, P. J. and Plesha, M. E. (2002). A porous elastoplastic compaction model for asphalt mixtures with parameter estimation algorithm. *15th ASCE Engineering Mechanics Conference.* June 2–5, Columbia University, New York, NY. pp.126–143.

Gurson, A.L. (1977). Continuum theory of ductile rupture by void nucleation and growth: part i - yield criteria and flow rules for porous ductile media. *Journal of Engineering Materials and Technology*, Vol. 99, pp.2–15.

Hertz, H. (1895). Über die berührung fester elastischer körper: gesammelte werke. *Bd. 1. Leipzig.*

Huerne, H.T. (2004). Compaction of asphalt road pavements using finite elements and critical state theory. *Ph.D. dissertation*, University of Twente, Netherlands.

Itasca (2003). 3 dimensional distinct element code manual, *Itasca Consulting Group*, Inc.

Kezdi, A., 1969. Handbuch der bodenmechanik (Handbook for soil mechanics). *Verlag für Bauwesen*, Berlin.

Kichenin, J., Van, K.D. and Boytard, K. (1996). Finite-element simulation of a new two-dissipative mechanisms model for bulk medium-density polyethylene. *Journal of Materials Science*, Vol. 31, No. 6, pp.1653–1661.

Koneru, S., Masad, E. and Rajagopal, K.R., (2008). A thermomechanical framework for modeling the compaction of asphalt mixes. *Mechanics of Materials*, Vol. 40, No. 10, pp.846–864.

Krishnan, J.M. and Rao, C.L. (2000). Mechanics of air voids reduction of asphalt concrete using mixture theory. *International Journal of Engineering Science*, Vol.38, No.12, pp.1331–1354.

Krishnan, J.M. and Rajagopal, K.R. (2004). Thermodynamic framework for the constitutive modeling of asphalt concrete: theory and applications. *Journal of Materials in Civil Engineering*, Vol.16, pp155–166.

Li, G., Li, Y., Metcalf, J.B. and Pang, S.-S. (1999). Elastic modulus prediction of asphalt concrete. *Journal of Materials in Civil Engineering*. Vol. 11, pp.236–241.

Li, Y.Q. and Metcalf, J.B. (2005). Two-step approach to prediction of asphalt concrete modulus from two-phase micromechanical models. *Journal of Materials in Civil Engineering*, Vol.17, No.4, pp. 407–415.

Lundberg, G. (1939). Elastische berührung zweier halbräume, *Forschung auf dem Gebiete des Ingenieurwesens*: Band 10, pp. 201–211, Göteborg.

Masad, E., Jandhyala, V.K., Dasgupta, N., Somadevan, N. and Shashidhar, N. (2002). Characterization of air void distribution in asphalt mixes using x-ray computed tomography. *Journal of Materials in Civil Engineering*. Vol.14, pp.122–129.

Masad, E.A., Kassem, E.A. and Chowdhury, A. (2009). Application of imaging technology to improve the laboratory and field compaction of HMA. *0-5261-1. Texas Transportation Institute*, College Station, TX.

Nijboer, L.W. (1948). Plasticity as a factor in the design of sense bituminous road carpets. *Elsevier*, New York.

Papagiannakis, A.T., Abbas, A. and Masad, E. (2002). Micromechanical analysis of viscoelastic properties of asphalt concretes. *Transportation Research Record*, No.1789, pp.113–120.

Partl, M.N., Alexander, F. and Jonsson, M. (2007a). Comparison of laboratory compaction methods using x-ray computer tomography. *Road Materials and Pavement Design*, Vol. 8, No. 2, pp.139–164.

Partl, M.N., Alexander, F. and Jonsson M. (2007b). Gytatory compaction analysis using computer tomograhy. *Road Materials and Pavement Design*, Vol. 8, No. 2, pp.401–422.

Renken, P., (1980). Verdichtbarkeit von asphaltbetongemischen und ihr einfluss auf die satndfestigkeit (Compactibility of asphalt mixtures and effect of it on bearing capacity). *Technische Universitat Carola-Wilhelmina*, Braunschweig.

Reuss, A. (1929). Berechnung der fliessgrenze von mischkristallen auf grund der plastizitatsbedingung fur einkristalle , *Zeitschrift fur Angewandte Mathematik und Mechanik*, Vol. 9, pp.49–58.

Shell International Petroleum Company Limited (1978). *Shell Pavement Design Manual-Asphalt Pavements and Overlays for Road Traffic*. London.

Tashman, L., Masad, E., Zbib, H., Little, D. and Kaloush, K. (2005). Microstructural viscoplastic continuum model for permanent deformation in asphalt pavements. *Journal of Engineering Mechanics*, Vol.131, No.1, pp.48–57.

Tvergaard, V. (1981). Influence of voids on shear band instabilities under plane strain compdition. *International Journal of Fracture Mechanics*, Vol. 17, pp.389–407.

Vizi, L. and Büttner, C. (1981). Verdichten von asphalt im strassenbau (Compaction of asphalt roads). *Werner Verlag*, Düsseldorf.

Voigt, W. (1889). Ueber die beziehung zwischen den beiden elasticitatsconstanten isotroper korper, *Annalen der physik und chemie*, Vol.38, pp.573–587.

Wang, D. (2007). Binder Film Thickness Effect on Aggregate Contact Behavior, *M.S. Thesis*, Department of Civil and Environment Engineering, Virginia Tech.

Wang, L.B., Frost, J.D. and Lai. J.S. (1999). Non-invasive measurement of permanent strain field resulting from rutting in asphalt concrete. *Transportation Research Record*, No.1687, pp.85–94.

Wang, L.B, Frost, J.D. and Lai, J.S. (2004). 3D digital representation of the microstructure of granular materials from x-ray tomography imaging. *Journal of Computing in Civil Engineering*, Vol.18, No.1, pp. 28–35.

Wang, L.B. and Frost, J.D. (2004). Dissipated strain energy method for determining preconsolidation pressure. *Canadian Geotechnical Journal*. Vol.41, No.4, pp.760–768.

Wang, L.B. and Lai, J.S. (1997). Estimate the moduli and ko of soil and soil-pile system using micromechanics. *Proceedings on Computer Methods and Advances in Geomechanics*, pp.2345–2347, October 11–13, Wuhan, China.

Wang,L.B., Wang, Y.P. and Mohammad, L.N. (2003). Application of mixture theory in the evaluation of mechanical properties of asphalt concrete. *Journal of Materials in Civil Engineering*, Vol.16, No.2, pp.167–174.

Wang, L.B., Zhang, B., Wang, D. and Yue, Z.Q. (2007a). Fundamental mechanics of asphalt compaction through FEM and DEM modeling. *ASCE Geotechnical Special Publication*. No. 176, pp.45–63.

Wang, Y.P., Wang, L.B., Li, Q.B. and Harman, T.P. (2007b). Noninvasive measurement of 3D permanent strains in asphalt concrete using x-ray tomography imaging. *Transportation Research Record*, No. 2005, pp.95–103.

Wood, D.M. (1990). Soil Behavior and Critical State Soil Mechanics. *Cambridge University Press*, Cambridge.

Xia K. and Chi L. (2008). A viscoplastic foam model for prediction of asphalt pavement compaction. *Geotechnical Special Publication*, No.182, pp.136–145.

You, Z. and Buttlar, W.G. (2004). Discrete Element Modeling to Predict the Modulus of Asphalt Concrete Mixtures. *Journal of Materials in Civil Engineering*, Vol.16, pp.140–146.

You, Z. and Buttlar, W.G. (2005). Application of discrete element modeling techniques to predict the complex modulus of asphalt–aggregate hollow cylinders subjected to internal pressure. *Transportation Research Board*, No.1929, pp.218–226.

Zhang, B. and Wang, L.B. (2007). Digital simple performance test of asphalt concrete. *Proceedings of the 7th International Conference of Chinese Transportation Professionals*, pp.98–107, May 20–21, Shanghai, China.

CHAPTER 12

Characterization and Modeling Anisotropic Properties of Asphalt Concrete

12.1   Introduction

Asphalt concrete (AC) is a bonded granular material. Its internal structure is anisotropic, which could be due to the anisotropic particle and void shape, particle orientation distribution, anisotropic compaction (restrain and force pattern applied during the compaction), and compaction-induced anisotropic distribution of contact normals. The characterization and modeling of the anisotropic properties of soils have been widely explored in geomechanics and geotechnical engineering (Kuwano et al., 2000; Pennington et al., 2001; Adu-Osei et al., 2001; Roy et al., 2002; Li and Dafalias, 2002). However, few studies have focused on the characterization and modeling of aggregate base (Tutumluer et al., 2001; Seyhan and Tutumluer, 2002), AC (Masad et al., 2002; Wang et al., 2005a; Li and Wang, 2008) and pavement analysis (Boulbibane et al., 1999; Wang et al., 2005b). It would be worthwhile to evaluate the degree of anisotropy and its effect on material response because the characterization and analysis methods for isotropic and anisotropic materials are quite different.

In general, whether a material is isotropic or anisotropic affects the selection of material characterization methods, response models, and distress models. For example, if the axial stiffness and the lateral stiffness of a gyratory specimen are significantly different from those of a field specimen, the deformation characteristics obtained using gyratory specimens may not well represent those of field specimens. This chapter presents a comparison between the stress field of an isotropic pavement and that of an anisotropic pavement through analytical solutions and finite element (FE) simulations, and a method to characterize the orthotropic material properties using a cubical device. Studies on other anisotropic properties of AC, including permeability and rutting, are also summarized in this chapter.

12.2   Orthotropic Elasticity

AC is a viscoelastoplastic material. In pavement analysis and especially stress analysis, it is often treated as an isotropic elastic material. This treatment has its advantages in simplicity. It also has some theoretical basis in that the initial stresses at $t = 0$ has significant influence on the subsequent viscous and plastic deformation and the stress analysis can be performed using elasticity theory through the elastic-viscoelastic correspondence principle. In this chapter, an anistropic elasticity analysis of a pavement under wheel load will be presented. Although AC may demonstrate general anisotropy, only cross anisotropy or orthotropy is considered for the purpose of simplification, and due to the experimental evidence. For the orthotropic case, AC is considered to have significant differences only in vertical and horizontal directions due to the anisotropic compaction, aggregate orientation, restraint conditions, and gravity direction. In orthotropic elasticity, there are only five material constants. However, the general anisotropic elasticity has 21 material constants and is not realistic for modeling and characterization. The five material constants of the orthotropic elasticity are E_v, E_h, v_{vh}, v_{hh}, and G_{vh}. Where E_v and E_h are elastic modulus in vertical and horizontal directions respectively; v_{vh} and v_{hh} are Poisson's ratios for vertical-horizontal and horizontal-horizontal responses, respectively; and G_{vh} is the shear modulus along the vertical plane. Hooke's Law for the orthotropic case can be expressed as follows:

$$\Delta \varepsilon_x = \frac{1}{E_h} \Delta \sigma_x - v_{hh} \frac{1}{E_h} \Delta \sigma_y - v_{vh} \frac{1}{E_v} \Delta \sigma_z \tag{12-1a}$$

$$\Delta \varepsilon_y = \frac{1}{E_h} \Delta \sigma_y - v_{hh} \frac{1}{E_h} \Delta \sigma_x - v_{vh} \frac{1}{E_v} \Delta \sigma_z \tag{12-1b}$$

$$\Delta \varepsilon_z = \frac{1}{E_v} \Delta \sigma_y - v_{vh} \frac{1}{E_h} \Delta \sigma_x - v_{vh} \frac{1}{E_v} \Delta \sigma_y \tag{12-1c}$$

$$\Delta \gamma_{yz} = \frac{1}{G_{vh}} \Delta \tau_{yz} \tag{12-1d}$$

$$\Delta \gamma_{zx} = \frac{1}{G_{vh}} \Delta \tau_{zx} \tag{12-1e}$$

$$\Delta \gamma_{xy} = \frac{(1 + v_{hh})}{E_h} \Delta \tau_{xy} \tag{12-1f}$$

Where $\Delta \varepsilon_x$, $\Delta \varepsilon_y$, $\Delta \varepsilon_z$ are normal strain increments and $\Delta \sigma_x$, $\Delta \sigma_y$, $\Delta \sigma_z$ are normal stress increments; $\Delta \gamma_{yz}$, $\Delta \gamma_{zx}$ and $\Delta \gamma_{xy}$ are shear strain increments; and $\Delta \tau_{yz}$, $\Delta \tau_{zx}$ and $\Delta \tau_{xy}$ are the corresponding shear stress increments. It should be noted that in the above formulations, the properties in tension and compression are considered the same.

12.3   Boussinesq's Solution for Orthotropic Materials

For a full-depth AC pavement of isotropic materials, if the distributed tire load could be approximated as a point load, the stress field could be approximated by the Boussinesq's solution, a half-space subjected to a concentrated load P as illustrated in Figure 12.1. The corresponding Boussinesq's solution for orthotropic material was obtained by Wolf (1935). The analytical expressions for two of the four stress components, σ_θ and τ_{rz}, are presented in Equations 12-2a and 12-2b. The solution is based on the cylindrical

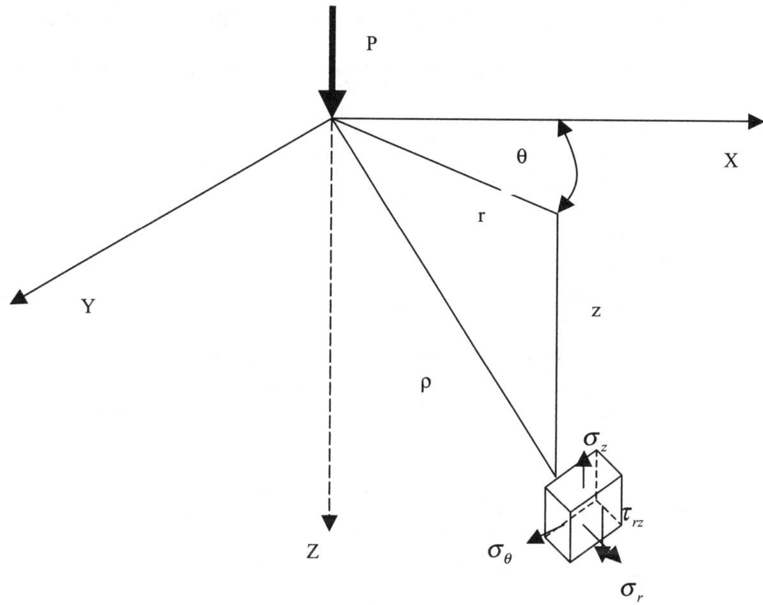

Figure 12.1   Illustration of the geometric and stress terms.

coordinate system shown in Figure 12.1. The parameters s_1, s_2, a, c, d, q_1, q_2, λ, μ are related to E_v, E_h, ν_{vh}, ν_{hh}, G_{vh}. Their implications can be found in Wolf (1935).

$$\sigma_\theta = -\frac{P}{2\pi}\left\{ \frac{\sqrt{d}}{ac-d}\frac{z}{s_1-s_2}\left[-\frac{s_1^2 q_2}{(r^2+s_1^2 z^2)^{3/2}} + \frac{s_2^2 q_1}{(r^2+s_2^2 z^2)^{3/2}} \right] - \frac{\lambda}{s_1-s_2}\frac{z}{r^2}\left[\frac{s_1^2 p_2}{(r^2+s_1^2 z^2)^{3/2}} - \frac{s_2^2 p_1}{(r^2+s_2^2 z^2)^{3/2}} \right] \right.$$

$$(12\text{-}2a)$$

$$\tau_{rz} = -\frac{P}{2\pi\sqrt{d}}\frac{z}{s_1-s_2}\left[\frac{1}{(r^2+s_1^2 z^2)^{3/2}} - \frac{1}{(r^2+s_2^2 z^2)} \right]$$

$$(12\text{-}2b)$$

To illustrate how these two stress distributions vary with the elastic material constants, the solutions for σ_θ and τ_{rz} are plotted in Figures 12.2a and b for $E_h = 1.0, 0.9, 0.7, 0.5, 0.2E_v$, ν_{vh}, $\nu_{hh} = 0.3$ and $G_{vh} = 0.384E_h$ (see the experimental data in the experiment section for the rationality of the assumed values). It can be seen that both stresses are larger than those of the isotropic cases if the horizontal stiffness is smaller than the vertical stiffness. This fact has an important implication for pavement design and analysis. The larger tensile stress σ_θ may imply that the stress level for fatigue cracking might be underestimated by using isotropic elasticity stress analysis; the larger shear stress τ_{rz} may imply that shear flow is underestimated using the isotropic elasticity stress analysis.

12.4   FEM Analysis of an Anisotropic Model Pavement

The above solution refers to the half-space case or to the full-depth asphalt pavement case. For the layered pavement, an FEM analysis is needed (the commercial software ADINA is used in this case). In the FEM analysis, a block of AC of $5 \times 5 \times 3$ in³ subjected to a 100-pound load distributed on a 0.5-inch strip is simulated (in simulation

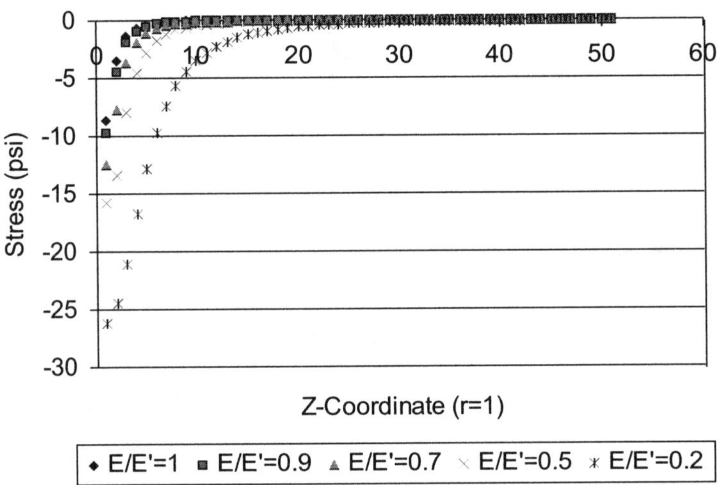

a. Shear Stress vs. Elastic Modulus Ratio

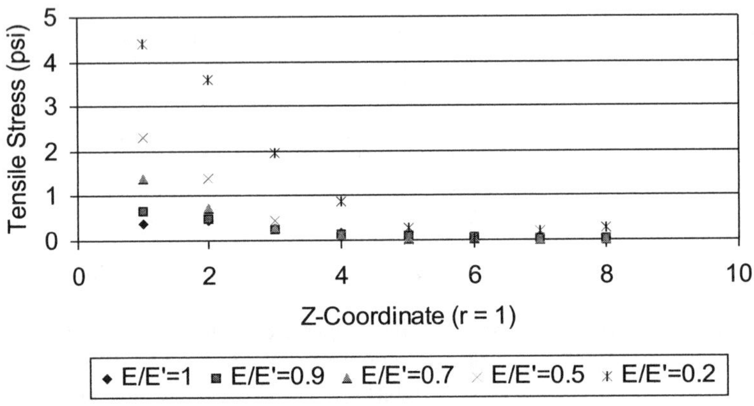

b. Tensile Stress vs. Elastic Modulus Ratio
(Note, $E=E_h$ and $E'=E_v$)

FIGURE 12.2   Stress distribution versus the ratio of the elastic modulus.

of the APA testing conditions). The block is discretized into 1000 3D 20-node solid elements after refinement analysis. Two sets of boundary conditions are applied. In set 1, the bottom surface is fixed in x, y, and z directions (Figure 12.3); the other four lateral surfaces are fixed in normal directions (horizontal directions). This set of boundary conditions is similar to an asphalt overlay placed on top of a rigid pavement (strong base) with excellent bonding. In Set 2, the bottom surface is fixed in z directions only and the other four lateral surfaces are fixed in normal directions (horizontal directions). This configuration is similar to the testing configuration of the APA. Because

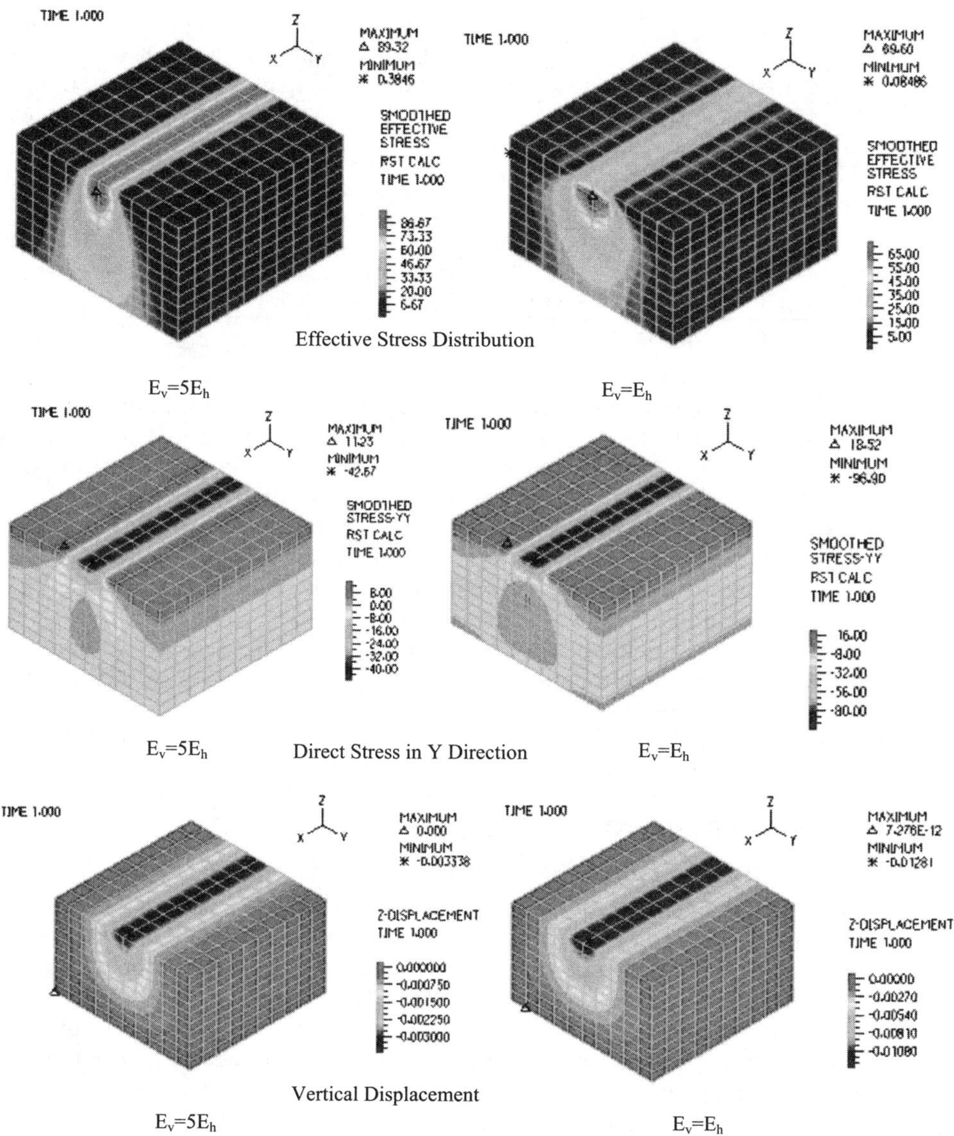

FIGURE 12.3 Typical stress and vertical displacement distribution for the case of fixed XYZ (set 1 conditions).

only linear elasticity is involved, the configuration can be proportionally scaled to larger sections. Table 12.1 presents the material inputs for the simulation. The simulation mainly evaluates how the ratio between horizontal and vertical stiffness affects the stress distribution and vertical displacements. The analysis is performed for five elastic modulus ratios (E_h/E_v = 1, 0.8, 0.5, 0.2, 0.1).

Figures 12.3 and 12.4 present the comparisons of stresses and displacements at the elastic modulus ratio of 0.2 for Set 1 and 2 boundary conditions, respectively. Other cases are similar and the general trend is that the more significant the anisotropy, the

Case	E_v (Psi)	E_{hx} (Psi)	E_{hy} (Psi)	Poisson's Ratio	Shear Modulus G_x (Psi)	Shear Modulus G_y (Psi)	Shear Modulus G_z (Psi)
1	100000	100000	100000	0.3	3846.15	3846.15	38461.54
2	100000	80000	80000	0.3	3846.15	3846.15	30769.23
3	100000	50000	50000	0.3	3846.15	3846.15	19230.77
4	100000	20000	20000	0.3	3846.15	3846.15	7692.31
5	100000	10000	10000	0.3	3846.15	3846.15	3846.15

Note: Different expressions but are consistent with the orthotropic model. E_v: Modulus in vertical direction; E_{hx}: Horizontal modulus in x-direction; E_{hy}: Horizontal modulus in y-direction

TABLE 12.1   Material properties for the simulation.

larger the differences in stress and displacement fields. Analysis of the data presented in Figures 12.3 and 12.4 and other cases indicates that with the above boundary conditions and materials properties, the largest effective stress (the von Mises stress) and the vertical displacement increases with the increase of anisotropy (about 13% larger for the effective stress); the largest tensile stress in y direction decreases with the increase of anisotropy. It should be noted that the horizontal stiffness (in y direction) decreases with the increase of anisotropy, and therefore the tensile stresses decrease (the normal direction is restrained).

Although a definitive conclusion (i.e., increase or decrease, safe or unsafe) about the trend due to the effects from anisotropy cannot be drawn based on this limited study, the analysis indeed indicates some significant differences. Further study to characterize the material constants and to apply rational boundary conditions is needed.

12.5   Analytical Anisotropic Model of Multi-Layered System Pavement

Figure 12.5 presents a multi-layered system of cross-anisotropic materials lying on a homogeneous half-space. Pan (1989) and Chen and Pan (2008) presented analytical solutions for the stresses, strains, and displacements of the layered system under vertical and horizontal loading applied on the surface. The general methodology of solving the boundary value problem of the multi-layered system is to make use of the Cartesian and cylindrical systems of vector functions and of the propagator matrix method.

They first introduced the following Cartesian system of vector functions (Equation 12-3) and expressed the surface loading in Equation 12-4.

$$L(x,y;\alpha,\beta) = e_z S(x,y;\alpha,\beta)$$
$$M(x,y;\alpha,\beta) = (e_x \partial_x + e_y \partial_y) S(x,y;\alpha,\beta) \tag{12-3}$$
$$N(x,y;\alpha,\beta) = (e_x \partial_y - e_y \partial_x) S(x,y;\alpha,\beta)$$

With $S(x,y;\alpha,\beta) = e^{-i(\alpha x + \beta y)} / (2\pi)$

$$P(x,y) = \int\int_{-\infty}^{+\infty} \left[P_L(\alpha,\beta)L(x,y) + P_M(\alpha,\beta)M(x,y) + P_N(\alpha,\beta)N(x,y) \right] d\alpha d\beta \tag{12-4}$$

Then, they expressed the displacement and surface traction in the vector format in Equations 12-5 and 12-6 and in component format in Equations 12-7 and 12-8.

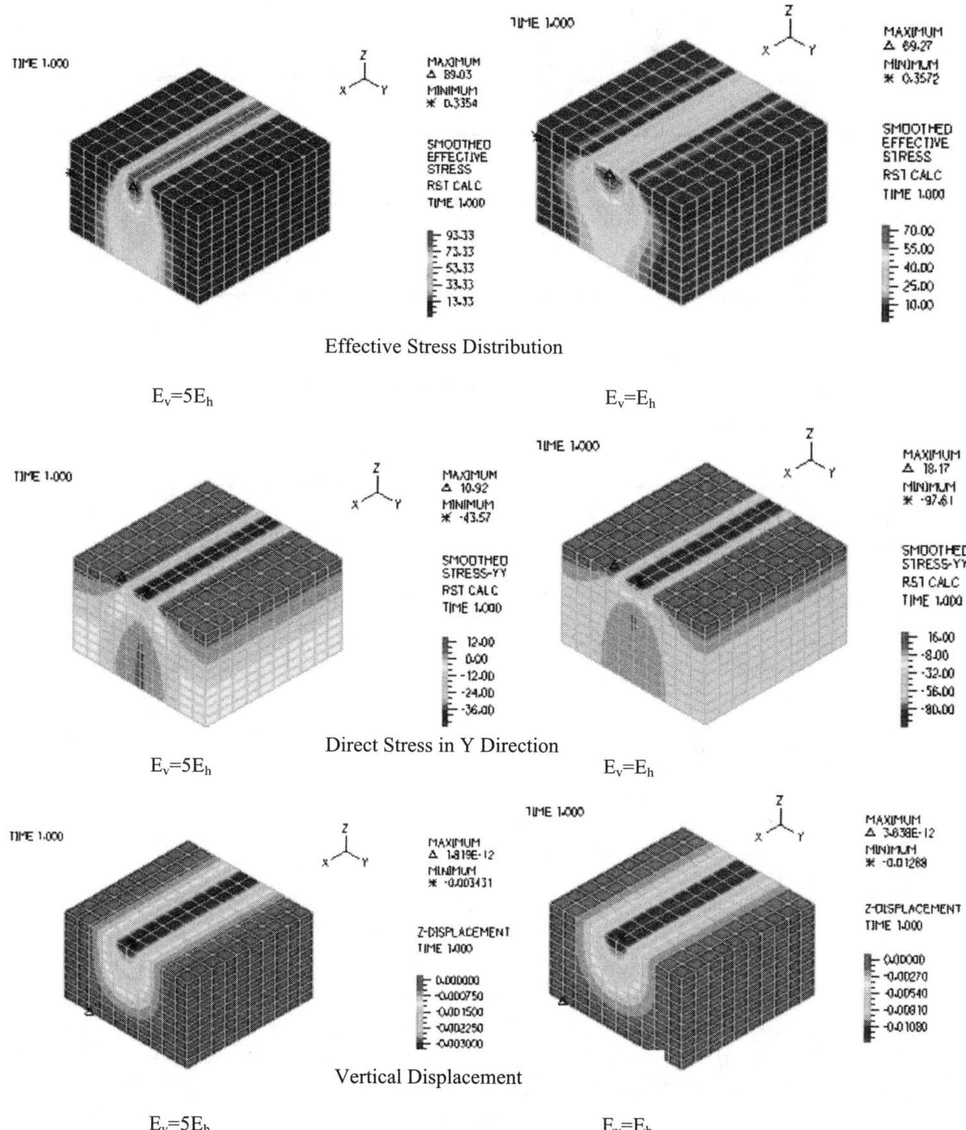

FIGURE 12.4 Typical stress and vertical displacement distribution for the case of fixed Z (set 2 conditions).

$$\boldsymbol{u}(x,y,z) = \int_{-\infty}^{+\infty}\int_{-\infty}^{+\infty} [U_L(z)\boldsymbol{L}(x,y) + U_M(z)\boldsymbol{M}(x,y) + U_N(z)\boldsymbol{N}(x,y)]\, d\alpha\, d\beta \tag{12-5}$$

$$\boldsymbol{t}(x,y,z) \equiv \sigma_{xz}\boldsymbol{e}_x + \sigma_{yz}\boldsymbol{e}_y + \sigma_{zz}\boldsymbol{e}_z$$
$$= \int_{-\infty}^{+\infty}\int_{-\infty}^{+\infty} [T_L(z)\boldsymbol{L}(x,y) + T_M(z)\boldsymbol{M}(x,y) + T_N(z)\boldsymbol{N}(x,y)]\, d\alpha\, d\beta \tag{12-6}$$

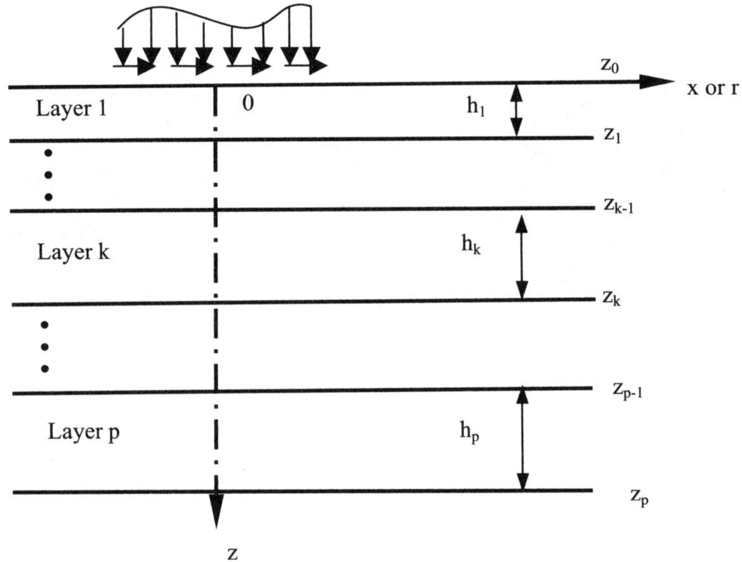

FIGURE 12.5   Illustration of the multiple layered system (courtesy Ernie Pan).

$$u(x,y,z) = u_x e_x + u_y e_y + u_z e_z$$

$$= \int\limits_{-\infty}^{+\infty}\int\limits_{-\infty}^{+\infty} \left\{ \left[U_M(z)\frac{\partial S}{\partial x} + U_N(z)\frac{\partial S}{\partial y} \right] e_x + \left[U_M(z)\frac{\partial S}{\partial y} - U_N(z)\frac{\partial S}{\partial x} \right] e_y + U_L(z)Se_z \right\} d\alpha d\beta \tag{12-7}$$

$$t(x,y,z) = \sigma_{xz} e_x + \sigma_{yz} e_y + \sigma_{zz} e_z$$

$$= \int\limits_{-\infty}^{+\infty}\int\limits_{-\infty}^{+\infty} \left\{ \left[T_M(z)\frac{\partial S}{\partial x} + T_N(z)\frac{\partial S}{\partial y} \right] e_x + \left[T_M(z)\frac{\partial S}{\partial y} - T_N(z)\frac{\partial S}{\partial x} \right] e_y + T_L(z)Se_z \right\} d\alpha d\beta \tag{12-8}$$

Finally, through the propagator matrix method, they solved for the stress, strain, and displacement fields under various loading and boundary conditions. A program (MultiSmart3D) has also been developed by this group to allow convenient computation of the stress, strain, and displacement fields. Through Laplace transform, the analytical solutions have been further derived for the corresponding viscoleastic materials (Chen et al., 2009). Pan's work placed a foundation for considering the anisotropic material properties in the flexible layered pavement design.

12.6   Characterization of the Anisotropic Properties Using a True Triaxial Testing Device

12.6.1   General State

While significant efforts have been devoted to the study of the fundamental properties of soils, limited efforts have been devoted to AC. Properties such as stress path dependency, anisotropy, yielding, plastic deformation rates, etc. are barely studied in AC.

True triaxial devices that have been widely used in soil mechanics have advantages for use in the characterization of the above fundamental properties.

Different types of true triaxial devices have been developed worldwide, and they can be classified into three categories: (1) rigid boundary (Hambley, 1969; Airey and Wood 1988); (2) flexible boundary (Ko and Scott, 1967; Sture and Desai, 1979), and (3) mixed boundary (Green, 1971; Lade and Duncan, 1973). The advantages and disadvantages of these three types of true triaxial devices have been discussed by Sture (1979) and Arthur (1988). The original development of the flexible boundary type of device used in this work was presented by Atkinson (1972) for multiaxial testing of rock materials. A detailed description of the original components is presented by Atkinson (1972), Sture (1979), and NeSmith (1997). The stress-controlled, computer-driven cubical testing device consists basically of six main components or modules: (1) a frame; (2) six wall assemblies; (3) a deformation measuring system; (4) a stress application and control system; (5) six rigid membranes; and (6) a data acquisition and process control system (DA/PCS). A detailed and illustrated description of these components follows.

12.6.2 The Cubical Device System

12.6.2.1 Steel Frame

A photograph of the true triaxial cubical frame is shown in Figure 12.6a. The frame supports the top and four lateral wall assemblies, the cubical AC specimen, and the bottom wall assembly. An inner square cavity was machined into each of the six faces of the

FIGURE 12.6 Top, bottom, and lateral wall assemblies.

frame to accommodate the membranes and to form the pressure cavities. Connection bolts were provided on each face of the frame to fix the wall assemblies. The function of the frame is twofold: (1) It forms the top, bottom, and lateral sides of the six pressure chambers (wall assemblies) that apply the external multiaxial load to the cubical AC specimen; and (2) It serves as the reaction structure for the application of the total normal stresses to the top, bottom, and lateral faces of the cubical AC specimen. The frame was machined from a solid billet of AISI 4140 heat treated, re-sulfurized forged steel. The outside of the frame was machined to a dimension of 23.0 cm (9.05 in), while the inner square cavities have a dimension of 10.35 cm (4.07 in). A 0.13 cm (0.05 in) radius is provided at the inner corners of the square cavities to reduce stress concentrations. The working pressure limits are 132.50 MPa (19200 psi) for uniaxial pressures, and 64.52 MPa (9350 psi) for a hydrostatic loading condition (Atkinson, 1972; NeSmith, 1997).

12.6.2.2   Top, Bottom, and Lateral Wall Assemblies

Figures 12.6b and c show a photograph of a wall assembly. The walls were machined from a 7075-T6 aluminum plate (Atkinson, 1972), and are fastened to the frame by steel studs mounted in tapped holes on the exterior of the frame. Each wall assembly consists mainly of three components: (1) a cover plate, which provides the wall seal for the interior pressure cavity, (2) two threaded fluid pressure inlet/outlet connections, and (3) three threaded holes to receive the stainless steel housing of the LVDTs (linear variable differential transducers). When the membranes are mounted on the frame, each assembly provides an effective seal against the leaking of the pressurized fluid into the atmosphere. An O-ring groove on the wall forms the pressure seal between the wall assembly and the reaction frame.

12.6.2.3   Deformation Measuring System

The deformation of the cubical asphalt concrete specimen is measured at three points on the top, bottom, and each of the four lateral faces, using 18 LVDTs (3 LVDTs/face). The core of each LVDT and its extension rod are thrust into contact with the rigid membranes by a low-stiffness spring (Figure 12.6b, c). Two threaded locknuts were used as the tip of each LVDT's extension rod. A zinc-plated steel flat washer allows for complete extension or compression of the spring. The output from the LVDTs is recorded by means of a DA/PCS. The 499 XS-C's Schaevitz series, high-pressure-sealed LVDT (from Lucas Control Systems Products, Inc.), is suitable for operating pressures up to 210 bars (3000 psi) in pressure-sealed chambers. The LVDTs have a nominal linear range of +/− 12.50 mm (+/−0.50 in). The EB-welded stainless steel housing is highly resistant to corrosive environments, and its internal magnetic and electrostatic shielding render the XS-C's LVDT insensitive to external magnetic influences. The sampling rate during a test can be easily selected from the menu of a data acquisition and process control computer software (LABTECH-NOTEBOOK from Laboratory Technologies Corp.). The computer software allows for the real-time display of the 18 LVDT channels. The analog input (volt) signal delivered by the lead wires of the LVDTs is converted into a digital output signal by an analog-to-digital converter (RTI-815 board from Analog Devices, Inc.) plugged into the CPU of the PC-based computer. The deformation of each side of the specimen along a particular direction is estimated by averaging the three LVDTs' outputs corresponding to the specimen's face, X(+/−), Y(+/−), or Z(+/−), perpendicular to that particular direction. The LVDTs are located at 120 degree spacing on a 3.18 cm (1.25 in) radius on each of the top, bottom, and lateral wall assemblies.

12.6.2.4 Stress Application and Control System

The hydraulic pressure applied through the six rigid membranes to the top, bottom, and lateral sides of the cubical AC specimen is generated and controlled by a computer-driven electro-hydraulic pressure system. The fluid is pressurized by a 20-gallon standard hydraulic power unit or pump (manufactured and installed by ORTON Industries, Inc.) that can deliver a variable output pressure up to 10.35 MPa (1500 psi) on each side of the specimen, depending on the initial setting and calibration of three computer-driven proportional pressure relief valves (manufactured by Parker Motion & Control, Inc.) connected to an equal number of outlet ports in the power unit (Figure 12.7). The RE series proportional relief valves provide variable hydraulic pressure control in response to a variable voltage or current command signal. The RE series' valves feature an on-board electronic driver with adjustments for the rate at which the pressure increases and decreases (Figure 12.7). In addition, there are adjustments to electronically set the minimum pressure and maximum pressure delivered to the specimen (valves' calibration process).

For this study, the proportional pressure relief valves were calibrated to work at a minimum pressure of 175 kPa (25 psi) and a maximum pressure of 1225 kPa (175 psi). The output lines from the three proportional pressure relief valves split into two lines to supply the pressures to the positive (+) and negative (−) faces of the cubical asphalt concrete specimen for each particular loading condition. These positive and negative faces are designated as X(+), X(−), Y(+), Y(−), accounting for the four lateral sides, and Z(+), Z(−), accounting for the top and bottom sides. The output lines from the valves are transmitted to the test cell by flexible hoses having a rated burst strength of 27.50 MPa (4000 psi).

A quick-disconnect coupling between the hose and the fluid pressure inlet/outlet connections of the wall assembly makes for easy assembly. Any principal stress combination path can be achieved by simultaneous control over the three computer-driven proportional relief valves. The proportional relief valves receive an analog input signal (volt) from an analog output signal-conditioning interface (IOB120-01 interface kit from Analog Devices, Inc.) connected to the digital-to-analog converter (RTI-815 board from Analog Devices, Inc.) plugged into the CPU of the computer. Low-viscosity Mobil-type (DTE 26) hydraulic fluid was used as the pressurizing medium.

FIGURE 12.7 Calibration of proportional valves of hydraulic power unit.

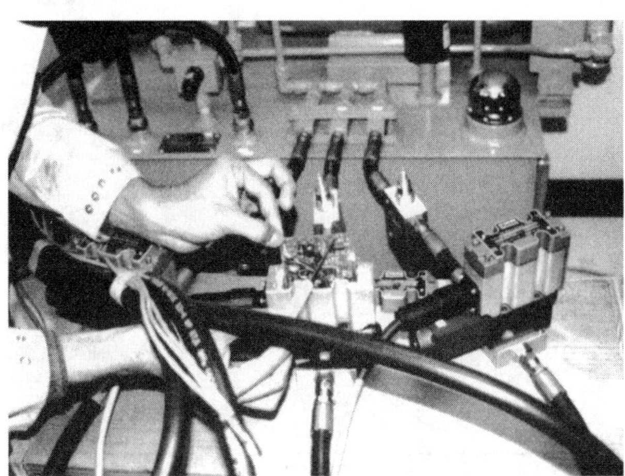

12.6.2.5 Rigid Membranes

Rigid membranes transmit the applied pressure uniformly to the top, bottom, and lateral faces of the cubical AC specimen. They form the actual fluid barrier of the hydraulic fluid acting against the cubical AC specimen. The membrane extends past the O-ring mounted in the inner face of the wall assemblies.

12.6.2.6 Data Acquisition and Process Control System (DA/PCS)

A DA/PCS was assembled to control the pressures applied to the cubical AC specimen, and to monitor and record its resulting deformations. Details of this system are presented by NeSmith (1997). A photograph of the complete testing setup is shown in Figure 12.8.

12.6.3 Tests

To investigate the capability of the cubical cell device in evaluating the properties of asphalt concrete, a multi-stage loading procedure was adopted. Multi-stage loading runs different tests such as triaxial compression, triaxial extension, and cyclic loading on the same specimen, eliminating the requirements for multiple specimens. This is especially useful when low-level stress is involved, causing little or no damage to the specimen. If numerical simulation can be used to account for specimen change, the results can be better interpreted.

A cubic specimen of 4-inch lateral length was cut from a block sample for the tests. The block sample was cored from the WesTrack project. The mix has targeted asphalt content of 5.7% and an air-void content of 8%. The loading procedure for the specimen is as follows: isotropic compression (IC) to 25 psi in each direction followed by triaxial compression (TC), triaxial extension (TE), simple shear (SS), conventional triaxial compression (CTC), conventional triaxial extension (CTE), and cyclic CTE tests. Figures 12.9a and 9b present the loading sequences in σ_x, σ_y, σ_z space and $\tau \sim p$ space respectively, where $\tau = \sqrt{(\sigma_x - \sigma_y)^2 + (\sigma_y - \sigma_z)^2 + (\sigma_z - \sigma_x)^2}$ $p = (\sigma_x + \sigma_y + \sigma_z)/3$. An initial stress of

FIGURE 12.8 Overview of the complete testing setup.

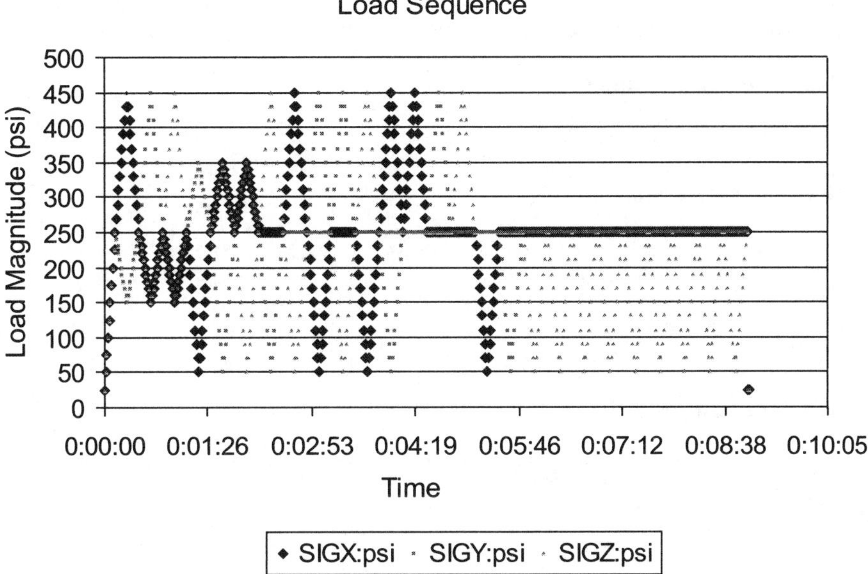

a. Load Sequence in the X, Y, and Z Space

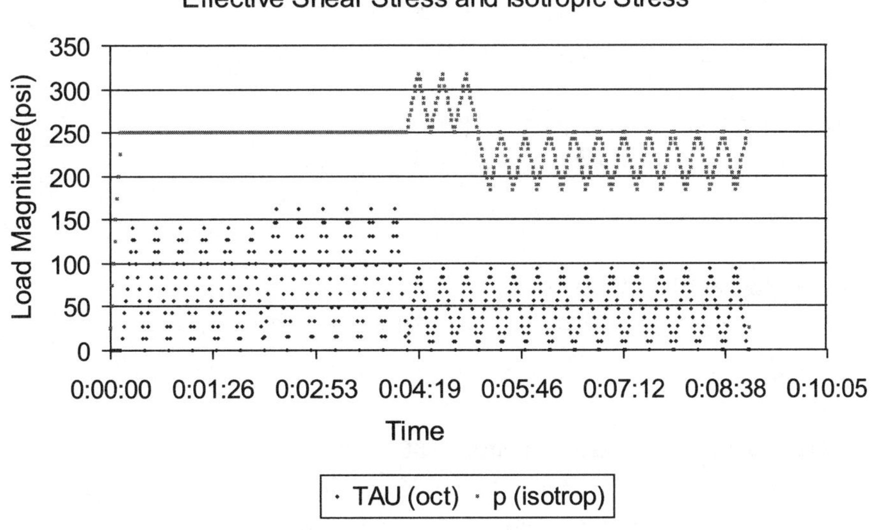

b. Load Sequence in the $\tau \sim p$ Space

FIGURE 12.9   Load sequence of the testing.

25 psi is necessary to keep the membranes in good contact with the specimen. Stress paths for IC, TC, TE, SS, CTC, and CTE tests were accomplished by simultaneous control of major, intermediate, and minor principal stresses as follows:

Isotropic compression (IC):

$$\Delta\sigma_x = \Delta\sigma_y = \Delta\sigma_z = \Delta\sigma$$

Triaxial compression (TC):

$$\Delta\sigma_x = \Delta\sigma, \Delta\sigma_y = \Delta\sigma_z = -\frac{1}{2}\Delta\sigma; \Delta\sigma_y = \Delta\sigma, \Delta\sigma_x = \Delta\sigma_z = -\frac{1}{2}\Delta\sigma$$

$$\Delta\sigma_z = \Delta\sigma, \Delta\sigma_x = \Delta\sigma_y = -\frac{1}{2}\Delta\sigma$$

Triaxial extension (TE):

$$\Delta\sigma_x = -\Delta\sigma, \Delta\sigma_y = \Delta\sigma_z = \frac{1}{2}\Delta\sigma; \Delta\sigma_y = -\Delta\sigma, \Delta\sigma_x = \Delta\sigma_z = \frac{1}{2}\Delta\sigma$$

$$\Delta\sigma_z = -\Delta\sigma, \Delta\sigma_x = \Delta\sigma_y = \frac{1}{2}\Delta\sigma$$

Simple shear (SS):

$$\Delta\sigma_x = 0, \Delta\sigma_y = \Delta\sigma, \Delta\sigma_z = -\Delta\sigma; \Delta\sigma_x = 0, \Delta\sigma_y = -\Delta\sigma, \Delta\sigma_z = \Delta\sigma$$

$$\Delta\sigma_y = 0, \Delta\sigma_x = \Delta\sigma, \Delta\sigma_z = -\Delta\sigma; \Delta\sigma_y = 0, \Delta\sigma_x = -\Delta\sigma, \Delta\sigma_z = \Delta\sigma$$

$$\Delta\sigma_z = 0, \Delta\sigma_y = \Delta\sigma, \Delta\sigma_x = -\Delta\sigma; \Delta\sigma_z = 0, \Delta\sigma_y = -\Delta\sigma, \Delta\sigma_x = \Delta\sigma$$

Conventional triaxial compression (CTC):

$$\Delta\sigma_x = \Delta\sigma, \Delta\sigma_y = 0, \Delta\sigma_z = 0; \Delta\sigma_x = 0, \Delta\sigma_y = \Delta\sigma, \Delta\sigma_z = 0$$

$$\Delta\sigma_y = 0, \Delta\sigma_x = 0, \Delta\sigma_z = \Delta\sigma$$

Conventional triaxial extension (CTE):

$$\Delta\sigma_x = -\Delta\sigma, \Delta\sigma_y = 0, \Delta\sigma_z = 0; \Delta\sigma_x = 0, \Delta\sigma_y = -\Delta\sigma, \Delta\sigma_z = 0$$

$$\Delta\sigma_y = 0, \Delta\sigma_x = 0, \Delta\sigma_z = -\Delta\sigma$$

Cyclic conventional triaxial extension (CCTE):

The CTE test ran for a number of cycles.

Where $\Delta\sigma_x$, $\Delta\sigma_y$, $\Delta\sigma_z$ are load increments and $\Delta\sigma > 0$.

It should be noted that TC, TE, and SS tests are run at $p = (\sigma_x + \sigma_y + \sigma_z)/3 = const$ or on the π planes, planes that are normal to $\sigma_x = \sigma_y = \sigma_z$. The stress paths of these three tests in the principal stress space are illustrated in Figure 12.10. There are three variations for TC and TE stress paths respectively, and six variations for SS stress paths. Table 12.2 listed all the stress paths that were followed in the test.

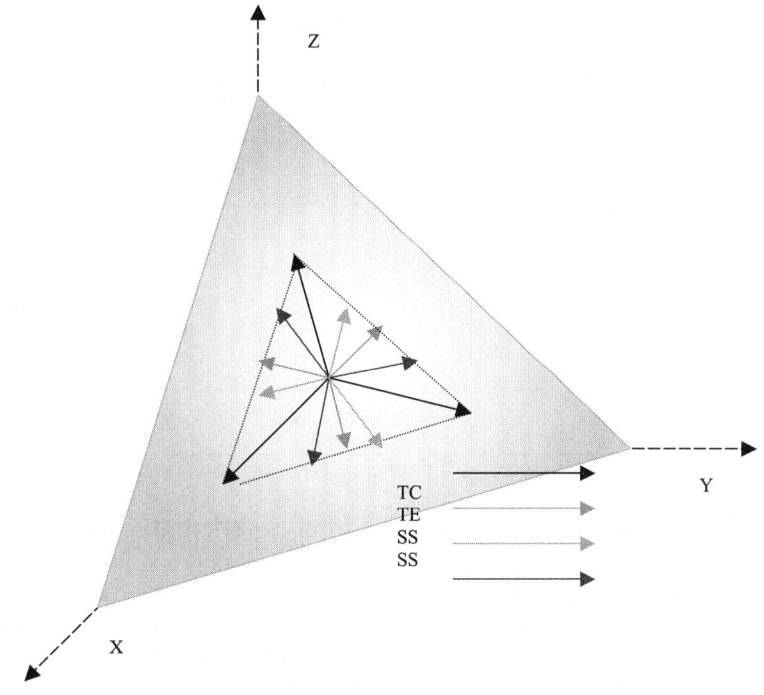

FIGURE 12.10   The TC, TE and SS stress paths on the π plane.

Sequence No.	Test Designation	X	Y	Z
1	IC (to 250 psi)	↑	↑	↑
2	TC	↑	↓	↑
3	TC	↓	↑	↓
4	TC	↓	↓	↑
5	TE	↓	↑	↑
6	TE	↑	↓	↑
7	TE	↑	↑	↓
8	SS	Constant	↑	↓
9	SS	↑	Constant	↓
10	SS	↑	↓	Constant
11	SS	Constant	↓	↑
12	SS	↓	Constant	↑
13	SS	↓	↑	Constant
14	CTC	↑	Constant	Constant
15	CTC	Constant	↑	Constant
16	CTC	Constant	Constant	↑
17	CTE	↓	Constant	Constant
18	CTE	Constant	↓	Constant
19	CTE	Constant	Constant	↓
20	CCTE	Constant	Constant	↓↑

↑ Stress increase; ↓ stress decrease

TABLE 12.2   Test sequence.

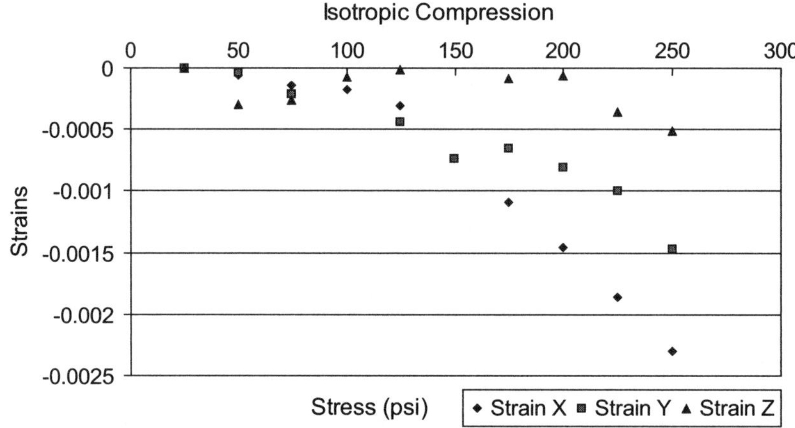

FIGURE 12.11   Stress-strain relation during isotropic compression.

12.6.4 Calculation of Modulus from Experimental Data

Since both stresses and strains in the three orthogonal orientations are monitored during each loading process, the information presented can be used to calibrate constitutive models. By eliminating the creeping strain (achieved using incremental stress and strain relations) the magnitudes of E_v, E_h can be calculated from CTC and CTE tests. Once E_v, E_h are obtained, v_{vh}, v_{hh} can be obtained from TC, TE, or IC tests. G_{vh} can be obtained from the SS test that is involved with the z direction (for example, $\Delta\tau_{yz} = \Delta\sigma, \Delta\gamma_{yz} = \Delta\varepsilon_y - \Delta\varepsilon_z$ and using Equation 12-1d). From the two tests performed, it is found that roughly $E_v = 200000$ psi, $E_{hh} = 60000$ psi, the Poisson's ratios v_{vh}, v_{hh} are 0.30, and $G_{vh} = 100000$ psi. Figure 12.11 presents the IC test results; it can be seen that AC compacted in the field follows a general anisotropy. The vertical modulus is usually two to five times larger than the horizontal modulus. The test is performed at room temperature (20°C). These relations are the basis for the FEM analysis. There are many other properties of AC that can be characterized from the test. For example, Figure 12.12 presents the volumetric strain changes during the entire test (please note the dilation effects), while Figure 12.13 presents the strain responses during the cyclic CTE test. The

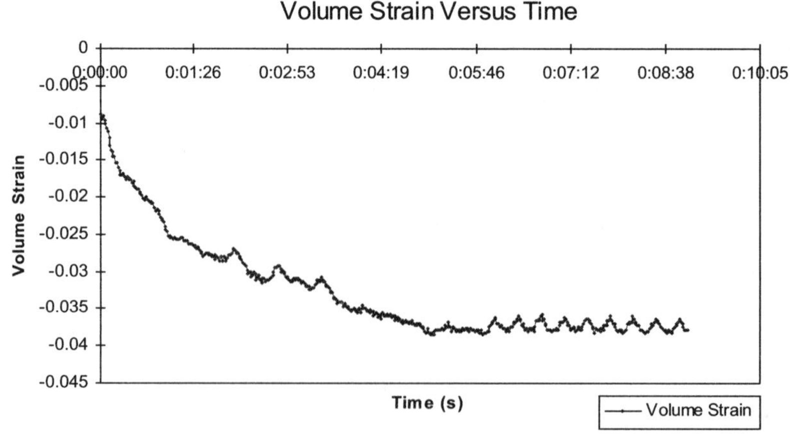

FIGURE 12.12   Volumetric strain during the entire test.

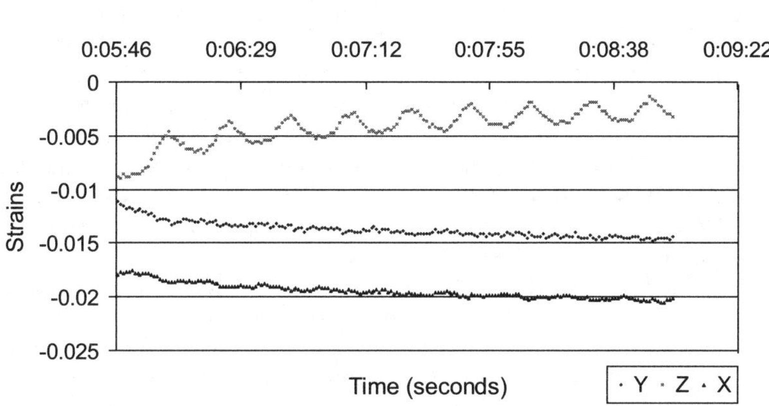

Figure 12.13   Strain plot of the cyclic CTE test.

local slopes of the loading-unloading cycles may present a resilient modulus measurement. Properties on compressibility, phase angles, stress-path dependency of elastic moduli, creeping, accumulative plastic deformation, and dynamic modulus can be evaluated.

12.7   Anisotropy in Compression and Tension

About 70 to 80% (by volume) of AC is aggregates, which form a skeleton with aggregate-aggregate (aggregates coated with thin asphalt film) contact. By contact mechanics (Johnson, 1985), the mixture should have different responses (stiffness) against compressive and tensile loading. To further complicate this issue, the material properties of asphalt mixture are loading-rate and temperature dependent. This is mainly due to the viscoelastic asphalt binder in the mixture. In other words, the difference in the mechanical response of asphalt materials in compression and tension could vary significantly as the properties of asphalt binder vary significantly with temperature and loading rate.

In the current pavement design methods, and even the anisotropy considered in the previous sections, the modulus in compression and tension are assumed to be the same to simplify the analysis of pavement structure. However, two of three major distresses considered in the pavement design, low-temperature cracking and fatigue cracking, are mainly referred to as Mode I fracture, or the so-called tension fracture, and thus are closely related to the mechanical response of asphalt materials in tension. Investigating the difference of moduli in compression and tension serves as the first step to understanding the different behavior of asphalt material in tension and compression and how significantly this feature of asphalt mixtures could affect the current design system.

Bi-modularity (of asphalt mixtures), a material behavior that shows two different moduli when subjected to tension and compression, has received attention from researchers in pavement engineering for a long time. Secor and Monismith (1965) noticed the difference in the strain of AC in tension and compression using a creep testing procedure. However, they did not estimate the modulus in tension and compression separately; instead they calculated a combined modulus to describe the mechanical response of asphalt material. Kallas (1970) conducted dynamic modulus tests on asphalt mixture specimens subjected to tension, tension-compression, and compression. He found that the measured values of dynamic modulus in tension and tension-compression were around 50% of that in compression at 37.8°C (100°F), and the difference decreases as the testing temperature decreases. Khanal and Mamlouk (1995) repeated all dynamic modulus tests that Kallas had done and static compression and tension tests, and they reached conclusions similar to those of Kallas. With different modulus in tension and compression, they found a reduction in the tensile strain at the bottom of the AC layer and the vertical compressive strain on the top of the subgrade. Christensen and Bonaquist (2004) reported that the compliance values in tension and compression are not the same, even at low temperatures of –20°C, –10°C, and 0°C.

All the above research efforts show that the difference between modulus in tension and modulus in compression is lower at lower-testing temperatures. Daniel and Lachance (2005) conducted dynamic modulus tests in tension and compression on asphalt mixtures with 15%, 25%, and 40% reclaimed asphalt pavement (RAP). They found at 20°C there was no difference between modulus in tension and compression for the control mixtures (with no RAP) and mixture with 40% RAP; while the modulus in compression was larger than that in tension for a mixture with 15% and 25% RAP. The data also indicates that when the loading frequency increases, both moduli in tension and compression increase, as does the difference between moduli in tension and compression. This means at the same loading frequency, the difference of moduli in tension and compression increases as temperature decreases, which is a different observation from other research. Note that this was observed on asphalt mixtures with RAP. Table 12.3 presents a summary of the research reviewed.

In addition, many research efforts (Christensen and Bonaquist, 2004; Von Quitas et al., 1982; Bonaquist et al., 1986) were devoted to the comparison of dynamic moduli from compression/tension to that of the indirect tensile test (IDT) (AASHTO T322). It is well known that the modulus calculated according to the IDT is based on the assumption of the same modulus in tension and compression. Once the bi-modularity of asphalt mixture is considered, the IDT modulus thus calculated is in fact neither a

Authors	Type of Test	Moduli Ratio (Compression to Tension)
Secor and Monisomith	Creep test on a bend beam	N/A (larger difference at higher temperature)
Kallas	Compression, tension, and alternate dynamic modulus test	1.3 at 25°C; 2 at 37 °C
Khanal and Mamlouk	Quasi-static compression and tension by a ramping load; compression, tension and alternate dynamic modulus test	0.86~1.37 at 5°C; 1.10~1.45 at 25°C; 1.50~2.54 at 40°C
Christensen and Bonaquist*	Uniaxial compression and tension creep test; the IDT test	Around 2 at temperatures from –20°C to 0°C (larger difference at higher temperature)
Daniel and Lachance	Compression and tension dynamic modulus test	N/A (larger difference at higher frequency)

* In this research, the compliance was actually measured, instead of modulus.

TABLE 12.3   Research on the ratio of compressive modulus to tensile modulus of asphalt mixture.

compressive modulus nor a tensile modulus. More sophisticated analysis is necessary for interpreting the stress and strain distribution and evaluating the modulus in tension and compression in the IDT configuration.

Most previous research has measured the moduli in tension and compression by applying tensile and compressive loads on different sets of asphalt mixture specimens and investigated the bimodularity by comparing the mean values from these sets. Considering the limited accuracy for dynamic modulus tests (Table 2 of AASHTO TP62), and the difficulty in conducting a dynamic modulus test in tension, a more convenient testing procedure is necessary. It is advantageous to develop an experiment measuring both moduli from the same specimen and at the same time to minimize experimental efforts and effects induced due to the use of different samples. This section describes a study on determining the moduli in tension and compression from a simply supported asphalt mixture beam.

12.7.1 Testing Setup and Procedure

Li and Wang (2008) developed a testing procedure to use the four-point beam fatigue test on asphalt mixtures (AASHTO T321) for the simultaneous determination of the tensile and compressive moduli of asphalt materials (Figure 12.14a). The beam is 381 mm (15 in) long (L), 50.4 mm (2 in) wide (w), and 63.5 mm (2.5 in) high (h) (Figure

FIGURE 12.14 Testing setup.

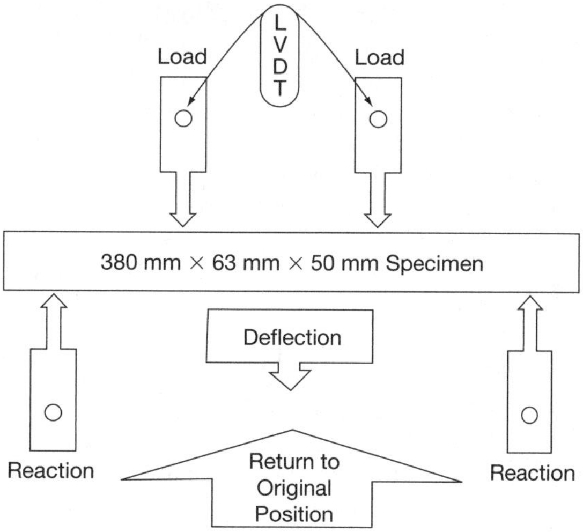

12.14b). Two extensometers were mounted on both the top and bottom of the beam to measure the compressive and tensile displacements (strains) simultaneously. A nominal 9.5-millimeter gradation asphalt mixture with PG64-22 binder and anti-stripping agent (0.5% by weight of binder Adhere HP+) was used for their study. The asphalt content is 5.7%. It should be noted that 15% of RAP is used in this mix. Testing at both 25°C and 40°C was performed. During the test, three parameters were recorded at every 0.1 s. The load was measured with a 2-kilonewton capacity load cell; the compressive deformation on the top of beam (Δ_c) and the tensile deformation at the bottom of the beam (Δ_t) were measured with strain-gage based extensometers with a gauge length of 25 mm. Both extensometers were attached at the mid-span of the beam. The measurement range of the two extensometers is ±2 mm.

A ramping load was applied to the beam at the rate of 5 N/s. This loading rate was chosen based on the experiences that the target peak load can be reached in a reasonable amount of time without introducing significant creep deformation at 25°C. Depending on the temperature and thus the modulus, the maximum load was varied such that the maximum incurred strain (tensile and compressive strains) was comparable to the previous study (Secor and Monismith, 1965). At 25°C, the maximum load was 200N corresponding to the data collection time of 40 s. The self-weight of the beam was also considered as it may induce significant strains at relatively high temperatures.

Beam specimens were kept at the target temperature in an environmental chamber for four hours before testing. This minimized the temperature gradient within the asphalt specimen and consequent modulus gradient, which can invalidate the following analysis of the stress and strain in a beam with bi-modulus.

12.7.2 Elastic Analysis of a Bernoulli-Euler Beam of Bi-Modulus

The model of testing is a simply supported beam subjected to two equal, concentrated loads symmetric to the center of the span, as shown in Figure 12.15. The span of the beam is L and the distance between the support and the location at which a concentrated load is applied is d. The cross-section of the beam is shown in Figure 12.15a, with h denoting the height and w the width. The self-weight of the beam is also considered and denoted as q with unit of N/m.

The moments from concentrated loads and uniformly distributed self-weight can be obtained with the classical beam theory as described in many structural mechanics books, e.g., Gere and Timoshenko (1984). The moment diagrams from both loads are plotted separately in Figure 12.15. The total momentum in the beam is obtained using the superposition principle. Let the longitudinal direction of the beam be the same as the direction of the x coordinate. For convenience, the origin of the coordinates is set at the mid-span of the beam in the following analysis.

The moment between two concentrated loads is described as:

$$M = \frac{Pd}{2} - \frac{q}{2}x^2 + \frac{qL}{2}x \text{ for } -\frac{s}{2} \leq x \leq \frac{s}{2} \tag{12-9}$$

Where s is the distance between the two concentrated loads.

The Bernoulli-Euler beam theory assumes that the transverse plane sections remain plane and normal to the longitudinal fibers after bending. Thus, the following relation on strain distribution always exists for a continuous beam.

$$\varepsilon_x = -\kappa y \tag{12-10}$$

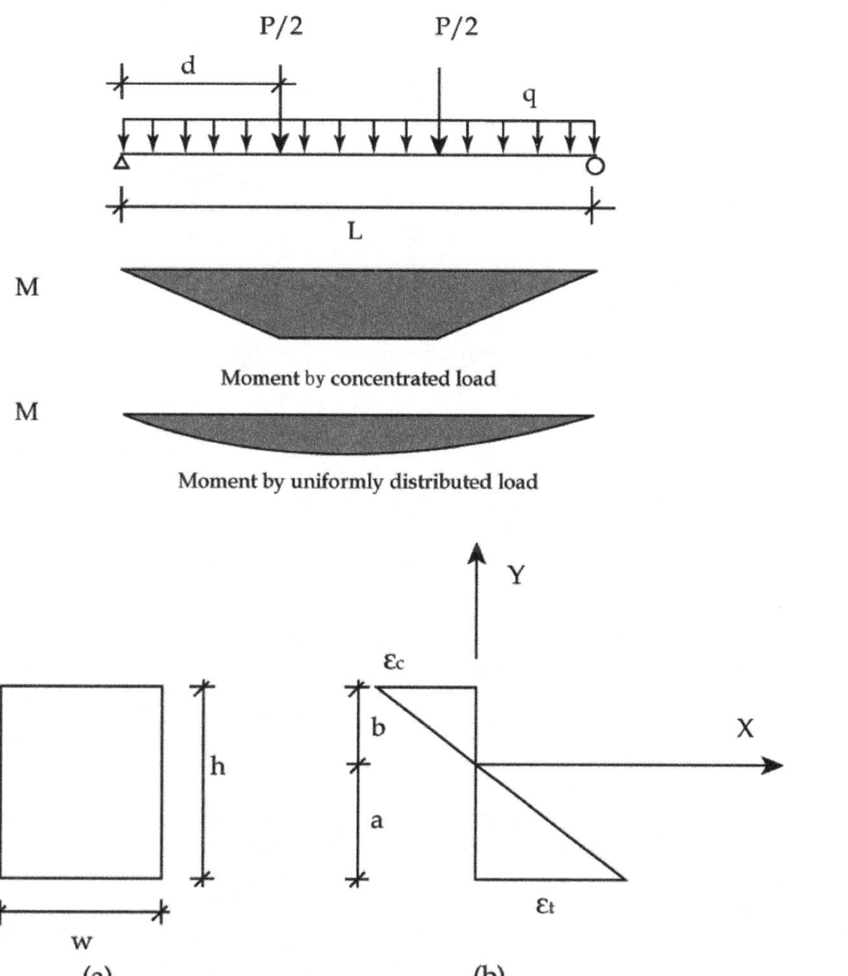

FIGURE 12.15   Analytical model of the beam testing.

Where κ is the curvature. For a beam with a single elastic modulus, i.e., the compressive modulus equals the tensile modulus, the location where strain equals to zero, or the so-called neutral axis, is at the middle of the height for a cross-section shown in Figure 12.15a. When the compressive modulus is different from the tensile modulus, Equation 12-10 still holds, but the neutral axis no longer coincides with the mid-height. In Figure 12.15b, the neutral axis is off the middle of the height and distance a away from the bottom fiber and b from the top fiber. The strain corresponding to distance a is the largest tensile strain and is denoted as ε_t. Similarly, ε_c is used to denote the largest compressive strain at distance b. With the geometry in Figure 12.15b, it is obtained that:

$$\frac{\varepsilon_t}{\varepsilon_c} = \frac{a}{b} = r \qquad (12\text{-}11)$$

Based on the balance of force at the cross-section, i.e., $\sum F = 0$, it can be written as:

$$\int_{-a}^{0} \sigma_t' dA + \int_{0}^{b} \sigma_c' dA = 0 \tag{12-12a}$$

Where σ_t' and σ_c' are tensile and compressive stresses distributed along the height respectively. Solving Equation 12-12, it is obtained that:

$$\frac{E_t}{E_c} = \frac{b^2}{a^2} = \frac{1}{r^2} \tag{12-12b}$$

Considering the resultant moment from the stress acting over the cross-section:

$$M = \int_{-a}^{0} \sigma_t' \, y \, dA + \int_{0}^{b} \sigma_c' \, y \, dA \tag{12-13}$$

With some simple integration one can obtain:

$$M = \frac{w}{3}(E_t \cdot a^2 \cdot \varepsilon_t + E_c \cdot b^2 \cdot \varepsilon_c) \tag{12-14}$$

Where E_t and E_c are the tensile and compressive moduli, respectively. With Equations 12-12b and 12-14, one can obtain:

$$M = \frac{w}{3}E_c b^2 \left(1 + \frac{a}{b}\right)\varepsilon_c = \frac{w}{3}E_t a^2 \left(1 + \frac{b}{a}\right)\varepsilon_t \tag{12-15}$$

In their study, two extensometers are used to measure the deformation of the outside fiber of the beam. One extensometer is attached on the top and the other at the bottom of the beam and along the longitudinal fiber. The center of the extensometer is aligned with the mid-span of the beam. Next, the deformation of gauges amounted on the beam is estimated. With self-weight, pure bending does not exist between the two concentrated loads. The deformation measured by the extensometers should be calculated by:

$$\Delta_t = \int_{GL} \varepsilon_t dx \tag{12-16}$$

$$\Delta_c = \int_{GL} \varepsilon_c dx \tag{12-17}$$

Where GL is the gauge length. With the relationship in Equation 12-15, Equation 12-16 can be expressed as:

$$\Delta_t = \int_{GL} \frac{a}{b}\varepsilon_c dx = \frac{a}{b}\Delta_c \tag{12-18}$$

Therefore:

$$\frac{\Delta_t}{\Delta_c} = \frac{\varepsilon_t}{\varepsilon_c} = \frac{a}{b} = r \tag{12-19}$$

With Equations 12-17 and 12-18, it can be derived that:

$$E_c = \frac{A_1 P + A_2 q}{\Delta_c} \tag{12-20}$$

Where $A_1 = \dfrac{3d \cdot GL}{2wh^2}(1+r)$

$$A_2 = \frac{GL}{8wh^2}(3L^2 - GL^2)(1+r)$$

P is the applied concentrated load and Δ_c is the compressive deformation measured by the extensometer attached on the top of the beam. When Equation 12-14 is applied, the tensile modulus, E_t, can also be derived.

12.7.3   Viscoelastic Analysis of a Beam of Bi-Modulus

The correspondence principle can be used to develop the solution considering asphalt mixture as a viscoelastic material. This principle (see Chapter 6) states that the solution to the problem for a viscoelstic continuum is closely related to the solution to a corresponding problem for an elastic continuum occupying the same geometric configuration and subjected to the same boundary conditions. The only difference between these two solutions is that the elastic constants and variables in the elastic problem are replaced by their appropriate variables in the transformed space in the viscoelastic problem. In this study, the Laplace transformation was used.

The viscoelastic solution was obtained through the following steps:

1. Perform the Laplace transform of two loading functions and one deformation function.

2. Replace variables with their corresponding Laplace transforms in Equation 12-20. That is $E(t) \rightarrow sE(s)$, $P(t) \rightarrow P(s)$, $\Delta(t) \rightarrow \Delta(s)$

3. Solve the equation in the Laplace transform space.

4. Invert the solution in the transform space back into the time domain.

The loading functions in the time domain are:

$$P(t) = p_0 \cdot t \quad ; \quad q(t) = q_0 \cdot U(t) \tag{12-21}$$

Where $U(t)$ is a unit step function. The form of the deformation function, $\Delta(t)$, can be obtained through regression analysis of the experimental data.

12.7.4   Data Analysis

The results for the six beams tested at 25°C are presented. The measured tensile deformation at the bottom of the beam was plotted against the compressive deformation on the top of the beam in Figure 12.16. Note the deformations plotted were the extension or compression measured by extensometers and they are Δ_t or Δ_c in Equations 12-16 and 12-17 and are not the corresponding strains, which vary within the gauge length. As expected, the compressive and tensile deformation changed proportionally and a linear regression typically resulted in a good fit.

The slope of the regression curve evaluated the ratio r used in the above derivations. The values of the slopes listed in Table 12.4 are an average of 1.28 with a standard deviation of 0.13. With Equation 12-12b, this indicates the compressive modulus, E_c, is

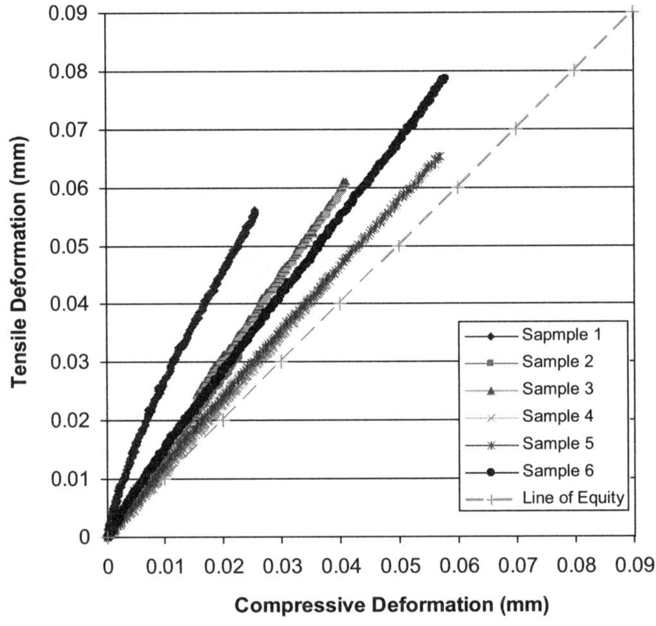

Figure 12.16 Deformations at 25°C.

around 1.6 times larger than the tensile modulus, E_t. This value agrees with the observations derived from previous research (Secor and Monismith, 1965).

The compressive deformation curve was used to obtain the deformation function, $\Delta(t)$ in Equation 12-21. The best fit for the typical compressive deformation process was regressed as a second-order polynomial:

$$\Delta(t)=2\times10^{-5}t^2+4\times10^{-4}t \tag{12-22}$$

A Laplace transform was performed on Equation 12-20 and 12-21 and the following transformed functions were obtained:

$$\overline{E}(s)=\frac{A_1\overline{P}(s)+A_2\overline{q}(s)}{s\overline{\Delta}(s)} \tag{12-23}$$

The inverse Laplace transform was then applied to obtain the relaxation modulus function in the time domain as:

$$E(t)=1.524\times10^{10}e^{-0.1t}-1.574\times10^{11}\delta(t) \tag{12-24}$$

Beam 1	Beam 2	Beam 3	Beam 4	Beam 5	Beam 6	Mean	Standard Deviation
1.97*	1.27	1.48	1.16	1.14	1.34	1.28	0.14

* This is an outlier and not included in the calculation

Table 12.4 The r ratio at 25°C.

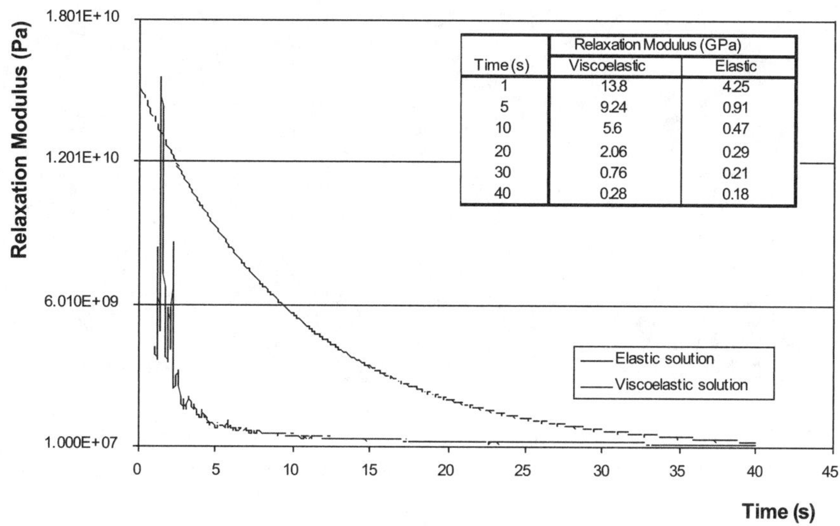

Time (s)	Relaxation Modulus (GPa)	
	Viscoelastic	Elastic
1	13.8	4.25
5	9.24	0.91
10	5.6	0.47
20	2.06	0.29
30	0.76	0.21
40	0.28	0.18

FIGURE 12.17   Modulus estimated by elastic and viscoelastic solutions.

Where $\delta(t)$ is the dirac delta function and this item in Equation 12-24 is related to the initial material response. The relaxation modulus is plotted in Figure 12.17 and the values at certain time points are calculated and listed in the same figure. These values of the relaxation modulus are comparable with the values reported in the literature (Secor and Monismith, 1965; Khanal and Mamlouk, 1995). The modulus evaluated with the elastic solution in Equation 12-20 is plotted in the same graph. The obvious difference between these two curves implies the necessity of a viscoelastic solution.

12.7.5   Conclusions

The measurement and analytical procedures developed for the four-point bending beam test present an effective method to evaluate the bimoduli of AC. The viscoelastic analysis developed allows the use of the above test for evaluating the relaxation modulus of asphalt materials in tension and compression. While it might be difficult to develop an analysis of pavements that considers the bimoduli, the measured difference that is consistent with literature data provides a basis for future pavement design and analysis that may consider the bimoduli effects.

12.8   Anisotropy in Permeability

Harris (2007) conducted a study to experimentally evaluate the anisotropy properties of AC. A horizontal permeability test device and a field permeability test device were developed for both laboratory and field permeability tests (Figure 12.18). An FEM simulation of the field permeability test was also performed. Both the vertical permeability calculated using Equation 12-25 and the horizontal lab permeability calculated using Equation 12-26 utilized the falling head method.

$$k_v = \frac{a \cdot L}{A \cdot \Delta t} \cdot \ln\left(\frac{h_1}{h_2}\right) \qquad (12\text{-}25)$$

(a)   (b)   (c)

a. Florida vertical permeability device; b. Virginia Tech Transportation Institute (VTTI) horizontal permeability device; c. VTTI field permeability device

FIGURE 12.18   Lab and field permeability test devices.

where k_v = vertical coefficient of permeability
$\quad$ a = area of standpipe
$\quad$ L = length of specimen
$\quad$ h_1 = water head at beginning of test
$\quad$ h_2 = water head at end of test
$\quad$ A = cross sectional area of specimen
$\quad$ t = time between reading h_1 and h_2

$$k_h = \frac{\left(\dfrac{\pi * r_p^{\,2}\left(h_o - h_t\right)}{t}\right)\ln\left(\dfrac{r_o}{r_i}\right)}{2 * \pi * L * \left(\dfrac{h_o - h_t}{2}\right)}$$

(12-26)

where r_p = radius of standpipe
$\quad$ r_o = outer radius of sample, cm
$\quad$ r_i = inner radius of sample, cm
$\quad$ h_o = initial head, cm
$\quad$ h_t = final head, cm
$\quad$ t = time, s
$\quad$ L = length of sample, cm

In the horizontal permeability formulation, only cross anisotropy is considered. Their experimental results (Figure 12.19) indicate that horizontal permeability is roughly two times that of the vertical permeability and this ratio decreases with the void content. When void content reaches the critical void content (Cooley et al., 2001), the ratio approaches 1.

FEM simulation (Figure 12.20) using $k_h/k_v = 5.0$ results correlate well with the data obtained from the field permeameter test. The permeability was largest with the 3.5-inch diameter contact area. Since the influence of horizontal permeability is greatest with this

Horizontal/Vertical Permeability

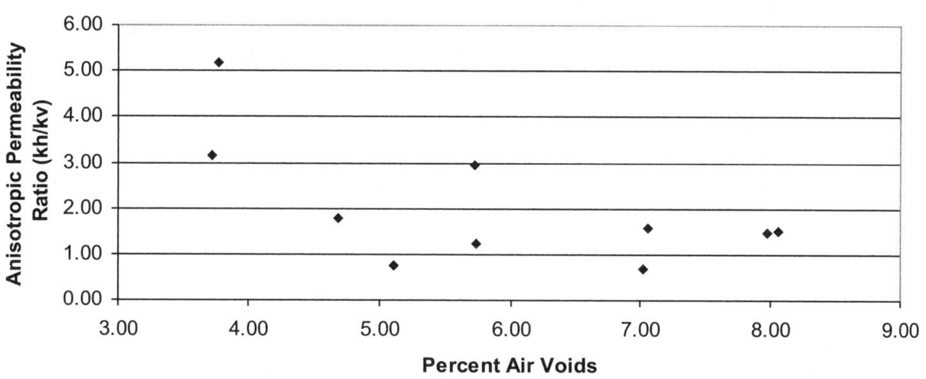

FIGURE 12.19   Vertical/horizontal permeability.

area, and the horizontal permeability is generally larger than vertical permeability, this would yield the largest permeability values. As the contact area increased, the influence of horizontal permeability decreased. Since the vertical permeability is lower than the horizontal permeability, the permeability values that were calculated based on larger contact areas were lower.

Kutay et al. (2006) presented a comprehensive study on the general anisotropy permeability behavior of AC. The generalized Darcy's Law (Equation 12-27) is utilized for his simulation using the Lattice Boltzman method. In his study, XCT was used to obtain the void structure, including the connectivity of the voids, and simulation results were

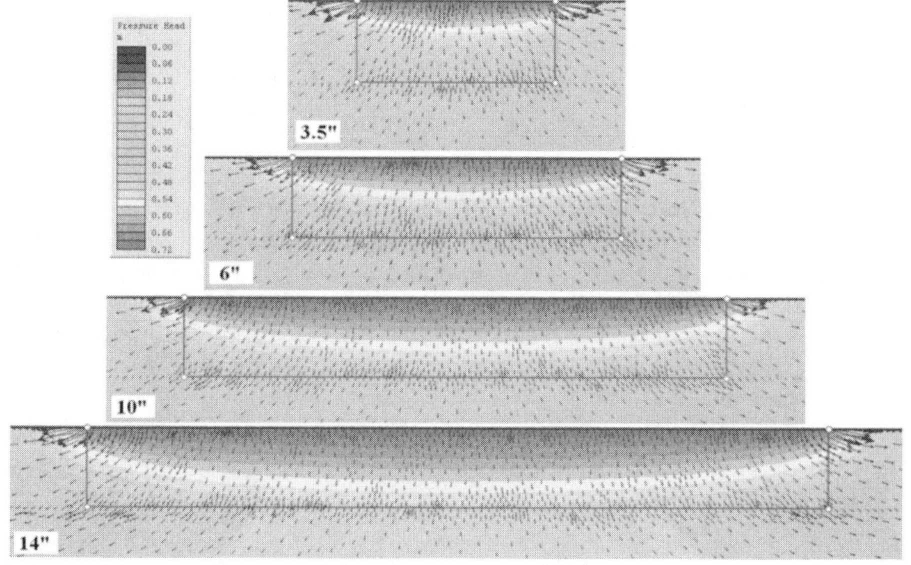

FIGURE 12.20   Flow lines/pressure head (permeable) for $k_h/k_v = 5$, $k = 5 \times 10^{-3}$ cm/s.

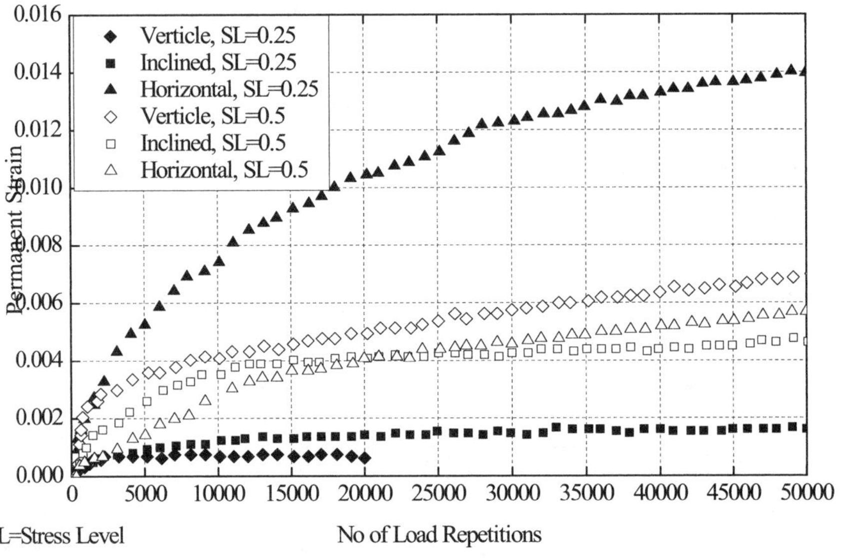

Figure 12.21 Permanent deformation of asphalt mixes at 10 psi confining stress (courtesy Robert Liang).

compared with experimental results. His simulation results and experimental results show a significant permeability difference between vertical (the compaction direction) and horizontal directions, while the horizontal direction permeability is statistically the same (k_{xx} and k_{yy}).

$$\begin{pmatrix} u_x \\ u_y \\ u_z \end{pmatrix} = -\frac{1}{\gamma n_{eff}} \begin{bmatrix} k_{xx} & k_{xy} & k_{xz} \\ k_{yx} & k_{yy} & k_{yz} \\ k_{zx} & k_{zy} & k_{zz} \end{bmatrix} \begin{bmatrix} \nabla P_x \\ \nabla P_y \\ \nabla P_z \end{bmatrix} \tag{12-27}$$

12.9   Anisotropic Behavior of Permanent Deformation

Liang et al. (2006) presented a study on evaluating the anisotropic properties of AC using regular binder and 5% SBS modified binder (PG-76-22). They discovered that the vertical modulus and strength (against permanent deformation) are larger than those in the horizontal and inclined (45° to the vertical direction) directions (Figure 12.21). Anisotropy is more significant for the modified binder mixture (based on their limited study).

References

AASHTO Designation T 322. (2007). Determining the creep compliance and strength of hot-mix asphalt (HMA) using the indirect tensile test device. AASHTO Standards, Washington, D.C., *American Association of State Highway Transportation Officials.*

AASHTO Designation TP 62. (2005). Determining dynamic modulus of hot-mix asphalt concrete mixtures. *AASHTO Provisional Standards, Washington, D.C., American Association of State Highway Transportation Officials.*

Adu-Osei, A, Little, D.N. and Lytton, R.L. (2001). Cross-anisotropic characterization of unbound granular materials. *Transportation Research Record*, No. 1757, pp.82–91.

Airey, D.W. and Wood, D.M. (1988). The cambridge true triaxial apparatus. advanced triaxial testing of soil and rock, *STP 977, ASTM, Philadelphia*, PA, pp.796–805.

Arthur, J.R.F. (1988). Cubical devices: versatility and constraints. advanced triaxial testing of soil and rock, *STP 977. ASTM, Philadelphia*, PA, pp.743–765.

Atkinson, R.H. (1972). A cubical test cell for multiaxial testing of materials. Ph.D. *Dissertation, University of Colorado at Boulder*, Boulder, CO.

Bonaquist, R., Anderson, D.A. and Fernando, E. (1986). Relationship between moduli measured in the laboratory and field deflection measurements. *Proc. Association of Asphalt Paving Technologist*, Vol. 55, pp.419–448.

Boulbibane, M., Weichert, D. and Raad, L. (1999). Numerical application of shakedown theory to pavements with anisotropic layer properties. *Transportation Research Record*, No.1687, pp.75–81.

Chen, Y.G., Pan, E. and Green, R. (2009). Surface loading of a multilayered viscoelastic pavement: semianalytical solution. *Journal of Engineering Mechanics*, Vol.135, pp.517–528

Chen, Y.G. and Pan, E. (2008). Responses of transversely isotropic layered half-space to multiple horizontal loads. *ASCE Geotechnical Special Publication, No. 182, Pavements and Materials: Characterization, Modeling and Simulation*, pp.63–77.

Christensen, D.W. and Bonaquist, R.F. (2004). Evaluation of indirect tensile test (IDT) procedures for low-temperature performance of hot mix asphalt. *NCHRP Report 530.*

Cooley, L.A. Jr, Brown, E.R. and Maghsoodloo, S. (2001). Development of critical field permeability and pavement density values for coarse-graded superpave pavements. *NCAT Report 01-03. National Center for Asphalt Technology.* Auburn, AL.

Daniel, J.S. and Lachance, A. (2005). Mechanical and volumetric properties of asphalt mixtures with recycled asphalt pavement. *Transportation Research Record*, No. 1929, pp.28–36.

Green, G.E. (1971). Strength and deformation of sand measured in an independent stress control cell. *In stress-strain behavior of soils. Proceedings of the Roscoe Memorial Symposium*, G.T. Foulis and Co. Ltd., Cambridge, England, pp.285–323.

Gere, J.M. and Timoshenko, S.P. (1984). *Mechanics of Materials*, 2nd Edition. PWS-KENT Publishing Company, Boston.

Hambley, E.C. (1969). A new triaxial apparatus. *Geotechnique*, Vol. 19, No.2, pp.307-309.

Harris, C.H. (2007). Hot Mix Asphalt Permeability: Tester Size Effects and Anisotropy. *Master Thesis*, Virginia Tech.

Johnson, K.L. (1985). Contact Mechanics. *Cambridge University Press,* Cambridge.

Kallas, B.F. (1970). Dynamic modulus of asphalt concrete in tension and tension-compression. *Proceedings of Association of Asphalt Paving Technologist*, Vol. 39, pp.1–20.

Khanal, P.P. and Mamlouk, M.S. (1995). Tensile versus compressive moduli of asphalt concrete. *Transportation Research Record*, No.1492, pp.144-150.

Ko, H.Y. and Scott, R.F. (1967). A new soil testing apparatus. *Geotechnique*, Vol.17, No.1, pp.40–57.

Kuwano, R., Connolly, T.M. and Jardine, R.J. (2000). Anisotropic stiffness measurements in a stress-path triaxial cell. *Geotechnical Testing Journal*, Vol.23, No.2, pp.141–157.

Kutay, M.E., Aydilek, A.H., Masad, E. and Harman, T. (2006). Computational and experimental evaluation of hydraulic conductivity anisotropy in hot-mix asphalt, *International Journal of Pavement Engineering*, Vol. 8, No. 1, pp.29–43.

Lade, P.V. and Duncan, J.M. (1973). Cubical triaxial tests on cohesionless soil. *ASCE Journal of Soil Mechanics*, Vol.99, SM10, pp.793–812.

Li, X. and Wang, L.B. (2008). Simultaneous determination of bimoduli of asphalt material with single viscoelastic beam. *International Journal of Pavement and Technology*, Vol.1, No.2, pp.57–63.

Li, X.S. and Dafalias, Y.F. (2002). Constitutive modeling of inherently anisotropic sand behavior. *Journal of Geotechnical and Geoenvironmental Engineering*, Vol.128, No. 10, pp. 868–880.

Liang, R.Y., Abu Alfoul, B.A. and Mohammad, K. (2006). Laboratory investigation of anisotropic behaviour of HMA. *2006 International Conference on Perpetual Pavement*. Hilton Columbus at Easton Columbus, Ohio

Masad, E., Tashman, L., Somedavan, N. and Little, D. (2002). Micromechanics-based analysis of stiffness anisotropy in asphalt mixtures. *Journal of Materials in Civil Engineering*, Vol.14, No.5, pp.374–383.

NeSmith, W. M. (1997). Development of a computer controlled multiaxial cubical testing apparatus. *M.Sc. Thesis, Georgia Institute of Technology*, Atlanta, GA.

Pan, E. (1989). Static response of a transversely isotropic and layered half space to general surface loads. *Physics of the Earth and Planetary Interiors*, Vol.54, pp.353–363.

Pennington, D.S., Nash, D.F.T. and Lings, M.L. (2001). Horizontally mounted bender elements for measuring anisotropic shear moduli in triaxial clay specimens. *Geotechnical Testing Journal*, Vol. 24, No.2, pp.133–144.

Roy, D., Campanella, R.G., Byrne, P.M. and Hughes, J. (2002). Undrained anisotropic monotonic behavior of sand from in situ tests. *Journal of Geotechnical and Geoenvironmental Engineering*, Vol.128, No.1, pp.85–91.

Secor, K.E. and Monismith, C.L. (1965). Viscoelastic response of asphalt paving slabs under creep loading. *Highway Research Record*, No. 67, pp.84–97

Seyhan, U. and Tutumluer, E. (2002). Anisotropic modular ratios as unbound aggregate performance indicators. *Journal of Materials in Civil Engineering*, Vol.14, No.5, pp 409–416.

Sture, S. and Desai, C.S. (1979). A fluid cushion truly triaxial or multiaxial testing device. *ASTM Geotechnical Testing Journal*, Vol. 2, No.1, pp.20–33.

Sture, S. (1979). Development of multiaxial cubical test device with pore-water pressure monitoring facilities. *Report No. VPI-E-79.18, Dept of Civil Engineering, Virginia Polytechnic Institute and State University*, Blacksburg, VA.

Tutumluer, E., Adu-Osei, A., Little, D.N. and Lytton, R.L. (2001). Field validation of the cross-anisotropic behavior of unbound aggregate bases. *ICAR-502-2*.

Von Quintas, H.L., Rauhut, J.B. and Kennedy, T.W. (1982). Comparisons of Asphalt concrete stiffness as measured by various testing techniques. *Proceeding of Association of Asphalt Paving Technologist*, Vol. 51, 1982, pp.35–49.

Wang, L.B., Hoyos, L.R., Mohammad, L. and Abadie, C. (2005a). Characterization of asphalt concrete by multi-stage true triaxial testing. *Journal of ASTM International*, Vol.2, No.10, pp.7–16.

Wang, L.B., Hoyos L.R., Wang, X., Voyiadjis, G. and Abadie, C. (2005b). Anisotropic properties of asphalt concrete: characterization and implications in pavement analysis and design. *ASCE Journal of Engineering Mechanics*, Vol.17, No.5, pp.535–543.

Wof, K. (1935). Ausbreitung der kraft in der halbebene und im halbraum bei anisotropem material. *Zeitschrift f. Angew. Math. Und Mech.*, B.15, H.5.

CHAPTER 13

Multiscale Modeling and Moisture Damage

13.1 Introduction

This chapter presents some recent developments on multiscale modeling and moisture damage mechanisms of asphalt concrete (AC). These recent developments, to a certain degree, are still at their inception stage.

13.2 Multiscale Characterization and Modeling

13.2.1 Needs for Multiscale Modeling

The mix design methods for AC are highly empirical. It usually involves testing a representative volume (a specimen) of different blends to evaluate their properties and applying macro-constitutive models and failure criteria to assess the behavior or performance of these materials placed in the field at certain assumed conditions. Due to the differences in size, pavement structure, environmental and loading conditions, and complex coupling, lab specimen behavior may not reflect the field performance of the mix. In addition, the mixing of these different components results in various complex physical interactions, chemical reactions, and electromagnetic interactions. Since these reaction processes have not been well understood, mixture properties are often not as expected. Mix design often involves understanding chemical reactions where quantum mechanics, chemomechanics, may play an important role in interpreting these complex reaction processes and guiding mix design.

Asphalt concrete is highly heterogeneous, with components of significantly distinct physical, chemical, thermal-mechanical, and electromagnetic properties. In addition, the sizes of the particles range from nanometer to centimeter; the inherent and induced defects have also a similar size range. The particle and defect at different sizes interact differently and in a complicated manner. Traditional approaches in homogeneous continuum mechanics, micromechanics, continuum damage mechanics, and fracture mechanics may be able to describe the behavior of these materials under complicated environmental and loading conditions in one scale, but fail to address the failure and interaction of components and defects at different scales with significant heterogeneity. Multiscale characterization, modeling, and simulation prove to be a useful tool. In addition, recent developments in supercomputers and parallel computation make it realistically feasible to perform these computational simulations for both material design and structural behavior prediction.

The critical point here is the failure modes and mechanism dominating the initiation and propagation of localized deformation and cracking at different length scales. While all the failures (cracking) or large plastic or permanent deformations are attributed ultimately to either the breakage of atomic bonds or dislocations of atoms, a continuum approach may be valid in many cases where failure mechanisms are well understood empirically, and the deformation gradient or strain rates, etc. are not so large. For example, microstructure defects far away from the crack tip may not assist in the arrest or speed up of cracking; nano-size defects or inhomogeneity may not affect a millimeter-size crack at high-speed fracturing. Therefore, determining where a smaller scale model is needed also requires multiscale modeling.

With high performance and even nano-composite materials, the failure mechanism has not been well understood, therefore modeling and simulation at the atomistic scale becomes critical. Mechanical properties of the thin films, etc., are so difficult to test. Macroscopic properties such as strength and strain relationships, modulus, strength, and toughness may not be able to be conveniently characterized. In addition, lack of understanding of the deformation and failure mechanisms of the structural materials under extreme loadings such as explosion, high-speed impact, and burning all require understanding the fundamental mechanisms across the different scales. Even friction between tires and pavement materials, and ice may require understanding the cohesion and adhesion of these materials at nanoscale.

Steel corrosion and alkaline silicon reaction both involve complex chemo-mechanical interactions. Even the hydration process has not been well understood. Recently, chemo-mechanics of bituminous materials has caught the attention of many researchers. Phenomena such as self-healing mechanisms of polymers require the use of quantum mechanics and nanoscale devices or techniques to understand. Furthermore, the use of nanosensors involves the interactions at nanoscale between the hosting materials and the sensor materials. Understanding of chemo-mechanical and electromagnetic coupling is also important for the design and interpretation of the sensing data.

Developments in nanoscale science and technology as well as high performance computation make it possible to perform such multiscale characterization, modeling, and simulation. Multiscale modeling techniques and their numerical implementation have reached a point where multiscale modeling techniques may help select materials, develop mix design, and characterize mixture properties more rationally.

13.2.2 A Brief Overview of Multiscale Modeling Methods

13.2.2.1 Continuum Modeling

Continuum modeling based on governing constitutive equations has dominated the realm of numerical modeling since its advent. The constitutive equations are actually a set of phenomenological relationships between cause and effect such as stress and strain. Continuum methods have been successful in characterizing large-scale structures and components, for which the detailed response of materials is not so critical. In rigorous mechanics sense, constitutive equations in continuum mechanics are built on the statistical description of atomic and molecular scale processes (average stress, strain, or velocity over time and space). In other words, constitutive equations represent the mechanical behavior of materials over long-time and large-length scales.

All materials are composed of atoms. Existing defects such as cracks can lead to stress concentrations under loads, which in turn lead to nucleation of dislocations or generation of new material surfaces due to strong deformation at highly localized regions such as the crack tips. These nano- and microscale defects typically include point defects, dislocations, interfaces, and grain boundaries that essentially determine the mechanical properties of materials and structures. According to the statistical law of large numbers (LLN), continuum analyses are valid only for large enough systems that include a substantial number of defects at the meso or above-meso scales. Continuum approaches cease to work as the system size approaches the average separation distance in-between defects.

13.2.2.2 Atomistic Modeling

The capability of continuum modeling of construction materials and structures is further limited by a number of factors. First, a wide variety of experimental observations on the mechanical behavior of materials cannot be explained satisfactorily within the paradigm of continuum mechanics. These observations include the dislocation patterns in fatigue and creep of metals, surface roughening and crack nucleation in fatigue of concretes, and some common problems in generic building materials such as the statistical nature of brittle failure, the inherent inhomogeneity of plastic deformation and plastic flow localization in shear bands, and the effects of size, geometry, and stress state on the yield properties. Second, constitutive equations in continuum models are usually phenomenological descriptions that are determined within a certain range of temperature, stress state, strain rate, and material conditions. Without a clear physico-chemical understanding, the behavior of materials under unanticipated conditions beyond the measured range becomes unknown, which in turn significantly limits the widespread use of experimentally-derived constitutive equations. Third, nowadays, transportation engineering has evolved to emphasize small length-scale phenomena, such as to develop high performance materials based on nanotechnologies. Equally imperative is the fact that the technology of micro-electro-mechanical systems (MEMS) has become mature for monitoring the reliability of structures, which points to the need for a more physically-based approach than continuum modeling.

Being faced with the challenges to the continuum modeling of construction materials, together with the recent advancement of computation technologies that enable large-scale computing, the development of more efficient numerical methods for modeling complex physical phenomena in materials has become imperative and attractive. The availability of large-scale computing capability has, in the past decade, greatly propelled the modeling of material phenomena into a more rational direction: the atomistic modeling and simulation, mainly quantum molecular dynamics (QMD) and classic molecular dynamics (MD), in which individual nucleons, electrons, and atoms are explicitly followed during their dynamic evolution. Opposite to the continuum-based analysis that attempts to reduce the complexity of a material system's behavior by a process of reduction of its degree of freedom (DOF), researchers now are trying to represent large numbers of DOFs and solve for them numerically. The last decade has seen tremendous advances on a number of computational and physical fronts in atomistic simulation techniques, with rigorous ways to approximate the quantum mechanical behavior of atoms and molecules. These advances in QMD have been parallel to the development of physically-based inter-atomic potentials for performing more accurate classical MD simulations. It is noteworthy that the quantum simulation methods are much more time expensive (10^5 to 10^6 times more) than classical atomistic simulations.

Currently, the simulated system size has increased significantly due to the fast increase in computing power and reduction in computing costs. The linkage between classical MD and *ab initio* QMD computations has become more clear and rigorous.

13.2.2.3 Multiscale Modeling

The explicit modeling of atomic/molecular structures through tracing the details of atomic-scale processes, although having been enthusiastically pursued, has involved a set of inherent spatial and temporal limitations. The spatial and temporal scale limitations come from both small and large directions. Time scales of atoms are on the order of femto-second and length scales on the order of angstroms. Typical atomistic simulations therefore are limited to very small systems over very short times due to the very small spatial and temporal scales. Although the computing power has grown significantly, atomistic modeling methods are incapable of describing systems much larger than billions of atoms (1μm geometric dimension) or longer than billions of femto-second time steps (1 ms simulation period).

Realizing that continuum and atomistic modeling methods are complementary, the idea of multiscale modeling came first to the mesoscopic modeling methods that are developed to bridge this critical gap between the extremes of length scales, i.e., the macroscale and atomic-scale. Continuum analyses tend to break down at the mesoscale, at which atomistic methods start to reach their inherent spatial and temporal limitations. There are two general solutions that have evolved so far to bridge the mesoscopic gap. The first one is to study the dynamics of defect ensembles in the system, instead of atomistically simulating the entire system. This approach has been successfully used in dynamic simulations of interacting cracks in brittle materials and dislocations in crystalline materials. The second solution to the mesoscale problem is based on statistical mechanics approaches, in which evolution equations for statistical averages are solved for a complete description of the analyzed problem (Walgraef and Aifantis, 1985; Hähner et al., 1998; Hähner, 1996; Zaiser and Hähner, 1997; El-Azab, 2000).

13.2.2.4 Classification of Multiscale Problems

The very first step in multiscale modeling is to design an efficient multiscale method. From this standpoint, multiscale problems can be divided into four general categories, each sharing the special features that can be used when designing an efficient multiscale method for the category.

Type A: problems for which a macroscopic model is valid across a domain except for localized region(s) in which the macroscopic model does not work spatially and/or temporally and therefore a more accurate microscopic description is needed. Typical type A problems include isolated defects or singularities such as cracks, dislocations, shocks, and contact lines. For these problems, the microscopic model is necessary near defects or singularities, whereas the continuum theory remains valid in the surrounding regions for establishing macroscopic model(s).

Type B: problems for which a closed macroscopic description is known to exist across a domain, however, the macroscopic model may not be explicitly expressed or is too expensive to obtain. Typical type B problems include transport of fluids and solutes through heterogeneous porous media, complex fluids, and plasticity.

Type C: problems for which the macroscopic model is not explicitly known and does not work correctly near defects. Type C problems have characteristics of both Type A and Type B.

Type D: problems that exhibit self-similarity in scales, such as the critical phenomena in fractals, turbulent transport, and statistical physics.

Type A and Type B problems feature the characteristic of scale separation, while Type D problems are characterized by the special feature of statistical self-similarity. The featured characteristics of each category generally guide the design of an efficient multiscale modeling approach to be used.

13.2.2.5 Hierarchical vs. Concurrent Modeling

One traditional multiscale modeling approach is the *hierarchical coupling method* (sequential or serial coupling method) in which an effective macroscale model is determined in pre-processing steps using microscale models. The macroscale model is usually targeted for applications. Hierarchical coupling methods are well known to pass parameters in sequence, i.e. from a lower hierarchy (smaller spatial or temporal scale) to an upper one (larger spatial or temporal scale) and vice versa. The scheme of hierarchical multiscale modeling is illustrated in Figure 13.1 (Ghoniem and Cho, 2002). The focus of multiscale modeling in current times, however, is more on the *concurrent coupling methods* (Abraham et al., 1998). The concurrent coupling methods entail the linkage of the macroscale and the microscale models in a domain based on a spatial decomposition of the system into regions dominated by different length scales. Both the macroscale and the microscale models are executed in each computation step (such as one time step). This is how the terminology *concurrent* comes into being.

For asphalt mixtures, one additional complexity is their meso-scale heterogeneity, the air-void, binder, and aggregate constituents. If these constituents are treated as micro-continuums, for each micro-continuum, the above scale classification may be

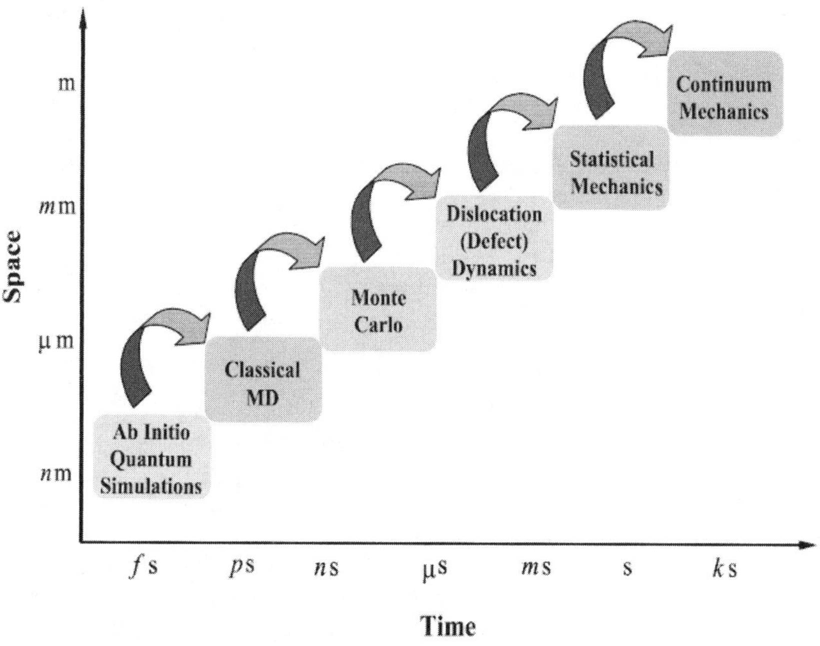

FIGURE 13.1 Schematic illustration of a typical hierarchical multiscale modeling (Ghoniem and Cho, 2002).

valid. The only difference would involve the interfaces among the different micro-continuums.

When the governing constitutive equation contains few variables, a hierarchical coupling method is desired that can provide a more viable solution to concurrent coupling techniques. In contrast, the concurrent coupling methods are preferred when the constitutive relationship depends on more variables, so it is difficult to extract it by pre-computing. From a numerical viewpoint, however, the two strategies are closely related in that for both hierarchical and concurrent coupling methods, the key issue is to design simulations at the micro-scale level that offer the needed macroscopic data. The two strategies can be combined to yield optimal efficiency. For example, a concurrent simulation can be used to pre-evaluate the constitutive equation for the hierarchical coupling method.

Recent years have seen a surge of interest in developing concurrent multiscale modeling methods that can obtain the same results as atomistic modeling at a lower computational price. The efficiency is accomplished by atomistically modeling selected small sub-domains of the problem, while analyzing the rest of the domain using continuum modeling. Table 13.1 summarizes the current multiscale modeling methods available in open literature.

13.2.3   General Philosophy of Multiscale Modeling of AC

Figure 13.2 presents the general philosophy to resolve the multiscale modeling problem for AC. AC typically consists of aggregates (including mineral fillers), asphalt binder, interfacial transition zones, and voids. These components can be considered as micro-continuums. Each of the micro-continuums and the interfaces can be modeled using the Reaction Force Theory (van Duin et al., 2003; Goddard et al., 2006; and Chenoweth et al., 2008) based on quantum mechanics, molecular dynamics, and thermal mechanics (statistical mechanics, and can be upscaled to continuum). Through the use of FEM and with the consideration of the microstructure and the techniques of homogenization, the material can be modeled as either a traditional continuum or 3M (micro-morphic, micro-stretch, and micro-polar, Eringen, 1999) continuum. A macro-constitutive model for the material can then be developed, represented by the upscaling process. The material structure can be represented as digital specimens, which may represent the scales in separate databases. For the downscaling process, solving a boundary value problem would involve the use of the continuum model for the completely homogenized continuum, triggering the micro-continuum and directly MD and RFF (reaction force field) calculations concurrently for zones of special interest. Numerically, the above problem can be resolved concurrently or hierarchically.

13.2.4   Reaction Force Field Theory

In molecular mechanics/molecular dynamics, a force field refers to the functional form and parameter sets used to describe the potential energy of a system of atoms. Force-field functions and parameter sets are derived from both experimental work and high-level quantum mechanical calculations.

The consistent-valence force field (CVFF) is a generalized valence force field. Parameters are provided for amino acids, water, and a variety of other functional groups. The augmented CVFF was developed for materials science applications. It includes additional atom types for aluminosilicates and aluminophosphates.

Concurrent Multiscale Modeling Method	Developers
Artificial Compressibility Method (Incompressible Flow)	Chorin, 1967
Atomistic to Continuum Coupling (Concurrent)	Fish et al., 2007
Bridging Domain Method	Xiao & Belytschko, 2004
Bridging Scale Method	Wagner and Liu, 2003
Composite Grid Atomistic Continuum Method	Datta et al., 2004
Coupled Atomistics and Discrete Dislocations	Shilkrot et al., 2004
Continuum - Molecular Dynamics Models for Fluids	Li et al., 1999
Continuum - Molecular Dynamics Models for Solids	E and Huang, 2002
Cluster-Energy Quasi-Continuum Method	Eidel & Stukowski, 2008
Cluster-Force Quasi-Continuum Method	Knap and Ortiz, 2001
Coarse-Grained Monte Carlo	Katsoulakis et al., 2003
Coarse-Grained Molecular Dynamics	Rudd & Broughton, 1998
Coupling of Length Scales Method	Rudd & Broughton, 2000
Computational Materials Design Facility Method	Buehler et al., 2005
Car-Parrinello Method	Car and Parrinello, 1985
Discrete Simulation Monte Carlo	Garcia et al., 1999
Finite-Element/Atomistics Method	Kohlhoff et al., 1991
Ghost-Force Corrected Quasi-Continuum Method	Shenoy et al., 1999
Ghost-Force Corrected Cluster-Energy Quasi-Continuum	Tadmor et al., 2002
Gas-Kinetic Scheme	Xu & Prendergast, 1994
Gap-Tooth Scheme	Kevrekidis et al., 2003
Heterogeneous Multiscale Modeling	E and Engquist, 2003
Hybrid Simulation Method	Luan, 2007
Local Quasi-Continuum Method	Ericksen, 1984
Coupled Local and Nonlocal Quasi-Continuum Method	Ortiz & Phillips, 1999
Macroscopic, Atomistic, Ab Initio Dynamics	Abraham et al., 1998
Non-Local Quasi-Continuum Method	Tadmor et al., 1996
Optimal Prediction	Chorin et al., 2002
Projective Dynamics for Kinetic Monte Carlo	Kolesik, 1998
Quasi-Continuum –Density Functional Theory Method	Tadmor et al., 2002
Quantum Mechanics - Molecular Dynamics	Zhang et al., 1999
Stochastic Ordinary Differential Equations	Vanden-Eijnden, 2003
Virtual Internal Bond - Cauchy-Born Based Techniques	Gao and Klein, 1998

TABLE 13.1 Summary of concurrent multiscale modeling methods.

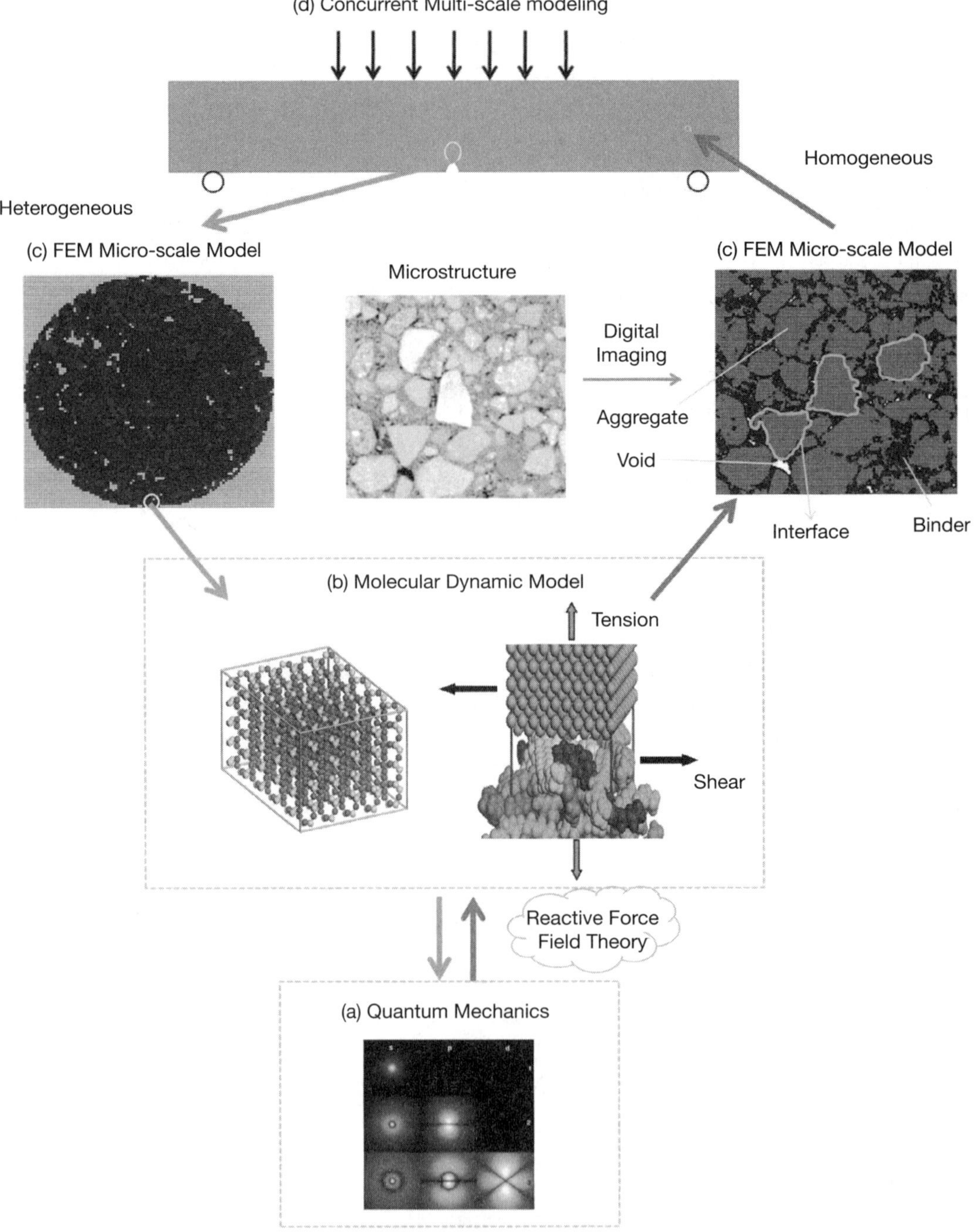

Figure 13.2   General philosophy of multiscale modeling of AC.

To understand the chemical reactions, mechanical behavior, and electromechanical coupling, an atomistic level of interaction and behavior is needed. This requires quantum mechanics, which is numerically very challenging (i.e., on only a few hundred atoms). The Reaction Force Field (ReaxFF) Theory bridges the length scale between quantum mechanics and molecular dynamics. The ReaxFF simulations (van Duin et al., 2003; Goddard et al., 2006; and Chenoweth et al., 2008) provide a full-physics, full-chemistry description of processes in materials, otherwise unreachable via quantum mechanics (QM), and constitute the ideal tool for studying reactivity on surfaces and condensed matter. The results of these simulations will provide an atomistic description of the processes and structure, including kinetic conditions, that will provide key insight into optimizing asphalt mixtures, as well as provide complementary information to the available experimental methodologies and continuum-level simulation of these materials.

13.2.5 Molecular Dynamics (MD)

13.2.5.1 MD Fundamentals

Traditional MD will increase the computational efficiency at the cost of accuracy. Under National Science Foundation (NSF) support for the Unified Approach for Multiscale Characterization, Modeling, and Simulation of Stone-based Infrastructure Materials, MD simulations of asphalt binder-aggregate interfaces have been developed (Lu and Wang, 2008; Lu and Wang, 2009, 2010; Lu, 2010), where the binder model developed by Zhang and Greenfield (2007a, 2007b and 2008). An atomistic FEM approach is being developed to bridge the length scale between MD and continuum mechanics. This is a hierarchal approach but may be more realistically achievable and provide better links between state-of-the-art research and current practice using the continuum and phenomenological models.

MD was originally used in thermodynamics and physical chemistry to calculate the collective or average thermochemical properties of various materials including gases, liquids, and solids. There are two basic assumptions made for the standard molecular dynamics simulations (Haile, 1992):

1. Molecules or atoms are described as a system of interacting material points, whose motions are described dynamically with vectors of instantaneous positions and velocities. The atomic interaction has a strong dependence on the spatial orientations of atoms and the distances among them.

2. The simulated system is usually treated as an isolated domain with conserved energy. Usually there is no mass change in the system. Equivalently, the number of atoms in the system remains the same. However, non-conservative techniques are also available, which can model the dissipation of the kinetic energy into the surrounding media.

As shown in Figure 13.3, the MD simulation can be represented as a loop using the finite difference method. This method involves the solution of the classical equations of motion for a system of N atoms. In classical statistical mechanics, the phase space trajectories (position vectors and velocity vectors) $\{r_i(t), v_i(t)\}$, are calculated by integrating Newton's second law:

$$m_i \ddot{r}_i = -\frac{\partial U\left(r_1, r_2, \ldots, r_N\right)}{\partial r_i} \equiv F_i, \quad i = 1, 2, \ldots, N \tag{13-1}$$

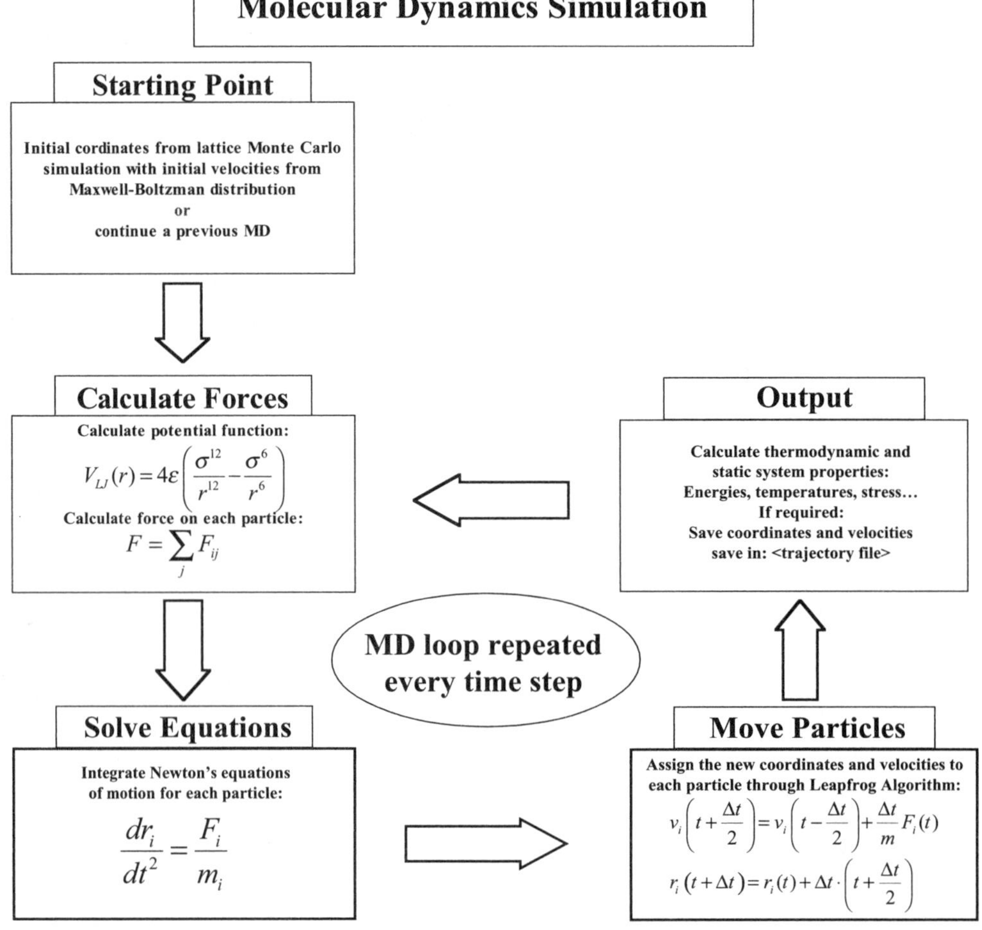

Figure 13.3 MD simulation procedures.

Where $U(r_1, r_2, \ldots r_N)$ is the interatomic potential; the force F_i is usually referred to as the internal force, e.g., the force exerted on atom i due to surrounding atoms. To solve the set of 3N equations in (13-1), the initial and boundary conditions need to be specified. In the case of a crystal, the initial conditions are the lattice positions; for an amorphous material, the initial configuration is obtained by quenching a molten state through constraining the molecule velocity fields. The initial set of atomic velocities is chosen randomly from a Maxwell-Boltzmann distribution. A number of robust algorithms (Allen and Tildesley, 1987) are available to integrate the Equation 13-1. It requires the use of the finite difference method at a suitable time step, Δt, that conserves the total Hamiltonian of the system (typically, Δt is on the order of a femtosecond).

13.2.5.2 Interatomic Potentials

The essential input to a MD simulation is the interatomic potential $U(r_1, r_2, \ldots r_N)$. The degree to which the results of MD simulation represent the properties of real materials is determined by how realistic the potential is. The degree of the complexity of an interatomic potential also determines the amount of time needed for the simulation. Over the years, a vast number of potentials have been developed that include pair potentials (Haile, 1992), embedded-atom potentials (Farkas and Duranduru 2001), and bond-order potentials (Dauber-Osguthorpe et al., 1988).

Potential energy functions are constructed as a function of atomic coordinates, making use of parameters for describing bond stretching and bond-bending and allowing for interactions between non-bonded atoms. Functional forms of potential energy functions are shown in Figure 13.4. Generally, force-field methods describe the potential energy of molecules as a sum of bond stretching (V_{str}), bond-angle bending (V_{bend}), out-of-plane bending (V_{oop}), internal rotation (torsion) about bonds (V_{tors}), interactions between these motions (V_{cross}), van der Waals attractions and repulsions between non-bonded atoms (V_{vdW}), and electrostatic interactions between non-bonded atoms (V_{el}):

$$V = V_{str} + V_{bend} + V_{oop} + V_{tors} + V_{cross} + V_{vdW} + V_{el} \qquad (13\text{-}2)$$

There are several excellent textbooks on MD simulations: Haile (1992), Leach (2001), and Rapaport (2004).

13.2.5.3 Virial Strain

The atomistic quantities (trajectory quantities) must be averaged in space and time in order to be compared with continuum concepts. Thus, the virial needs to be averaged over space and time to converge to the Cauchy stress tensor.

The strain field is a measure of geometric deformation of the atomic lattice. The local atomic strain is calculated by comparing the local deviation of the lattice from a reference configuration. Usually, the reference configuration is taken to be the undeformed lattice.

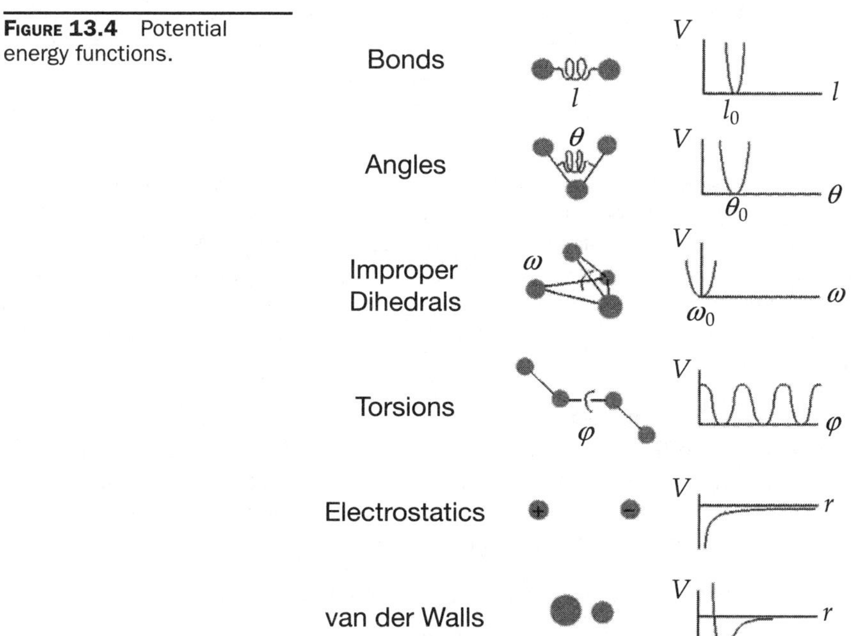

Figure 13.4 Potential energy functions.

In the atomistic simulations, the information about the position of every atom is readily available. The virial strain is relatively straightforward. The following is the general formulation.

The following tensor is defined for atom l:

$$q_{ij}^l = \frac{1}{N} \sum_{k=1}^{N} \left(\frac{\Delta x_i^{kl} \Delta x_j^{kl}}{r_0^2} \right) \tag{13-3}$$

Where $\Delta x_i^{kl} = x_i^l - x_i^k$ and $\Delta x_j^{kl} = x_j^l - x_j^k$. The quantity N refers to the number of nearest neighbors considered. The left Cauchy-Green strain tensor is given by:

$$b_{ij}^l = \frac{N}{\lambda} q_{ij}^l = \frac{1}{\lambda} \sum_{k=1}^{N} \left(\frac{\Delta x_i^{kl} \Delta x_j^{kl}}{r_0^2} \right) \tag{13-4}$$

Where λ is a factor depending on the lattice being considered.

This definition provides an expression for a measure of deformation defined using continuum mechanics and in terms of atomic positions. The Eulerian strain tensor of atom l is obtained from equation, $e_{ij}^l = \frac{1}{2}\left(\delta_{ij} - b_{ij}^l\right)$. Unlike the virial stress, the atomic strain is valid instantaneously in space and time.

13.2.5.4   Statistical Definition of Stress for Macro Continuum Model

Uncoupled simple constitutive laws including linear elasticity, viscoelasticity, and placidity can be derived from thermomechanics through free energy formulations and can be derived from the state variables in the MD simulations. For example, the average stress of a specific region in the atomistic simulation, σ_{ij}, on the j plane and in the i direction is calculated via virial formula (Allen and Tildesley, 1987; Haile, 1992; Chen, 2006):

$$\sigma_{ij} = -\frac{1}{N} \sum_{i=1}^{N_s} \left(\frac{m_i v_i \otimes v_i}{V_i} + \frac{1}{2V_i} \sum_{j=1}^{N_s} r_{ij} \otimes F_{ij} \right) \tag{13-5}$$

Where m_i is the mass of the atom i, V_i is the volume assigned to atom i, N_s is the number of atoms contained in the region of atomic interaction, r_{ij} is the vector from atom j to atom i, and F_{ij} is that force acting on atom i due to interaction with atom j. The first term of the right-hand side of Equation 13-5 represents the kinetic effect associated with atomic motion, and is related to temperature. The second term represents the effect of atomic interaction and is related to the interatomic forces and the separation between the atoms.

13.2.5.5   Free Energy and Elasticity

Elasticity stems from the interactions of atoms, and thus it is intimately linked to electro-chemistry. Depending on what material is being considered, these chemical interactions are more or less complex. In amorphous bitumen materials, it is vital to consider various types of chemical interactions, including:

- Ionic interactions (electro-static columbic interactions)
- Covalent bonds (due to overlap of electron orbital)

- Hydrogen-bonds
- Weak or dispersive van der Waals (vdW) interactions.

In general, the elastic properties of materials can be expressed as the partial derivative of the free energy density with respect to the strain tensor that characterizes the deformation.

$$G = U - TS \qquad (13\text{-}6)$$

Where G, U, T, and S represent free energy, internal energy, absolute temperature, and entropy, respectively.

One can define the free energy density as:

$$\Phi = \frac{G}{V} \qquad (13\text{-}7)$$

In the simplest 1D situation, the (scalar) stress is then is given by:

$$\sigma = \frac{\partial \Phi}{\partial \varepsilon} \qquad (13\text{-}8)$$

The stress and strain can be related by Hooke's law, $\sigma = E\varepsilon$. The above relationships can be conveniently extended to 3D formulations.

One can use the Energetic Elasticity Theory to link molecular or atomistic properties with overall elastic modulus. Energetic elasticity is characterized by the stretching of atomic, metallic, covalent, or ionic bonds that leads to a change in potential energy in the material.

13.2.6 Digital Specimen and Digital Test for Meso-scale Simulations

Clearly, at the meso-scale the digital specimen and digital test techniques can be used for either upscaling or downscaling. A gradual homogenization approach at such scale has been developed by Lutif et al. (2008). No further detailed discussions will be offered for this section.

13.2.7 Multiscale Structure Characterization

Modeling without characterization is not complete. Characterization of the 3D multi-scale structure using nano CT and micro CT is very important. Traditional nano devices such as Atomic Force Microscope (AFM) and Scanning Electron Microscope (SEM) can only characterize surface structure. A recent grant by NSF allows Virginia Tech to purchase a Nano CT at 50-nanometer resolution. This is a very important step. The Virginia Tech Institute for Critical Technology and Applied Science (ICTAS) has various micro CTs, the nano CT and the nano indentation system. They enable the characterization of AC structure at various scales and mechanical properties at various scales for modeling.

13.2.7.1 Interface Characterization

In previous chapters, microstructure characterization at mesoscale (0.1-millimeter resolution) has been discussed and the major technology presented is XCT. For features smaller than this scale, micro CT at very high resolutions (0.5 µm) and nano CT is necessary.

In addition, other characterization methods such as AFM, 3D digital optical microscope, and environmental scanning electron microscope (ESEM) can also provide microstructure characterizations. Figure 13.5 shows an AFM image of the aggregate-binder interface taken by the height mode (non-contact mode) after scanning is made. The image in Figure 13.6 is taken by a 3D digital optical microscope with a magnification of 100X. Figure 13.7 presents an interface image at 500x acquired with an ESEM. ESEM images can provide a useful chemistry map to characterize the chemical components that can be employed in the atomistic modeling stage. Figure 13.8 presents the micropores identified by focused ion beams. Figure 13.9 presents the asphalt-aggregate interface structure characterized by the tunneling electron microscope (TEM). Images by TEM indicate that the thickness of the interface transition zone is roughly 3 to 5 nm.

13.2.8 LAMMPS and Massive Parallel Computation

Large-scale molecular dynamics simulations require a significant amount of computing resources. Classical molecular dynamics can be quite efficiently implemented on supercomputers using parallelized computing strategies. Such supercomputers consist of hundreds of individual computers, which are called clusters. Concurrent computing using modern parallel computers enables the small computers to work simultaneously on different parts of the same problem. Information between these small computers is shared by communicating, which is achieved by message-passing procedures, implemented by software libraries such as the message passing interface (MPI) (Gropp et al., 1994; Gropp et al., 1999).

A large-scale parallel classical molecular dynamics code (LAMMPS, Plimpton and Hendrickson, 1994; Plimpton, 1995) is widely employed for MD simulations. LAMMPS implementation is based on spatial domain decomposition. Its parallel MD mechanism reaches linear scaling, that is the total execution time scales linearly with the number of particles $\sim N$, and scales inversely proportional with the number of processors used to solve the numerical problem, $\sim 1/P$ (P is the number of processors) (Kadau et al., 2004).

With a parallel computer, whose number of processors increases with the number of cells (the number of particles per cell does not change), the computational burden remains constant. To achieve this, the computational space is divided into cells such that in searching for neighbors interacting with a given particle, only the cell in which it is located and the next-nearest neighbors have to be considered. Thus, the domain decomposition scheme can treat huge systems with several billion particles (Kadau et al., 2004).

13.2.9 Mineral Crystal's Elastic Modulus

The MD method can be used to calculate the elastic constants of a crystal structure such as quartz. In general, it may be used to calculate the elasticity constants of aggregates. Table 13.2 presents the results showing the elastic constants (the 6 × 6 elastic constant matrix) of a perfect quartz lattice structure, calculated using a static method. After an initial energy minimization, a very small strain (remain within elastic limits, ± 0.001) is applied to the system and energy re-minimization is applied. More details can be found in Lu (2010).

13.2.10 Modeling of Interface Behavior

Interfaces between aggregates and asphalt binder are a zone whose behavior and structure have not been well understood. The multiscale characterization would allow the

FIGURE 13.5 AFM image of bitumen-rock interface region.

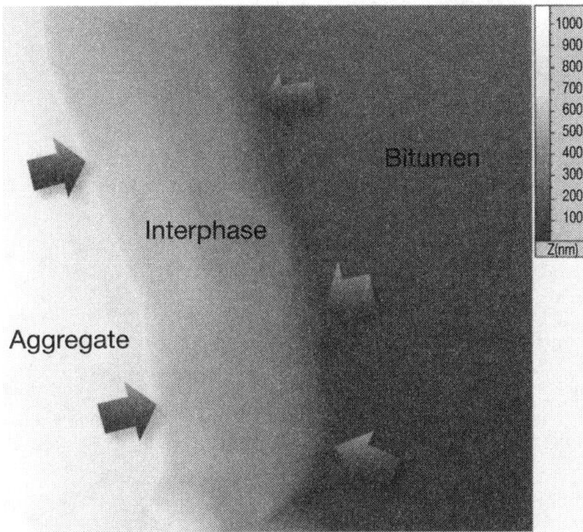

FIGURE 13.6 Interface image by digital optical microscope.

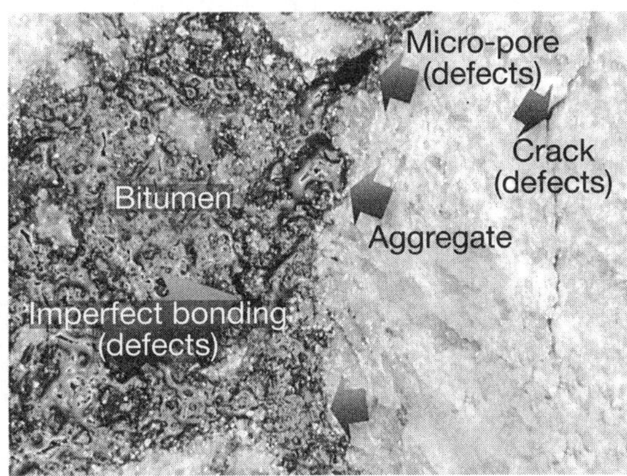

FIGURE 13.7 Chemistry map by ESEM.

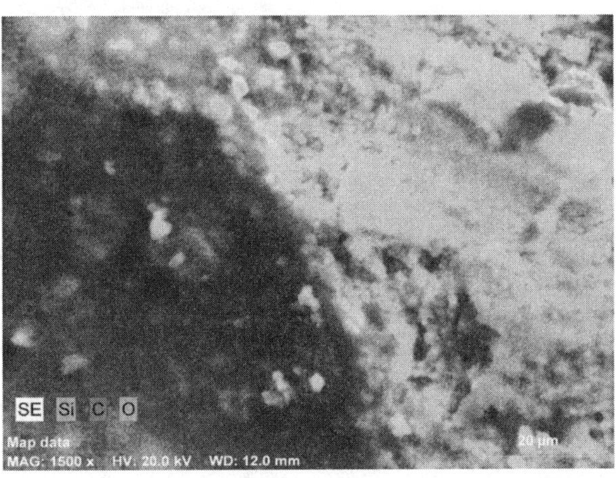

Figure 13.8 Micropores located in the asphalt phase along the asphalt-rock interface identified by focused ion beam (FIB).

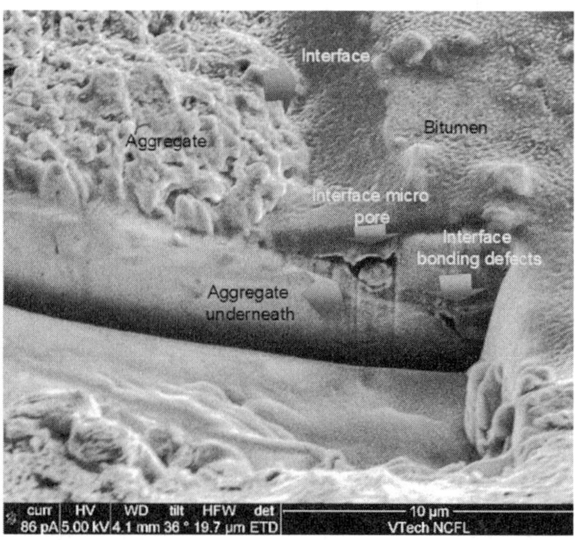

Figure 13.9 Atomistic structure at the interface by high resolution TEM (Nanoscale Resolution), by TITAN at VT.

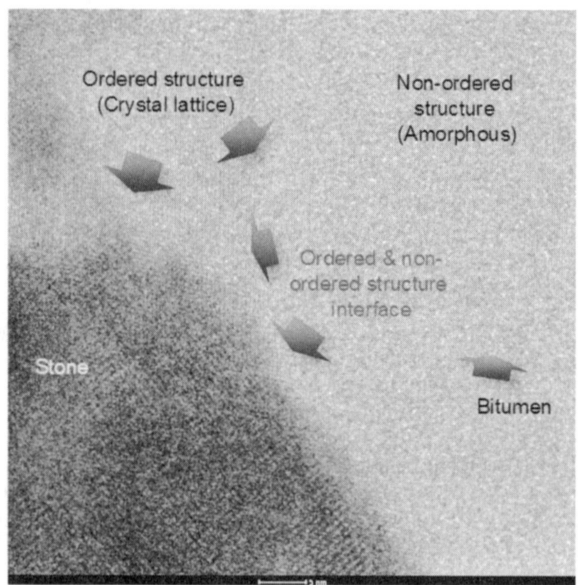

8.6886	−4.5084	0.7276	−0.7584	2.8529	2.3534
−4.5084	4.2377	−2.2465	0.0784	0.1619	2.2883
0.7276	−2.2465	3.7855	3.1904	4.1923	0.4572
−0.7584	0.0784	3.1904	−4.1273	−6.1776	2.2486
2.8529	0.1619	4.1923	−6.1776	2.5741	0.5005
2.3534	2.2883	0.4572	2.2486	0.5005	3.2891

Table 13.2 Elastic stiffness constants C_{ij} (GPa) (Lu, 2010).

interface structure to be characterized at different scales and modeled using MD and RFF. Using MD modeling the tensile and shear behavior (including modulus and viscosity) can be calculated using the statistical mechanics from the state variables in the MD simulations.

13.2.10.1 Modeling the Stress-Strain-Strain Rate Behavior of Interfaces

In order to develop an efficient model that represents bitumen-rock interfacial interaction, an in-depth study of bitumen molecular structure is needed. Chemical methods separate asphalt into multiple parts such as asphaltene, resin, and maltene. Solubility differences and chromatography subdivide an asphalt into asphaltene, polar aromatic, naphthene aromatic, and saturate components (Roberts et al., 1996). Asphaltenes are the most viscous and polar components, maltenes are the least viscous and mostly non-polar, and resins are in-between. In the bitumen model adopted, asphalt mixtures of reasonable compositions are with different compounds representing each constituent. Each component is represented using one molecule type. Molecule choice is based on the measurements by Storm et al. (1994). It contains a moderate-size aromatic core with small branches. It is taken from NMR studies by Artok et al. (1999). They proposed a collection of sample molecules that in total represent the statistics of molecular asphaltenes. This model is chosen because it represents one of the bonding patterns present in asphaltenes.

Once the asphalt mixture structure is built, CVFF-aug force field parameters can be used to characterize the interatomic interaction. A commonly used potential function for characterizing intra-connected organic molecules is the CHARMM potential. Bond energy and torsion energy are built based on a harmonic angle style. Dihedral and improper dihedral energy calculation use CHARRM and CVFF type, respectively. The atomistic model of asphalt-quartz interface is depicted in Figure 13.10.

At a fixed temperature, static calculations are performed to minimize energy. This process also ensures that convergence is achieved, thus relieving any potential overlap in vdW interactions. In the second step, the molecules system is annealed after heating

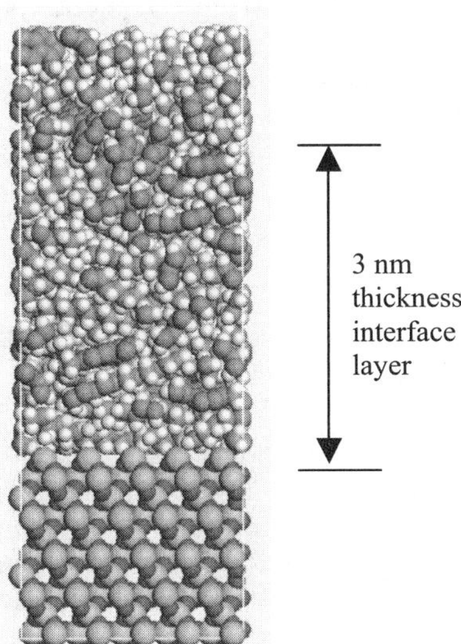

FIGURE 13.10 An interface model for MD simulations.

it to a temperature $T = 300K$. The heat up rate is $\Delta T = 25K$ every 25 steps. The temperature is fixed after the final temperature $T = 300K$ is achieved. Then the temperature is controlled in an NVT ensemble. It is also ensured that the energy remains constant after the annealing procedure.

Depending on loading conditions, the mechanical forces can be applied with various constraints. For example, the shear and tensile loading can be applied to the 3D asphalt-quartz interface structure for investigating the stress-strain relationships. To impose shear loading conditions, the solid bounding wall parallel to the asphalt-rock interface is moving in tangential direction at a constant velocity. Uniaxial tension boundary conditions involve the application of deformation at a constant strain rate normal to the interface plane, while the lateral boundaries are prescribed as stress free and thus allowed to contract during the deformation process.

The interfacial asphalt layer stress-strain relationship is analyzed by using statistical mechanics with the stress and strain definitions presented in 13.2.5.3 and 13.2.5.4. The interface width, statistics, and radial distribution function are measured as a function of distance from the interface. The MD simulations allow state variables to be analyzed in terms of dynamic properties including radial distribution function (RDF), mean squared displacement (MSD), space-time correction function, structural properties (atomistic configuration), and energetic properties (energy evolution, temperature profile).

Using the mechanisms presented in the previous section, Figures 13.11a and b present the shear stress-strain curve and uniaxial tensile stress-strain curve for the interface model. These simulations provide insight into the interface shear/tension strength of the molecule and the critical strains for onset of different modes of deformation. Figure 13.12a depicts a snapshot as nano-pores occur. Figure 13.12b shows a viscosity-time step relationship of asphalt-aggregate interface under shear loading. For more details of these simulations and other studies, please refer to the three papers (Lu and Wang, 2008; Lu and Wang, 2009; Lu and Wang, 2010).

13.2.11 3M Continuum and Electromagnetic-Mechanical Coupling

13.2.11.1 3M Continuum

Stone-based materials are highly heterogeneous. Due to the significant difference in stiffness of the component materials (for example, aggregates versus binder), aggregate may be subjected to a significant degree of rigid (relatively) rotation. Due to the close packing of the aggregate particles, the strain gradient is also important and therefore the non-local-type continuum theory should be adopted. By considering the electromagnetic-mechanical coupling, the 3M (*Micromorphic, Microstretch, and Micropolar*) micro-continuum, non-local, and coupled theory proposed by Eringen (1999) may have advantages.

13.2.11.2 Multi-Physics Coupled Constitutive Relation

The effect of electromagnetic-mechanical coupling can be evaluated using reaxFF. In molecular dynamics, particles have a point charge representation used to compute the Coulomb interactions (this is generally presented as a static limit of Maxwell's equations valid after all transients have decayed to zero). Yet, in the case of reaxFF, the charges on each atom are distributed over the size of the atom so that Coulomb interactions between atoms are shielded as the atoms come together, allowing electrostatic interactions between bonded atoms (conventionally shielded in other force fields). The energy of each atom depends on the net charge in such a way as to properly describe

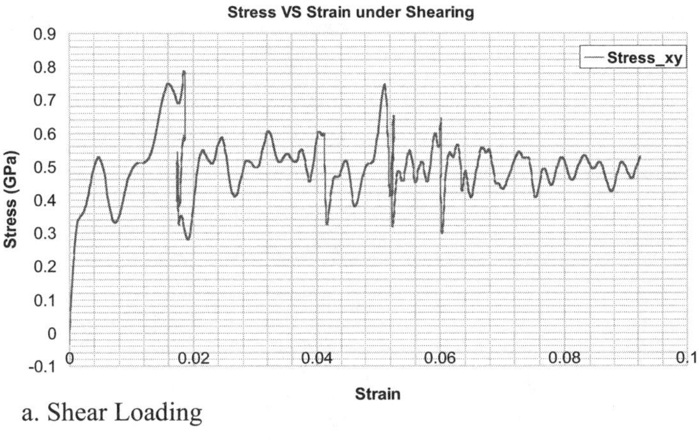

a. Shear Loading

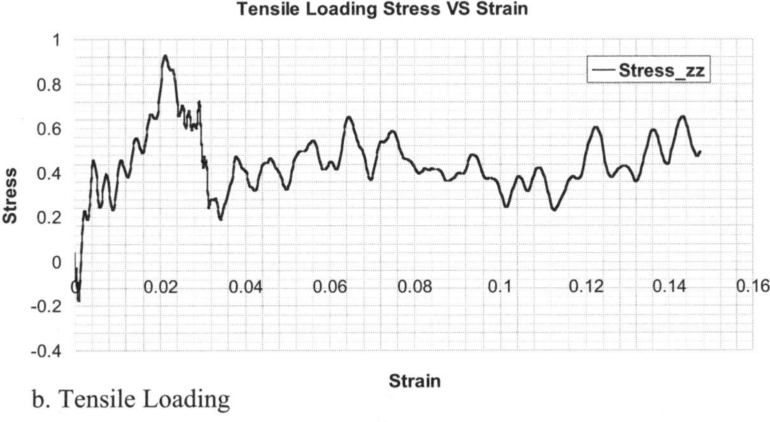

b. Tensile Loading

FIGURE 13.11   Stress-strain relationships based on MD simulations.

the ionization potential and electronic affinity in the molecular environment (i.e., environmentally dependent charge distributions). This way charge transfer processes under non-equilibrium conditions, whence the effect of the polarization phenomena from an external (electromagnetic) field, can be modeled. The results from reaxFF-MD could then be coupled to, for example, a finite-difference time-domain method (in both the spatial and temporal dimensions) to predict the electromechanical coupling of the material under the influence of, say, an external electromagnetic field at the atomic level.

At continuum levels, electromagnetic-mechanical coupling can be handled through adopting a specific format of the free-energy functions. There are a few research publications in this area specifically dealing with piezoelectric material. The constitutive equations (Jayachandran and Guedes et al., 2009) which correlate the stress (T_{ij}), strain (ε_{kl}), electric field (E_k), and electric displacement (D_i) for the piezoelectric medium are cited here as examples:

$$\begin{cases} \varepsilon_{ij} = S^E_{ijkl}T_{kl} - d_{kij}E_k \\ D_i = d_{ijk}T_{jk} + \in^T_{ij} E_j \end{cases} \tag{13-9}$$

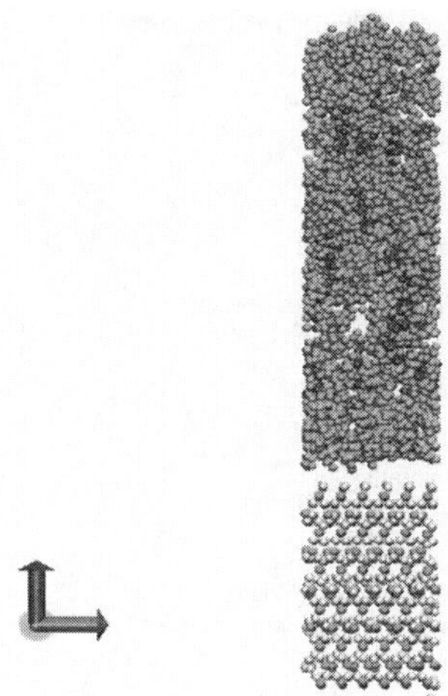

a. Nano-Pore Initiation under Tensile Loading

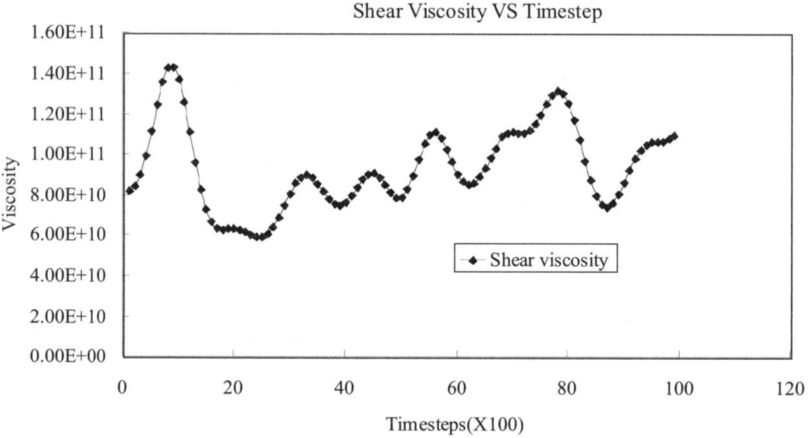

b. Viscosity at Interface under Shear Loading

FIGURE 13.12   Nano pore initiation and viscosity modeling based on MD simulations.

Where the tensor S^E_{ijkl}, d_{ijkl}, and $\in^T_{ij}$ are respectively the compliance at $E = 0$ and the piezoelectric strain coefficient and unclamped dielectric coefficient; E is an external electric field.

A linearly non-homogeneous magnetoelectroelastic properties formulation, which coupled magnetic potential, electric potential, and continuum mechanics field (Lee and Li, 2008) is also listed here:

$$\begin{pmatrix} \nabla^2 w & w_{,y} \\ \nabla^2 \phi & \phi_{,y} \\ \nabla^2 \varphi & \varphi_{,y} \end{pmatrix} \begin{pmatrix} 1 + \beta y \\ \beta \end{pmatrix} = \begin{pmatrix} 0 \\ 0 \\ 0 \end{pmatrix} \tag{13-10}$$

Where $w(x,y)$, $\tau_{yz}(x,y)$, $\phi(x,y)$, $D_y(x,y)$, $\varphi(x,y)$, and $\beta_y(x,y)$ are the anti-plane mechanical displacement, stress, in-plane electric potential, electric displacement, in-plane magnetic potential, and magnetic induction.

Some other theories, including the 3M continuum (Eringen, 1999), are briefly described as follows. Eringen (1999) assumes the Helmholtz free energy Ψ in the following format for micro continuum (or continuum for a single material):

$$\psi = \Psi\left\{\zeta_{KL}, \Gamma_{KLM}, \partial_{KL}, \theta, \varepsilon_K, B_K; X\right\} \tag{13-11}$$

$$\zeta_{KL} \equiv x_{k,K}\chi_{Lk}, \ \partial_{KL} \equiv x_{kK}x_{kL} = \partial_{LK}, \ \Gamma_{KLM} \equiv \chi_{Kk}x_{kL,M}$$

$x_{k,K}$ micro-deformation tensor, χ_{Kk} inverse micro-deformation tensor

Here ζ_{KL} is called the deformation tensor, ∂_{KL} the microdeformation tensor, and Γ_{KL} the wryness tensor; θ absolute temperature, X coordinate in the reference configuration, x coordinate in the deformed configuration, and B in the magnetic flux vector, $\varepsilon_K = \varepsilon_k x_{k,k}$, $B_K x_{k,k}$.

This way, the mechanical-electromagnetic coupling can be considered.

While the electronic-mechanical coupling may not have direct applications in asphalt mechanics, it may help characterize the electron-magnetic properties of asphalt concrete that can enhance the understanding of electron-magnetic wave propagation in asphalt concrete.

13.3 Moisture Damage of AC

13.3.1 Overall Review

Among asphalt pavement distresses, moisture damage still remains a primary cause of its premature failure (Fromm, 1974; Graf, 1986; Curtis, 1993; Kandhal, 1994; Scholz et al., 1994; Roberts et al., 1996; Alam et al., 1998; Mohammad et al., 2005) as water or moisture penetrates and settles within its layers (Hicks, 1991; Epps et al., 2000; Solaimanian et al., 2007). Moisture damage also strongly influences other types of distresses such as rutting, raveling, and cracking, which significantly reduce the performance and service life of HMA pavements resulting in high maintenance costs for state and federal highways (Izzo and Tahmoressi, 1999; Bhasin, 2006).

In the Superpave volumetric mix design SP-2 (2001) process, evaluation of a mixture's moisture susceptibility is the final step and is accomplished using the AASHTO T-283 standard. Research performed in the last three decades or so showed that moisture susceptibility is influenced by the aggregate mineralogy and surface texture, asphalt binder chemistry, and the interaction between the aggregate and asphalt binder. Although bitumen properties and aggregate characteristics are determinant factors (Mohammad et al., 2003), aspects such as hot mix processing, mixture characteristics, quality control during construction, water in the aggregate-asphalt interface, dynamic effect of traffic, and type and properties of anti-stripping additives also influence the hot mix asphalt (HMA) moisture susceptibility.

Generally, moisture damage occurs in AC pavements due to a loss of adhesion and/or cohesion that accelerates structural degradation of the mixtures in conjunction with cracking and plastic deformation. Moisture in pavements typically reduces strength or stiffness of the binder and mastic through diffusion and weakens the adhesive bond between the mastic and aggregates (Stuart, 1990; Maupin, 1999). It has also been shown through micro-mechanics that mixtures with thin asphalt films are prone to adhesive failure, while those having thick asphalt films tend to fail cohesively (Lytton et al., 2005). The adhesive failure, mostly known as stripping, takes place when water gets between the aggregate surface and asphalt film and breaks the bonding, while the cohesive failure occurs within the asphalt film due to its combination with water. Moisture usually infiltrates the asphalt pavement from the rise of underground water, rainfall, and/or water vapor absorption/adsorption (Arambula et al., 2007a).

Some other factors influencing the loss of adhesion are poor mechanical interlocking of aggregate and chemical interaction between the minerals on the aggregate surface and the asphalt cement. Even though cohesive failure has been regarded as a less important factor of moisture damage of HMA, work done by Kanitpong and Bahia (2005) showed that the failure surfaces in asphalt mixtures obtained from the tensile strength ratio (TSR) test were visually observed within the binder film without evidence of apparent loss of adhesion to the aggregate particles. Fromm (1974) proved that water could penetrate the asphalt film and form a water-in-asphalt emulsion (emulsification) that causes asphalt particles to separate from the asphalt film (cohesive failure). Ultimately, this leads to an adhesive failure when this emulsification boundary progresses to the aggregate surface.

Over the years, numerous laboratory tests have been developed to identify the moisture susceptibility of asphalt mixtures (Epps et al., 2000; Caro et al., 2008a). They can be classified into two main categories: (1) tests performed on loose mixes, such as the static immersion and boil tests, and (2) tests performed on compacted mixes, such as the indirect tensile strength, immersion compression, and modulus tests (Roberts et al., 1996; Solaimanian et al., 2007). It has been agreed, as a general consensus in the industry, that laboratory tests performed on compacted mixtures are capable of conveying better information on moisture sensitivity than tests on loose mixtures or their components. As numerous types of unmodified and modified asphalt binders and a great number of different aggregate minerals are used, combined with various construction practices, traffic levels, and environmental conditions, testing to accurately predict HMA moisture susceptibility has become quite a difficult task. Also, several mechanisms have been recognized as sources of moisture damage: displacement, detachment (debonding), spontaneous emulsification, film rupture, pore pressure, hydraulic scouring (Hammons et al., 2006; Solaimanian et al., 2007; Caro et al., 2008a), degradation or fracture of aggregate (Cheng et al., 2003), and mastic dispersion and desorption (Kringos and Scarpas, 2005, 2008; Caro et al., 2008b). Even though these mechanisms are held responsible for moisture damage, moisture damage is much more complex and an identification of the damage mechanisms of various asphalt-aggregate systems in the presence of water is still difficult to achieve. Usually, a synergistic interaction of the identified mechanisms remains the best justification of the moisture damage process. Table 13.3 presents these mechanisms in more detail.

13.3.2 Mechanics Method

The mechanisms in the previous section are mainly empirically identified. A new approach for evaluating moisture-related damage in asphalt mixes is based on the fundamental properties of component materials. Possible moisture-related damage prediction is sought from these properties by comparing mechanical behaviors of dry and conditioned (water-immersed) asphalt mixtures. In this respect, Cheng et al. (2002a,b;

Moisture Damage Mechanism	Description	Process Nature
Displacement	Removal of the asphalt film from the aggregate surface through a break in the asphalt film and/or possible separation of the aggregate/mastic	Mechanical
Detachment (debonding)	Asphalt film separation from an aggregate surface by a thin film of water without a visible break in the binder film	Chemical, mechanical
Pore pressure	Development of high pore pressures in the saturated flow condition high percentage of air-voids with low connectivity	Mechanical
Spontaneous emulsification	Asphalt particles separate from the asphalt film in the presence of water	Chemical
Film rupture/microcracks	Microcracks generation from pore pressure that leads to asphalt film/aggregate fracture making new paths for moisture transport	Mechanical, thermodynamic
Degradation/aggregate fracture	Mixture is subjected to freeze/thaw cycles and repetitive wheel loads	Mechanical, thermodynamic
Mastic dispersion	Asphalt binder or mastic cohesion weakening due to long-term diffusion and loss of material due to the water flow	Mechanical, thermodynamic
Mastic desorption	Removal of the outer layers of mastics due to the water flow	Chemical, mechanical

TABLE 13.3 Moisture damage mechanisms.

2003) investigated the moisture effects on asphalt mixes by employing an adhesion failure model based on the surface energy theory and a moisture diffusion model based on universal sorption device (USD) testing. For this purpose they used the viscoelastic fracture law as follows:

$$2\Gamma_f = E_R D_f(t_\alpha) J_v \tag{13-12}$$

where Γ_f = surface energy of crack surface (FL^{-1}); F, L = force and length units
$\quad$ E_R = reference modulus (FL^{-2})
$\quad$ $D_f(t_\alpha)$ = tensile creep compliance at time of loading (t_α)
$\quad\quad$ J_v = viscoelastic J-integral representing the change in dissipated pseudostrain energy/unit of crack area

In an asphalt-aggregate system the surface energies are mainly composed of a non-polar component and an acid-base component (Cheng et al., 2003). Equation 13-13 describes the total surface energy and its components:

$$\Gamma = \Gamma^{LW} + \Gamma^{AB} \tag{13-13}$$

where Γ = surface energy of asphalt/aggregate (FL/L^2)
$\quad$ Γ^{LW} = Lifshitz-Van der Waals component of surface energy (FL/L^2), and
$\quad$ Γ^{AB} = acid-base component of surface energy (FL/L^2)

To calculate the surface energy of adhesion between two different materials (i and j) without the presence of water, the following equation was used (Cheng et al., 2003):

$$\Delta G^a{}_{ij} = \Delta G^{aLW}{}_{ij} + \Delta G^{aAB}{}_{ij}$$
(13-14)

where $\Delta G^{aLW}{}_{ij}$ = non-polar part of the surface energy of adhesion
$\Delta G^{aAB}{}_{ij}$ = polar part of the surface energy of adhesion

In his research, Kim et al. (2004) used the dynamic mechanical analysis (DMA) to characterize the fatigue damage behavior and fracture of asphalt binders and mastics by measuring the viscoelastic properties and damage characteristics. The model employed was used to study the effects of moisture on mastic sensitivity to fatigue damage. The model equation is presented as follows:

$$N_f = \frac{f(S_f)^k}{k(0.5 I_p C_1 C_2)^\alpha}(\gamma_0)^{-2\alpha}$$
(13-15)

where N_f = number of loading cycles to failure
I_p = initial pseudostiffness
$k = 1 + (1 - C_2)\alpha$
S_f = damage parameter for fatigue failure
α = viscoelastic-related material parameter
C_1, C_2 = regression constants from stiffness versus damage parameter relation

The measure of the free energy using suction in AC materials with different air-void contents and gradations was studied by Kassem et al. (2006). The suction measurements were used to calculate the moisture diffusion coefficient and relate it with the moisture damage. The total suction under isothermal conditions was calculated using the following formula:

$$h_t = (u_a - u_w) + h_\pi$$
(13-16)

where h_t = total suction (kPa)
$(u_a - u_w)$ = matric (capillary) suction
u_a = pore-air pressure
u_w = pore-water pressure
h_π = osmotic suction

Bhasin et al. (2006, 2007) looked into the strength of physical adhesion between the asphalt and the aggregate in a dry condition, which is quantified in terms of the adhesive bond energy. A high level of adhesive bond energy (in dry state) combined with a low level of reduction in free energy in the presence of moisture are beneficial for HMA to resist debonding in the presence of water. The total adhesive bond energy was calculated using the following formula, based also on Lifshitz-Van der Waals and acid-base components:

$$\Delta G_{AS} = \gamma_A + \gamma_S - \gamma_{AS}$$
(13-17)

where γ_A, γ_S = surface energy of unit area (asphalt and aggregate)
γ_{AS} = interfacial surface energy

Birgisson et al. (2003, 2004, 2007) employed the HMA fracture mechanics model based on dissipated creep strain energy (DCSE) as a fundamental model for evaluating asphalt mixes moisture damage. Properties evaluated were indirect tensile strength, resilient modulus, creep compliance, creep rate, fracture energy limit, and dissipated creep strain energy limit and results showed that neither one of these properties consistently reveals the effects of moisture damage on the mixtures. To measure the fracture resistance of mixtures, the energy ratio (ER) was introduced:

$$ER = \frac{DCSE_f}{DCSE_{min}} = \frac{a \bullet DCSE_f}{0.98 \bullet m^2 \bullet D_1} \qquad (13\text{-}18)$$

where $DCSE_f$ = dissipated creep strain energy (kJ/m^3)
$\qquad DCSE_{min}$ = minimum DCSE for adequate cracking performance (kJ/m^3)
$\qquad D_1, m$ = creep parameters (1/psi)
$\qquad\quad a = 0.0299\ \sigma_t^{-3.1}\ (6.36 - S_t) + 2.46 \times 10^{-8}$
$\qquad\quad \sigma_t$ = tensile stress of asphalt layer (psi)
$\qquad\quad S_t$ = tensile strength (MPa)

Fracture model (crack growth model) based on dissipated pseudostrain energy (DPSE) was investigated by Masad et al. (2006) and Arambula et al. (2007a,b) for moisture susceptibility of asphalt mixes. The crack growth index for assessing the damage was calculated using an equation derived from Paris's law expressed in terms of the J-integral (Lytton et al., 2005):

$$\frac{dr}{dN} = A(J_R)^n \qquad (13\text{-}19)$$

where r = average crack radius
$\qquad N$ = number of load cycles
$\qquad A, n$ = material constants
$\qquad J_R$ = J integral

Further, J_R can be expressed as a function of the change in pseudostrain energy/unit volume (W_R) of the intact material over the change in crack surface area (CSA) using the equation:

$$J_R = \frac{\partial W_R}{\partial(CSA)},\ \text{with}\ W_R = \frac{DPSE}{(S_i / S_1)} \qquad (13\text{-}20)$$

where $DPSE$ = dissipated pseudostrain energy/load cycle
$\qquad S_i$ = pseudostiffness for each load cycle
$\qquad\quad$ (max. applied stress/pseudostrain)
$\qquad S_1$ = maximum pseudostiffness in the first load cycle

Using a linear regression model based on ultrasonic energy, McCann and Sebaaly (2001) showed that the rate at which the asphalt binder is removed from the surface of an aggregate can be quantified. Also, Pinzon and Such (2004) used a complex modulus approach for assessing the damage by moisture in AC cores having various dimensions. Via this approach, an empirical model for evaluating the extent of deterioration was under development based on classifying the modulus loss levels before and after immersion for each mix design under given temperature and immersion time conditions. They concluded that the stiffness modulus alone cannot provide reasonable means for evaluating the moisture sensitivity of asphaltic mixtures, nor does it allow to evidently determine the effect of mix design parameters on the weakening of the binder-aggregate bond.

Kringos et al. (2007) and Kringos and Scarpas (2008) proposed a 3D elasto-visco-plastic FE model for mastic to evaluate moisture-induced damage in asphaltic mixes from quantifying the physical and mechanical processes which lead to moisture damage. Also, a finite element tool that enables the simulation of physical damage due to water flow and moisture diffusion through asphaltic mixes was developed. The 1-D schematic of the proposed elasto-visco-plastic material model consists of a single elasto-plastic component in parallel with a random number of viscoelastic ones. To simulate the water flow through the asphalt mix, the water velocity field was expressed using the following formula, assuming the validity of Darcy's Law:

$$div(v) = -\tilde{\theta}\frac{\partial p}{\partial t} \quad \text{with } \tilde{\theta} = \frac{dS}{dp} \tag{13-21}$$

where v = water velocity

p = water pressure

$\tilde{\theta}$ = water capacity

S = pressure dependent saturation

A diffusion flux similar to the Fick's law of diffusion was assumed to simulate the moisture diffusion through the AC material.

$$J = -D\nabla(C_d) \tag{13-22}$$

where J = diffusive-dispersive mass flux (M/L²T)

D = diffusion-dispersion tensor (L²/T)

C_d = concentration of desorbed mastic (M/L³)

Numerical models used to simulate fluid flow in 3D microstructures of porous materials were developed by Kutay et al. (2007) and Masad et al. (2007). From these models it is possible to create realistic simulations of the material distresses considering the presence of moisture. The model by Masad et al. (2007) is based on the governing equations (continuity and Navier-Stokes) in 3D space, for the steady incompressible fluid flow using finite difference techniques. This way the fluid flow equations are solved within the actual boundary conditions of the 3D air-void structure.

To establish the fluid flow field within the microstructure, Kutay et al. (2007) used the Lattice Boltzmann (LB) approach. Through these proposed models, the flow speed and pressure are determined at all points within the void structure, which means that at the microstructural level they represent essential inputs for numerical simulations of moisture damage.

13.3.3 Mesoscale Moisture Damage Mechanisms

13.3.3.1 General Concept

Pore water in the asphalt pavement will produce excess pore water pressure. The excess pore water pressure will produce tensile stress in the surrounding medium. This additional tensile stress may reduce the fatigue life and increase the deformation rate. Figure 13.13 illustrates this mechanism through a simple model: a plate containing a small circular hole.

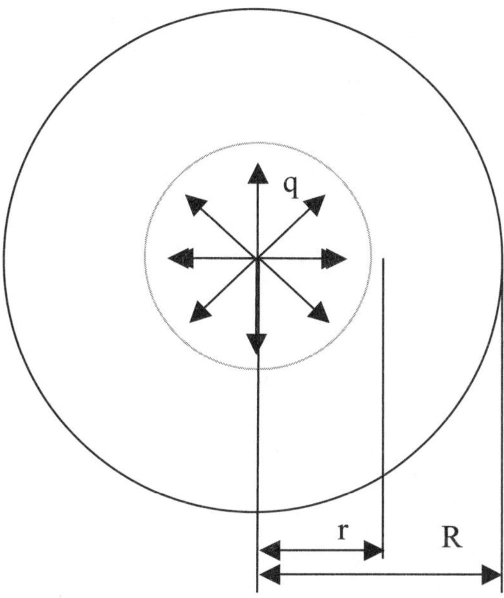

FIGURE 13.13 Circular hole in a plate.

The radius of the hole is r. The radius of the plate is R. The inside pressure is q. The tangential stress distribution along the edge of the hole under the polar coordinates is given in Equation 13-23:

$$\sigma_\theta = \frac{\dfrac{R^2}{\rho^2} - 1}{\dfrac{R^2}{r^2} - 1} q \tag{13.23}$$

Which means that tensile stress is always produced along the edge of the hole due to the inside pressure and causes the failure or premature cracks of the asphalt concrete mixture. For any incremental pressure Δq, the incremental hoop stress will be $\Delta\sigma_\theta = \dfrac{\dfrac{R^2}{\rho^2} - 1}{\dfrac{R^2}{r^2} - 1} \Delta q$. The average of such stress in the specimen is related to the void/pore structure, the permeability, and the stress increment rates.

In recent years, Cui et al. (2009) used the Lagrangian finite difference method to study the dynamic deflection of the pavement with existence of pore water pressure under traffic loading. Kutay (2009) analyzed the in-depth variations of pore water pressures and viscous shear stresses caused by water flow based on the lattice Boltzmann Method. Zhou et al. (2008) used the Galerkin Weighted Residual Method to compute the pore water pressure in an asphalt pavement. Dong et al. (2008) compared modified indirect tension tests and uniaxial compression test results with the FEM computation to study the coupled water/loading action on AC specimens. Yang et al. (2006) discussed the use of a Stiffness Matrix Method to calculate pore pressure within a saturated pavement. Kringos and Scarpas (2005) developed an FEM to model to investigate the raveling of asphalt mixes due to water damage from the point of microscopic view. Masad et al. (2003) used 2D SEEP/W FE code to study the relationship between the water flow patterns and the gradients of percent air voids in HMA. Novak et al. (2002) used the theory of mixture to describe the saturated asphalt pavement and studied the effects of vehicle speed and permeability on pore pressure with a 2D finite element program—PlasFEM.

This section presents a brief outline of a recent study on excess pore water pressure development in asphalt pavement based on the Poroelasticity Theory (Wang, 2000) and ABAQUS Theory manual. More details can be found in Wang et al. (2010a, 2010b).

13.3.3.2   Constitutive Model

The poroelasticity model treats the pavement (AC) with porous water as a continuum, assuming that the total stress at each point is the sum of an "effective stress" carried by the aggregate skeleton and a pore pressure in the fluid permeating the structure. This fluid pore pressure can change with time, and the gradient of the pressure through the pavement (or material) that is not balanced by the weight of fluid between the points will cause the fluid to flow. The flow velocity is proportional to the pressure gradient in the fluid according to Darcy's Law. In essence, the fundamental concept is based on mixture theory.

The model considers the medium as a multi-phase material and adopts an effective stress principle to describe its behavior. Two fluids are considered. The "wetting liquid" is assumed to be relatively incompressible. The other is gas, which is compressible. The total stress at any point, σ, is assumed to be made up of an average pressure stress in the wetting liquid, u_w, an average pressure stress in the other fluid, u_a, and an "effective stress," $\bar{\sigma}^*$, which is defined by:

$$\bar{\sigma}^* = \sigma - (\chi u_w + (1-\chi)u_a)I \tag{13-24}$$

In which χ is a factor that depends on the level of saturation and on the surface tension of the liquid/solid system.

Equilibrium is expressed by writing the principle of virtual work for the volume under consideration in its current configuration at any time:

$$\int_V \sigma : \delta\varepsilon \, dV = \int_S t \cdot \delta v \, dS + \int_V \hat{f} \cdot \delta v \, dV \tag{13-25}$$

Where δv is a virtual velocity field, $\delta\varepsilon \overset{def}{=} sym(\partial\delta v / \partial x)$ is the virtual rate of deformation, σ is the true stress, t is surface traction per unit area, and $\hat{f}$ is body forces per unit volume.

The mechanical behavior of the porous medium consists of the response of the liquid and solid to local pressure and the response of the overall material to the effective stress.

For the liquid in the system:

$$\frac{\rho_w}{\rho_w^0} \approx 1 + \frac{u_w}{K_w} - \varepsilon_w^{th} \tag{13-26}$$

Where ρ_w is the density of the liquid, ρ_w^0 is its density in the reference configuration, K_w is the liquid's bulk modulus, and ε_w^{th} is the volumetric expansion of the liquid caused by temperature change.

For the solid in the system:

$$\frac{\rho_g}{\rho_g^0} \approx 1 + \frac{1}{K_g}(su_w + \frac{\overline{P}}{1-n-n_t}) - \varepsilon_g^{th} \tag{13-27}$$

Where K_g is the bulk modulus of the solid, s is the saturation in the wetting fluid, and ε_g^{th} is its volumetric thermal strain.

The constitutive behavior of pore fluid flow is governed by Darcy's Law. Under saturated conditions, the volumetric flow rate of the wetting liquid through a unit area of the medium, snv_m, is proportional to the negative of the gradient of the piezometric head.

$$snv_m = -\hat{k} \cdot \frac{\partial \phi}{\partial x} \tag{13-28}$$

Where $\hat{k}$ is the permeability of the medium and ϕ is the piezometric head, defined as:

$$\varphi = z + \frac{u_w}{g\rho_w} \tag{13-29}$$

Where z is the elevation above some datum and g is the magnitude of the gravitational acceleration, which acts in the direction opposite to z.

13.3.3.3 Finite Element Modeling

This section presents results on how the poroelasticity model predicts the excess pore water pressure accumulation and dissipation. A 2D finite element mesh is shown in Figure 13.14. A body of pavement 3 m long and 60 cm thick is confined by impermeable,

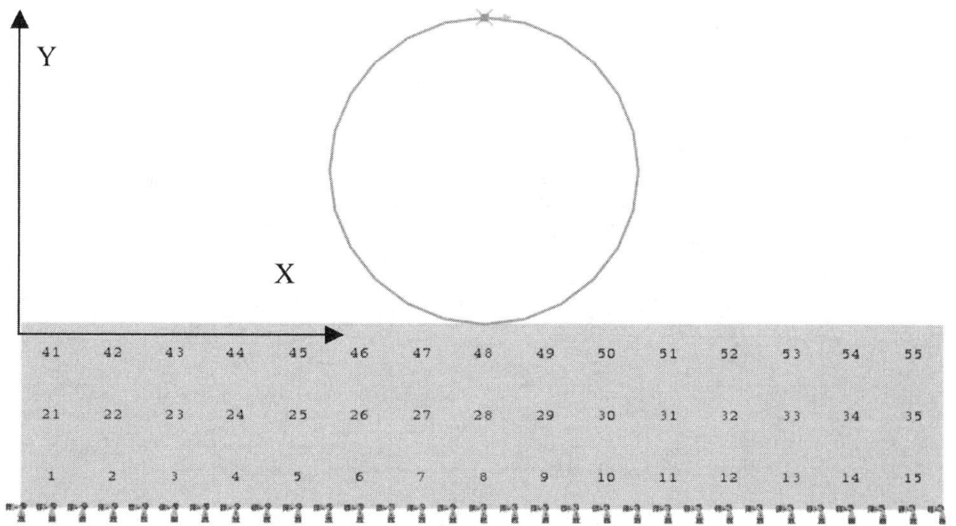

FIGURE 13.14 2D finite element mesh.

smooth, rigid walls on all but the top surface. On that surface, perfect drainage boundary is defined and a tire loading is applied. The eight-node biquadratic, reduced integration element is used, which is comparatively accurate and less memory consuming. Gravity is neglected.

Vehicle loading of 2500N, which is approximately ¼ of a car's weight, is constantly applied on the wheel in the $-y$ direction and the wheel is analytically rigid: the deformation of which is not considered. The speed of the wheel is set toward the x direction. The contact between the wheel and pavement surface is tangentially frictional with a friction coefficient of 0.3. The material of the pavement is elastic with Young's Modulus of 1000 MPa, Poisson's ratio of 0.3. The specific weight of the pore fluid is assumed to be $276.8 \times 10^3 \, \text{N/m}^3$ (1 lb/in³).

The boundary conditions are as follows. On the bottom, the displacements in both X and Y directions are fixed. On the side walls, there are no constraints in any directions. No flow of pore fluid through these walls is permitted. The top surface is where the vehicle loading is applied and allows perfect drainage so that the excess pore pressure is always zero on this surface.

Development of pore water pressure under several situations has been modeled (Wang et al., 2010a). Due to a moving load, the excess pore water pressures change with time. Figure 13.15 is the pore water pressure history of nodes along a vertical line at the middle. It can be found that the deeper the nodes are located, the higher the pore water pressure is generated. This is related to the drainage boundary conditions: the surface is fully drained and the bottom is fully impermeable.

Permeability is another key influencing factor of the pavement to affect the pore water pressure generation and dissipation. Figure 13.16 shows the pore water pressure history of the element below the tire with different permeability. The results indicate that the peak value of the pore water pressure decreases as the permeability increases. Smaller permeability allows longer time for pore water pressure to decay. The results are very similar to the work by Zhou et al. (2007).

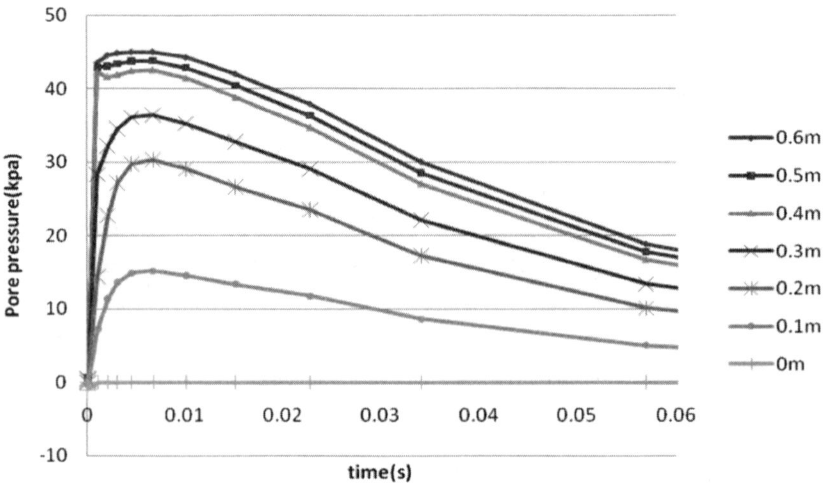

Figure 13.15 Pore pressure-time relationship.

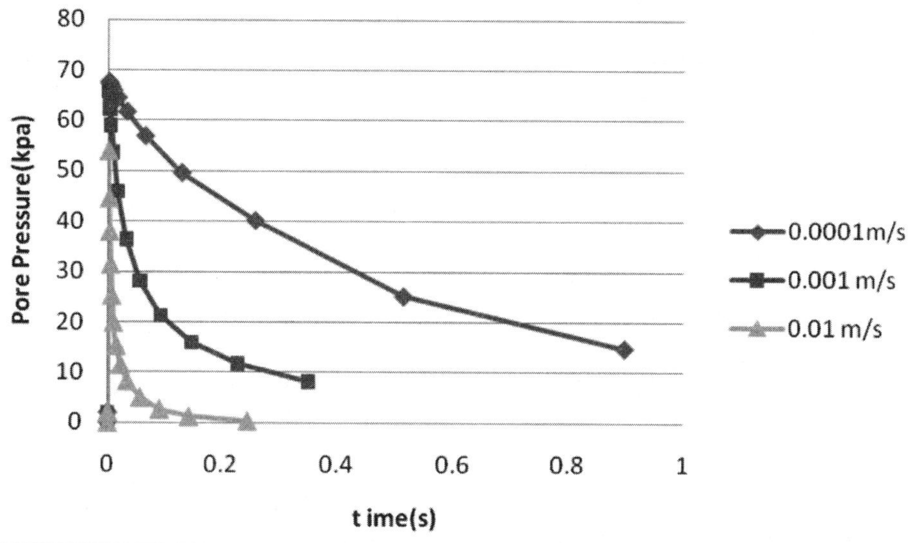

Figure 13.16 Pore water pressure dissipation with varied permeability.

The results of these simulations qualitatively indicate the rationality of the poro-elasticity model in predicting the pore water pressure generation and dissipation process. Wang et al. (2010b) extended this model to the study of erosion of the bases of cement concrete and provided a theoretical background for erosion design.

13.3.3.4 MD Simulation for Interface Moisture Sensitivity

Moisture-induced damage in the form of stripping is very much related to the interface bonding between aggregates and bond (Huang and Wu et al., 2008; Gorkem and Sengoz, 2009). Interactions between asphalt and aggregate have several mechanisms, including physical and chemical aspects of asphalt-aggregate bonding that promote adhesion.

Most publications (Kringos and Scarpas, 2008; Kringos and Scarpas et al., 2008) attribute stripping to effects like bitumen properties, aggregate properties, pH at the interface, traffic-related load, construction, and temperature. Although available research based upon experimental testing (Bhasin and Little, 2007a; Kim et al., 2008; Bhasin and Little, 2009) considered bitumen and aggregates as influencing factors for stripping (Little and Petersen, 2005; Hefer et al., 2006), the dependence of moisture damage on the fundamental composition of the components (bitumen and aggregate) is not yet well understood. There is a need for fundamental research to investigate how water, bitumen, and rock minerals may interact at the interface region; and how water may induce moisture damage at the bonding area. For the bitumen rock composite matrix to function durably, a high-bond energy between the aggregate and bitumen is desirable. This adhesion also ensures proper wetting of the rock mineral surface by the bitumen during the mixing process, ensuring the mechanical interlock between the rock surface and the binder. The surface energy of different components may serve this purpose.

The comparison between the surface free energy of aggregate and asphalt interface when dry, and its corresponding value when wet, gives an assessment of propensity of the interfaces to moisture damage. Previous studies have demonstrated the correlation of these two parameters with the moisture sensitivity of asphalt mixtures. These two parameters are computed from the surface free energy components of the asphalt binder, aggregate, and water using Equations 13-30 and 13-31.

Dry interface condition (Bhasin et al., 2007):

$$\Delta G_{AS}^{abs\,dry} = 2\sqrt{\gamma_A^{LW}\gamma_S^{LW}} + 2\sqrt{\gamma_A^{+}\gamma_S^{-}} + 2\sqrt{\gamma_A^{-}\gamma_S^{+}}$$
$$= \left|\gamma_A + \gamma_S - \gamma_{AS}\right| \qquad (13\text{-}30)$$

Moisture interface condition (Bhasin and Little, 2007):

$$\Delta G_{AS}^{abs\,wet} = \left|\gamma_{AW} + \gamma_{SW} - \gamma_{AS}\right| \qquad (13\text{-}31)$$

Where $\Delta G_{AS}^{abs\,wet}$ represents free energy of an interface system moisture conditioned, $\Delta G_{AS}^{abs\,dry}$ represents free energy of a dry interface system. The subscripts A, S, and W represent asphalt binder, aggregate, and water, respectively; γ_A denotes the free surface energy of asphalt, γ_S denotes the free surface energy of aggregate, and γ_{ij} denotes interfacial surface free energy between any two materials i and j. The terms γ^{LW}, γ^{+}, and γ^{-} refer to Lifshitz-v$_{dW}$, acid, and base components of surface free energy, respectively.

A higher value of $\Delta G_{AB}^{abs\,dry}$ indicates better adhesion between the aggregate and the asphalt binder, whereas a lower value of $\Delta G_{AB}^{abs\,wet}$ indicates that a lower potential drives water to displace the asphalt binder from its interface with the aggregate.

A simple moisture damage interface model (Lu, 2010) with 100 water molecules sandwiched between bitumen and a quartz layer was developed and illustrated in Figure 13.17. This model is constructed for the undeformed structure of interfaces that may lead to either adhesive or cohesive failure.

Lu (2010) investigated the interface atomistic models with different minerals such as quartz and calcite. The adhesion energy and cohesion energy based on the surface energy method were used to assess the sensitivity of aggregate-binder interface to moisture

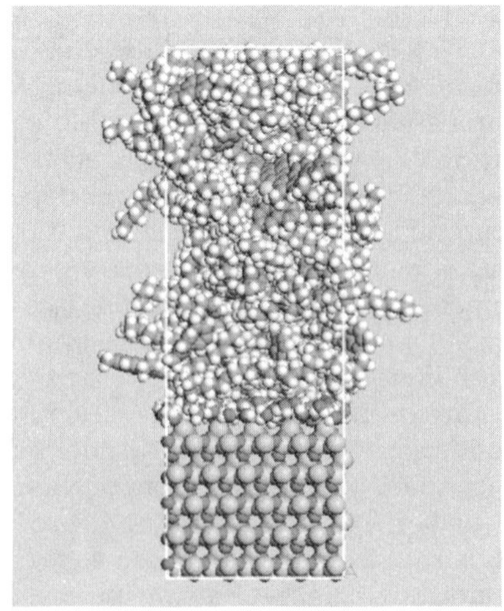

FIGURE 13.17   Interface model for moisture damage simulations.

damage. The simulations results indicate that the quartz surface is more susceptible to moisture than Calcite. This is in agreement with macroscale experimental observations. The major reason is that quartz is an acid substrate, whereas calcite has an alkaline inclination which increases its affinity to asphalt and makes it less susceptible to water.

13.4 A Two-Scale Homogenization Method

Computational simulation based on 3D microstructure is very time expensive. Lutif et al. (2009) proposed a multiscale approach in which a separate scale computation is performed for each of the smaller structural scales. If statistical homogeneity at any smaller length scale has been satisfied, a homogenization principle may be adopted to produce the field equations for the next larger length scale. In addition, damage can also be modeled at each length scale by incorporating appropriate types of fracture mechanics models to the analysis.

In this multiscale model, a global scale computation is performed concurrently with the local scale analyses. Figure 13.18 illustrates a schematic depiction of a two-scale problem. The global scale is considered to be statistically homogeneous, whereas the local scale is heterogeneous. Note that the local scale size needs to meet the required dimensions of the representative volume element (RVE), because the heterogeneous mixture RVE is homogenized to produce its effective properties that are sequentially updated to the global scale constitutive relations. In the local scale object, various sources of heterogeneity can be considered, such as particles, voids, and cracks. Cracks in the local scale are modeled by implementing the cohesive zones, which are fictitious zones embedded in the local scale body to represent the initiation and the propagation of physical cracks.

The method developed by Lutif et al. (2009) and Souza (2008), which is based on the finite element method, can be summarized in the following steps.

1. Input data for global and local scales.
2. Solve the global scale problem.
3. Apply global scale solution to the local scale problem.
4. Obtain local scale solution from the local scale RVE.
5. Homogenize local scale field variables.
6. Update homogenized local scale results to the global scale problem.

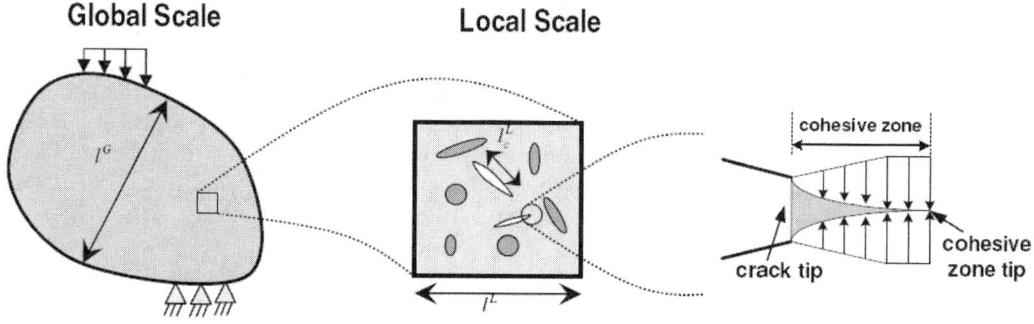

Figure 13.18 A schematic depiction of a global-local multiscale problem (Courtesy Y. R. Kim).

The above procedure is actually a two-way coupling method. The state of deformation observed on the global structure is passed to the boundary of the local scale RVE, and the solution obtained from the local scale is homogenized and returned to the global scale problem for the next time step. The two-way coupling is essentially necessary when the problem is for viscoelasticity and evolved microstructure. This type of problem involves viscoelasticity and cracking and is typically time, history, and space dependent. In the multiscale modeling process, the global scale can be discretized with a homogeneous finite element mesh, which may dramatically reduce the time required for computation. The local scale, considering mixture heterogeneity, may take more complicated meshes. Furthermore, a local scale RVE is attached to each global scale integration point of selected global scale finite elements.

It is important to note that the results from the local scale analysis are homogenized and linked to the global scale problem. The concept of homogenization replacement (Allen and Searcy, 2006) of a heterogeneous medium with a macroscopically equivalent homogeneous one is applicable whenever the heterogeneous medium satisfies statistical homogeneity. Homogenization is central to the idea of multiscale simulation and is typically created through averaging local fields within the heterogeneous medium. Equation 13-32 represents such an average:

$$C_{ijkl}^{\mu+1}(t) = \frac{1}{V^\mu} \int_{V^\mu} C_{ijkl}^{\mu}(t-\tau)dV \tag{13-32}$$

Where $C_{ijkl}^{\mu+1}$ is the averaged global scale modulus, C_{ijkl}^{μ} is the local scale modulus; μ denotes the local length scale, $\mu + 1$ indicates the next larger length scale, that is the global scale, V^μ denotes the volume of the local scale object.

It is also noteworthy that the local scale RVE could include cracking. Souza (2008) included a nonlinear viscoelastic cohesive zone model to enable the modeling of local scale viscoelastic fracture. The cohesive zone model can predict damage evolution in the global scale. Hence, the model can capture the formation of micro-cracking and the propagation in the heterogeneous local scale RVE and update the local scale fracture process to the global scale damage-dependent performance.

13.5   Future Development

While it is hard to accurately predict and evaluate the behavior of AC due to the reasons outlined in Chapter 1, multiscale modeling of asphalt behavior presents a promising potential to realistically model its behavior more accurately.

Research and education in the mechanics of asphalt have gained momentum in recent years due mainly to the need to understand and predict the mixture behavior for building long-lasting AC pavements. Nevertheless, significant efforts are still needed in validating, refining, and integrating various existing models. Special efforts should go to the exploration of complicated coupling phenomena such as rutting (large plastic deformation) induced fatigue and vice versa. Development of a generalized FEM code for these specific applications is a task that will benefit all researchers, educators, and practicing engineers.

References

Abraham, F.F., Broughton, J.Q., Bernstein, N. and Kaxiras, E. (1998). Spanning the continuum to quantum length scales in a dynamic simulation of brittle fracture. *Europhys. Lett.* Vol.44, No.6, pp.783–787.

Alam, M.M., Tandon, V., Nazarian, S. and Tahmoressi, M. (1998). Identification of moisture-susceptible asphalt concrete mixes using modified environmental conditioning system. *Transportation Research Record*, No.1630, pp.106–116.

Allen, D.H. and Searcy, C.R.(2006). A model for predicting the evolution of multiple cracks on multiple length scales in viscoelastic composites.*Journal of Materials Science*, Vol.41, pp.6510–6519

Allen, M.P. and D.J. Tildesley (1987). Computer Simulation of Liquids. *Clarendon Press; Oxford University Press.* Oxford [England], New York.

Arambula, E., Masad, E. and Martin, A.E. (2007a). Moisture susceptibility of asphalt mixtures with known field performance, Evaluated with dynamic analysis and crack growth model. *Transportation Research Record*, No. 2001, pp.20–28.

Arambula, E., Masad, E. and Martin, A.E. (2007b). Influence of air void distribution on the moisture susceptibility of asphalt mixes. *Journal of Materials in Civil Engineering*, Vol. 19, No. 8, pp.655–664.

Artok, L., Su, Y., Hirose, Y., Hosokawa, M., Murata, S. and Nomura, M (1999). Structure and reactivity of petroleum-derived asphaltene. *Energy Fuels.* Vol.13, pp.287–296.

Bhasin, A. and Little, D.N. (2007). Characterization of aggregate surface energy using the universal sorption device. *Journal of Materials in Civil Engineering*, Vol.19, No.8, pp.634–641.

Bhasin, A. and Little. D.N. (2007). Methods to quantify interfacial adhesion and debonding in bitumen-aggregate systems in dry and wet conditions. *International Conference on Advanced Characterisation of Pavement and Soil Engineering Materials*, Athens, Greece.

Bhasin, A. and Little, D.N. (2009). Application of microcalorimeter to characterize adhesion between asphalt binders and aggregates. *Journal of Materials in Civil Engineering*, Vol.21, No.6, pp.235–243.

Bhasin, A., Masad, E., Little, D. and Lytton, R. (2006). Limits on adhesive bond energy for improved resistance of hot-mix asphalt to moisture damage. *Transportation Research Record.* No.1970, pp.3–13.

Bhasin, A., Little, D.N., Vasconcelos, K.L and Masad, E. (2007). Surface free energy to identify moisture sensitivity of materials for asphalt mixes. *Transportation Research Record*, No.2001, pp.37–45.

Birgisson, B., Roque, R. and Page, G.C. (2003) Evaluation of water damage using hot mix asphalt fracture mechanics. *Journal of the Association of Asphalt Paving Technologists*, Vol.72, pp.424–462.

Birgisson, B., Roque, R. and Page, G.C. (2004) Performance-based fracture criterion for evaluation of moisture susceptibility in hot-mix asphalt. *Transportation Research Record*, No.1891, pp.55-61.

Birgisson, B., Roque, R. and Page, G.C. and Wang, J. (2007) Development of new moisture-conditioning procedure for hot-mix asphalt. *Transportation Research Record*, No. 2001, pp.46–55.

Buehler, M.J., Dodson, J., van Duin, A.C.T., Meulbroek, P. and Goddard III, W. A. (2005). The Computational materials design facility (CMDF): a powerful framework for multi-paradigm multi-scale simulations. *MRS Proceedings,*Vol.894.

Caro, S., Masad, E., Bhasin, A. and Little,D.N. (2008a). Moisture susceptibility of asphalt mixtures, part 1: mechanisms. *International Journal of Pavement Engineering*, Vol. 9, No.2, pp.81–98.

Caro, S., Masad, E., Bhasin, A. and Little, D.N. (2008b). Moisture susceptibility of asphalt mixtures, part 2: characterisation and modelling. *International Journal of Pavement Engineering,* Vol. 9, No.2, pp.99–114.

Car, R. and Parrinello, M. (1985). Unified approach for molecular dynamics and density-functional theory. *Physical Review Letters.* Vol.55, pp.2471–2474.

Chenoweth, K., van Duin, A. and Goddard, W.A. (2008). Reactive force field for molecular dynamics simulations of hydrocarbon oxidation. *J. Phys. Chem. A.,* Vol. 112, pp.1040–1053.

Chen, Y.P. (2006). Local stress and heat flux in atomistic systems involving three-body forces. *Journal of Chemical Physics,* Vol.124, No.5.

Cheng, D.X., Little, D. N., Lytton, R.L. and Holste, J.C. (2002a). Surface energy measurement of asphalt and its application to predicting fatigue and healing in asphalt mixtures. *Transportation Research Record,* No. 1810, pp.44–53.

Cheng, D.X., Little, D.N., Lytton, R.L. and Holste, J.C. (2002b). Use of surface free energy of asphalt-aggregate system to predict moisture damage potential. *Journal of the Association of Asphalt Paving Technologists,* Vol. 71, pp.59–88.

Cheng, D.X., Little, D.N., Lytton, R.L. and Holste. J.C. (2003). Moisture damage evaluation of asphalt mixtures by considering both moisture diffusion and repeated-load conditions. *Transportation Research Record,* No. 1832, pp.42–49.

Chorin, A.J. (1967). A numerical method for solving incompressible viscous flow problems. *J. Comput. Phys.,* Vol.2, pp.12–26.

Chorin, A. J., Hold, O. and Kupferman R. (2002). Optimal prediction with memory. *Physica D,* Vol.166, 239–257.

Curtis, C.W., Ensley, K. and Epps, J. (1993). Fundamental properties of asphalt-aggregate interactions including adhesion and absorption. *SHRP-A-341. Strategic Highway Research Program,* TRB, National Research Council, Washington, D.C.

Cui, X.Z. (2009). Numerical simulation of dynamic pore pressure in asphalt pavement, *Journal of Southeast University,* Vol. 25, pp.79–82.

Datta, D.K., Picu, R.C. and Shephard. M.S. (2004). Composite grid atomistic continuum method: an adaptive approach to bridge continuum with atomistic analysis. *Intl. J. Multiscale Computational Engineering,* Vol.2, No.3, pp.71–90.

Dauber-Osguthorpe, P., Roberts, V.A., Osguthorpe, D.J., Wolff, J., Genest, M., and Hagler, A.T. (1988). Structure and energetics of ligandbinding to proteins: Escherichia coli dihydrofolate reductase trimethoprim, a drug-receptor system. *Proteins: Structure, Function, and Genetics,* Vol.4, pp.31–47.

Dong, Z.J., Tan, Y.Q. and Cao, L.P. (2008). The effect of water on pavement response based on 3D FEM simulation and experiment evaluation. *Geotechnical Special Publication, No.182,* pp.4–15.

E,W. and Huang, Z. (2002). A dynamic atomistic-continuum method for the simulation of crystalline materials, *J. Comput. Phys.,* Vol.182, pp.234–261.

E,W. and Engquist, B. (2003). The heterogeneous multi-scale methods. *Comm. Math. Sci.,* Vol.1, pp.87–132.

Eidel, B. and Stukowski, A. (2008). A variational formulation of the quasi-continuum method based on energy sampling of clusters. *J. Mech. Phys. Sol.,* Vol.57, No.1, pp87–108.

El-Azab,A., (2000), Statistical mechanics treatment of the evolution of dislocation distributions in single crystals. *Phys. Rev. B* 61, pp.11956–11966

Epps, J.A., Sebaaly, P.E., Penaranda, J., Maher, M.R., McCann, M.B. and Hand, A.J. (2000). NCHRP Report 444: Compatibility of a test for moisture-induced damage with superpave Volumetric Mix Design. *TRB, National Research Council.* Washington, D.C.

Ericksen, J.L., Gurtin, M. (1984), The Cauchy-born hypotheses for crystals. *Phase Transformations and Material Instabilities in Solids,* Academic Press, New York. pp.61–77.

Eringen, A.C. (1999). Microcontinuum Field Theories. *Springer.*

Farkas, D. and Duranduru, M. (2001). Multiple-dislocation emission from the crack tip in the ductile fracture of Al. *Philosophical Magazine a-Physics of Condensed Matter Structure Defects and Mechanical Properties,* Vol.81, No.5, pp.1241–1255.

Fish, J., Nuggehally, M.A., Shephard, M.S., Picu, C.R., Badia, S., Parks, M.L. and Gunzburger, M. (2007). Concurrent AtC coupling based on a blend of the continuum stress and the atomistic force. *Computer Methods in Applied Mechanics and Engineering.* Vol.196, No.45–48, pp.4548–4560.

Fromm, H.J. (1974). The mechanisms of asphalt stripping from aggregate surfaces. Journal of the *Association of Asphalt Paving Technologists,* Vol. 43, pp.191–223.

Gao, H. and Klein, P. (1998), Numerical simulation of crack growth in an isotropic solid with randomized internal cohesive bonds. *Journal of the Mechanics and Physics of Solids,* Vol. 46, pp.187-218.

Garcia, A.L., Bell, J.B., Crutchfield, W.Y. and Alder, B.J. (1999). Adaptive mesh and algorithm refinement using direct simulation. Monte Carlo. *J. Comput. Phys.,* Vol.154, pp.134–155.

Ghoniem, N.M. and Cho, K. (2002). The Emerging role of multiscale modeling in nano- and micro-mechanics of materials. *Comp. Modeling in Engr. & Sci.,* Vol.3, pp.147–174.

Goddard, W.A., Van Duin, A., Chenoweth, K., Cheng, M., Pudar, S., Oxgaard, J., Merinov, B., Jang, Y.H. and Persson, P. (2006). Development of the FF reactive force field for mechanistic studies of catalytic selective oxidation processes on BiMoOx. *Topics in Catalysis,* Vol. 38, Nos.1–3, pp.93–103.

Gorkem, C. and Sengoz, B. (2009). Predicting stripping and moisture induced damage of asphalt concrete prepared with polymer modified bitumen and hydrated lime. *Construction and Building Materials,* Vol.23, No.6, pp.2227–2236.

Graf, P. (1986). Factors affecting moisture susceptibility of asphalt concrete mixes. Journal of the *Association of Asphalt Paving Technologists,* Vol.55, pp.175–212.

Gropp, W., Lusk, E. and Skjellum, A. (1994). Using MPI : portable parallel programming with the message-passing interface. *Cambridge, Mass, MIT Press.*

Gropp, W., Lusk, E. and Thakur, R. (1999). Using MPI-2 : advanced features of the message-passing interface. *Cambridge, Mass., MIT Press.*

Hähner, P., Bay,K., and Zaiser,M. (1998). Fractal dislocation patterning during plastic deformation. *Phys. Rev. Lett.,* Vol.81, pp.2470–2473.

Hähner, P., (1996), On the foundation of stochastic dislocation dynamics. *Applied Physics A Materials Science & Processing,* Vol.62, Issue. 5, pp.473–481.

Haile, J.M. (1992). Molecular Dynamics Simulation : Elementary Methods. *Wiley.* New York.

Hammons, M.I., Quintus, V., Geary, H., Wu, G.P.Y. and Jared, D.M. (2006) Detection of stripping in Hot-Mix Asphalt. *Transportation Research Record,* No. 1949, pp.20–31.

Hefer, A.W., Bhasin, A. and Little, D.N. (2006). Bitumen surface energy characterization using a contact angle approach. *Journal of Materials in Civil Engineering,* Vol.18, No.6, pp.759–767.

Hicks, R.G. (1991). Moisture damage in asphalt concrete. *NCHRP Synthesis of Practice 175: Moisture Damage in Asphalt Concrete,* TRB, National Research Council, Washington, D.C.

Huang, J.F., Wu, S.P. Ma, L.X., and Liu, Z.F. (2008). Material selection and design for moisture damage of HMA Pavement. *6th International Forum on Advanced Material Science and Technology.* Hong Kong, China.

Izzo, R.P. and Tahmoressi, M. (1999). Use of the Hamburg wheel-tracking device for evaluating moisture susceptibility of Hot-Mix Asphalt. *Transportation Research Record,* No. 1681, pp.76–85.

Jayachandran, K.P., Guedes, J.M., and Rodrigues, H.C. (2009). Effect of Microstructure and Texture on the Macroscopic Piezoelectric Response of Ferroelectric Barium Titanate and PZN-PT Films. *Journal of Intelligent Material Systems and Structures.* Vol.20, No. 2, pp.193–204.

Kadau, K., Germann, T.C. , Lomdahl, P.S. (2004). Large-scale molecular-dynamics simulation of 19 billion particles. *International Journal of Modern Physics* C, Vol.15, No.1, pp.193–201.

Kandhal, P.S. (1994). Field and laboratory investigation of stripping in asphalt pavements: state of the art report. *Transportation Research Record*, No.1454, pp.36–47.

Kanitpong and Bahia, H. (2005). Relating adhesion and cohesion of asphalts to the effect of moisture on laboratory performance of asphalt mixtures. *Transportation Research Record*, No.1901, pp.33–43.

Kassem, E., Masad, E., Bulut, R. and Lytton. R. (2006). Measurements of moisture suction and diffusion coefficient in hot-mix asphalt and their relationships to moisture damage. *Transportation Research Record*, No.1970, pp.45–54.

Katsoulakis, M., Majda, A.J. and Vlachos, D.G. (2003). Coarse-grained stochastic processes and monte carlo simulations in lattice systems. *J. Comp. Phys.*, Vol.186, pp.250–278

Kevrekidis, I., Gear, C.W. , Hyman, J.M., Kevrekidis, P.G., Runborg, O. and Theodopopoulos, C.(2003). Equation-free coarse-grained multiscale computation: enabling microscopic simulators to perform system-level analysis. *Comm. Mathematical Sciences.* Vol. 1, No. 4, pp.715–762.

Kim, Y.R., Lutif, J.S., Bhasin, A. and Little, D.N. (2008). Evaluation of moisture damage mechanisms and effects of hydrated lime in asphalt mixtures through measurements of mixture component properties and performance testing. *Journal of Materials in Civil Engineering.* Vol. 20, No.10, pp.659–667.

Kim, Y.R., Little, D.N. and Lytton, R.L. (2004). Effect of moisture damage on material properties and fatigue resistance of asphalt mixtures. *Transportation Research Record*, No. 1891, pp.48–54.

Knap, J., and Ortiz M. (2001). An analysis of the quasi-continuum method. *J. Mech. Phys. Sol.*, Vol.49, pp.1899–1923.

Kohlhoff, S., Gumbsch P., and Fischmeister, H.F. (1991). Crack propagation in BCC crystals studied with a combined finite-element and atomistic model. *Phil. Mag. A,* Vol. 64, No.4, pp.851–878.

Kolesik, M., Novotny, M. A. and Rikvold, P. (1998). Projection method for statics and dynamics of lattice spin systems. *Phys. Rev. Lett.*, Vol.80, pp.3384–3387

Kringos, N. and Scarpas, A. (2008). Physical and mechanical moisture susceptibility of asphaltic mixtures. *International Journal of Solids and Structures,* Vol.45, No.9, pp.2671–2685.

Kringos, N., Scarpas, A. and deBondt, A. (2008). Determination of moisture susceptibility of mastic-stone bond strength and comparison to thermodynamical properties. *Journal of the Association of Asphalt Paving Technologists.* Vol.77, pp.435–478.

Kringos, N. and Scarpas. A. (2005). Raveling of asphaltic mixes due to water damage: computational identification of controlling parameters. *Transportation Research Record,* No. 1929, pp.79–87.

Kringos, N., Scarpas, A. and Kasbergen,C. (2007). Three dimensional elasto-visco-plastic finite element model for combined physical-mechanical moisture induced damage in asphaltic mixes. *Journal of the Association of Asphalt Paving Technologists,* Vol. 76, 2007, pp.495–524.

Kutay, M.E., Aydilek, A.H. and Masad, E. (2007). Computational and experimental evaluation of hydraulic conductivity anisotropy in hot-mix asphalt. *International Journal of Pavement Engineering,* Vol. 8, No.1, pp.29–43.

Kutay, E. (2009) Pore pressure and viscous shear stress distribution due to water flow within asphalt pore structure, *Computer-Aided Civil and Infrastructure Engineering,* Vol. 24, pp.212–224.

Leach, A.R. (2001). Molecular Modelling: Principles and Applications, 2nd Edition. *Pearson Education EMA.*

Lee, C.L. and Li, S. (2008). The size effect of thin films on the Peierls stress of edge dislocations. *Mathematics and Mechanics of Solids*, Vol.13, pp.316–335.

Li, J., Liao, D. and Yip, S. (1999). Nearly exact solution for coupled continuum/md fluid simulation. *J. Computer-Aided Materials Design*, Vol.6, pp.95–102.

Little, D.N. and Petersen, J. C. (2005). Unique effects of hydrated lime filler on the performance-related properties of asphalt cements: Physical and chemical interactions revisited. *Journal of Materials in Civil Engineering*, Vol.17, No.2, pp.207–218.

Lu, Y. (2010). Atomistic modeling and simulation of asphalt-rock interface deformation and failure behaviors. Ph.D. *Dissertation*. Civil and Environmental Engineering. Blacksburg, Virginia Tech.

Lu, Y. and Wang, L. (2008). Molecular dynamics simulation to characterize asphalt-aggregate interfaces. *Research Symposium on the Characterization and Behavior of Interfaces (CBI)*. Atalanta, Georgia, USA.

Lu, Y. and Wang, L. (2009). Atomistic modeling of asphalt-aggregate interface behavior under tensile loading. *Proceedings of 2009 NSF Engineering Research and Innovation Conference*, Honolulu, Hawaii. USA.

Lu, Y. and Wang, L. (2009). Nanoscale modeling of mechanical properties of bitumen-rock interface under tensile loading. *IJPE Special Issue on Multiscale Modeling (in Press)*.

Luan, B.Q., Hyun, S., Molinari, J.F., Bernstein, N. and Robbins M.O. (2007). Multiscale modeling of two-dimensional contacts. *Phys. Rev. E*, Vol.74.

Lutif, J., Souza, F., and Kim, Y-R. (2009). A multiscale model for predicting mechanical behavior of geomaterials. *Proceedings of the NSF CMMI Engineering Research and Innovation Conference*, Honolulu, HI, USA. National Science Foundation.

Lytton, R. L., Masad, E., Zollinger, C., Bulut, R. and Little, D. (2005). Measurements of surface energy and its relationship to moisture damage. *FHWA/TX-05 Rep. No. 0-4524-2, Texas Transportation Institute*, College Station, TX.

Masad, E., Zollinger, C. Bulut, R., Little, D. and Lytton, R. (2006). Characterization of HMA moisture damage using surface energy and fracture properties. *Journal of the Association of Asphalt Paving Technologists*, Vol. 75, pp.713–754.

Masad, E. and Al-Omari, A. and Chen, H.C.(2007). Computations of permeability tensor coefficients and anisotropy of asphalt concrete based on microstructure simulation of fluid flow. *Computational Material Science*, Vol. 40 (4), pp.449–459.

Masad, E., Birgisson, B., Al-Omari, A. and Cooley, A. (2003). Analysis of permeability and fluid flow in asphalt mixes. *The 82nd Annual Meeting of the Transportation Research Board*, TRB, Washington D.C., CD.

Maupin, Jr., G.W. (1999). Evaluation of stripping in Virginia's pavements. *Transportation Research Record*, No. 1681, pp.37–42.

McCann and Sebaaly, P. (2001). Quantitative evaluation of stripping potential in Hot-Mix Asphalt, using ultrasonic energy for moisture-accelerated conditioning. *Transportation Research Record*, No.1767, pp.48–59.

Mohammad, L.N., Herath,A. and Huang,B. (2003). Evaluation of permeability of superpave asphalt mixtures. *Transportation Research Record*, No. 1832, pp.50–58.

Mohammad, L.N., A. Herath, Z. Wu, and S. Cooper. (2005). A comparative study of factors influencing the permeability of hot-mix asphalt Mixtures. *Journal of the Association of Asphalt Paving Technologists*. Vol. 74, pp.1–25.

Novak, M., Birgisson, B. and McVay, M. (2002). Effects of permeability and vehicle speed on pore pressure in hot mix asphalt pavements. *The 81th Annual Meeting of the Transportation Research Board*, TRB, Washington D.C. (CD).

Ortiz, M. and Phillips, R. (1999). Nanomechanics of Defects in Solids. *Adv. Appl. Mech.* Vol.36, pp.1–79.

Pinzón, E.C. and Such, C. (2004). Evaluation of moisture sensitivity of bituminous mixtures by a complex modulus Approach. *Transportation Research Record,* No. 1891, pp.62–67.

Plimpton, S. (1995). Fast parallel algorithms for short-range molecular-dynamics. *Journal of Computational Physics.* Vol.117, No.1, pp.1–19.

Plimpton, S. and Hendrickson, B. (1994). Parallel molecular-dynamics algorithms for simulation of molecular-systems. *Symposium on Parallel Computing in Computational Chemistry.* The 207th National Meeting of the American-Chemical-Society, San Diego, CA.

Rapaport, D.C. (2004). The Art 0f Molecular Dynamics Simulation, *Cambridge University Press.*

Roberts, F.L., Kandhal, P.S., Brown, E.R., Lee, D.Y. and Kennedy, T.W. (1996). Hot Mix Asphalt, Mixture Design, and Construction. 2nd Edition, *National Asphalt Pavement Association Education Foundation Inc.* Lanham, MD.

Rudd, R.E. and Broughton, J.Q. (1998). Coarse-grained molecular dynamics and the atomic limit of finite elements. *Phys. Rev. B,* Vol.58, pp.5893–5896.

Rudd, R.E. and Broughton, J.Q. (2000). Concurrent coupling of length scales in solid state systems. *Phys. Stat Solid B,* Vol.217, pp.251–291.

Scholz, T.V., Terrel, R.L. Al-Joaib, A. and Bea, J. (1994). Water sensitivity: binder validation. *Report. SHRP-A-402. Strategic Highway Research Program,* National Research Council, Washington, D.C.

Shenoy, V.B., Miller, R. Tadmor, E.B. Rodney, D. Phillips, R. and Ortiz, M. (1999). An adaptive methodology for atomic scale mechanics: the quasi-continuum method. *J. Mech. Phys. Sol.,* Vol.47, pp.611–642.

Shilkrot, L.E. Miller, R.E. and Curtin, W.A. (2004). Multiscale plasticity modeling: coupled atomistic and discrete dislocation mechanics. *J. Mech. Phys. Sol.,* Vol. 52, No.4, pp.755–787.

Solaimanian, M., Bonaquist, R.F. and Tandon, V.(2007). NCHRP Report 589: Improved conditioning and testing procedures for HMA moisture susceptibility. *TRB,* National Research Council, Washington, D.C.

Souza, F.V., Allen, D.H. and Kim, R.K.(2008).Multiscale model for predicting damage evolution in composites due to impact loading.*Composites Science and Technology,* Vol.68, No.13, pp.2624–2634.

Storm, D.A., Edwards, J.C., Decanio, S.J. and Sheu, E.Y. (1994). Molecular representations of Ratawi and Alaska north slope asphaltenes based on liquid- and solid-state NMR. Energy & Fuels, Vol. 8, No. 3, pp.561–566.

Stuart, K.D. (1990). Moisture damage in asphalt mixtures: A State-of-the-Art Report. *Report FHWA-RD-90-019.* FHWA, McLean, VA.

Tadmor, E. B., Ortiz, M. and Phillips, R. (1996). Quasi-continuum analysis of defects in crystals. *Phil. Mag. A,* Vol.73, pp.1529–1563.

Tadmor, E.B., Waghmare, U.V. Smith, G.S. and Kaxiras, E. (2002). Polarization switching in PbTiO3: An ab initio finite element simulation. *Acta Materialia,* Vol.50, pp.2989–3002.

Vanden-Eijnden, E. (2003). Numerical techniques for multiscale dynamical systems with stochastic effects. *Comm. Math. Sci.,* Vol.1, pp.385–391.

Van Duin, A., Strachan,A., Stewman,S., Zhang, Q., Xu,X. and Goddard,W.A.(2003). ReaxFFSiO reactive force field for silicon and silicon oxide systems. *J. Phys. Chem. A,* Vol.107, pp.3803–3811

Wagner, G.J. and Liu W.K., (2003). Coupling of atomistic and continuum simulations using a bridging scale decomposition. *J. Comput. Phys.,* Vol.190, pp.249–274.

Walgraef, D. and Aifantis, E.C. (1985). On the formation and stability of dislocation patterns, I–III. *Int. J. Eng. Sci.,* Vol.12, pp.1351–1372

Wang, D., and Wang,L.B. (2010a). Understand the moisture damage of the asphalt pavement using poro-elasticity and FEM. *Paper submitted to International Journal of Pavement research technology.*

Wang, D., Wang, L.B., Sheng, Y.P., and Chen, S.F. (2010b). Analysis of excess pore water pressure and base erosion of rigid pavements using porous elasticity and FEM. *Paper submitted to International Journal of Pavement Research Technology.*

Wang, H.F. (2000). Theory of Linear Poroelasticity with Applications to Geomechanics and Hydrogeology, *Princeton University Press,* Princeton, NJ, Vol.287, pp.2000.

Xiao, S.P. and Belytschko, T. (2004). A bridging domain method for coupling continua with molecular dynamics. *Computer Methods in Applied Mechanics and Engineering,* Vol. 193, pp.1645–69.

Xu, K. and Prendergast, K.H. (1994). Numerical Navier-stokes solutions from gas kinetic theory. *J. Comput. Phys.,* Vol.114, pp.9–17.

Yang, Z., Litao, G. and Linzhou, F. (2006). Computing the express pore fluid stress of flexible pavement by stiffness matrix method. *Journal of ShenyangJianzhu University,* Vol.22, No.1, pp.25–29.

Zaiser,V. and Hähner, P. (1997). Oscillatory modes of plastic deformation: theoretical concept. *Physica Status Solidi (b),* Vol. 199, pp.267–330.

Zhang, L.Q. and Greenfield, M. L. (2007a). Molecular orientation in model asphalts using molecular simulation. Energy & Fuels, Vol.21, No.2, pp.1102–1111.

Zhang, L. and Greenfield, M.L. (2007b). Analyzing properties of model asphalts using molecular simulation. Energy & Fuels, Vol.21, No.3, pp.1712–1716.

Zhang, L. and Greenfield, M. L. (2008). Effects of polymer modification on properties and microstructure of model asphalt systems. Energy & Fuels, Vol.22, No.5, pp.3363–3375.

Zhang, Y., Lee, T.-S. and Yang, W. (1999). A pseudo-bond approach to combining quantum mechanical and molecular mechanical methods. *J. Chemical Phys.,* Vol.110, pp.46–54.

Zhou, C.H., Wang, Z.R., Chen, J.Y. and Qiao, Y.J .(2007). Numerical computation and analysis on dynamic pore water pressure in asphalt pavement. *International Conference on TransportaionEngineering (ICTE 2007),* pp.489–495.

Eshelby's Tensor (S) for Special Cases

This appendix is abstracted from (Nemat-Nasser and Hori, 1998). It refers to the Eshelby tensor components for some specific cases. The general case is the ellipsoids embedded with semi-principal axes (a_i) coincide with the coordinate axes (x_1, x_2, x_3) (Figure A.1).

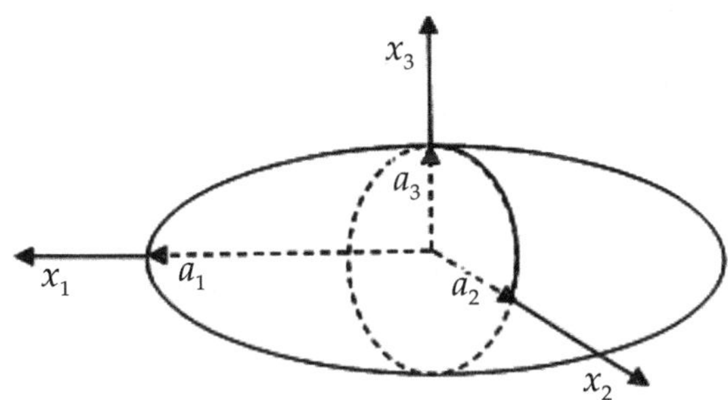

Figure A.1 Ellipsoid coaxial with the cartesian coordinates

General form of the components for $a_1 > a_2 > a_3$ are:

$$S_{1111} = \frac{3}{8\pi(1-v)}a_1^2 I_{11} + \frac{1-2v}{8\pi(1-v)}I_1$$

$$S_{1122} = \frac{1}{8\pi(1-v)}a_2^2 I_{12} - \frac{1-2v}{8\pi(1-v)}I_1$$

$$S_{1212} = \frac{1}{16\pi(1-v)}(a_1^2 + a_2^2)I_{12} + \frac{1-2v}{16\pi(1-v)}(I_1 + I_2)$$

The I_i and I_{ij} integrals are given by:

$$I_1 = \frac{4\pi a_1 a_2 a_3}{(a_1^2 - a_2^2)(a_1^2 - a_3^2)^{1/2}}\{F(\theta,k) - E(\theta,k)\}$$

$$I_3 = \frac{4\pi a_1 a_2 a_3}{(a_2^2 - a_3^2)(a_1^2 - a_3^2)^{1/2}}\{\frac{a_2(a_1^2 - a_3^2)^{1/2}}{a_1 a_3} - E(\theta,k)\}$$

$$I_1 + I_2 + I_3 = 4\pi$$

and

$$3I_{11} + I_{12} + I_{13} = \frac{4\pi}{a_1^2}, \quad 3a_1^2 I_{11} + a_2^2 I_{12} + a_3^2 I_{13} = 3I_1, \quad I_{12} = \frac{I_2 - I_1}{a_1^2 - a_2^2}$$

Where F and E are the elliptic integrals of the first and second kind, and

$$\theta = \arcsin\{\frac{a_1^2 - a_3^2}{a_1^2}\}^{1/2}, \ k = \arcsin\{\frac{a_1^2 - a_2^2}{a_1^2 - a_3^2}\}^{1/2}$$

For sphere ($a_1 = a_2 = a_3 = a$)

$$S_{ijkl} = \frac{5v - 1}{15(1 - v)}\delta_{ij}\delta_{kl} + \frac{4 - 5v}{15(1 - v)}(\delta_{ik}\delta_{jl} + \delta_{il}\delta_{jk})$$

For elliptical cylinder ($a_3 \rightarrow \infty$)

$$S_{1111} = \frac{1}{2(1 - v)}\{\frac{a_2^2 + 2a_1 a_2}{(a_1 + a_2)^2} + (1 - 2v)\frac{a_2}{a_1 + a_2}\}$$

$$S_{2222} = \frac{1}{2(1 - v)}\{\frac{a_1^2 + 2a_1 a_2}{(a_1 + a_2)^2} + (1 - 2v)\frac{a_1}{a_1 + a_2}\}, \ S_{3333} = 0$$

$$S_{1122} = \frac{1}{2(1 - v)}\{\frac{a_2^2}{(a_1 + a_2)^2} - (1 - 2v)\frac{a_2}{a_1 + a_2}\}$$

$$S_{2233} = \frac{1}{2(1 - v)}\frac{2va_1}{a_1 + a_2}$$

$$S_{2211} = \frac{1}{2(1 - v)}\{\frac{a_1^2}{(a_1 + a_2)^2} - (1 - 2v)\frac{a_1}{a_1 + a_2}\}$$

$$S_{1133} = \frac{1}{2(1 - v)}\frac{2va_2}{a_1 + a_2}, \ S_{3311} = S_{3322} = 0$$

$$S_{1212} = \frac{1}{2(1 - v)}\{\frac{a_1^2 + a_2^2}{2(a_1 + a_2)^2} + \frac{1 - 2v}{2}\}$$

$$S_{2323} = \frac{a_1}{2(a_1 + a_2)}, \ S_{3131} = \frac{a_2}{2(a_1 + a_2)}$$

For penny shape ($a_1 = a_2 \gg a_3$)

$$S_{1111} = S_{2222} = \frac{\pi(13 - 8v)}{32(1 - v)}\frac{a_3}{a_1}, \ S_{3333} = 1 - \frac{\pi(1 - 2v)}{4(1 - v)}\frac{a_3}{a_1}$$

$$S_{1122} = S_{2211} = \frac{\pi(8v - 1)}{32(1 - v)}\frac{a_3}{a_1}, \ S_{1133} = S_{2233} = \frac{\pi(2v - 1)}{8(1 - v)}\frac{a_3}{a_1}$$

$$S_{1122} = S_{2211} = \frac{\pi(8v - 1)}{32(1 - v)}\frac{a_3}{a_1}, \ S_{1133} = S_{2233} = \frac{\pi(2v - 1)}{8(1 - v)}\frac{a_3}{a_1}$$

$$S_{1212} = \frac{\pi(7 - 8v)}{32(1 - v)}\frac{a_3}{a_1}, \ S_{3131} = S_{2323} = \frac{1}{2}\{1 + \frac{\pi(v - 2)}{4(1 - v)}\frac{a_3}{a_1}\}$$

Reference

Nemat-Nasser, S. and Hori, M. (1998). *Micromechanics: Overall Properties of Heterogeneous Materials*. North-Holland.

Laplace Transform

TABLE A2.1 Typical operations and relations.

Laplace transform	$L\{f(t)\} \equiv F(s) \equiv \int_0^\infty f(t)e^{-st}\,dt$
Convolution	$\int_0^t f(t-\xi)g(\xi)\,d\xi$
Convolution for Laplace	$L\{\int_0^t f(t-\xi)g(\xi)\,d\xi\} = L\{f(t)\}L\{g(t)\}$
Derivative theorem for Laplace	$L\{\dfrac{df(t)}{dt}\} = sL\{f(t)\} - f(0)$
Linearity (superposition)	$L\left\{a_1 f_1(t) + a_2 f_2(t)\right\} = a_1 F_1(s) + a_2 F_2(s)$
Derivative property	$L\left\{\dfrac{df(t)}{dt}\right\} = sF(s) - f(0)$
Integral property	$L\left\{\int_s^t f(\tau)\,d\tau\right\} = \dfrac{F(s)}{s}$
Multiplication by time	$L\{tf(t)\} = -\dfrac{d}{ds}F(s)$
Division by time	$L\left\{\dfrac{f(t)}{t}\right\} = \int_s^\infty F(u)\,du$
Multiplication by an exponential	$L\{e^{-at}f(t)\} = F(s+a)$
Time shift	$L\{f(t-T)H(t-T)\} = e^{-Ts}F(s)$
Scale change	$L\{f(at)\} = \dfrac{1}{a}F(\dfrac{s}{a})$
Convolution theorem	$L\left\{\int_0^t f(\tau)g(t-\tau)\,d\tau\right\} = F(s) \bullet G(s)$

TABLE A2.2 Laplace transformations of typical functions.

$f(t)$	$F(s)$
$\delta(t)$	1
$\delta(t-a)$	e^{-as}
$H(t)$	$1/s$
$H(t-a)$	$\dfrac{1}{s}e^{-as}$
$tH(t)$	$1/s^2$
e^{-at}	$\dfrac{1}{s+a}$
$\dfrac{t^{n-1}e^{-at}}{(n-1)!}$	$\dfrac{1}{(s+a)^n}$
$\dfrac{1}{a}\left(1-e^{-at}\right)$	$\dfrac{1}{s(s+a)}$
$\dfrac{\left[be^{bt}-ae^{at}\right]}{(b-a)}$	$\dfrac{1}{(s-a)(s-b)}$ for $a \neq b$
$\dfrac{\left[e^{bt}-e^{at}\right]}{(b-a)}$	$\dfrac{s}{(s-a)(s-b)}$ for $a \neq b$
$\cosh(at)$	$\dfrac{s}{s^2-a^2}$
$\sinh(at)$	$\dfrac{a}{s^2-a^2}$
$\cos(at)$	$\dfrac{s}{s^2+a^2}$
$\sin(at)$	$\dfrac{a}{s^2+a^2}$
$\dfrac{\sin at}{t}$	$\tan^{-1}(a/s)$
$t\cos(at)$	$\dfrac{s^2-a^2}{\left(s^2+a^2\right)^2}$
$t\sin(at)$	$\dfrac{2as}{s^2+a^2}$
$e^{-at}\sin\omega t$	$\dfrac{\omega}{\left(s+a\right)^2+\omega^2}$
$e^{-at}\cos\omega t$	$\dfrac{s+a}{\left(s+a\right)^2+\omega^2}$
t^n	$\Gamma(n+1)s^{-n-1}$
$t^n e^{-qt}$	$\Gamma(n+1)(s+q)^{-n-1}$

TABLE A2.3   Convolution laws.

Convolution definition	$C(t) = f(t) * g(t) = \int\limits_{0}^{t} f(\xi)g(t - \xi)d\xi$
Commutative property	$f(t) * g(t) = g(t) * f(t)$
Associativity	$f(t) * \left[g(t) * h(t) \right] = \left[f(t) * g(t) \right] * h(t)$
Distributivity	$f(t) * \left[g(t) + h(t) \right] = f(t) * g(t) + f(t) * h(t)$

References

Sneddon, I. N. (1951). *Fourier Transforms*, McGraw-Hill, NY.

Spiegl, M. R. (1968). *Mathematical Handbook*, McGraw-Hill, NY.

Swisher, G. M. (1976). *Introduction to Linear Systems Analysis*, Matrix, Cleveland.

Isotropic Elastostatics Fundamental Solution

This appendix is abstracted from Gaul et al (2003). It documents the fundamental solution to the following equation as the basis solution for BEM applications.

$$\frac{1}{1-2v} u^*_{ik,kj} + u^*_{ij,kk} = -\frac{1}{\mu}\delta(x,\xi)\delta_{ij}$$

Where v is Poisson's ratio and μ is the shear modulus. By introducing the tensorial potential G_{ij}, one obtains:

$$u^*_{ij} = G_{ij,mm} - \frac{1}{2(1-v)}G_{im,jm}$$

For the 3D case, the tensorial potential function is:

$$G_{ij}(r) = \frac{1}{8\pi\mu}r\delta_{ij}$$

Where r is the position vector. Some of the useful derivatives include:

$$G_{ij,m} = \frac{1}{8\pi\mu}\delta_{ij}r_{,m}$$

$$G_{ij,mn} = \frac{1}{8\pi\mu r}\delta_{ij}(\delta_{mn} - r_{,m}r_{,n})$$

The fundamental solution is:

$$u^*_{ij} = \frac{1}{16\pi\mu(1-v)r}((3-4v)\delta_{ij} + r_{,i}r_{,j})$$

The traction field is:

$$t^*_{ij} = \frac{1}{8\pi(1-v)r^2}\left(((1-2v)\delta_{ij} + 3r_{,i}r_{,j})\frac{\partial r}{\partial n} + (1-2v)(r_{,j}n_i - r_{,i}n_j)\right)$$

The solution represented in X, Y and Z coordinates is presented in Becker (1992).

References

Becker, A.A. (1992). *The Boundary Element Method in Engineering.* McGraw-Hill Book Company.

Gaul, L., Kogl, M. and Wagner, M. (2003). *Boundary Element Methods for Engineers and Scientists- An Introductory Course with Advanced Topics,* Springer.

Index